Non-Native and Invasive Ticks

UNIVERSITY PRESS OF FLORIDA

Florida A&M University, Tallahassee
Florida Atlantic University, Boca Raton
Florida Gulf Coast University, Ft. Myers
Florida International University, Miami
Florida State University, Tallahassee
New College of Florida, Sarasota
University of Central Florida, Orlando
University of Florida, Gainesville
University of North Florida, Jacksonville
University of South Florida, Tampa
University of West Florida, Pensacola

# Non-Native and Invasive Ticks

Threats to Human and Animal Health
in the United States

Michael J. Burridge

UNIVERSITY PRESS OF FLORIDA

Gainesville · Tallahassee · Tampa · Boca Raton
Pensacola · Orlando · Miami · Jacksonville · Ft. Myers · Sarasota

Copyright 2011 by Michael J. Burridge
Printed in the United States of America on acid-free paper
All rights reserved

16 15 14 13 12 11   6 5 4 3 2 1

Library of Congress Cataloging-in-Publication Data
Burridge, M. J.
Non-native and invasive ticks : threats to human and animal health in the
United States / Michael J. Burridge.
p. cm.
Includes bibliographical references and index.
ISBN 978-0-8130-3537-6 (alk. paper)
 1. Ticks as carriers of disease—United States. 2. Nonindigenous
pests—United States. 3. Tick-borne diseases—United States. 4. Tick-borne
diseases in animals—United States. 5. Ticks. I. Title.
RA641.T5B87   2011
614.4'33—dc22   2010042209

The University Press of Florida is the scholarly publishing agency for the
State University System of Florida, comprising Florida A&M University, Florida Atlantic University, Florida Gulf Coast University, Florida
International University, Florida State University, New College of Florida,
University of Central Florida, University of Florida, University of North
Florida, University of South Florida, and University of West Florida.

University Press of Florida
15 Northwest 15th Street
Gainesville, FL 32611-2079
http://www.upf.com

I dedicate this book to my wife, Karen, and daughter, Christina,
for their love and encouragement.

# Contents

List of Illustrations xi

## 1. Introduction 1

1.1. Layout of Book 1
1.2. Tick Taxonomy 2
1.3. Acknowledgments 3

## 2. Invasive Species 4

2.1. International Trade in Live Animals 4
2.2. International Trade in Live Reptiles 5
2.3. Invasive Animal Species in Florida 6
2.4. Non-native and Invasive Ticks 9
    2.4.1. Biology of ticks 9
    2.4.2. Invasive ticks 10
    2.4.3. Modes of introduction of invasive ticks 11
    2.4.4. Role of migratory birds in dissemination of ticks 12

## 3. Ticks from Africa 14

3.1. *Amblyomma chabaudi* Rageau 14
3.2. *Amblyomma compressum* Macalister 14
3.3. *Amblyomma exornatum* Koch 16
3.4. *Amblyomma falsomarmoreum* Tonelli-Rondelli 18
3.5. *Amblyomma flavomaculatum* (Lucas) 18
3.6. *Amblyomma gemma* Donitz 20
3.7. *Amblyomma hebraeum* Koch 21
3.8. *Amblyomma latum* Koch 22
3.9. *Amblyomma lepidum* Donitz 24
3.10. *Amblyomma marmoreum* Koch 25
3.11. *Amblyomma nuttalli* Donitz 27
3.12. *Amblyomma pomposum* Donitz 29
3.13. *Amblyomma rhinocerotis* (de Geer) 29
3.14. *Amblyomma sparsum* Neumann 30
3.15. *Amblyomma splendidum* Giebel 32
3.16. *Amblyomma sylvaticum* (de Geer) 32
3.17. *Amblyomma tholloni* Neumann 33
3.18. *Amblyomma transversale* (Lucas) 34
3.19. *Dermacentor rhinocerinus* (Denny) 35
3.20. *Haemaphysalis elongata* Neumann 36
3.21. *Haemaphysalis hoodi* Warburton & Nuttall 36
3.22. *Haemaphysalis leachi* (Audouin) 37
3.23. *Haemaphysalis muhsamae* Santos Dias 38
3.24. *Hyalomma albiparmatum* Schulze 39
3.25. *Hyalomma impressum* Koch 39
3.26. *Hyalomma truncatum* Koch 40
3.27. *Ixodes pilosus* Koch 41
3.28. *Ixodes schillingsi* Neumann 42
3.29. *Rhipicephalus appendiculatus* Neumann 43
3.30. *Rhipicephalus capensis* Koch 45
3.31. *Rhipicephalus compositus* Neumann 46
3.32. *Rhipicephalus evertsi* Neumann 47
3.33. *Rhipicephalus kochi* Donitz 49
3.34. *Rhipicephalus muehlensi* Zumpt 50
3.35. *Rhipicephalus pulchellus* (Gerstacker) 51
3.36. *Rhipicephalus senegalensis* Koch 52
3.37. *Rhipicephalus simus* Koch 52
3.38. *Rhipicephalus sulcatus* Neumann 54
3.39. *Rhipicephalus (Boophilus) decoloratus* (Koch) 55

## 4. Ticks from the Afro-Caribbean Region 57

4.1. *Amblyomma variegatum* (Fabricius) 57

## 5. Ticks from the Afro-Asian Region 62

5.1. *Hyalomma dromedarii* Koch 62

## 6. Ticks from the Afro-European Region 64

6.1. *Hyalomma lusitanicum* Koch 64

## 7. Ticks from the Afro-Eurasian Region 66

7.1. *Haemaphysalis punctata* Canestrini & Fanzago 66
7.2. *Hyalomma aegyptium* (Linnaeus) 67
7.3. *Hyalomma anatolicum* Koch spp. 68
7.4. *Hyalomma detritum* Schulze 70
7.5. *Hyalomma marginatum* Koch group 71
7.6. *Ixodes ricinus* (Linnaeus) 74
7.7. *Rhipicephalus bursa* Canestrini & Fanzago 77
7.8. *Rhipicephalus turanicus* Pomerantsev 78

## 8. Ticks from Europe 81

8.1. *Ixodes hexagonus* Leach 82

## 9. Ticks from Eurasia 83

9.1. *Dermacentor reticulatus* (Fabricius) 83

## 10. Ticks from Asia 85

10.1. *Amblyomma clypeolatum* Neumann 86
10.2. *Amblyomma crassipes* Neumann 87
10.3. *Amblyomma fuscolineatum* (Lucas) 87
10.4. *Amblyomma geoemydae* (Cantor) 88
10.5. *Amblyomma javanense* (Supino) 89
10.6. *Amblyomma komodoense* (Oudemans) 89
10.7. *Amblyomma kraneveldi* (Anastos) 91
10.8. *Amblyomma testudinarium* Koch 91
10.9. *Amblyomma varanense* (Supino) 92
10.10. *Dermacentor auratus* Supino 93
10.11. *Dermacentor nuttalli* Olenev 94
10.12. *Haemaphysalis hystricis* Supino 94

## 11. Ticks from the Australasian-Asiatic Region 96

11.1. *Amblyomma fimbriatum* Koch 96
11.2. *Amblyomma helvolum* Koch 97
11.3. *Amblyomma trimaculatum* (Lucas) 98
11.4. *Haemaphysalis longicornis* Neumann 99

## 12. Ticks from Australia 100

12.1. *Amblyomma moreliae* (Koch) 100
12.2. *Amblyomma triguttatum* Koch 101
12.3. *Bothriocroton concolor* (Neumann) 101
12.4. *Bothriocroton hydrosauri* (Denny) 102

## 13. Ticks from the Americas 105

13.1. *Amblyomma albopictum* Neumann 105
13.2. *Amblyomma argentinae* Neumann 106
13.3. *Amblyomma auricularium* (Conil) 108
13.4. *Amblyomma cajennense* (Fabricius) 109
13.5. *Amblyomma calcaratum* Neumann 112
13.6. *Amblyomma coelebs* Neumann 113
13.7. *Amblyomma dissimile* Koch 114
13.8. *Amblyomma geayi* Neumann 116
13.9. *Amblyomma humerale* Koch 117
13.10. *Amblyomma incisum* Neumann 117
13.11. *Amblyomma longirostre* (Koch) 118
13.12. *Amblyomma multipunctum* Neumann 119
13.13. *Amblyomma nodosum* Neumann 120
13.14. *Amblyomma oblongoguttatum* Koch 120
13.15. *Amblyomma ovale* Koch 121
13.16. *Amblyomma parvum* Aragao 122
13.17. *Amblyomma pictum* Neumann 123
13.18. *Amblyomma pseudoconcolor* Aragao 123
13.19. *Amblyomma quadricavum* (Schulze) 124
13.20. *Amblyomma rotundatum* Koch 125
13.21. *Amblyomma sabanerae* Stoll 126
13.22. *Amblyomma scutatum* Neumann 127
13.23. *Amblyomma varium* Koch 127
13.24. *Dermacentor nitens* Neumann 128
13.25. *Haemaphysalis juxtakochi* Cooley 130
13.26. *Ixodes luciae* Senevet 130

## 14. Ticks with Widespread Distributions 132

14.1. *Rhipicephalus (Boophilus) annulatus* (Say) 132
14.2. *Rhipicephalus (Boophilus) microplus* (Canestrini) 134

## 15. Tickborne Diseases 137

15.1. Anaplasmosis 137
  15.1.1. *Anaplasma marginale* infection 137
  15.1.2. *Anaplasma centrale* infection 137
  15.1.3. *Anaplasma bovis* infection 138
  15.1.4. *Anaplasma ovis* infection 138
  15.1.5. *Anaplasma phagocytophilum* infection 138
15.2. Babesiosis 139
  15.2.1. *Babesia bigemina* infection 139

- 15.2.2. *Babesia bovis* infection 139
- 15.2.3. *Babesia divergens* infection 140
- 15.2.4. *Babesia major* infection 140
- 15.2.5. *Babesia motasi* infection 140
- 15.2.6. *Babesia ovis* infection 141
- 15.2.7. *Babesia trautmanni* infection 141
- 15.2.8. *Babesia canis* and *Babesia rossi* infections 141
- 15.2.9. *Babesia gibsoni* infection 141
- 15.2.10. *Babesia microti* infection 142
- 15.3. Borreliosis 142
  - 15.3.1. Lyme borreliosis 142
  - 15.3.2. Bovine borreliosis 143
- 15.4. Crimean-Congo Hemorrhagic Fever 144
- 15.5. Dermatophilosis 145
- 15.6. Heartwater 146
- 15.7. Hemogregarine Infections 147
- 15.8. Louping Ill 147
- 15.9. Nairobi Sheep Disease 148
- 15.10. Omsk Hemorrhagic Fever 148
- 15.11. Piroplasmosis 149
  - 15.11.1. *Babesia caballi* infection 149
  - 15.11.2. *Theileria equi* infection 150
- 15.12. Q Fever 150
- 15.13. Rickettsiosis 151
  - 15.13.1. *Rickettsia africae* infection 151
  - 15.13.2. *Rickettsia conorii* infection 152
  - 15.13.3. *Rickettsia honei* infection 152
  - 15.13.4. *Rickettsia japonica* infection 152
  - 15.13.5. *Rickettsia rickettsii* infection 152
  - 15.13.6. *Rickettsia sibirica sibirica* infection 153
  - 15.13.7. *Rickettsia sibirica mongolitimonae* infection 153
  - 15.13.8. *Rickettsia slovaca* infection 153
- 15.14. Theileriosis 153
  - 15.14.1. *Theileria parva* infection 154
  - 15.14.2. *Theileria annulata* infection 155
  - 15.14.3. *Theileria lestoquardi* infection 155
  - 15.14.4. *Theileria buffeli/Theileria orientalis* infection 156
  - 15.14.5. *Theileria mutans* infection 156
  - 15.14.6. *Theileria taurotragi* infection 156
- 15.15. Tick Paralysis 156
- 15.16. Tick Toxicosis 157
  - 15.16.1. Brown tick toxicosis 157
  - 15.16.2. Sweating sickness 157
- 15.17. Tickborne Encephalitis 157
- 15.18. Tularemia 158

## 16. Risks of Invasive Ticks to the United States 160

- 16.1. Establishment of Invasive Ticks in the United States 160
- 16.2. Tickborne Diseases Transmissible to Humans 168
- 16.3. Tickborne Diseases Transmissible to Cattle 170
  - 16.3.1. Risk of introduction of heartwater 170
  - 16.3.2. Risk of introduction of bovine babesiosis 173
- 16.4. Tickborne Diseases Transmissible to Sheep and Goats 175
- 16.5. Tickborne Diseases Transmissible to Other Mammals 175
- 16.6. Ticks and Tickborne Diseases Associated with Reptiles 176

## 17. Measures Used to Combat Invasive Ticks 178

- 17.1. Regulation of Global Trade in Animals 178
  - 17.1.1. The U.S. Department of Agriculture 179
  - 17.1.2. The U.S. Fish and Wildlife Service 180
  - 17.1.3. The Centers for Disease Control and Prevention 181
  - 17.1.4. The National Marine Fisheries Service 181
  - 17.1.5. Overview of U.S. regulatory agencies 181
  - 17.1.6. National Reptile Improvement Plan 182
  - 17.1.7. Reptile smuggling 182
- 17.2. Eradication of Infestations of Invasive Ticks 183
  - 17.2.1. Eradication of a *Rhipicephalus evertsi* infestation 183

17.2.2. Eradication of an *Amblyomma sparsum* infestation  184
  17.2.3. Eradication of an *Amblyomma komodoense* infestation  186

# 18. Actions Needed to Minimize Introduction of Invasive Ticks  187

  18.1. Regulatory Actions  187
    18.1.1. CITES regulations  187
    18.1.2. U.S. regulations  187
  18.2. Working Groups  187
  18.3. Reptile Importations  188
  18.4. USDA Reports  189
  18.5 Eradication of the Tropical Bont Tick from the Caribbean  189
  18.6. Prevention of Reestablishment of Cattle Fever Ticks in the United States  193

Appendix 1. Invasive Ticks Introduced into the Continental United States  195

Appendix 2. Invasive Ticks by Host  217

Glossary  263

References  267

Index  289

# Illustrations

**Maps**

3.1. Africa  15
4.1. The Caribbean  58
8.1. Europe  81
10.1. Asia  85
11.1. Australasia and the Pacific  96
13.1. The Americas  105
16.1. Florida counties in which non-native vertebrate species have established breeding populations  167

**Figures**

2.1. Spider tortoises (*Pyxis arachnoides*) imported from Madagascar  6
2.2. Green iguana (*Iguana iguana*) in southern Florida  7
2.3. Burmese python (*Python molurus*) captured in southern Florida  8
2.4. Stages in the life cycle of the Cayenne tick *Amblyomma cajennense*  9
3.1. Nile monitor (*Varanus niloticus*)  16
3.2. Leopard tortoise (*Stigmochelys pardalis*)  19
3.3. *Amblyomma latum* tick feeding in the eye socket of a ball python (*Python regius*)  23
3.4. *Amblyomma marmoreum* ticks feeding under shell of tortoise  26
3.5. Black rhinoceros (*Diceros bicornis*)  30
3.6. African buffalos (*Syncerus caffer*)  32
3.7. African elephant (*Loxodonta africana*)  33
3.8. *Rhipicephalus appendiculatus* ticks feeding on the ear of a cow  44
3.9. Nyala (*Tragelaphus angasii*)  50
4.1. Engorged *Amblyomma variegatum* ticks taken from cattle  58
5.1. Camel (*Camelus dromedarius*)  63
10.1. Indian star tortoise (*Geochelone elegans*)  86
10.2. Adult Komodo dragon (*Varanus komodoensis*) with *Amblyomma komodoense* ticks  90
10.3. Water monitor (*Varanus salvator*)  92
12.1. Short-beaked echidna (*Tachyglossus aculeatus*)  102
12.2. Shingleback skink (*Trachydosaurus rugosus*)  103
13.1. Cayman Islands ground iguana (*Cyclura nubila*)  106
13.2. Male and female *Amblyomma argentinae* ticks  107
13.3. *Amblyomma argentinae* ticks on the leg of an Argentine tortoise (*Chelonoidis chilensis*)  107
13.4. Female *Amblyomma auricularium* tick  108
13.5. Six-banded armadillo (*Euphractus sexcinctus*)  108
13.6. Male and female Cayenne ticks (*Amblyomma cajennense*)  110
13.7. Giant anteater (*Myrmecophaga tridactyla*)  112
13.8. South American tapir (*Tapirus terrestris*) and newborn calf  113
13.9. Boa constrictor (*Boa constrictor*)  115
13.10. Female *Amblyomma parvum* tick  122
13.11. Female *Amblyomma pseudoconcolor* tick  123
13.12. Engorged female and larvae of the tick *Amblyomma pseudoconcolor* on a larger hairy armadillo (*Chaetophractus villosus*)  123
13.13. Giant toad (*Bufo marinus*)  125
13.14. Southern two-toed sloth (*Choloepus didactylus*)  128
13.15 Female *Haemaphysalis juxtakochi* tick  130
15.1. Cow suffering from acute dermatophilosis and cow that recently died from the disease  145
15.2. White-tailed deer (*Odocoileus virginianus*) exhibiting incoordination in terminal stages of heartwater  146

15.3. Froth in trachea of a cow that died from heartwater  147
15.4. Frothy fluid exuding from the nostrils of a cow that died from East Coast fever  154
16.1. Gopher tortoise (*Gopherus polyphemus*)  163
16.2. Cattle egrets (*Bubulcus ibis*) with a grazing steer in Florida  172
16.3. White-tailed deer buck (*Odocoileus virginianus*)  174
16.4. Adult Gulf Coast tick (*Amblyomma maculatum*)  174
17.1. *Amblyomma marmoreum* ticks feeding on an Aldabra giant tortoise (*Aldabrachelys gigantea*)  184
17.2. Treating a tortoise with an acaricide spray  185
17.3. Treating tortoise housing with an acaricide spray  185
17.4. Treating a snake container with an acaricide spray  186
18.1. Spraying cattle with acaricides manually and with a spray race  190
18.2. Tick decoy attached to tail switch of a cow  191
18.3. AppliGator self-medicating applicator, and cattle feeding from trough with necks rubbing on AppliGator  192

**Tables**

3.1. Hosts of the tick *Amblyomma chabaudi*  15
3.2. Hosts of the tick *Amblyomma compressum*  15
3.3. Hosts of the tick *Amblyomma exornatum*  17
3.4. Hosts of the tick *Amblyomma falsomarmoreum*  18
3.5. Hosts of the tick *Amblyomma flavomaculatum*  19
3.6. Hosts of the tick *Amblyomma gemma*  20
3.7. Hosts of the tick *Amblyomma hebraeum*  21
3.8. Hosts of the tick *Amblyomma latum*  23
3.9. Hosts of the tick *Amblyomma lepidum*  24
3.10. Hosts of the tick *Amblyomma marmoreum*  26
3.11. Hosts of the tick *Amblyomma nuttalli*  28
3.12. Hosts of the tick *Amblyomma pomposum*  29
3.13. Hosts of the tick *Amblyomma rhinocerotis*  30
3.14. Hosts of the tick *Amblyomma sparsum*  31
3.15. Hosts of the tick *Amblyomma splendidum*  32
3.16. Hosts of the tick *Amblyomma sylvaticum*  33
3.17. Hosts of the tick *Amblyomma tholloni*  34
3.18. Hosts of the tick *Amblyomma transversale*  35
3.19. Hosts of the tick *Dermacentor rhinocerinus*  35
3.20. Hosts of the tick *Haemaphysalis elongata*  36
3.21. Hosts of the tick *Haemaphysalis hoodi*  37
3.22. Hosts of the tick *Haemaphysalis leachi*  37
3.23. Hosts of the tick *Haemaphysalis muhsamae*  38
3.24. Hosts of the tick *Hyalomma albiparmatum*  39
3.25. Hosts of the tick *Hyalomma impressum*  39
3.26. Hosts of the tick *Hyalomma truncatum*  40
3.27. Hosts of the tick *Ixodes pilosus*  42
3.28. Hosts of the tick *Ixodes schillingsi*  42
3.29. Hosts of the tick *Rhipicephalus appendiculatus*  43
3.30. Hosts of the tick *Rhipicephalus capensis*  45
3.31. Hosts of the tick *Rhipicephalus compositus*  46
3.32. Hosts of the tick *Rhipicephalus evertsi*  48
3.33. Hosts of the tick *Rhipicephalus kochi*  49
3.34. Hosts of the tick *Rhipicephalus muehlensi*  50
3.35. Hosts of the tick *Rhipicephalus pulchellus*  51
3.36. Hosts of the tick *Rhipicephalus senegalensis*  52
3.37. Hosts of the tick *Rhipicephalus simus*  53
3.38. Hosts of the tick *Rhipicephalus sulcatus*  54
3.39. Hosts of the tick *Rhipicephalus (Boophilus) decoloratus*  55
4.1. Hosts of the tick *Amblyomma variegatum*  59
5.1. Hosts of the tick *Hyalomma dromedarii*  63
6.1. Hosts of the tick *Hyalomma lusitanicum*  64
7.1. Hosts of the tick *Haemaphysalis punctata*  67
7.2. Hosts of the tick *Hyalomma aegyptium*  68
7.3. Hosts of *Hyalomma anatolicum* spp. ticks  69
7.4. Hosts of the tick *Hyalomma detritum*  71
7.5. Hosts of ticks of the *Hyalomma marginatum* group  73
7.6. Hosts of the tick *Ixodes ricinus*  75
7.7. Hosts of the tick *Rhipicephalus bursa*  78
7.8. Hosts of the tick *Rhipicephalus turanicus*  79
8.1. Hosts of the tick *Ixodes hexagonus*  82
9.1. Hosts of the tick *Dermacentor reticulatus*  83
10.1. Hosts of the tick *Amblyomma clypeolatum*  86
10.2. Hosts of the tick *Amblyomma crassipes*  87
10.3. Hosts of the tick *Amblyomma fuscolineatum*  87
10.4. Hosts of the tick *Amblyomma geoemydae*  88
10.5. Hosts of the tick *Amblyomma javanense*  89
10.6. Hosts of the tick *Amblyomma komodoense*  90
10.7. Hosts of the tick *Amblyomma kraneveldi*  91
10.8. Hosts of the tick *Amblyomma testudinarium*  91

10.9. Hosts of the tick *Amblyomma varanense*  93
10.10. Hosts of the tick *Dermacentor auratus*  93
10.11. Hosts of the tick *Dermacentor nuttalli*  95
10.12. Hosts of the tick *Haemaphysalis hystricis*  95
11.1. Hosts of the tick *Amblyomma fimbriatum*  97
11.2. Hosts of the tick *Amblyomma helvolum*  97
11.3. Hosts of the tick *Amblyomma trimaculatum*  98
11.4. Hosts of the tick *Haemaphysalis longicornis*  98
12.1. Hosts of the tick *Amblyomma moreliae*  100
12.2. Hosts of the tick *Amblyomma triguttatum*  101
12.3. Hosts of the tick *Bothriocroton concolor*  102
12.4. Hosts of the tick *Bothriocroton hydrosauri*  103
13.1. Hosts of the tick *Amblyomma albopictum*  106
13.2. Hosts of the tick *Amblyomma argentinae*  107
13.3. Hosts of the tick *Amblyomma auricularium*  109
13.4. Hosts of the tick *Amblyomma cajennense*  110
13.5. Hosts of the tick *Amblyomma calcaratum*  113
13.6. Hosts of the tick *Amblyomma coelebs*  114
13.7. Hosts of the tick *Amblyomma dissimile*  115
13.8. Hosts of the tick *Amblyomma geayi*  116
13.9. Hosts of the tick *Amblyomma humerale*  117
13.10. Hosts of the tick *Amblyomma incisum*  118
13.11. Hosts of the tick *Amblyomma longirostre*  119
13.12. Hosts of the tick *Amblyomma multipunctum*  119
13.13. Hosts of the tick *Amblyomma nodosum*  120
13.14. Hosts of the tick *Amblyomma oblongoguttatum*  120
13.15. Hosts of the tick *Amblyomma ovale*  121
13.16. Hosts of the tick *Amblyomma parvum*  122
13.17. Hosts of the tick *Amblyomma pictum*  123
13.18. Hosts of the tick *Amblyomma pseudoconcolor*  124
13.19. Hosts of the tick *Amblyomma quadricavum*  124
13.20. Hosts of the tick *Amblyomma rotundatum*  125
13.21. Hosts of the tick *Amblyomma sabanerae*  126
13.22. Hosts of the tick *Amblyomma scutatum*  127
13.23. Hosts of the tick *Amblyomma varium*  128
13.24. Hosts of the tick *Dermacentor nitens*  129
13.25. *Dermacentor nitens* infestations on horses in Florida and Texas, 1958–1989  129
13.26. Hosts of the tick *Haemaphysalis juxtakochi*  131
13.27. Hosts of the tick *Ixodes luciae*  131
14.1. Hosts of the tick *Rhipicephalus (Boophilus) annulatus*  133
14.2. *Rhipicephalus (Boophilus) annulatus* infestations on native animals in Texas, 1962–1989  134
14.3. Hosts of the tick *Rhipicephalus (Boophilus) microplus*  135
14.4. *Rhipicephalus (Boophilus) microplus* infestations on native animals in Texas, 1962–1989  136
16.1. Invasive ticks that have spread to exotic host species from different geographical regions  161
16.2. Invasive ticks that have spread to host species native to the United States  164
16.3. Invasive ticks that have, or may have, become established in limited foci in the United States  165
16.4. Non-native vertebrate species established in Florida since 1900  166
16.5. Invasive ticks reported feeding on humans  168
16.6. Diseases caused by invasive ticks  169
16.7. Diseases transmitted to humans naturally by invasive ticks  169
16.8. Diseases of cattle for which invasive ticks are vectors  171
16.9. Selected hosts of invasive tick species that are vectors of heartwater  173
16.10. Diseases of sheep and goats for which invasive ticks are vectors  176
16.11. Diseases of other mammals for which invasive ticks are vectors  176

# 1

# Introduction

Several books have been written on ticks, but the majority have focused on tick identification and taxonomy. Few have provided comprehensive information on the diseases that ticks cause or transmit even though ticks are considered second only to mosquitoes in importance as vectors of diseases of humans (Parola & Raoult, 2001) and are the most important arthropod vectors of diseases of domestic and wild animals. Fewer still have mentioned the ticks of reptiles, amphibians, birds, or wild mammals such as marsupials, armadillos, sloths, and anteaters, all of which are continuously transported around the world by the international trade in live animals. Finally, no book has addressed the issue of non-native and invasive ticks and the risks their introduction poses to countries such as the United States. This book will focus on all 100 non-native ticks that have been introduced into the continental United States over the past half-century, documenting details of their introduction, their hosts, their geographical distribution, their life cycle and habitat, their disease associations, the risks they pose to the health of humans and animals, and methods for their control and eradication. Unlike other books, which typically mention merely the diseases associated with ticks, this book includes a separate chapter with short reviews of every tickborne disease discussed, providing the reader an appreciation of the relevance of each disease to human and/or animal health without the need to consult other publications. Tick identification and taxonomy have been discussed in detail in many publications and, therefore, will be considered in this book only when there has been past confusion regarding the identity of the ticks reported in earlier publications or changes in scientific names of ticks.

## 1.1. Layout of Book

Chapter 2 introduces the reader to the nature of invasive species in general and non-native and invasive ticks in particular. The problem of non-native and invasive ticks has increased in recent years as the international trade in live animals has increased, and both the legal and illegal aspects of this important trade are reviewed.

Chapters 3–14 provide detailed information, by geographical origin of the tick species, on each of the 100 non-native tick species introduced into the continental United States. For each tick species, the following information is given, where available: details of introductions, hosts, geographical distribution, life cycle and habitat, disease associations, and selected references for those who wish to explore a tick species in greater depth. Data for these chapters were gathered from an extensive search of the world scientific literature using databases such as Web of Science, BIOSIS Previews, CAB Abstracts, MEDLINE and Zoological Record. One challenging task was to standardize the names of hosts. This was accomplished by using only the scientific and common names of host species as found in the United Nations Environment Programme/World Conservation Monitoring Centre Species Database (http://www.unep-wcmc.org/isdb/Taxonomy/index.cfm). Data on tick introductions are few, and they fall primarily into three categories:

(1) U.S. Department of Agriculture (USDA) reports, primarily the national tick surveillance reports for the years 1962–1989 and the cooperative tick eradication reports for the years 1962–1974; (2) scattered publications in the scientific literature on introductions of ticks on animals; and (3) a series of publications by the author and colleagues on recent introductions of ticks on reptiles. Unfortunately, since 1989, data on non-native ticks introduced into the United States have been collected by USDA but rarely published and made available to the scientific community; therefore, the extent of the non-native tick problem in the United States is far greater than that depicted in this book.

The nature of the diseases transmitted and/or caused by non-native ticks is discussed in chapter 15 as a prerequisite to an understanding of the risks they pose to public and animal health of the United States. These risks are analyzed in chapter 16. Methods for the regulation of global trade in animals and for the control and eradication of non-native ticks are discussed in chapter 17. The final chapter provides a summary of current problems in efforts to prevent establishment of non-native ticks in the United States and offers actions that, if implemented, would help to minimize the risks posed by these ticks to public and animal health.

Finally, the book ends with a detailed listing of the 669 references cited from the total 2,127 articles reviewed and that provided data for this book, followed by two appendixes, a glossary, and an index. Among the references cited are "USDA (unpublished data)" and "UF (unpublished data)"; the former relates to ticks identified at the USDA National Veterinary Services Laboratory and the latter to ticks identified in the author's laboratory at the University of Florida. Appendix 1 is a comprehensive list of all introductions of non-native ticks into the continental United States that have been documented in the scientific literature, but it is realized that this is only a fraction of the true number of actual introductions. Appendix 2 is a comprehensive list of all known hosts of the 100 non-native tick species. Hosts are listed by class, order, family, genus, and species in alphabetical order for ease of review. The common names of classes, families, and species are included. Beside each host are listed the tick species found to infest them, together with, where available, the stages of the tick and an indication of which tick species are common parasites of that host. The glossary contains primarily medical and herpetological words that might be unfamiliar to some readers.

This book is written for several audiences. It is targeted particularly to two subject areas that have received little to no coverage in previous publications, namely non-native and invasive ticks and the diseases with which they are associated, and the ticks of reptiles. The book provides detailed information that will be invaluable to those involved in controlling ticks disseminated globally by the international trade in animals and, in particular, to those interested in the trade in reptiles and other wild animals. As such, the book is of value to parasitologists, to wildlife biologists, to government and international agencies concerned with control of diseases, to physicians, veterinarians and other scientists involved with tickborne diseases, and to those involved in the reptile industry. This book is the only source of data on non-native and invasive ticks for the United States, a concise source of information on tickborne diseases (including several emerging diseases) of both humans and animals, and the only reference book on ticks of reptiles.

## 1.2. Tick Taxonomy

All tick names used in this book are those listed in the recent taxonomic revision of ticks by Horak et al. (2002). This revision was necessitated by the confusion in tick nomenclature. Several recent studies have redescribed certain tick species after having discovered misidentifications in earlier publications. Where problems with the taxonomic status of a given tick species or group of ticks raised questions about the identity of ticks in past publications, the author has discussed these problems in the introduction to the given tick species and, where appropriate, excluded references where the identity of the tick remains uncertain or is in error. The stages of ticks found on specific hosts has been included where the information was available in the reviewed literature; however, it must be remembered that the immature stages (larvae and nymphs) of tick species are often difficult to identify and, therefore, the data on immatures in this book should be taken only as a guide to potential host species.

## 1.3. Acknowledgments

The author's studies on invasive ticks of reptiles were made possible only by the excellent technical assistance of Petey Simmons, whose contributions to the research are noted with deep gratitude. Preparation of this book was possible only due to the invaluable assistance of Janice Kahler, who obtained innumerable scientific articles for me through interlibrary loan services, of Debra Couch, who provided advice on computer technologies, and of Kia Hendrix, who typed various drafts of the manuscript with such patience and expertise. The support of each of them has been greatly appreciated. In addition, several scientific colleagues provided the author with generous technical assistance. Alberto Guglielmone critically reviewed the manuscript and provided the author with insights into issues related to tick taxonomy. Santiago Nava, Ray Ashton, Kenny Krysko, Kevin Enge, Scott Citino, Darryl Heard, Mariano Mastropaolo, and Andres Pautasso kindly provided images used to illustrate the book. Nicholas Campiz used his superior cartographic skills to prepare the maps in the book. Jim Mertins of the U.S. Department of Agriculture provided confirmation of tick identifications throughout the author's studies. Again, the assistance of each of these colleagues has been much appreciated.

# 2

# Invasive Species

Recent years have seen an increase in attention given to the introduction, establishment, and spread of non-native (alien, nonindigenous, or exotic) species in new environments. Many of these non-native species can cause economic damage, irreversible ecological changes, and significant impacts on human, animal, and plant health in their new environments and, thus, have been reported as harmful invasive species (Andersen et al., 2004). In 1999, President Clinton signed Executive Order 13112 establishing the National Invasive Species Council, in which an invasive species was defined as "an alien species whose introduction does or is likely to cause economic or environmental harm or harm to human health" (Clinton, 1999). In 2006, The National Invasive Species Council clarified the definition of an invasive species as "a non-native species whose introduction does or is likely to cause economic or environmental harm or harm to human, animal, or plant health" (http://www.invasivespecies-info.gov/docs/council/isacdef.pdf). The council further pointed out that, for a non-native species to be considered an invasive species, the negative effects the species causes or is likely to cause must be deemed to outweigh any beneficial effects, and it called this the concept of causing harm. Examples of direct negative effects of invasive species on native species cited by the council included the causing or vectoring of diseases.

## 2.1. International Trade in Live Animals

Many of the non-native animal species have been introduced into the United States through the thriving international trade in animals. For example, 710 of 799 species (89%) of reptiles, 172 of 195 species (88%) of amphibians, 559 of 653 species (86%) of birds and 263 of 308 species (85%) of mammals imported into the United States between 2000 and 2004 were non-native (Jenkins et al., 2007). The sheer size of this global trade in live animals is staggering. For example, more than 1.07 billion individually counted live animals were imported into the United States from 163 countries during the five-year period 2000–2004 according to U.S. Fish and Wildlife Service data cited by Jenkins et al. (2007) which is, on average, an incredible 588,382 animals per day for five years. These imported animals included vertebrates (amphibians, birds, fishes, mammals, and reptiles) and invertebrates (annelids, arachnids, cnidarians, crustaceans, insects, mollusks, and sponges), but not accompanying parasites such as ticks. They were imported for food and products, for exhibitions at zoos, for scientific education, research, and conservation programs, and, increasingly, for the pet and aquarium industries. Within the thriving commercial pet trade, the fastest-growing sector is the international trade in live reptiles, which is responsible for the introduction of many invasive animal species into the United States.

A recent study of the international trade in live amphibians has confirmed that this trade has introduced into the United States two major disease threats to amphibian populations (Schloegel et al., 2009): chytridiomycosis, caused by the fungus *Batrachochytrium dendrobatidis*, and ranaviral disease. Chytridiomycosis has been associated with high mortality rates in adult anurans and implicated in a series of declines and extinction events globally, and ranaviral disease is known to cause amphibian mortality, often with large die-offs. Current evidence strongly suggests that the international trade in live amphibians has spread chytridiomycosis to pet, zoo, and laboratory animals, that the trade in U.S. salamanders as fish bait has resulted in the geographic expansion of ranaviruses, and that the bullfrog trade has introduced both diseases into the United States. These findings have prompted the World Organization for Animal Health (see section 17.1) to list chytridiomycosis and ranaviral disease of amphibians as notifiable diseases. This situation exemplifies the inherent dangers of the international trade in live animals as a means of global dissemination of dangerous diseases.

## 2.2. International Trade in Live Reptiles

The international trade in live reptiles has been active for many years, with the pet trade recently becoming a significant consumer of a wide variety of live tortoises, turtles, lizards, and snakes. This trade has grown dramatically over the past two decades, with the United States responsible for more than 80% of the total world trade in live reptiles listed by the Convention on International Trade in Endangered Species of Wild Fauna and Flora (CITES). More than 18.3 million live reptiles, representing more than 600 taxa, were imported into the United States from 1989 through 1997 (Telecky, 2001). In 1997, lizards were the most commonly imported reptiles (70%), followed by tortoises and turtles (15%), snakes (13%), and crocodilians (1%). Most live reptiles are imported through the ports of Los Angeles (California) and Miami (Florida), which are also home to the largest reptile importers.

The United States is the largest importer of live reptiles in the world, with about 9 million reptiles kept as pets in the United States in 2000 (Telecky, 2001). The sheer scale of this trade can be appreciated from the statistics of a surveillance program of imported reptiles conducted by the U.S. Department of Agriculture (USDA) at Miami International Airport in Florida from November 1994 to January 1995. During that three-month period, a total of 349 shipments, comprising 117,690 animals from 142 reptilian and two amphibian species, were imported into this one airport from 22 different countries (Clark & Doten, 1995).

In addition to the import trade, the United States is also a major exporter of reptiles, exporting more live reptiles than it imports. For example, while more than 18.3 million reptiles were imported into the United States from 1989 through 1997, more than 57.8 million were exported during that same time period (Telecky, 2001). These exports included reexports, reptiles that had been imported into the United States and later exported, as opposed to reptiles merely shipped through the United States in transit to their final destinations. The United States is playing an expanding role as a reexporter in the live reptile trade, largely because of its geographical location, whereby it acts as a supplier to Asia of imported African species, to Europe as a supplier of imported Asian and Pacific species, and to Asia and Europe as a supplier of imported Central and South American species.

Once in the United States, imported reptiles can be acquired through traditional sources such as pet stores and superstores; however, alternative sources are flourishing. Reptiles are bought and sold over the Internet via breeder Web pages, classified advertisements, e-mails and online auctions. Expos (reptile breeders' expositions), shows, and swap meets are other places to buy reptiles, with some events occurring on a monthly basis in some states. Some of the large reptile expos can have attendance of up to 10,000 people, with more than 500 vendor tables and over 100,000 reptiles on display. Swap meets are small informal gatherings of reptile owners and/or collectors who come together to buy or trade reptiles; they are usually sponsored by local herpetological groups.

Along with the dramatic increase in the legal trade in live reptiles, there has been an apparent increase in the illegal trade in live reptiles both to and from the United States. This illegal trade continues to flourish (Hoover, 1998), particularly in Australia and Madagascar (Fig. 2.1). Australia has long been a focus of illegal trade because it prohibits the export for commercial purposes of all its reptiles and because so many of these reptiles are found nowhere else on earth; however, despite these legal prohibitions, a substantial number of Australian species of reptiles are available on the U.S. market, showing that the illegal trade is alive and well.

Fig. 2.1. Spider tortoises (*Pyxis arachnoides*) imported from Madagascar at a reptile dealer's premises in southern Florida.

## 2.3. Invasive Animal Species in Florida

The thriving international trade in live animals has introduced many invasive species into the United States. Although the problems caused by invasive species exist throughout the country, Florida has been identified as one of two states with the most severe problems (Sementelli et al., 2008). Much of the focus on invasive animal species in Florida has centered on some vertebrates that, because of their size and uniqueness, have captured the attention of both the media and the public. Examples are the green iguana (*Iguana iguana*), the Nile monitor (*Varanus niloticus*), the red lionfish (*Pterois volitans*), the Gambian rat (*Cricetomys gambianus*) and the Burmese python (*Python molurus*).

The green iguana is native to Latin America (from Mexico south to Bolivia and Paraguay) and the Caribbean. It was first reported in the wild in southern Florida in the 1960s (Townsend et al., 2003). Since that time, the green iguana has become a widely distributed and well-established exotic species in southern Florida, primarily because of a lack of predators and an abundance of food. It is usually found near water, often in trees or on embankments bordering canals and lakes, or even basking on lawns or pavements in urban and suburban areas (Fig. 2.2). A recent report by Sementelli et al. (2008) discusses some negative impacts of this invasive species. They found that the green iguana population has increased dramatically in the past few years along canals and levees in southern Florida, with their burrowing habits leading to both bank erosion and instability of the canals and levees. They also point out that green iguanas have been observed impacting endangered species such as the Florida burrowing owl (*Athene cunicularia*) and the gopher tortoise (*Gopherus polyphemus*) by usurping their burrows. Finally, they mention that green iguanas have been identified as dispersal agents, consuming the fruits of invasive plants and depositing seeds throughout their ranges, to the extent that the expansion and distribution of some invasive plants mirror the expansion and distribution of green iguana populations in southern Florida.

The Nile monitor is native to sub-Saharan Africa, where it is the longest lizard, attaining a maximum length of 2.4 m (see Fig. 3.1). It was first observed free in Cape Coral in southwestern Florida about 1990 (Enge et al., 2004). Nile monitors are found in residential areas, canal banks and surrounding pine flatwoods, mangroves, and barrier islands in the Cape Coral area.

Even though the Nile monitor is the second-most commonly sold African monitor species in the United States, with most individuals imported from Africa, there are few lizards less suited to life in captivity because of their large adult size and nervous disposition. It is well known that in Florida, pet monitors often escape or are intentionally released by owners when the novelty of the exotic pet wanes, when they outgrow their cages, when they become too expensive to feed, or when their temperament and size make them difficult to handle. Furthermore, monitors may be illegally released by reptile dealers when they are too damaged or sick to sell, or when the dealers wish to establish a local breeding population for later capture and resale. Enge et al. (2004) have discussed possible ecological impacts of Nile monitors from knowledge of this species in Africa. They concluded that Nile monitors have the potential to significantly reduce native wildlife species by preying upon them or competing for habitat or food, specifically mentioning colonial waterbirds such as the brown pelican (*Pelecanus occidentalis*), sea turtles, diamondback terrapins (*Malaclemys terrapin*), American crocodiles (*Crocodylus acutus*), burrowing owls, and gopher tortoises as native species that would be affected.

The red lionfish is native to Asia, Australia, and the Pacific, with its range extending from southern Japan through Indonesia to Australia, Micronesia, and French Polynesia. It is a strikingly colorful fish with distinctive red, maroon, and white stripes, fleshy tentacles above the eyes and below the mouth, a fan-like pectoral fin, and long separated dorsal, anal, and pelvic spines. The red lionfish is a valued aquarium fish; its first documented release in the eastern United States took place in Florida in 1992, when six lionfish were freed when Hurricane Andrew destroyed a large marine aquarium in Biscayne Bay (Ruiz-Carus et al., 2006). The red lionfish is the first exotic marine fish to have become established off the coast of Florida and now appears to be widely distributed along the shallow continental shelf of the eastern United States, from the Florida Keys north to Cape Hatteras in North Carolina (Schofield, 2009). The ecological impacts of red lionfish on the reefs off the southeastern United States are unknown, but it will pose a threat to fishermen, divers, and others because it is venomous, capable of injecting venom through its dorsal, anal, and pelvic fin spines.

The Gambian rat is native to western and central Africa and among the largest members of the rodent family Muridae, with males achieving weights as much as 2.8 kg. Eight Gambian rats escaped from the pen of an exotic pet breeder on Grassy Key in southern Florida in 1999 (Perry et al., 2006). In 2003, monkey-

Fig. 2.2. A green iguana (*Iguana iguana*) in southern Florida. (Photograph courtesy of Kevin Enge.)

Fig. 2.3. A Burmese python (*Python molurus*) captured in southern Florida. (Photograph courtesy of Kenneth Krysko.)

pox virus was introduced into the United States by a pet-trade shipment of infected Gambian rats imported from Ghana by an exotic animal distributor in Illinois. The distributor kept the rats in close proximity to prairie dogs (*Cynomys* sp.) that subsequently were sold to private individuals through an animal distributor in Wisconsin (Ligon, 2004). Within a few months, 72 people in six states became ill with monkeypox through bites or contact with body fluids from prairie dogs that had become infected with the monkeypox virus through contact with the Gambian rats. The presence of a breeding population of Gambian rats on Grassy Key was confirmed in 2004. Its large size, high fecundity, omnivorous diet, and associations with zoonotic diseases such as monkeypox make the Gambian rat a threat to the native ecological communities within the Florida Keys and beyond, should it spread to the U.S. mainland (Engeman et al., 2006).

The Burmese python is native to southeastern Asia. It is one of the largest snakes in the world, as an adult reaching 8 m in length and 90 kg in weight. From 1995 to 2005, 160 Burmese pythons were found in Everglades National Park in southern Florida (Krysko et al., 2008); it is believed this wild population originated from unwanted pets released in the park (Rodda et al., 2009). The Burmese python is now well established in southern Florida (Fig. 2.3) and spreading northward. The enormous range of body sizes of Burmese pythons from hatchlings to adults allows them at some life stage to eat most terrestrial vertebrate species found in Florida, and, in fact, animals ranging in size from wrens to white-tailed deer have been removed from the stomachs of Burmese pythons in Florida. Furthermore, even modest-sized Burmese pythons are capable of killing humans; in 1993, a female captive Burmese python measuring only 3.36 m in length and weighing just 24 kg killed a 15-year-old boy weighing 43 kg in Colorado (Chiszar et al., 1993). Following the killing of a 2-year-old girl in Florida in July 2009 by a pet Burmese python, the Florida Fish and Wildlife Conservation Commission began considering a ban on possession of Burmese pythons and several other large non-native reptiles such as Nile monitors. Furthermore, recent studies by the U.S. Geological Survey have estimated that ecological risks associated with establishment of Burmese pythons in the United States are high (Reed & Rodda, 2009) and that areas of the United States climatically suitable for establishment of Burmese pythons range up the Atlantic and Pacific coasts and across the south to include most of California, Texas, Oklahoma, Arkansas, Louisiana, Mississippi, Alabama, Florida, Georgia, and South and North Carolina (Rodda et al., 2009). It is evident, therefore, that the Burmese python is a very dangerous invasive species whose continued northward spread must be stopped as a matter of urgency.

As dramatic as these invasive vertebrate species are, a potentially greater threat is posed by invasive invertebrates able to introduce and transmit diseases pathogenic to both humans and animals. Invasive ticks are a prime example of such invertebrates since they are second only to mosquitos as vectors of human diseases and the most important arthropod vectors of diseases of domestic and wild animals.

## 2.4. Non-native and Invasive Ticks

### 2.4.1. Biology of ticks

Ticks are obligate hematophagous arthropod ectoparasites that infest every class of vertebrates in almost every region of the world, including Antarctica. Ticks are arachnids belonging to the class Arachnida and the subclass Acari. There are three families of ticks: the Ixodidae or hard ticks, which have a sclerotized dorsal scutal plate, the Argasidae or soft ticks, which have a flexible leathery cuticle, and the Nuttalliellidae, which contains only a single species. All the ticks discussed in this book are ixodid ticks belonging to the family Ixodidae.

Ticks comprise an anterior capitulum and a posterior body or idiosoma. The capitulum bears the mouthparts (the chelicerae and hypostome) and the palps. Ticks are pool feeders, tearing the membranes and capillaries of the dermis of skin with the chelicerae. They attach themselves to host skin using their toothed hypostome to suck the blood and other fluids that exude into the bite wound. The body bears the walking legs, four pairs in adults and nymphs and three pairs in larvae. Ixodid ticks have a hard sclerotized plate on the dorsal body surface called the scutum. The scutum covers the entire dorsum in male ticks, so that males exhibit only slight growth of their bodies during feeding. In contrast, in the other stages of the life cycle, the scutum covers only the anterior part of the body with the remainder of the body covered by an expansible cuticle, allowing these stages to increase in size enormously during feeding. Cuticle growth takes time; thus these stages in the life cycle are by necessity slow feeders, remaining attached and engorging for many days while the process of cuticle growth is completed.

Ticks have three feeding stages in their life cycle, namely the larva which hatches from the egg, the nymph, and the adult, either male or female. Figure 2.4 depicts the stages in the life cycle of the Cayenne

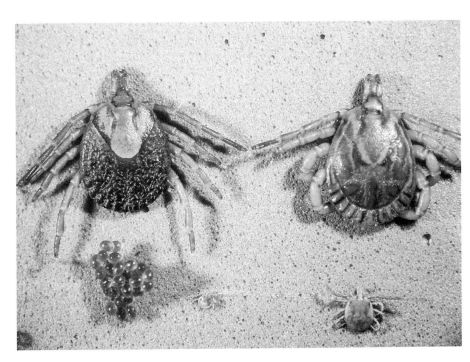

Fig. 2.4. Stages in the life cycle of the Cayenne tick *Amblyomma cajennense*. Counterclockwise from bottom left: eggs, larva, nymph, male and female. (Photograph courtesy of Santiago Nava.)

tick *Amblyomma cajennense*, showing a small batch of eggs, a small six-legged larva, a larger eight-legged nymph, and a much larger pair of unengorged adults, a male (4.6 x 2.9 mm) with a scutum covering the entire body, and a female (4.9 x 3.2 mm) with a scutum covering only the anterior part of the body. Ixodid ticks have one of three types of life cycle, a one-host life cycle in which all three parasitic stages feed on the same host, a two-host life cycle in which both the larval and nymphal stages feed on one host with the adult stage feeding on a second host, or a three-host life cycle in which each parasitic stage feeds on a different host. Each parasitic stage of the tick feeds only once but for a relatively long period of several days. Mating of adults occurs typically on the host; when the female is fully engorged, she detaches from the host and drops to the ground to digest her blood meal and lay from several hundred to more than 20,000 eggs (depending on tick species) in a sheltered environment, after which she dies. The life cycle of ixodid ticks takes anywhere from several months to a few years to complete, depending on a number of factors, including environmental conditions such as temperature, relative humidity, and photoperiod.

The two primary methods by which ixodid ticks transmit infectious pathogens are transstadial or transovarial. Transstadial transmission is stage-to-stage transmission within a generation, whereby a larva (or nymph) picks up an infectious agent from an infected host and then transmits it to a susceptible host as a nymph (or adult). Transovarial transmission, on the other hand, is transmission between generations through the egg, whereby a female picks up an infectious agent and then transmits it through its eggs to larvae that infect susceptible hosts.

For a detailed account of the biology of ticks, the reader is referred to Sonenshine (1991).

### 2.4.2. Invasive ticks

Ixodid ticks are enormously important parasites of both humans and animals. They cause direct damage by prolonged feeding with large mouthparts, some cause tick paralysis, tick toxicosis, or blockage of nasal passages, and, most significantly, many transmit serious and often fatal diseases. Ticks possess many unusual features that contribute to their remarkable success as parasites and as vectors of diseases (Sonenshine, 1991). They are long lived (with life cycles often measured in years), they consume large volumes of blood (up to 4–5 ml/tick), they are prodigious egg producers (with up to 22,891 produced by a single female), and they surpass all other arthropods including insects in the variety of pathogenic organisms they can transmit (including fungi, protozoa, bacteria, and viruses). It is evident, therefore, that non-native ticks fit perfectly into the definition of invasive species provided by the National Invasive Species Council, since they are introduced species that are purely parasitic on their hosts, causing harm to both human and animal health with no noticeable beneficial effects. **Consequently, hereinafter, all non-native ticks introduced into the continental United States will be described as invasive species in this book.**

During the past half-century a total of 100 species of invasive ticks have been introduced into the continental United States from many countries in Africa, Asia, Europe, Australasia, the Pacific, the Caribbean, Mexico, Central America, and South America. These invasive ticks comprise 56 species from the genus *Amblyomma* (*A. albopictum, A. argentinae, A. auricularium, A. cajennense, A. calcaratum, A. chabaudi, A. clypeolatum, A. coelebs, A. compressum, A. crassipes, A. dissimile, A. exornatum, A. falsomarmoreum, A. fimbriatum, A. flavomaculatum, A. fuscolineatum, A. geayi, A. gemma, A. geoemydae, A. hebraeum, A. helvolum, A. humerale, A. incisum, A. javanense, A. komodoense, A. kraneveldi, A. latum, A. lepidum, A. longirostre, A. marmoreum, A. moreliae, A. multipunctum, A. nodosum, A. nuttalli, A. oblongoguttatum, A. ovale, A. parvum, A. pictum, A. pomposum, A. pseudoconcolor, A. quadricavum, A. rhinocerotis, A. rotundatum, A. sabanerae, A. scutatum, A. sparsum, A. splendidum, A. sylvaticum, A. testudinarium, A. tholloni, A. transversale, A. triguttatum, A. trimaculatum, A. varanense, A. variegatum* and *A. varium*), two species from the genus *Bothriocroton* (*B. concolor* and *B. hydrosauri*), five species from the genus *Dermacentor* (*D. auratus, D. nitens, D. nuttalli, D. reticulatus* and *D. rhinocerinus*), eight species from the genus *Haemaphysalis* (*Ha. elongata, Ha. hoodi, Ha. hystricis, Ha. juxtakochi, Ha. leachi, Ha. longicornis, Ha. muhsamae* and *Ha. punctata*), nine species from the genus *Hyalomma* (*Hy.*

*aegyptium, Hy. albiparmatum, Hy. anatolicum* spp., *Hy. detritum, Hy. dromedarii, Hy. impressum, Hy. lusitanicum, Hy. marginatum* group and *Hy. truncatum*), five species from the genus *Ixodes* (*I. hexagonus, I. luciae, I. pilosus, I. ricinus* and *I. schillingsi*), 12 species from the genus *Rhipicephalus* (*R. appendiculatus, R. bursa, R. capensis, R. compositus, R. evertsi, R. kochi, R. muehlensi, R. pulchellus, R. senegalensis, R. simus, R. sulcatus* and *R. turanicus*) and three species from the subgenus *Rhipicephalus Boophilus* (*R. (B.) annulatus, R. (B.) decoloratus* and *R. (B.) microplus*). Of these 100 species of invasive ticks, 94 are truly exotic to the continental United States, while two (*A. dissimile* and *A. rotundatum*) have become established in southern Florida, one (*A. cajennense*) has become established in southern Texas, one (*A. auricularium*) appears to have become established in Florida, and two (*R. (B.) annulatus* and *R. (B.) microplus*) periodically cross the U.S.-Mexican border.

## 2.4.3. Modes of introduction of invasive ticks

Invasive ticks have been introduced into the continental United States in a variety of ways. As would be expected, most were introduced attached to their hosts; however, others were introduced by more unusual means, such as on animal products, on plant material, and on inanimate objects. Several ticks have entered the continental United States by multiple methods; *A. cajennense*, for example, has been introduced on humans, on domestic animals, on wild mammals, on frozen meat, on plant material, and on baggage. The most common methods of introduction were on wild mammals (48 invasive tick species), followed by reptiles (32), animal products (22), domestic animals (17), on plant material (13), birds (11), humans (11), amphibians (2), and baggage (2). The ticks introduced on animal hides and trophies were sometimes alive. The invasive tick species introduced by each method are listed below, demonstrating the diversity of ways in which ticks can cross international borders.

### Introduction on humans

Humans: *A. cajennense, A. hebraeum, A. latum, A. oblongoguttatum, A. parvum, A. triguttatum, A. variegatum, D. auratus, Hy. marginatum* group, *I. ricinus, R. pulchellus*

### Introduction on domestic animals

Cattle: *A. cajennense, A. variegatum, D. nitens, R. (B.) annulatus, R. (B.) microplus*
Sheep: *A. variegatum*
Pigs: *Ha. punctata*
Horses: *A. cajennense, A. variegatum, D. nitens, D. reticulatus, Ha. longicornis, Hy. anatolicum* spp., *Hy. detritum, Hy. marginatum* group, *I. ricinus, R. bursa, R. evertsi, R. (B.) annulatus, R. (B.) microplus*
Donkeys: *D. nitens, I. ricinus, R. evertsi*
Camels: *Hy. dromedarii, R. evertsi*
Dogs: *A. ovale, Ha. leachi*

### Introduction on wild mammals

Wild goats: *R. bursa*
Equids (ass, onager, zebra, zebra × horse): *A. gemma, A. hebraeum, A. variegatum, Hy. albiparmatum, Hy. anatolicum* spp., *Hy. marginatum* group, *Hy. truncatum, R. appendiculatus, R. evertsi, R. pulchellus, R. (B.) decoloratus*
Antelope (blesbok, bontebok, eland, gemsbok, hartebeest, impala, kudu, nilgai, nyala, oryx, sable antelope, topi, wildebeest): *A. hebraeum, A. pomposum, A. variegatum, D. reticulatus, Hy. truncatum, R. evertsi, R. muehlensi, R. pulchellus, R. (B.) decoloratus*
African buffalos: *A. hebraeum*
Giraffes: *A. gemma, A. hebraeum, A. variegatum, Hy. albiparmatum, R. evertsi, R. (B.) decoloratus*
Deer: *D. nitens, R. (B.) annulatus*
Tapirs: *A. incisum, A. multipunctum, Ha. hystricis*
Rhinoceroses: *A. gemma, A. hebraeum, A. rhinocerotis, A. sparsum, A. testudinarium, A. variegatum, D. rhinocerinus, Hy. albiparmatum, Hy. marginatum* group, *Hy. truncatum, R. pulchellus*
Elephants: *A. hebraeum, A. tholloni*
Hyraxes: *I. schillingsi*
Carnivores (cougar, fox, jackal, lion, ocelot): *A. ovale, D. nitens, Ha. leachi, Ha. muhsamae, R. simus*
Pangolins: *A. compressum, A. javanense*
Xenarthrans (anteater, armadillo, sloth): *A. cajennense, A. geayi, A. nodosum, A. parvum, A. pictum, A. pseudoconcolor, A. varium, R. (B.) microplus*
Rodents (capybara, porcupine): *A. cajennense, A. longirostre*

Insectivores (hedgehog, tenrec): *A. nuttalli, D. reticulatus, Ha. elongata, Ha. leachi, Ha. muhsamae, I. hexagonus, R. sulcatus*

Marsupials (opossum): *I. luciae* (on opossum in shipment of bananas)

Monotremes (echidna): *B. concolor, B. hydrosauri*

Introduction on reptiles

Alligators: *A. dissimile*

Lizards: *A. albopictum, A. crassipes, A. dissimile, A. exornatum, A. falsomarmoreum, A. fimbriatum, A. flavomaculatum, A. helvolum, A. komodoense, A. latum, A. moreliae, A. nuttalli, A. rotundatum, A. scutatum, A. sparsum, A. trimaculatum, A. varanense, A. variegatum, I. ricinus*

Snakes: *A. dissimile, A. exornatum, A. fimbriatum, A. flavomaculatum, A. fuscolineatum, A. helvolum, A. kraneveldi, A. latum, A. moreliae, A. nuttalli, A. quadricavum, A. rotundatum, A. sabanerae, A. sparsum, A. transversale, A. varanense*

Tortoises: *A. chabaudi, A. clypeolatum, A. falsomarmoreum, A. humerale, A. marmoreum, A. nuttalli, A. rotundatum, A. sabanerae, A. sparsum, A. sylvaticum, Hy. aegyptium*

Turtles: *A. dissimile, A. geoemydae, A. rotundatum, A. sabanerae*

Unidentified reptile: *A. argentinae*

Introduction on amphibians

Toads: *A. dissimile, A. rotundatum*

Introduction on birds

Ostriches: *A. gemma, A. lepidum, A. variegatum, Ha. punctata, Hy. albiparmatum, Hy. lusitanicum, Hy. marginatum* group, *Hy. truncatum, R. turanicus*

Parrots: *Ha. hoodi*

Migratory birds: *A. longirostre*

Introduction on animal products

Animal hides and trophies: *A. gemma, A. hebraeum, A. lepidum, A. splendidum, A. tholloni, A. variegatum, D. nuttalli, Hy. albiparmatum, Hy. impressum, Hy. lusitanicum, Hy. marginatum* group, *Hy. truncatum, R. appendiculatus, R. evertsi, R. muehlensi, R. pulchellus, R. senegalensis, R. simus, R. (B.) annulatus, R. (B.) decoloratus, R. (B.) microplus*

Frozen beef: *A. cajennense*

Refrigerated beef: *R. (B.) microplus*

Hair: *R. (B.) microplus*

Wool: *R. appendiculatus*

Introduction on plant material

Cut flowers: *I. pilosus, R. capensis, R. kochi*

Plants (dracaena, fern, orchid): *A. cajennense*

Palm fronds: *A. cajennense, Ha. juxtakochi*

Fruits (banana, lime, pineapple, plantain): *A. cajennense, A. nodosum, I. luciae* (on opossum in shipment of bananas)

Seeds: *A. coelebs*

Cork: *Hy. marginatum* group, *I. hexagonus*

Unidentified plant material: *D. auratus, I. pilosus, R. compositus, R. (B.) microplus*

Introduction on inanimate objects

Baggage: *A. cajennense, Hy. anatolicum* spp.

### 2.4.4. Role of migratory birds in dissemination of ticks

Birds are important hosts for several of the invasive tick species, and those that migrate play a significant role in carrying and dispersing ticks and the infectious agents they harbor. In the Western Hemisphere, migrating birds have been implicated in the international dissemination of four invasive tick species: *A. longirostre, A. humerale, A. sabanerae,* and *A. variegatum*.

The Latin American tick *A. longirostre* has been introduced into the United States on a migrating white-eyed vireo (*Vireo griseus*), a migrating summer tanager (*Piranga rubra*), and a migrating northern cardinal (*Cardinalis cardinalis*) (Eads et al., 1956; Snetsinger, 1968; Durden & Kollars, 1992). Furthermore, *A. longirostre* was introduced into Canada on a migrating yellow-bellied flycatcher (*Empidonax flaviventris*), on migrating willow flycatchers (*Empidonax trailii*), on a migrating red-eyed vireo (*Vireo olivaceus*), and on a migrating Canada warbler (*Wilsonia canadensis*) (Scott et al., 2001; Morshed et al., 2005). Although porcupines and other rodents are the preferred hosts for *A. longirostre* adults, 99 species of birds are known to be hosts for the larval and nymphal stages of this tick. It is clear, therefore, that birds play a role in the international dispersal of *A. longirostre*.

The Latin American ticks *A. sabanerae* and *A. humerale* have been introduced into Canada on a migrating veery (*Catharus fuscescens*) and a migrating grey-cheeked thrush (*Catharus minimus*), respectively (Scott et al., 2001; Morshed et al., 2005); however, both are ticks of reptiles and, since these are the only documented records of *A. sabanerae* and *A. humerale* infesting birds, their spread by birds would appear to be a rare occurrence.

The tropical bont tick *A. variegatum* spread from Africa to the eastern Caribbean on infested cattle around 1828, where it has since become well established, with larval and nymphal stages parasitizing local birds such as cattle egrets (*Bubulcus ibis*), Carib grackles (*Quiscalus lugubris*), and black-faced grassquits (*Tiaris bicolor*) (Barre et al., 1988, 1991). There is strong circumstantial evidence that much of the recent inter-island spread of *A. variegatum* in the Caribbean has occurred through the movement of infested migratory birds, in particular cattle egrets. Furthermore, the potential for cattle egrets to introduce *A. variegatum* into the United States was graphically demonstrated in 1992 when a cattle egret banded on Guadeloupe, where *A. variegatum* is plentiful, was found on Long Key in Florida (Corn et al., 1993).

In the Eastern Hemisphere, migrating birds have been implicated in the international dissemination of six invasive tick species, *A. lepidum*, *A. nuttalli*, *A. variegatum*, *Ha. punctata*, the *Hy. marginatum* group, and *I. ricinus*.

The African tick *A. lepidum* has been introduced into Azerbaijan on a migrating stone curlew (*Burhinus oedicnemus*) and into Cyprus on a migrating blackbird (*Turdus merula*) and migrating European turtle doves (*Streptopelia turtur*) (Kaiser et al., 1974). Furthermore, Yeruham et al. (1996) believed that *A. lepidum* was probably introduced into Israel by migrating birds from Africa. Although *A. lepidum* is a tick of ungulates, it has been found infrequently on 12 avian species; migrating birds thus appear to play some role in the spread of this tick to Europe and Asia.

Another African tick, *A. nuttalli*, has been introduced into Cyprus on a migrating thrush nightingale (*Luscinia luscinia*) (Kaiser et al., 1974). *Amblyomma nuttalli* is a tick of reptiles but has been found as larvae or nymphs on 16 avian species; thus, migrating birds may play some role in the spread of this tick as far as the Mediterranean.

The African tick *A. variegatum* has been introduced into Cyprus on a migrating tree pipit (*Anthus trivialis*) (Kaiser et al., 1974). In addition, *A. variegatum* adults have been found on a dog in France, on animals in Sicily, and on cattle in Russia, and, in each case, the authors (Hoogstraal, 1967; Albanese et al., 1971; Voltzit & Keirans, 2003) speculated that the infestations had been introduced into these European regions by migrating birds carrying *A. variegatum* nymphs. Even though it is primarily a tick of ungulates, various stages of *A. variegatum* have been found on 47 species of birds. Given this avian predisposition and previously cited reports from the Caribbean and Europe, it is evident that migrating birds are involved in the dissemination of this important tick species over considerable distances.

Three important tick species of ungulates are the *Hy. marginatum* group, *I. ricinus* and *Ha. punctata*, with each having a very wide host range that includes 262, 161, and 121 species, respectively, of birds as hosts for larval and nymphal stages. Several studies of birds migrating between Africa, Europe, and Asia have found many avian species infested with immature stages of one or more of these tick species (Hoogstraal & Kaiser, 1958, 1961a, b; Kaiser & Hoogstraal, 1958; Hoogstraal et al., 1961, 1963, 1964; Nuorteva & Hoogstraal, 1963; Brinck et al., 1965; Saikku et al., 1971; Kaiser et al., 1974; Siuda & Dutkiewicz, 1979; Nosek & Balat, 1982; Mehl et al., 1984; Olsen et al., 1995; Alekseev et al., 2001; Bjoersdorff et al., 2001; Papadopoulos et al., 2001; Hubalek, 2004; Comstedt et al., 2006; Poupon et al., 2006; Waldenstrom et al., 2007), demonstrating the role played by migratory birds in the spread of the *Hy. marginatum* group, *I. ricinus*, and *Ha. punctata*.

# 3

# Ticks from Africa

A total of 51 invasive tick species from Africa (Map 3.1) have been introduced into the continental United States. The 39 species whose distribution is limited to Africa are discussed in this chapter. The other 12 species with more extensive distributions are discussed in chapter 4 (*Amblyomma variegatum*), chapter 5 (*Hyalomma dromedarii*), chapter 6 (*Hy. lusitanicum*), chapter 7 (*Haemaphysalis punctata*, *Hy. anatolicum* spp., *Hy. detritum*, *Hy. marginatum* group, *Ixodes ricinus*, *Rhipicephalus bursa* and *R. turanicus*) and chapter 14 (*Rhipicephalus (Boophilus) annulatus* and *R. (B.) microplus*).

## 3.1. *Amblyomma chabaudi* Rageau

The tick *Amblyomma chabaudi* was introduced into the continental United States on one occasion in 2001 on spider tortoises (*Pyxis arachnoides*) imported into Florida from Madagascar (Simmons & Burridge, 2002).

### Hosts

*Amblyomma chabaudi* is a rarely reported tick of spider tortoises (see Fig. 2.1), the only host on which all three stages of the tick have been found. The only other hosts known to have been parasitized by *A. chabaudi* are radiated tortoises (*Astrochelys radiata*) and flat-backed spider tortoises (*Pyxis planicauda*) (Table 3.1). On spider tortoises, *A. chabaudi* has been found attached to both forelegs and hindlegs.

### Distribution

The geographical distribution of *A. chabaudi* is restricted to the island of Madagascar.

### Disease Associations

None reported.

### References

Uilenberg (1965, 1967), Uilenberg et al. (1979), Voltzit & Keirans (2003).

## 3.2. *Amblyomma compressum* Macalister

The tick *Amblyomma compressum* was introduced into the continental United States on one occasion on a pangolin (*Manis* sp.) imported into Ohio from Togo (Keirans & Durden, 2001).

### Hosts

*Amblyomma compressum* is a tick of pangolins and in particular the tree pangolin (*Manis tricuspis*) (Table 3.2). Occasional hosts include ungulates, birds and hyraxes.

### Distribution

The geographical distribution of *A. compressum* ranges over much of sub-Saharan Africa. It has been reported from western Africa (Gambia, Ghana, Guinea, Ivory Coast, Liberia, Nigeria, Senegal and Togo), central

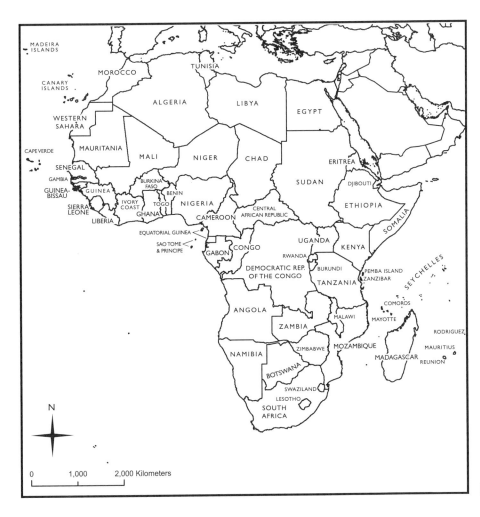

Map 3.1. Africa

Table 3.1. Hosts of the tick *Amblyomma chabaudi*

| Hosts | Stages[a] | No. of reports |
|---|---|---|
| **Tortoises** | | |
| *Pyxis arachnoides* (spider tortoise) | A  N  L | 9 |
| *Astrochelys radiata* (radiated tortoise) | A | 4 |
| *Pyxis planicauda* (flat-backed spider tortoise) | - | 2 |

[a] A = adult, N = nymph, L = larva, - = not recorded

Table 3.2. Most commonly reported hosts of the tick *Amblyomma compressum*

| Hosts | Stages[a] | No. of reports |
|---|---|---|
| **Pangolins** | | |
| *Manis tricupsis* (tree pangolin) | A  N  L | 10 |
| *Manis tetradactyla* (long-tailed pangolin) | A     L | 4 |
| **Ungulates** | | |
| *Tragelaphus scriptus* (bushbuck) | A | 4 |
| *Cephalophus niger* (black duiker) | A | 2 |
| **Birds** | | |
| *Tauraco persa* (Guinea turaco) | A | 3 |

[a] A = adult, N = nymph, L = larva

Fig. 3.1. A Nile monitor (*Varanus niloticus*), the preferred host of the tick *Amblyomma exornatum* and a common host of the tick *A. flavomaculatum*. (Photograph courtesy of iStockphoto.com.)

Africa (Cameroon, Central African Republic, Congo, Democratic Republic of the Congo, Equatorial Guinea and Gabon), eastern Africa (Kenya and Uganda) and southern Africa (Angola, Botswana, Mozambique, Namibia and South Africa). The distribution of *A. compressum* is similar to that of its pangolin hosts, with the tree pangolin, the long-tailed pangolin (*Manis tetradactyla*) and the giant pangolin (*Manis gigantea*) the principal hosts in western and central Africa and the Cape pangolin (*Manis temminckii*) in eastern and southern Africa.

### Life Cycle

The life cycle of *A. compressum* has been studied in the Ivory Coast by Aeschlimann (1963, 1967). He reported a preoviposition period of 4–15 days, an egg incubation period of 32 days, a larval prefeeding period of 20 days, a larval feeding period of 9 days, a larval premolting period of 10–15 days, a nymphal prefeeding period of 30 days, a nymphal feeding period of 12 days, a nymphal premolting period of 30 days, a female prefeeding period of 30 days and a female feeding period of 4–14 days, giving a generation period of 181–207 days for *A. compressum*.

### Disease Associations

None reported.

### References

Aeschlimann (1963), Matthysse & Colbo (1987), Voltzit & Keirans (2003).

### 3.3. *Amblyomma exornatum* Koch

The taxonomic status of *Amblyomma exornatum* and *Amblyomma arcanum* has been a matter of debate for several decades, with Kaufman (1972) considering *A. arcanum* a synonym of *A. exornatum* and later authors (Santos Dias, 1993; Horak et al., 2002) considering them two different species. In this book, *A. exornatum* is treated as a separate species from *A. arcanum*.

The tick *A. exornatum*, previously known as *Aponomma exornatum*, has been reported to have been introduced into the continental United States on at least 19 occasions between 1969 and 2001 on lizards and snakes imported into California, Florida, Maryland, Nebraska and New Jersey from African countries such as Tanzania and Togo (see Appendix 1). The saurian hosts were Nile monitors (*Varanus niloticus*) (Fig.

Table 3.3. Most commonly reported hosts of the tick *Amblyomma exornatum*

| Hosts | Stages[a] | No. of reports |
|---|---|---|
| LIZARDS | | |
| *Varanus niloticus* (Nile monitor) | A  N  L | 33 |
| *Varanus exanthematicus* (African savanna monitor) | A  N  L | 20 |
| *Varanus albigularis* (white-throated monitor) | A  N  L | 13 |
| *Acontias plumbeus* (giant lance skink) | A | 8 |
| SNAKES | | |
| *Python sebae* (African rock python) | A | 8 |

[a]A = adult, N = nymph, L = larva

3.1), white-throated monitors (*Varanus albigularis*), African savanna monitors (*Varanus exanthematicus*) and a great plated lizard (*Gerrhosaurus major*), and the ophidian hosts, ball pythons (*Python regius*) and a mamba (*Dendroaspis* sp.).

## Hosts

*Amblyomma exornatum* is a one-host reptilian tick. The preferred hosts for all stages are lizards, in particular the Nile, African savanna and white-throated monitors (Table 3.3). Other reptilian hosts of adults include snakes such as the African rock python (*Python sebae*). Occasional hosts include ungulates, carnivores, squirrels, birds and tortoises.

Although *A. exornatum* is an African tick, it has infested hosts native to Asia such as the water monitor (*Varanus salvator*) and Gray's monitor (*Varanus olivaceus*) while they were in captivity (Burridge et al., 2000a; Reeves et al., 2006), demonstrating the potential for *A. exornatum* to spread outside its native geographical range to new reptilian hosts.

Sites of attachment of *A. exornatum* on monitor lizards are the head, including the nasal passage, the legs including the axillae, the elbow joint, between the toes and the cloacal region. Sites of attachment on snakes include the head.

## Distribution

The geographical distribution of *A. exornatum* covers much of sub-Saharan Africa. It has been reported from western Africa (Gambia, Ghana, Guinea-Bissau, Ivory Coast, Mali, Niger, Nigeria and Senegal), from central Africa (Cameroon, Congo, Democratic Republic of the Congo, Equatorial Guinea and Gabon), from eastern Africa (Burundi, Djibouti, Eritrea, Ethiopia, Kenya, Rwanda, Somalia, Sudan, Tanzania and Uganda) and from southern Africa (Angola, Botswana, Malawi, Mozambique, Namibia, South Africa, Swaziland, Zambia and Zimbabwe). It has also been found on the islands of Sao Tome and Principe off the western coast of Africa. *Amblyomma exornatum* has been reported also from Algeria in northern Africa (Elbl & Anastos, 1966c). The distribution of *A. exornatum*, of which all stages feed on the same host, is similar to that of its principal hosts, the African monitor lizards.

Recently in 2005 *A. exornatum* was found on a native injured gopher tortoise in Hillsborough County in central Florida during a statewide tick survey (Berube, 2006). This is the first report of *A. exornatum* on native wildlife in the United States, suggesting that *A. exornatum* might have become established in Florida.

## Disease Associations

Heavy infestations of *A. exornatum* in the nasal passages can cause suffocation and death in monitor lizards (Young, 1965; Norval, 1985), with as many as 56 adults removed from the nostrils of a single Nile monitor exhibiting severe dyspnea. Arthur (1962) reported that *A. exornatum* was 'alleged' to harbor natural infections of *Coxiella burnetii*, the causative agent of Q fever, in Guinea-Bissau, but presented no data to demonstrate any role for *A. exornatum* in the transmission of the disease.

## References

Hoogstraal (1956), Theiler (1962), Yeoman et al. (1967), Kaufman (1972), Walker (1974), Matthysse & Colbo (1987), Burridge (2001).

## 3.4. *Amblyomma falsomarmoreum* Tonelli-Rondelli

The tick *Amblyomma falsomarmoreum* was introduced into the continental United States on two occasions in 1974, once on a leopard tortoise (*Stigmochelys pardalis*) imported into Arizona from Kenya and again on an unidentified lizard also imported into Arizona (USDA, 1975).

### Hosts

*Amblyomma falsomarmoreum* is a tick of reptiles and in particular tortoises (Table 3.4). Its preferred host is the leopard tortoise (Fig. 3.2). Other reptilian hosts include Bell's hinged tortoise (*Kinixys belliana*), the pancake tortoise (*Malacochersus tornieri*), the African savanna monitor and the Nile monitor. In addition, *A. falsomarmoreum* has been collected in Somalia from three species of domestic animals, cattle (*Bos taurus*), camels (*Camelus dromedarius*) and sheep (*Ovis aries*).

### Distribution

The geographical distribution of *A. falsomarmoreum* is limited to semiarid regions of eastern Africa, extending from Ethiopia and Somalia in the north through Kenya and Uganda to Tanzania in the south.

### Disease Associations

None reported.

### References

Theiler (1962), Walker (1974), van der Borght-Elbl (1977), Burridge & Simmons (2003), Voltzit & Keirans (2003).

## 3.5. *Amblyomma flavomaculatum* (Lucas)

The tick *Amblyomma flavomaculatum*, previously known as *Aponomma flavomaculatum*, was introduced into the continental United States on at least 27 occasions between 1978 and 2005 on lizards and snakes imported into California, Florida, Georgia, New York and Ohio from African countries such as Benin, Ghana, Tanzania and Togo (see Appendix 1). The saurian hosts were African savanna monitors, white-throated monitors and Nile monitors, and the ophidian hosts, ball pythons and African rock pythons.

### Hosts

*Amblyomma flavomaculatum* is a one-host reptilian tick of lizards and snakes. The preferred hosts for all stages are the African savanna and Nile monitors (see Fig. 3.1) and ball pythons (Table 3.5). Other reptilian hosts include African rock pythons, Gold's tree cobras (*Pseudohaje goldii*) and black-necked spitting cobras (*Naja nigricollis*). Occasional hosts include crocodiles. Reported infestations of pangolins and warthogs (*Phacochoerus africanus*) need to be confirmed.

Although *A. flavomaculatum* is an African tick, it has infested a host native to Asia, the water monitor, while it was in captivity (Wilson & Barnard, 1985), demonstrating the potential for *A. flavomaculatum* to spread outside its native geographical range to new reptilian hosts.

Attachment sites of adult *A. flavomaculatum* on monitors are the head, including inside the nasal cavities and the ear canals, on the legs and toes, on the belly and around the cloacal orifice. Nymphs are

Table 3.4. Most commonly reported hosts of the tick *Amblyomma falsomarmoreum*

| Hosts | Stages[a] | No. of reports |
|---|---|---|
| **Tortoises** | | |
| *Stigmochelys pardalis* (leopard tortoise) | A | 10 |
| *Kinixys belliana* (Bell's hinged tortoise) | - | 2 |
| *Malacochersus tornieri* (pancake tortoise) | - | 2 |
| **Lizards** | | |
| *Varanus exanthematicus* (African savanna monitor) | A | 4 |
| **Ungulates** | | |
| *Ovis aries* (sheep) | - | 4 |

[a] A = adult, - = not recorded

Fig. 3.2. A leopard tortoise (*Stigmochelys pardalis*), the preferred host of the ticks *Amblyomma falsomarmoreum*, *A. marmoreum* and *A. sparsum*. (Photograph courtesy of Ray Ashton.)

found engorging on the head, legs and axillae of monitors, while larvae attach preferentially on lateral body sites and the neck, where they are protected under the posterior edges of scales. Adults have been found attached to both the head and body of snakes.

### Distribution

The geographical distribution of *A. flavomaculatum* ranges across the northern half of sub-Saharan Africa. It has been reported from western Africa (Benin, Burkina Faso, Ghana, Guinea, Guinea-Bissau, Ivory Coast, Mali, Mauritania, Niger, Nigeria, Senegal, Sierra Leone and Togo), from central Africa (Cameroon, Central African Republic, Chad and Democratic Republic of the Congo) and from eastern Africa (Ethiopia, Kenya, Sudan, Tanzania and Uganda). Recently *A. flavomaculatum* was reported in Yemen in Asia across the Red Sea from Ethiopia (Ueckermann et al., 2006).

### Life Cycle

The life cycle of *A. flavomaculatum* has been studied in Guinea by Saratsiotis (1972). He reported a larval feeding period of 9–10 days, a nymphal feeding period of 6–8 days and a female feeding period of 21–28 days.

### Disease Associations

None reported.

### References

Kaufman (1972), Saratsiotis (1972), Matthysse & Colbo (1987), Burridge (2001).

Table 3.5. Most commonly reported hosts of the tick *Amblyomma flavomaculatum*

| Hosts | Stages[a] | No. of reports |
|---|---|---|
| LIZARDS | | |
| *Varanus exanthematicus* (African savanna monitor) | A  N  L | 25 |
| *Varanus niloticus* (Nile monitor) | A  N  L | 20 |
| SNAKES | | |
| *Python sebae* (African rock python) | A | 11 |
| *Python regius* (ball python) | A  N  L | 10 |
| *Pseudohaje goldii* (Gold's tree cobra) | - | 6 |

[a] A = adult, N = nymph, L = larva, - = not recorded

### 3.6. *Amblyomma gemma* Donitz

The tick *A. gemma* has been introduced into the continental United States on 16 occasions between 1956 and 1989 on wild ungulates, ostriches (*Struthio camelus*) and animal trophies imported into Florida, Illinois, Louisiana, Maryland, Michigan, New Jersey, New York, North Carolina and Texas from African countries such as Kenya and Tanzania (see Appendix 1). The ungulate hosts were zebras, giraffes (*Giraffa camelopardalis*) and a black rhinoceros (*Diceros bicornis*), and the animal trophies included a giraffe hide and a zebra hide. These findings clearly demonstrate that even dead hosts, in the form of animal hides, can introduce invasive ticks such as *A. gemma* into the United States.

### Hosts

*Amblyomma gemma* is a three-host tick of large ungulates. It has been collected from 25 species of ungulates, with cattle the preferred host (Table 3.6). Other ungulate hosts include goats (*Capra hircus*), sheep, camels, horses (*Equus caballus*), donkeys (*Equus asinus*) and mules (*Equus asinus* × *Equus caballus*) among domestic animals and giraffes, elands (*Taurotragus oryx*), African buffalos (*Syncerus caffer*) and warthogs among wild animals. Occasional hosts include carnivores, birds, primates including humans (*Homo sapiens*), reptiles, elephants and hares.

The sites of attachment of *A. gemma* to cattle are primarily the ventral part of the body, especially the axillae, groin, udder and scrotum, and the perineum. The tick is found most commonly on the escutcheon, groin and brisket of African buffalos, and on the back of and behind the ears of warthogs.

### Distribution

The geographical distribution of *A. gemma* has been limited to eastern Africa. It has been reported from Djibouti, Eritrea, Ethiopia, Kenya, Somalia, Tanzania (including the island of Zanzibar) and Uganda. Some reports have indicated that *A. gemma* has crossed the Red Sea to become established in Saudi Arabia on local native livestock (Abou-Elela et al., 1981; Al-Khalifa et al., 1987; Ueckermann et al., 2006) and in Yemen (Kolonin, 2009).

Apart from its introduction into the United States, *A. gemma* has been introduced on imported cattle and camels into other countries such as Egypt, Kuwait, Saudi Arabia, Ukraine and Yemen (Klushkina, 1972; Abou-Elela et al., 1981; Hoogstraal et al., 1981; Al-Khalifa et al., 1987; Ueckermann et al., 2006). Hoogstraal et al. (1981) even reported the introduction of *A. gemma* into the Seychelles on a tourist from eastern Africa.

### Life Cycle and Habitat

The life cycle of *A. gemma* has been studied in Kenya by Theiler et al. (1956). They reported a preoviposition period of 12 days, an egg incubation period of 69 days, a larval feeding period of 5 days, a larval premolting period of 17 days, a nymphal feeding period of 6 days, a nymphal premolting period of 28 days and a female feeding period of 12 days. In eastern Africa, *A. gemma* is confined to the drier parts of the region, with annual rainfall of 40–75 cm.

### Disease Associations

*Amblyomma gemma* is a tick of minor significance to animal health since it is a vector of *Theileria mutans* (Paling et al., 1981), a protozoan causing theileriosis in

Table 3.6. Most commonly reported hosts of the tick *Amblyomma gemma*

| Hosts | Stages[a] | No. of reports |
|---|---|---|
| **Ungulates** | | |
| *Bos taurus* (cattle) | A  N  L | 37 |
| *Giraffa camelopardalis* (giraffe) | A | 17 |
| *Capra hircus* (goat) | A | 14 |
| *Ovis aries* (sheep) | A | 14 |
| *Camelus dromedarius* (camel) | A | 11 |
| *Syncerus caffer* (African buffalo) | A | 11 |
| *Taurotragus oryx* (eland) | A | 11 |

[a]A = adult, N = nymph, L = larva

cattle. Additionally, *A. gemma* is an experimental vector of *Ehrlichia ruminantium* (Bezuidenhout, 1987), the rickettsia causing heartwater in domestic and wild ruminants. Theiler et al. (1956) stated that *A. gemma* transmitted Nairobi sheep disease without comment; Hoogstraal et al. (1981) later considered this incrimination of *A. gemma* as a vector of Nairobi sheep disease to be equivocal. Sang et al. (2006) found *A. gemma* infected with Dugbe and Thogoto viruses, the latter virus reported associated with abortions in sheep and clinical disease in humans (Labuda & Nuttall, 2004).

### References

Robinson (1926), Theiler (1962), Yeoman et al. (1967), Walker (1974), van der Borght-Elbl (1977), Matthysse & Colbo (1987), Voltzit & Keirans (2003).

## 3.7. *Amblyomma hebraeum* Koch

The "South African bont tick" *Amblyomma hebraeum* was introduced into the continental United States on 29 occasions between 1962 and 2001 on wild animals, humans and animal trophies imported into or entering Arizona, California, Colorado, Connecticut, District of Columbia, Florida, Georgia, Maine, Maryland, Mississippi, Missouri, New York, North Carolina, Oklahoma, Texas and Virginia from African countries such as Namibia and South Africa (see Appendix 1). The wild animal hosts were white rhinoceroses (*Ceratotherium simum*), African buffalos, black rhinoceroses, an eland, an African elephant (*Loxodonta africana*), a giraffe and a zebra, and the animal trophies included rhinoceros and zebra hides.

### Hosts

*Amblyomma hebraeum* is an important three-host tick with a wide host range. It has been found on 115 species of animals, including 36 species of ungulates, 33 species of birds, 23 species of carnivores, 8 species of reptiles, 5 species of rodents, 4 species of primates, 2 species of lagomorphs, and one species each of pangolins, elephants, hyraxes and insectivores (see Appendix 2). The preferred hosts for all stages are cattle (Table 3.7). Other ungulate hosts infested by all stages include goats and sheep among domestic animals and impalas (*Aepyceros melampus*), African buffalos, black and white rhinoceroses, greater kudus (*Tragelaphus strepsiceros*), warthogs, elands, giraffes, blue wildebeests (*Connochaetes taurinus*) and grey duikers (*Sylvicapra grimmia*) among wildlife. In addition

Table 3.7. Most commonly reported hosts of the tick *Amblyomma hebraeum*

| Hosts | Stages[a] | No. of reports |
|---|---|---|
| **Ungulates** | | |
| *Bos taurus* (cattle) | A  N  L | 59 |
| *Capra hircus* (goat) | A  N  L | 25 |
| *Aepyceros melampus* (impala) | A  N  L | 19 |
| *Syncerus caffer* (African buffalo) | A  N  L | 19 |
| *Ceratotherium simum* (white rhinoceros) | A  N  L | 18 |
| *Ovis aries* (sheep) | A  N  L | 18 |
| *Tragelaphus strepsiceros* (greater kudu) | A  N  L | 14 |
| *Giraffa camelopardalis* (giraffe) | A  N  L | 13 |
| *Phacochoerus africanus* (warthog) | A  N  L | 13 |
| *Taurotragus oryx* (eland) | A  N  L | 13 |
| *Diceros bicornis* (black rhinoceros) | A  N  L | 12 |
| *Connochaetes taurinus* (blue wildebeest) | A  N  L | 10 |
| *Sylvicapra grimmia* (grey duiker) | A  N  L | 10 |
| **Carnivores** | | |
| *Canis familiaris* (dog) | A  N  L | 17 |
| **Lagomorphs** | | |
| *Lepus saxatilis* (scrub hare) | N  L | 12 |
| **Primates** | | |
| *Homo sapiens* (human) | A  N  L | 10 |

[a] A = adult, N = nymph, L = larva

to ungulates, all stages of *A. hebraeum* will feed on carnivores such as dogs (*Canis familiaris*), jackals, lions (*Panthera leo*) and leopards (*Panthera pardus*), on ostriches and on humans. Larvae and nymphs are also found on scrub hares (*Lepus saxatilis*), leopard tortoises and helmeted guineafowl (*Numida meleagris*). Occasional hosts include non-human primates, rodents, pangolins, elephants, hyraxes, hedgehogs, lizards and snakes.

The preferred sites of attachment of *A. hebraeum* adults are the hairless areas under the tail, the perineum and axillae, the udder and around the genitalia of cattle as well as around the feet of sheep and goats. Nymphs are found on the feet, legs, groin, sternum and neck, and larvae on the feet, legs and muzzle. On impalas, all stages are found on the legs, with larvae also on the pinna of the ear. On elands, adults are found on the hindquarters, groin, abdomen, axillae and front legs. On ground-frequenting birds such as francolins, the larvae are found mainly on the head.

### Distribution

The geographical distribution of *A. hebraeum* is limited to southern Africa. It has been reported from Botswana, Mozambique, South Africa, Swaziland, Zambia and Zimbabwe. Recently it was collected from Aldabra giant tortoises (*Aldabrachelys gigantea*) in a nature park on the coast of Kenya in East Africa (Okanga & Rebelo, 2006; Wanzala & Okanga, 2006).

### Life Cycle and Habitat

*Amblyomma hebraeum* is a three-host tick, with each stage feeding on a different host. Larvae feed for 7–14 days, nymphs for 7–14 days and females for 7–9 days, after which the females will detach and lay up to 20,000 eggs (Walker et al., 2003). In contrast, *A. hebraeum* males will remain attached to cattle for as long as eight months for two reasons: (1) to engage in multiple matings and (2) as a source of attraction-aggregation-attachment pheromone to allow unfed adults to locate hosts and discriminate between suitable hosts with males attached and potentially unsuitable hosts on which there are no males (Norval et al., 1991). The life cycle usually takes one year to complete but can expand to three years in the cooler areas of southern South Africa. Larvae are active during summer and autumn, nymphs during winter and early spring, and adults during summer. *Amblyomma hebraeum* requires a moist warm environment with brush and bush for cover and does not survive in open grassland.

### Disease Associations

*Amblyomma hebraeum* is a tick of importance to both animal and human health because it is:

- the major vector in southern Africa of *Ehrlichia ruminantium* (Lounsbury, 1900), the rickettsia causing heartwater in domestic and wild ruminants;
- a vector of *Theileria mutans* (Lawrence et al., 1981), a protozoan causing theileriosis in cattle;
- a vector of *Theileria velifera* (Uilenberg, 1983a), a protozoan causing benign theileriosis in cattle;
- a vector of *Rickettsia africae* (Kelly & Mason, 1991; Kelly et al., 1996), the bacterium causing African tick-bite fever in humans;
- a cause of sores that can form foot abscesses in sheep and goats attractive to the blowfly *Chrysomya bezziana*, whose larvae can cause severe myiasis and which can lead to secondary bacterial infections and loss of udder quarters in cattle (Walker et al., 2003; Jongejan & Uilenberg, 2004).

### References

Robinson (1926), Theiler (1962), Norval (1974, 1977), Horak et al. (1987), Petney et al. (1987), Petney & Horak (1988), Norval et al. (1991), Voltzit & Keirans (2003), Walker et al. (2003).

## 3.8. *Amblyomma latum* Koch

The tick *Amblyomma latum*, previously known as *Aponomma latum* (syn. *Aponomma ochraeum*), has been introduced into the continental United States on at least 59 occasions between 1969 and 2005 on snakes, lizards and a human imported into or entering Arizona, California, Colorado, Connecticut, Florida, Georgia, Illinois, Indiana, Kansas, Kentucky, Louisiana, Maryland, Michigan, Mississippi, New Hampshire, New York, North Carolina, Ohio, Pennsylvania, South Carolina, Tennessee and Texas from African countries such as Benin, Ghana, Tanzania and Togo (see Appendix 1). The ophidian hosts were ball pythons, African rock pythons, puff adders (*Bitis arietans*), Gabon vipers (*Bitis gabonica*), forest cobras

Table 3.8. Most commonly reported hosts of the tick *Amblyomma latum*

| Hosts | Stages[a] | No. of reports |
|---|---|---|
| **SNAKES** | | |
| *Python regius* (ball python) | A N L | 51 |
| *Python sebae* (African rock python) | A N L | 23 |
| *Naja nigricollis* (black-necked spitting cobra) | A N L | 22 |
| *Bitis arietans* (puff adder) | A N | 21 |
| *Causus rhombeatus* (common night adder) | A N L | 18 |
| *Naja melanoleuca* (forest cobra) | A N L | 17 |
| *Rhamphiophis oxyrhynchus* (western beaked snake) | A N L | 16 |
| *Bitis gabonica* (Gabon viper) | A | 15 |
| *Dendroaspis polylepis* (black mamba) | A N L | 14 |
| *Naja haje* (Egyptian cobra) | A N L | 14 |
| *Psammophis sibilans* (African beauty racer) | A N L | 14 |
| *Dendroaspis angusticeps* (common mamba) | A | 12 |
| *Lamprophis fuliginosus* (brown house snake) | A N L | 12 |
| *Dendroaspis viridis* (green mamba) | A N L | 11 |
| *Pseudaspis cana* (mole snake) | A | 11 |
| *Crotaphopeltis hotamboeia* (herald snake) | A N | 10 |

[a] A = adult, N = nymph, L = larva

(*Naja melanoleuca*), an African burrowing python (*Calabaria reinhardtii*), a black mamba (*Dendroaspis polylepis*) and a black-necked spitting cobra, and the saurian hosts, African savanna monitors, great plated lizards, white-throated monitors and a Nile monitor.

## Hosts

*Amblyomma latum* is a one-host reptilian tick, with snakes the preferred host. Of the 65 ophidian species infested by *A. latum*, the ball python is the most commonly reported host (Table 3.8), in all likelihood due to its popularity in the pet reptile trade. Other less common reptilian hosts are lizards including monitors. Occasional hosts include shrews, rodents, pangolins, tortoises, humans and toads.

Although *A. latum* is an African tick, it has infested hosts native to Asia such as the reticulated python (*Python reticulatus*) (USDA, 1994; Burridge et al., 2000a), hosts native to South America such as the boa constrictor (*Boa constrictor*) and yellow-footed tortoise (*Chelonoidis denticulata*) (USDA, 1977b, 1980, 1988; Santos Dias, 1993; Burridge, 2001) and hosts native to Europe such as the asp viper (*Vipera aspis*) (Aeschlimann & Buttiker, 1975) while they were in captivity, clearly demonstrating the potential for *A. latum* to spread outside its native geographical range to new reptilian hosts.

Sites of attachment of *A. latum* adults and nymphs include all parts of the body of snakes. The preferred site is the head, with some ticks even found partially hidden feeding in the eye socket (Fig. 3.3). Adults are usually found under the dorsal and lateral scales, whereas immatures are most commonly found under the first belt of lateral scales above the wide ventral scales. On tortoises, *A. latum* attaches to the head and, on monitor lizards, in the nostrils and ear canals.

## Distribution

The geographical distribution of *A. latum* covers most of Africa. It has been reported from northern Africa

Fig. 3.3. *Amblyomma latum* tick feeding in the eye socket of a ball python (*Python regius*), with arrow locating partially hidden tick.

(Egypt), from western Africa (Benin, Burkina Faso, Guinea, Guinea-Bissau, Ivory Coast, Liberia, Mali, Niger, Nigeria, Senegal, Sierra Leone and Togo), from central Africa (Cameroon, Central African Republic, Chad, Congo and Democratic Republic of the Congo), from eastern Africa (Burundi, Ethiopia, Kenya, Rwanda, Sudan, Tanzania including the island of Zanzibar, and Uganda) and from southern Africa (Angola, Malawi, Mozambique, Namibia, South Africa, Zambia and Zimbabwe). Some reports have indicated that *A. latum* has crossed the Red Sea to become established in Yemen and Saudi Arabia (Hoogstraal & Kaiser, 1959b; Theiler, 1962; Kaufman, 1972; Hoogstraal et al., 1981; Santos Dias, 1993; Al-Khalifa et al., 2006; Ueckermann et al., 2006).

**Disease Associations**

None reported.

**References**

Hoogstraal (1956), Theiler (1962), Yeoman et al. (1967), Kaufman (1972), Saratsiotis (1972), Walker (1974), Matthysse & Colbo (1987), Burridge (2001).

## 3.9. *Amblyomma lepidum* Donitz

The tick *Amblyomma lepidum* has been introduced into the continental United States on three occasions in 1965 and 1989, on a zebra hide imported into Maryland and later on ostriches imported into New York and Texas from Tanzania (see Appendix 1).

**Hosts**

*Amblyomma lepidum* is a three-host tick of ungulates, with cattle the preferred host (Table 3.9). Other common hosts include sheep, goats, camels and dogs among domestic animals and African buffalos and ostriches among wild animals. Adults and nymphs have been found attached to humans. Occasional hosts include aardvarks (*Orycteropus afer*) and birds other than ostriches.

The preferred sites of attachment of *A. lepidum* adults are the ventral surface of the body from the dewlap to the sternum, including the axillae, the genital areas and the udder. Nymphs favor the heels for feeding.

**Distribution**

*Amblyomma lepidum* is primarily a tick of eastern Africa, reported frequently in Eritrea, Ethiopia, Kenya, Somalia, Sudan, Tanzania and Uganda. There are isolated reports of *A. lepidum* west, north and south of eastern Africa in the Central African Republic, Chad, Egypt, Malawi and Zambia.

Apart from its introduction into the United States, *A. lepidum* has been introduced into Cyprus and Asian countries such as Azerbaijan, Iraq, Israel, Kuwait, Saudia Arabia and Syria (Feldman-Muhsam, 1955; Feldman-Muhsam & Saturen, 1961; Kohler et al., 1967; Robson et al., 1969; Kaiser et al., 1974; Abou-Elela et al., 1981; Hoogstraal et al., 1981; Yeruham et al., 1996; Voltzit & Keirans, 2003). At least four of these introductions involved imported or migrating animals;

Table 3.9. Most commonly reported hosts of the tick *Amblyomma lepidum*

| Hosts | Stages[a] | | | No. of reports |
|---|---|---|---|---|
| **Ungulates** | | | | |
| *Bos taurus* (cattle) | A | N | L | 39 |
| *Ovis aries* (sheep) | A | N | | 20 |
| *Capra hircus* (goat) | A | N | L | 18 |
| *Syncerus caffer* (African buffalo) | A | | | 15 |
| *Camelus dromedarius* (camel) | A | | | 12 |
| **Birds** | | | | |
| *Struthio camelus* (ostrich) | A | | | 11 |
| **Carnivores** | | | | |
| *Canis familiaris* (dog) | A | N | | 10 |

[a]A = adult, N = nymph, L = larva

adults were found on a migrating stone curlew in Azerbaijan (Kaiser et al., 1974), nymphs and larvae on a migrating blackbird (*Turdus merula*) and migrating European turtle doves, respectively, in Cyprus (Kaiser et al., 1974), adults on cattle imported into Kuwait from Kenya (Hoogstraal et al., 1981) and adults on sheep imported into Saudi Arabia from Somalia (Abou-Elela et al., 1981). Furthermore, Yeruham et al. (1996) believed that *A. lepidum* was probably introduced into Israel by migrating birds from Africa.

### Habitat

In eastern Africa, *A. lepidum* is most common in arid habitats with 250–750 mm of annual rainfall. It is unable to tolerate rainfall that exceeds 900 mm annually. The adults of *A. lepidum* are most abundant from October to February south of the equator, beginning around the onset of the rainy season, whereas they are most active from May to June north of the equator during the rainy season.

### Disease Associations

*Amblyomma lepidum* is a tick of minor significance to animal health since it is:

- a vector in Sudan of *Ehrlichia ruminantium* (Karrar, 1965), the rickettsia causing heartwater in domestic and wild ruminants;
- a vector of *Theileria velifera* (Uilenberg, 1983a), a protozoan causing benign theileriosis in cattle;
- a cause of direct damage to the udders of cattle, resulting in the loss of one or more quarters of the udder (Jongejan et al., 1987).

Parola (2006) reported that *Rickettsia africae*, the bacterium causing African tick-bite fever in humans, was detected in *A. lepidum* in Sudan. Sang et al. (2006) found *A. lepidum* infected with Dugbe and Thogoto viruses which, on rare occasions, may infect humans and cause disease.

### References

Robinson (1926), Hoogstraal (1956), Theiler (1962), Yeoman et al. (1967), Walker (1974), Matthysse & Colbo (1987), Voltzit & Keirans (2003), Walker et al. (2003).

## 3.10. *Amblyomma marmoreum* Koch

The "South African tortoise tick" *Amblyomma marmoreum* has often been confused in the earlier literature with morphologically similar ticks such as *A. sparsum*. This confusion was resolved by Theiler and Salisbury (1959) and, consequently, all references to *A. marmoreum* prior to 1959 have been excluded because of uncertainty of the identity of the ticks in those earlier publications.

*Amblyomma marmoreum* has been introduced into the continental United States on at least 20 occasions between 1965 and 2000 on tortoises imported into Florida, Maryland, New York and Texas from African countries such as Mozambique, South Africa and Zambia (see Appendix 1). The chelonian hosts were leopard tortoises, African spurred tortoises (*Geochelone sulcata*), Bell's hinged tortoises and Karoo Cape tortoises (*Homopus femoralis*).

### Hosts

*Amblyomma marmoreum* is a three-host reptilian tick whose principal hosts are tortoises, in particular leopard tortoises (Fig. 3.2) (Table 3.10). Other reptilian hosts infested by adults and immatures are snakes, in particular puff adders, and monitor lizards. Other common hosts of larvae and nymphs are scrub hares and helmeted guineafowl. Cape rock hyraxes (*Procavia capensis*) and a variety of ungulates and carnivores have also been found infested, including goats, cattle, sheep and dogs among domestic animals. Occasional hosts include birds (other than guineafowl), rats, elephant shrews, hedgehogs and, rarely, humans.

Although *A. marmoreum* is an African tick, it has infested hosts native to other geographical regions while they were in captivity, including the yellow-footed tortoise from South America, the Aldabra giant tortoise from the Aldabra Islands in the Indian Ocean and the Galapagos giant tortoise (*Chelonoidis nigra*) from the Galapagos Islands in the Pacific Ocean (Allan et al., 1998b; Burridge et al., 2000a). Also, during an *A. marmoreum* eradication program in Florida in which Hermann's tortoises (*Testudo hermanni*) from Europe were used as sentinel animals, several Hermann's tortoises became infested with *A. marmoreum* (UF, unpublished data). These findings demonstrate

Table 3.10. Most commonly reported hosts of the tick *Amblyomma marmoreum*

| Hosts | Stages[a] | No. of reports |
|---|---|---|
| **REPTILES** | | |
| *Stigmochelys pardalis* (leopard tortoise) | A N L | 32 |
| *Bitis arietans* (puff adder) | A N | 11 |
| *Kinixys belliana* (Bell's hinged tortoise) | A N L | 9 |
| **BIRDS** | | |
| *Numida meleagris* (helmeted guineafowl) | N L | 13 |
| **LAGOMORPHS** | | |
| *Lepus saxatilis* (scrub hare) | N L | 13 |
| **UNGULATES** | | |
| *Capra hircus* (goat) | N L | 9 |
| **CARNIVORES** | | |
| *Caracal caracal* (African caracal) | N L | 9 |

[a] A = adult, N = nymph, L = larva

the potential for *A. marmoreum* to spread outside its native geographical range to new chelonian hosts.

*Amblyomma marmoreum* is a very large tick, with many engorging females becoming enormous in size, measuring up to 2.7 cm long by 1.4 cm wide and weighing as much as 4.66 g. Preferred attachment sites on tortoises are areas of soft skin that allow for easy penetration and engorgement and are protected from physical abrasion by the carapace. Consequently, adults are found primarily on the hind legs and around the cloaca and on the tail. Attachment to posterior regions of tortoises by adults is particularly important to *A. marmoreum* females since it provides sufficient space and protection for these large ticks to reach full engorgement. Unlike adults, *A. marmoreum* immatures are found mainly on the anterior part of tortoises, with both nymphs and larvae feeding predominantly on the neck and forelegs (Fig. 3.4). On monitor lizards, *A. marmoreum* adults are commonly found around the cloaca. On mammals, sites of attachment have been studied only on impalas, where *A. marmoreum* were found on the ears and the fore and hind legs. On birds, *A. marmoreum* larvae favor the body and wings of crested francolins (*Francolinus sephaena*).

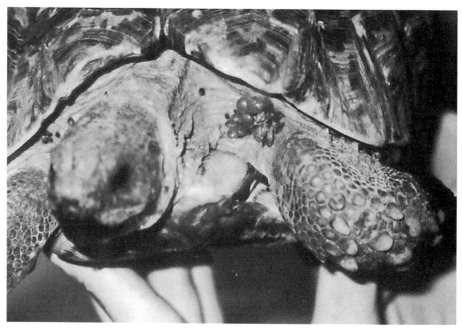

Fig. 3.4. *Amblyomma marmoreum* ticks feeding on base of foreleg under the shell of a tortoise.

### Distribution

The geographical distribution of *A. marmoreum* is limited to southern Africa. It has been reported from Botswana, Lesotho, Mozambique, Namibia, South Africa, Swaziland, Zambia and Zimbabwe, with the vast majority of reports from South Africa.

### Life Cycle and Habitat

The life cycle of *A. marmoreum* can take 160–373 days to complete on leopard tortoises, with larval feeding taking 8–104 days, nymphal feeding 4–77 days and adult feeding 1–111 days (Norval, 1975; Dower et al., 1988; Fielden et al., 1992). The slow rate of engorgement on tortoises is attributed to the low body temperatures of these poikilothermic hosts. The life cycle of *A. marmoreum* can be completed in either one or two years. Fielden et al. (1992) suggested that host selection by immature stages might be important in determining whether the life cycle is completed in a one- or two-year period under natural conditions. If the immature *A. marmoreum* feed on homoiothermic mammals or birds, the life cycle is more likely to be completed in one year than if all stages feed on poikilothermic tortoises. Furthermore, larvae that attach to tortoises in late autumn may not complete feeding until spring because of the low body temperature of tortoises in winter when they enter a state of semihibernation.

*Amblyomma marmoreum* exhibits a preference for grasslands with scrub or tree cover. It also requires moisture, being found in areas with 250–900 mm of annual rainfall but not in very arid habitats. Adult *A. marmoreum* are most numerous on tortoises in midsummer, nymphs in late summer and early spring, and larvae in late summer.

### Disease Associations

Peter et al. (2000) showed *A. marmoreum* to be a very capable experimental vector of *Ehrlichia ruminantium*, the rickettsia causing heartwater in domestic and wild ruminants. Even though *A. marmoreum* is primarily a tick of tortoises, all stages will feed on cattle and sheep, and nymphs and larvae will feed on goats, cattle and sheep, the major domestic animals susceptible to heartwater. Furthermore, African buffalos, black wildebeests (*Connochaetes gnou*), elands, giraffes, greater kudus, springboks (*Antidorcas marsupialis*) and steenboks (*Raphicerus campestris*) are all susceptible to heartwater (Peter et al., 2002), and all are known to be hosts of *A. marmoreum*. Consequently, *A. marmoreum* has the potential to feed on ruminants infected with *E. ruminantium* and to transmit heartwater to other susceptible ruminants, although such a scenerio has not yet been documented in nature.

### References

Theiler & Salisbury (1959), Theiler (1962), Norval (1975), Horak et al. (1987, 2006b), Petney & Horak (1988), Fielden & Rechav (1994), Burridge (2001), Voltzit & Keirans (2003).

## 3.11. *Amblyomma nuttalli* Donitz

The tick *Amblyomma nuttalli* has been introduced into the continental United States on at least 30 occasions between 1967 and 2005 on lizards, hedgehogs, snakes and tortoises imported into California, Florida, Massachusetts, Mississippi, Nevada, New Jersey, New York and Oklahoma from African countries such as Benin, Ghana and Togo (see Appendix 1). The saurian hosts were African savanna monitors, Nile monitors and common agamas (*Agama agama*), the ophidian hosts, ball pythons and Gabon vipers, and the chelonian hosts, Bell's hinged tortoises, a Home's hingeback tortoise (*Kinixys homeana*) and a Speke's hinged tortoise (*Kinixys spekii*).

### Hosts

*Amblyomma nuttalli* is a three-host reptilian tick. It is found most commonly on Bell's hinged tortoises and leopard tortoises among chelonians, African savanna and Nile monitors among lizards, and African rock pythons and Gabon vipers among snakes (Table 3.11). It is also found on a variety of ungulates, including cattle and goats, carnivores including dogs, rodents, insectivores and birds. All stages will feed on humans. Occasional hosts include bats and non-human primates.

Although *A. nuttalli* is an African tick, it has been found on a boa constrictor imported into Florida from Latin America (Burridge et al., 2000a), demonstrating the potential for *A. nuttalli* to spread outside its native geographic range to new reptilian hosts.

*Amblyomma nuttalli* has been found attached to the

Table 3.11. Most commonly reported hosts of the tick *Amblyomma nuttalli*

| Hosts | Stages[a] | | | No. of reports |
|---|---|---|---|---|
| **REPTILES** | | | | |
| *Kinixys belliana* (Bell's hinged tortoise) | A | N | | 24 |
| *Varanus exanthematicus* (African savanna monitor) | A | N | | 23 |
| *Python sebae* (African rock python) | A | N | | 10 |
| *Varanus niloticus* (Nile monitor) | A | N | | 8 |
| *Bitis gabonica* (Gabon viper) | A | N | | 8 |
| *Stigmochelys pardalis* (leopard tortoise) | A | | | 8 |
| **PRIMATES** | | | | |
| *Homo sapiens* (human) | A | N | L | 10 |
| **UNGULATES** | | | | |
| *Bos taurus* (cattle) | A | N | | 8 |
| **RODENTS** | | | | |
| *Atherurus africanus* (African brush-tailed porcupine) | | N | L | 8 |

[a] A = adult, N = nymph, L = larva

forelegs and axillae of tortoises, the belly and hindlegs of monitors and the heads of snakes. Furthermore, *A. nuttalli* nymphs have been reported to infest the nasal passages of humans.

## Distribution

The geographical distribution of *A. nuttalli* covers much of sub-Saharan Africa. It has been reported from western Africa (Benin, Burkina Faso, Ghana, Guinea, Guinea-Bissau, Ivory Coast, Niger, Nigeria, Senegal, Sierra Leone and Togo), from central Africa (Cameroon, Central African Republic, Chad and Democratic Republic of the Congo), from eastern Africa (Burundi, Ethiopia, Kenya, Rwanda, Somalia, Sudan, Tanzania and Uganda) and from southern Africa (Angola, Botswana, Malawi, Mozambique, Namibia, South Africa, Zambia and Zimbabwe).

An *A. nuttalli* nymph was found on a thrush nightingale in Cyprus during its fall migration (Kaiser et al., 1974), demonstrating that *A. nuttalli* can be disseminated outside its normal range by migrating birds.

## Life Cycle

The life cycle of *A. nuttalli* can take 134–296 days to complete (Hoogstraal, 1956). Adults take 9–50 days to complete engorgement on tortoises (Walker & Parsons, 1964), with a female capable of laying up to 22,891 eggs (Hoogstraal, 1956). This is the largest egg mass ever recorded for an individual tick, according to Oliver (1974).

## Disease Associations

Natural infections of *A. nuttalli* with the rickettsia *Coxiella burnetii*, the causative agent of the zoonotic disease Q fever, have been reported from Guinea-Bissau by Tendeiro and cited by Babudieri (1959) and Arthur (1962), even though Babudieri (1959) considered the report dubious. Theiler (1962) states that '*Am. nuttalli* has been proved to transmit the rickettsia *Coxiella burnetii*' but without providing any reference or supporting documentation. In 1978, an outbreak of Q fever was reported in New York among persons involved in unpacking and deticking a shipment of ball pythons imported from Ghana and infested with *A. nuttalli* and other ticks (Kim et al., 1978). Attempts to isolate *C. burnetii* rickettsiae from *A. nuttalli* were unsuccessful, and no causative association was made between the Q fever outbreak and the tick *A. nuttalli*. A later publication (Johnson-Delaney, 1996) discussed this Q fever outbreak and, without supportive references, called *A. nuttalli* a vector of *C. burnetii*.

These reports concerning Q fever provide no scientific evidence for an association between the disease and *A. nuttalli*; however, though the primary method of transmission of *C. burnetti* is through inhalation of infectious aerosols (Babudieri, 1959), the rickettsia is maintained in nature by ticks (Wilcocks & Manson-Bahr, 1972), with natural infections reported from 32 species of ixodid ticks, including *A. nuttalli* (Waag et al., 1991). Therefore, considering that all stages of *A. nuttalli* will feed on humans (Hoogstraal, 1956; Theiler

& Salisbury, 1959; Morel, 1961; Morel & Finelle, 1961; Aeschlimann, 1967; van der Borght-Elbl, 1977; Cornet, 1995; Estrada-Pena & Jongejan, 1999; Burridge, 2001), it is important that the role of *A. nuttalli* in the transmission of *C. burnetii* be clarified.

### References

Robinson (1926), Hoogstraal (1956), Theiler (1962), Elbl & Anastos (1966a), Walker (1974), van der Borght-Elbl (1977), Matthysse & Colbo (1987), Burridge (2001), Voltzit & Keirans (2003).

## 3.12. *Amblyomma pomposum* Donitz

The tick *Amblyomma pomposum* has been introduced into the continental United States on one occasion on a topi (*Damaliscus lunatus*) imported into New York (Becklund, 1968).

### Hosts

*Amblyomma pomposum* is a three-host tick of large ungulates. Its preferred hosts are cattle among domestic livestock and African buffalos, sable antelopes (*Hippotragus niger*), elands, warthogs, roan antelopes (*Hippotragus equinus*), hartebeests (*Alcelaphus buselaphus*) and topis among wild animals (Table 3.12). Occasional hosts include carnivores, monkeys, lizards, pangolins, elephants and rodents, with one report of a human infestation (Elbl & Anastos, 1966a). Adults attach to the underside of cattle.

### Distribution

The geographical distribution of *A. pomposum* is restricted to the wet highland areas of savanna and forest in southern Africa that extend from Angola in the west through Democratic Republic of the Congo and Zambia to western Tanzania in the east.

### Disease Associations

*Amblyomma pomposum* is a tick of minor significance to animal health since it is:

- a vector in Angola of *Ehrlichia ruminantium* (Walker & Olwage, 1987), the rickettsia causing heartwater in domestic and wild ruminants;
- a cause of severe skin damage to cattle, resulting in loss of productivity (Walker et al., 2003).

### References

Robinson (1926), Hoogstraal (1956), Theiler (1962), Elbl & Anastos (1966a), van der Borght-Elbl (1977), Voltzit & Keirans (2003), Walker et al. (2003).

## 3.13. *Amblyomma rhinocerotis* (de Geer)

The tick *Amblyomma rhinocerotis* has been introduced into the continental United States on two occasions, in 1967 and in 1989 on rhinoceroses imported into Alabama and Texas, respectively, from African countries such as Namibia (USDA, 1967b, 1994).

### Hosts

*Amblyomma rhinocerotis* is a tick whose preferred hosts are black rhinoceroses (Fig. 3.5) and white rhinoceroses (Table 3.13). Occasional hosts include cattle and other ungulates, lions, pythons and tortoises.

### Distribution

The geographical distribution of *A. rhinocerotis* has been limited to the eastern part of Africa from Sudan

Table 3.12. Most commonly reported hosts of the tick *Amblyomma pomposum*

| Hosts | Stages[a] | | No. of reports |
|---|---|---|---|
| **Ungulates** | | | |
| *Bos taurus* (cattle) | A | N | 19 |
| *Syncerus caffer* (African buffalo) | A | N | 17 |
| *Hippotragus niger* (sable antelope) | A | | 15 |
| *Phacochoerus africanus* (warthog) | A | N | 15 |
| *Taurotragus oryx* (eland) | A | | 14 |
| *Alcelaphus buselaphus* (hartebeest) | A | | 12 |
| *Hippotragus equinus* (roan antelope) | A | | 12 |
| *Damaliscus lunatus* (korrigum/topi) | A | | 10 |

[a] A = adult, N = nymph

Fig. 3.5. A black rhinoceros (*Diceros bicornis*), the preferred host of the ticks *Amblyomma rhinocerotis* and *Dermacentor rhinocerinus* and a common host of many other African ticks. (Photograph courtesy of iStockphoto.com.)

Table 3.13. Most commonly reported hosts of the tick *Amblyomma rhinocerotis*

| Hosts | Stages[a] | No. of reports |
|---|---|---|
| **Ungulates** | | |
| *Diceros bicornis* (black rhinoceros) | A | 22 |
| *Ceratotherium simum* (white rhinoceros) | A | 9 |
| *Bos taurus* (cattle) | A | 6 |
| *Hippopotamus amphibius* (hippopotamus) | A | 3 |
| *Taurotragus oryx* (eland) | A | 3 |

[a]A = adult

and Ethiopia in the north to South Africa in the south. It has been reported from central Africa (Central African Republic and Democratic Republic of the Congo), from eastern Africa (Ethiopia, Kenya, Somalia, Sudan, Tanzania and Uganda) and from southern Africa (Malawi, Mozambique, South Africa, Zambia and Zimbabwe). However, populations of rhinoceroses in Africa have declined markedly over the past 40 years, primarily as a result of poaching for their horns, which are prized in the Far East as fever-reducing medicinal agents and in north Yemen as prestigious traditional dagger handles (Hillman-Smith & Groves, 1994). As a result of the drastic declines in rhinoceros populations in Africa, the historical distribution of *A. rhinocerotis* outlined above has almost certainly also shrunk. Consequently, the present distribution of *A. rhinocerotis* is not known with any certainty.

**Disease Associations**

None reported.

**References**

Hoogstraal (1956), Theiler (1962), Elbl & Anastos (1966a), Yeoman et al. (1967), Walker (1974), Matthysse & Colbo (1987), Voltzit & Keirans (2003).

### 3.14. *Amblyomma sparsum* Neumann

The tick *Amblyomma sparsum* was introduced into the continental United States on at least 18 occasions between 1969 and 2001 on reptiles and rhinoceroses imported into Arizona, California, Florida, Minnesota, New York, Oregon, Pennsylvania, South Carolina, Tennessee and Texas from African countries such as

Kenya, South Africa, Tanzania and Zambia (see Appendix 1). The reptilian hosts included leopard tortoises, ball pythons, African spurred tortoises and a monitor lizard.

## Hosts

*Amblyomma sparsum* is an unusual tick in its host preferences since adults prefer two widely diverse groups of animals, reptiles and large wild ungulates (Table 3.14). This unusual host relationship led Matthysse & Colbo (1987) to speculate that *A. sparsum* might be more than one species. Among reptiles, *A. sparsum* adults are found on tortoises (especially leopard tortoises), snakes (especially puff adders and African rock pythons) and lizards (especially Nile monitors). Among ungulates, adults are found most commonly on African buffalos and black rhinoceroses. Adults are also found on lions and African elephants; however, all three parasitic stages of *A. sparsum* have been found only on leopard tortoises and Bell's hinged tortoises, indicating the importance of these chelonian species in maintenance of *A. sparsum* populations in nature. Occasional hosts include hares, birds, porcupines, dogs, aardvarks and elephant shrews. There is one report of an *A. sparsum* adult feeding on a human (Gomes, 1993).

On reptiles, *A. sparsum* has been found in the axillae and caudal folds and on the feet and lower legs of tortoises and on the neck of monitors. On ungulates, *A. sparsum* has been found on the brisket, groin and escutcheon of African buffalos.

## Distribution

The geographical distribution of *A. sparsum* covers much of the eastern half of sub-Saharan Africa. It has been reported from central Africa (Cameroon, Chad and Democratic Republic of the Congo), from eastern Africa (Burundi, Eritrea, Ethiopia, Kenya, Rwanda, Somalia, Sudan, Tanzania and Uganda) and from southern Africa (Angola, Malawi, Mozambique, Namibia, South Africa, Zambia and Zimbabwe).

## Life Cycle and Habitat

The larvae of *A. sparsum* take 6–10 days and the nymphs 6–11 days to engorge (Walker & Parsons, 1964). The adults take 23–26 days to feed on ungulates such as sheep, but up to 62 days on reptiles such as tortoises. Unfed adults can survive up to 2.5 years in nature. *Amblyomma sparsum* is found predominantly in arid and semiarid habitats containing areas of wooded savanna.

## Disease Associations

*Amblyomma sparsum* has been shown to transmit *Ehrlichia ruminantium*, the rickettsia causing heartwater in ruminants, to sheep (Norval & Mackenzie, 1981). Since *A. sparsum* seldom feeds on domestic livestock, it has not been reported as an important vector of heartwater; however, it does feed frequently on African buffalos and other wild ungulates susceptible to heartwater and thus could be significant in the transmission of the disease to wildlife.

## References

Theiler (1962), Elbl & Anastos (1966a), Yeoman et al. (1967), Walker (1974), van der Borght-Elbl (1977), Matthysse & Colbo (1987), Burridge (2001), Voltzit & Keirans (2003).

Table 3.14. Most commonly reported hosts of the tick *Amblyomma sparsum*

| Hosts | Stages[a] | | | No. of reports |
|---|---|---|---|---|
| REPTILES | | | | |
| *Stigmochelys pardalis* (leopard tortoise) | A | N | L | 20 |
| *Bitis arietans* (puff adder) | A | | | 10 |
| *Python sebae* (African rock python) | A | N | | 8 |
| *Varanus niloticus* (Nile monitor) | A | N | | 8 |
| UNGULATES | | | | |
| *Diceros bicornis* (black rhinoceros) | A | | | 19 |
| *Syncerus caffer* (African buffalo) | A | N | | 19 |

[a] A = adult, N = nymph, L = larva

## 3.15. Amblyomma splendidum Giebel

The tick *Amblyomma splendidum* was introduced into the continental United States on one occasion, in 1972, on an African elephant trophy imported into Arizona from Africa (USDA, 1973a).

### Hosts

*Amblyomma splendidum* is a tick of ungulates whose preferred hosts are cattle and African buffalos (Fig. 3.6) (Table 3.15). Other ungulate hosts include bay duikers (*Cephalophus dorsalis*), Maxwell's duikers (*Philantomba maxwellii*) and royal antelope (*Neotragus pygmaeus*). Occasional hosts include dogs, African elephants, rats and birds.

Fig. 3.6. African buffalos (*Syncerus caffer*), common hosts of *Amblyomma splendidum* and many other African ticks. (Photograph courtesy of iStockphoto.com.)

### Distribution

The geographical distribution of *A. splendidum* involves mainly western and central Africa. It has been reported from western Africa (Benin, Ghana, Guinea, Guinea-Bissau, Ivory Coast, Liberia, Nigeria, Sierra Leone and Togo), from central Africa (Cameroon, Central African Republic, Congo, Democratic Republic of the Congo and Gabon) and from southern Africa (Angola). It has also been found on the islands of Sao Tome and Principe off the western coast of Africa.

### Disease Associations

None reported.

### References

Robinson (1926), Theiler (1962), Elbl & Anastos (1966a), van der Borght-Elbl (1977), Voltzit & Keirans (2003).

## 3.16. Amblyomma sylvaticum (de Geer)

The tick *Amblyomma sylvaticum* has been introduced into the continental United States on one occasion, in 1990 on a tortoise believed to have been smuggled from South Africa (USDA, 1990; Keirans & Durden, 2001).

### Hosts

*Amblyomma sylvaticum* is a tick of reptiles whose preferred hosts are tortoises, in particular angulated tortoises (*Chersina angulata*), areolated tortoises (*Homopus areolatus*) and African tent tortoises (*Psammobates tentorius*) (Table 3.16). Other reptilian hosts include one species of snakes, the mole snake (*Pseudaspis cana*), and two species of lizards, the southern spiny agama (*Agama hispida*) and Knox's desert lizard

Table 3.15. Most commonly reported hosts of the tick *Amblyomma splendidum*

| Hosts | Stages[a] | | | No. of reports |
|---|---|---|---|---|
| UNGULATES | | | | |
| *Bos taurus* (cattle) | A | N | | 23 |
| *Syncerus caffer* (African buffalo) | A | N | | 22 |
| *Cephalophus dorsalis* (bay duiker) | A | N | L | 7 |
| *Neotragus pygmaeus* (royal antelope) | | | L | 7 |
| *Philantomba maxwellii* (Maxwell's duiker) | | | L | 7 |

[a]A = adult, N = nymph, L = larva

Table 3.16. Most commonly reported hosts of the tick *Amblyomma sylvaticum*

| Hosts | Stages[a] | | | No. of reports |
|---|---|---|---|---|
| TORTOISES | | | | |
| *Chersina angulata* (angulated tortoise) | A | N | L | 11 |
| *Homopus areolatus* (areolated tortoise) | A | N | L | 7 |
| *Psammobates tentorius* (African tent tortoise) | A | N | L | 6 |
| SNAKES | | | | |
| *Pseudaspis cana* (mole snake) | A | N | | 7 |
| LIZARDS | | | | |
| *Agama hispida* (southern spiny agama) | | | L | 4 |
| *Meroles knoxii* (Knox's desert lizard) | | | L | 4 |

[a] A = adult, N = nymph, L = larva

(*Meroles knoxii*). The only non-reptilian host found infested is the four-striped grass mouse (*Rhabdomys pumilio*).

### Distribution

The geographical distribution of *A. sylvaticum* is limited to western South Africa, where it is found only in the Eastern and Western Cape Provinces and Namaqualand.

### Disease Associations

None reported.

### References

Theiler (1962), Burridge & Simmons (2003), Voltzit & Keirans (2003).

## 3.17. *Amblyomma tholloni* Neumann

The tick *Amblyomma tholloni* has been introduced into the continental United States on seven occasions between 1965 and 1974 on African elephants and elephant trophies imported into Alaska, California, Colorado, Florida, Louisiana, Tennessee and Texas from African countries such as Uganda and Zimbabwe (see Appendix 1).

### Hosts

*Amblyomma tholloni* is a three-host tick of African elephants (Fig. 3.7) on which all stages feed (Table 3.17). Other hosts include ungulates, in particular hippopotamuses (*Hippopotamus amphibius*), African buffalos and black rhinoceroses. Larvae and nymphs have been found on domestic livestock such as cattle, horses, goats and sheep. Adults and nymphs will feed on humans. Occasional hosts include carnivores, birds and reptiles.

Fig. 3.7. An African elephant (*Loxodonta africana*), the preferred host of the tick *Amblyomma tholloni*. (Photograph courtesy of iStockphoto.com.)

The sites of attachment of *A. tholloni* on African elephants are behind the ears, in the axillae, on the abdomen and between the hind legs and even on the

Table 3.17. Most commonly reported hosts of the tick *Amblyomma tholloni*

| Hosts | Stages[a] | | | No. of reports |
|---|---|---|---|---|
| **ELEPHANTS** | | | | |
| *Loxodonta africana* (African elephant) | A | N | L | 52 |
| **UNGULATES** | | | | |
| *Hippopotamus amphibius* (hippopotamus) | A | N | L | 12 |
| *Syncerus caffer* (African buffalo) | A | N | | 12 |
| *Diceros bicornis* (black rhinoceros) | A | | | 10 |
| **BIRDS** | | | | |
| *Pitta reichenowi* (green-breasted pitta) | A | N | L | 7 |
| **PRIMATES** | | | | |
| *Homo sapiens* (human) | A | N | | 7 |

[a] A = adult, N = nymph, L = larva

roof of the mouth and the underside of the tongue. In contrast, *A. tholloni* was found attached in the nostrils of hippopotamuses.

## Distribution

The geographical distribution of *A. tholloni* covers those areas of sub-Saharan Africa inhabited by its preferred host, the African elephant. It has been reported from western Africa (Burkina Faso, Ghana, Ivory Coast, Liberia, Mali, Nigeria and Sierra Leone), from central Africa (Cameroon, Central African Republic, Chad, Congo, Democratic Republic of the Congo, Equatorial Guinea and Gabon), from eastern Africa (Burundi, Kenya, Rwanda, Somalia, Sudan, Tanzania and Uganda) and from southern Africa (Angola, Malawi, Mozambique, South Africa, Zambia and Zimbabwe).

## Life Cycle

The life cycle of *A. tholloni* was studied in Zimbabwe by Norval et al. (1980). They reported a preoviposition period of 14–29 days, an oviposition period of 24–44 days, an egg incubation period of 50–59 days, a larval feeding period of 4–11 days, a larval premolting period of 16–28 days, a nymphal feeding period of 4–11 days, a nymphal premolting period of 23–28 days and a female feeding period of 8–18 days. After engorgement and detachment, *A. tholloni* females laid up to 10,347 eggs.

## Disease Associations

*Amblyomma tholloni* has been shown to be an experimental vector of *Ehrlichia ruminantium* (Mackenzie & Norval, 1980), the rickettsia causing heartwater in domestic and wild ruminants.

## References

Robinson (1926), Hoogstraal (1956), Theiler (1962), Elbl & Anastos (1966a), Yeoman et al. (1967), Walker (1974), van der Borght-Elbl (1977), Matthysse & Colbo (1987), Voltzit & Keirans (2003).

## 3.18. *Amblyomma transversale* (Lucas)

The tick *Amblyomma transversale*, previously known as *Aponomma transversale*, has been introduced into the continental United States on four occasions, in 1973 on a ball python imported into New York from Africa (USDA, 1974b) and on pythons imported into New York, Ohio and Texas from African countries such as Togo (Keirans & Durden, 2001).

## Hosts

*Amblyomma transversale* is a tick of pythons, in particular ball pythons and African rock pythons (Table 3.18). Apart from pythons, *A. transversale* has been found on only one other species, an ungulate, the sable antelope. Sites of attachment of *A. transversale* on pythons are primarily around the eyes.

Although *A. transversale* is an African tick, it was found infesting a host native to Asia, the Burmese python, in a circus in Massachusetts in 1972 (Kaufman, 1972), demonstrating the potential for *A. transversale* to spread outside its native geographical range to new reptilian hosts.

Table 3.18. Hosts of the tick *Amblyomma transversale*

| Hosts | Stages[a] | No. of reports |
|---|---|---|
| SNAKES | | |
| *Python sebae* (African rock python) | A  N | 20 |
| *Python regius* (ball python) | A  N  L | 13 |
| *Python natalensis* (southern African python) | N | 1 |
| UNGULATES | | |
| *Hippotragus niger* (sable antelope) | N | 6 |

[a]A = adult, N = nymph, L = larva

## Distribution

The geographical distribution of *A. transversale* covers a wide range of sub-Saharan Africa from Senegal to South Africa. It has been reported from western Africa (Burkina Faso, Guinea-Bissau, Ivory Coast, Mali, Senegal and Togo), from central Africa (Cameroon and Democratic Republic of the Congo), from eastern Africa (Tanzania and Uganda) and from southern Africa (Mozambique, South Africa, Zambia and Zimbabwe).

## Disease Associations

None reported.

## References

Theiler (1962), Kaufman (1972), Matthysse & Colbo (1987), Burridge & Simmons (2003).

## 3.19. *Dermacentor rhinocerinus* (Denny)

The tick *Dermacentor rhinocerinus* has been introduced into the continental United States on three occasions, in 1969 and 1989 on rhinoceroses imported into Florida and Texas, respectively (USDA, 1970a, 1970b, 1994) and in 1973 on a white rhinoceros imported into Virginia (USDA, 1974a).

## Hosts

*Dermacentor rhinocerinus* is a tick whose preferred hosts are black rhinoceroses (Fig. 3.5) and white rhinoceroses (Table 3.19) on which it attaches around the genitalia. Other hosts for adults include African buffalos, elands, cattle, roan antelopes and African elephants. Nymphs and larvae have been found on three species of rodents, red rock rats (*Aethomys chrysophilus*), Natal multimammate mice (*Mastomys natalensis*) and bushveld gerbils (*Gerbilliscus leucogaster*). Occasional hosts include monitors, jackals and humans.

## Distribution

The geographical distribution of *D. rhinocerinus* is similar to that of the other tick of African rhinoceroses, *A. rhinocerotis*. It has been reported from central Africa (Cameroon, Chad and Democratic Republic of the Congo), from eastern Africa (Eritrea, Ethiopia, Kenya, Somalia, Sudan, Tanzania and Uganda) and from southern Africa (Angola, Malawi, Mozambique, Namibia, South Africa, Zambia and Zimbabwe). However, the historical distribution of *D. rhinocerinus* outlined above continues to be reduced by the ongoing destruction of rhinoceroses by poachers (Keirans, 1993).

Table 3.19. Most commonly reported hosts of the tick *Dermacentor rhinocerinus*

| Hosts | Stages[a] | No. of reports |
|---|---|---|
| UNGULATES | | |
| *Diceros bicornis* (black rhinoceros) | A | 20 |
| *Ceratotherium simum* (white rhinoceros) | A | 15 |
| *Syncerus caffer* (African buffalo) | A | 9 |
| *Taurotragus oryx* (eland) | A | 7 |
| ELEPHANTS | | |
| *Loxodonta africana* (African elephant) | A | 7 |

[a]A = adult

## Disease Associations

During nymphal feeding experiments, *D. rhinocerinus* caused paralysis in rabbits (Keirans, 1993); however, since rabbits are not natural hosts of *D. rhinocerinus*, these observations are of no more than academic interest.

## References

Hoogstraal (1956), Arthur (1960), Theiler (1962), Elbl & Anastos (1966c), Yeoman et al. (1967), Walker (1974), Matthysse & Colbo (1987), Keirans (1993).

## 3.20. *Haemaphysalis elongata* Neumann

The tick *Haemaphysalis elongata* was introduced into the continental United States on one occasion on a tenrec imported into Ohio from Madagascar (Keirans & Durden, 2001).

## Hosts

*Haemaphysalis elongata* is a tick of tenrecs, insectivorous mammals peculiar to Madagascar and nearby islands (Table 3.20). Five species of tenrecs have been found infested with *Ha. elongata*: tailless tenrecs (*Tenrec ecaudatus*), lowland streaked tenrecs (*Hemicentetes semispinosus*), greater hedgehog-tenrecs (*Setifer setosus*), lesser hedgehog-tenrecs (*Echinops telfairi*) and Talazac's shrew-tenrecs (*Microgale talzaci*). All three stages of *Ha. elongata* have been found on lowland streaked tenrecs. Occasional hosts are humans, striped mongooses (*Galidictis* sp.) and black rats (*Rattus rattus*).

## Distribution

The geographical distribution of *Ha. elongata* is restricted to the island of Madagascar.

## Disease Associations

None reported.

## References

Hoogstraal (1953), Hoogstraal et al. (1974), Uilenberg et al. (1979).

## 3.21. *Haemaphysalis hoodi* Warburton & Nuttall

The tick *Haemaphysalis hoodi* was introduced into the continental United States on one occasion in 1984 on a grey parrot (*Psittacus erithacus*) imported into New York (USDA, 1985b).

## Hosts

*Haemaphysalis hoodi* is a tick of ground-frequenting birds, in particular coucals, francolins, chickens (*Gallus gallus*) and guineafowl (Table 3.21). A total of 59 avian species have been found infested with *Ha. hoodi* (see Appendix 1). Occasional hosts include rodents, carnivores and ungulates.

Sites of attachment of *Ha. hoodi* are on the head and neck. In heavy infestations, the affected areas can be almost devoid of feathers.

## Distribution

The geographical distribution of *Ha. hoodi* covers much of sub-Saharan Africa, from Senegal to South Africa and Madagascar. It has been reported from western Africa (Benin, Burkina Faso, Gambia, Ghana, Guinea, Guinea-Bissau, Ivory Coast, Mali, Nigeria, Senegal and Sierra Leone), from central Africa (Cameroon, Central African Republic, Chad, Congo and Democratic Republic of the Congo), from eastern Africa (Ethiopia, Kenya, Sudan and Uganda), from southern Africa (Angola, Botswana, Malawi, Mozam-

Table 3.20. Most commonly reported hosts of the tick *Haemaphysalis elongata*

| Hosts | Stages[a] | | | No. of reports |
|---|---|---|---|---|
| **INSECTIVORES** | | | | |
| *Tenrec ecaudatus* (tailless tenrec) | A | | L | 4 |
| *Erinaceus* sp. (hedgehog) | | N | | 3 |
| *Hemicentetes semispinosus* (lowland streaked tenrec) | A | N | L | 3 |
| *Setifer setosus* (greater hedgehog-tenrec) | A | N | | 3 |
| **PRIMATES** | | | | |
| *Homo sapiens* (human) | | N | | 4 |

[a] A = adult, N = nymph, L = larva

bique, South Africa, Zambia and Zimbabwe) and from the island of Madagascar.

### Disease Associations

There is one report of a heavy infestation of *Ha. hoodi* causing fatal anemia in chickens (Lucas, 1954).

### References

Hoogstraal (1953, 1956), Theiler (1962), Elbl & Anastos (1966c), Matthysse & Colbo (1987).

## 3.22. *Haemaphysalis leachi* (Audouin)

The taxonomy of the African *Haemaphysalis leachi* group of ticks has been problematic for many years. Apanaskevich et al. (2007) recently conducted an exhaustive study of many collections of *Haemaphysalis* ticks that had been identified as *Ha. leachi* and found that many southern and eastern African ticks were actually *Ha. elliptica*. Consequently, data on *Ha. leachi* relying only on references published prior to 2007 have been excluded because of the uncertainty of the identity of the ticks in those earlier publications.

The "yellow dog tick" *Ha. leachi* was reported introduced into the continental United States on five occasions between 1965 and 1983 on hedgehogs, a dog, a jackal and a lion imported into Connecticut, Missouri, New York and Texas (see Appendix 1); however, given the fact that many specimens identified as *Ha. leachi* in the past have been found to be *Ha. elliptica*, there is some uncertainty about the true identity of the *Ha. leachi* ticks introduced into the United States between 1965 and 1983.

### Hosts

*Haemaphysalis leachi* is a three-host tick of carnivores. The preferred hosts for adults are domestic dogs and cats (*Felis catus*) and wild carnivores such as African civets (*Civettictis civetta*), leopards, lions, side-striped jackals (*Canis adustus*) and black-backed jackals (*Canis mesomelas*) (Table 3.22). Occasional hosts include mice, wild hogs and blue wildebeests.

### Distribution

The geographical distribution of *Ha. leachi* covers much of Africa. It has been reported from northern Africa (Egypt), western Africa (Guinea, Liberia, Mali and Senegal), from central Africa (Cameroon, Central African Republic, Chad, and Democratic Republic of the Congo), from eastern Africa (Burundi, Ethiopia, Kenya, Sudan, Tanzania and Uganda) and from southern Africa (Zambia and Zimbabwe).

Table 3.21. Most commonly reported hosts of the tick *Haemaphysalis hoodi*

| Hosts | Stages[a] | No. of reports |
|---|---|---|
| **BIRDS** | | |
| *Centropus senegalensis* (Senegal coucal) | A  N | 16 |
| *Francolinus bicalcaratus* (double-spurred francolin) | A  N  L | 11 |
| *Gallus gallus* (chicken) | A  N  L | 10 |
| *Numida meleagris* (helmeted guineafowl) | A | 9 |
| *Centropus monachus* (blue-headed coucal) | A | 6 |
| *Centropus superciliosus* (white-browed coucal) | A  N  L | 6 |

[a] A = adult, N = nymph, L = larva

Table 3.22. Most commonly reported hosts of the tick *Haemaphysalis leachi*

| Hosts | Stages[a] | No. of reports |
|---|---|---|
| **CARNIVORES** | | |
| *Canis familiaris* (dog) | A  N  L | 46 |
| *Civettictis civetta* (African civet) | A  N  L | 30 |
| *Panthera pardus* (leopard) | A     L | 21 |
| *Felis catus* (cat) | A  N | 18 |
| *Panthera leo* (lion) | A | 17 |
| *Canis adustus* (side-striped jackal) | A | 10 |
| *Canis mesomelas* (black-backed jackal) | A | 10 |

[a] A = adult, N = nymph, L = larva

In the older literature there are some records of *Ha. leachi* var *indica* from the Indian subcontinent. These refer to another species, *Haemaphysalis indica* (Hoogstraal, 1970).

### Disease Associations

*Haemaphysalis leachi* has been reported to be a tick of significance to both human and animal health as a vector of *Rickettsia conorii* (Gear & de Meillon, 1941), the bacterium causing tick typhus in humans, and the vector of *Babesia rossi* (Shaw et al., 2001), a protozoan causing babesiosis in dogs; however, Apanaskevich et al. (2007) now consider these two diseases to be transmitted by *Ha. elliptica* rather than *Ha. leachi*.

### References

Apanaskevich et al. (2007).

## 3.23. *Haemaphysalis muhsamae* Santos Dias

As previously stated, the taxonomy of the African *Haemaphysalis leachi* group of ticks has been problematic for many years, and that group includes *Haemaphysalis muhsamae*. Apanaskevich et al. (2007) recently addressed the taxonomic status of *Ha. leachi* and indicated they will soon conduct a similar study of *Ha. muhsamae*. Consequently, the data presented below on *Ha. muhsamae* should be considered tentative until the taxonomic status of this tick is clarified.

The tick *Ha. muhsamae*, previously known as *Haemaphysalis leachi muhsami* and *Haemaphysalis muhsami*, was reported introduced into the continental United States on two occasions, in 1966 on a bat-eared fox (*Otocyon megalotis*) imported into New York (USDA, 1966b, 1967a) and on a hedgehog imported into New York from Kenya (Becklund, 1968).

### Hosts

*Haemaphysalis muhsamae* is a tick of carnivores with a preference for small wild carnivores such as African civets, white-tailed mongooses (*Ichneumia albicauda*), small-spotted genets (*Genetta genetta*) and large-spotted genets (*Genetta tigrina*) (Table 3.23). Other hosts of adults are four-toed hedgehogs (*Atelerix albiventris*) and Cape hares (*Lepus capensis*). Occasional hosts include rodents, birds, hyraxes, ungulates, aardvarks and elephant shrews.

### Distribution

The geographical distribution of *Ha. muhsamae* is reported to cover much of sub-Saharan Africa. It has been reported from western Africa (Benin, Burkina Faso, Guinea, Ivory Coast, Nigeria, Senegal and Sierra Leone), from central Africa (Cameroon, Central African Republic, Chad and Democratic Republic of the Congo), from eastern Africa (Ethiopia, Kenya, Rwanda, Somalia, Sudan, Tanzania and Uganda) and from southern Africa (Angola, Botswana, Lesotho, Malawi, Mozambique, Namibia, South Africa, Zambia and Zimbabwe). Some reports have indicated that *Ha. muhsamae* has crossed the Red Sea to become established in Yemen (Hoogstraal, 1956; Hoogstraal & Kaiser, 1959b).

### Disease Associations

None reported.

### References

Hoogstraal (1956), Theiler (1962), Elbl & Anastos (1966c), Yeoman et al. (1967).

Table 3.23. Most commonly reported hosts of the tick *Haemaphysalis muhsamae*

| Hosts | Stages[a] | No. of reports |
|---|---|---|
| **Carnivores** | | |
| *Civettictis civetta* (African civet) | A | 9 |
| *Ichneumia albicauda* (white-tailed mongoose) | A  N | 9 |
| *Genetta genetta* (small-spotted genet) | A | 6 |
| *Genetta tigrina* (large-spotted genet) | A | 6 |
| *Panthera pardus* (leopard) | A | 6 |
| **Insectivores** | | |
| *Atelerix albiventris* (four-toed hedgehog) | A  N | 6 |

[a]A = adult, N = nymph

## 3.24. *Hyalomma albiparmatum* Schulze

The tick *Hyalomma albiparmatum* was introduced into the continental United States on five occasions between 1969 and 1989 on wild ungulates, ostriches and a giraffe hide imported into New Jersey, New York and Texas from African countries such as Kenya and Tanzania (see Appendix 1). The wild ungulate hosts were giraffes, a rhinoceros and a zebra.

### Hosts

*Hyalomma albiparmatum* is a tick of ungulates, with cattle, goats and sheep the most common hosts among domestic livestock and giraffes and warthogs among wild animals (Table 3.24). Occasional hosts include carnivores, hares, ostriches and, rarely, humans.

Preferred sites of attachment for adults on cattle are the heels, abdomen, perianal region and tail tip, with some also found on the neck and legs.

### Distribution

The geographical distribution of *Hy. albiparmatum* is limited to eastern Africa. It has been reported from Ethiopia, Kenya, Somalia, Tanzania and Uganda, but Apanaskevich & Horak (2008b) believe its true distribution is confined to just Kenya and Tanzania.

### Disease Associations

*Hyalomma albiparmatum* has been reported to be a vector of *Rickettsia conorii* (Apanaskevich & Horak, 2008b), the bacterium causing tick typhus in humans.

### References

Hoogstraal (1956), Theiler (1962), Yeoman et al. (1967), Walker (1974), Apanaskevich & Horak (2008b).

## 3.25. *Hyalomma impressum* Koch

The tick *Hyalomma impressum* was introduced into the continental United States on one occasion in 1984 on a rhinoceros trophy hide imported into Maine (USDA, 1985b).

### Hosts

*Hyalomma impressum* is a three-host tick whose preferred hosts are cattle (Table 3.25). Other hosts of adults are sheep, horses, camels, warthogs and pigs (*Sus scrofa*). Occasional hosts include rats, hares, hedgehogs, aardvarks and birds.

### Distribution

The geographical distribution of *Hy. impressum* extends from western Africa east to Sudan. It has been

Table 3.24. Most commonly reported hosts of the tick *Hyalomma albiparmatum*

| Hosts | Stages[a] | No. of reports |
|---|---|---|
| UNGULATES | | |
| *Bos taurus* (cattle) | A | 14 |
| *Giraffa camelopardalis* (giraffe) | A | 9 |
| *Capra hircus* (goat) | A | 8 |
| *Ovis aries* (sheep) | A | 7 |
| *Phacochoerus africanus* (warthog) | A | 7 |

[a] A = adult

Table 3.25. Most commonly reported hosts of the tick *Hyalomma impressum*

| Hosts | Stages[a] | No. of reports |
|---|---|---|
| UNGULATES | | |
| *Bos taurus* (cattle) | A | 28 |
| *Ovis aries* (sheep) | A | 9 |
| *Camelus dromedarius* (camel) | A | 7 |
| *Equus caballus* (horse) | A | 7 |
| *Phacochoerus africanus* (warthog) | A | 7 |

[a] A = adult

Table 3.26. Most commonly reported hosts of the tick *Hyalomma truncatum*

| Hosts | Stages[a] | | | No. of reports |
|---|---|---|---|---|
| **UNGULATES** | | | | |
| *Bos taurus* (cattle) | A | N | L | 108 |
| *Capra hircus* (goat) | A | N | L | 42 |
| *Ovis aries* (sheep) | A | N | L | 38 |
| *Phacochoerus africanus* (warthog) | A | | | 29 |
| *Taurotragus oryx* (eland) | A | | | 25 |
| *Syncerus caffer* (African buffalo) | A | | | 23 |
| *Camelus dromedarius* (camel) | A | | | 21 |
| *Equus burchelli* (Burchell's zebra) | A | | | 18 |
| *Giraffa camelopardalis* (giraffe) | A | | | 18 |
| *Equus caballus* (horse) | A | N | | 15 |
| *Hippotragus equinus* (roan antelope) | A | | | 15 |
| *Connochaetes taurinus* (blue wildebeest) | A | | | 14 |
| *Tragelaphus scriptus* (bushbuck) | A | | | 14 |
| *Tragelaphus strepsiceros* (greater kudu) | A | | | 14 |
| *Hippotragus niger* (sable antelope) | A | | | 13 |
| *Diceros bicornis* (black rhinoceros) | A | | | 11 |
| *Oryx gazella* (gemsbok) | A | | | 11 |
| *Equus zebra* (mountain zebra) | A | | | 10 |
| **LAGOMORPHS** | | | | |
| *Lepus saxatilis* (scrub hare) | | N | L | 18 |
| *Lepus capensis* (Cape hare) | A | N | L | 12 |
| **CARNIVORES** | | | | |
| *Canis familiaris* (dog) | A | N | | 17 |
| *Panthera leo* (lion) | A | N | L | 14 |
| **PRIMATES** | | | | |
| *Homo sapiens* (human) | A | | | 14 |
| **BIRDS** | | | | |
| *Struthio camelus* (ostrich) | A | | | 12 |

[a]A = adult, N = nymph, L = larva

reported from western Africa (Benin, Burkina Faso, Ghana, Guinea, Guinea-Bissau, Ivory Coast, Mali, Mauritania, Niger, Nigeria and Senegal), from central Africa (Cameroon, Central African Republic, Chad and Democratic Republic of the Congo) and from eastern Africa (Ethiopia and Sudan).

**Disease Associations**

None reported.

**References**

Theiler (1962), Elbl & Anastos (1966c), Walker et al. (2003), Apanaskevich & Horak (2007).

*3.26. Hyalomma truncatum* Koch

The tick *Hyalomma truncatum*, known in the past as *Hyalomma transiens* and colloquially known as the "shiny hyalomma," was introduced into the continental United States on 11 occasions between 1973 and 1989 on wild ungulates, ostriches and animal trophies imported into Arizona, California, Colorado, Mississippi, New York, North Carolina and Texas from African countries such as Angola and Namibia (see Appendix 1). The wild ungulate hosts were rhinoceroses, an oryx and a zebra, and the animal trophies included hides.

**Hosts**

*Hyalomma truncatum* is a two-host tick of great importance to both animal and human health. It has a wide host range, having been found on 109 species, including 41 species of ungulates, 27 species of birds, 13 species of rodents, 10 species of carnivores, 6 species of reptiles, 4 species of lagomorphs, 2 species each of elephant shrews and primates, and 1 species each of insectivores, aardvarks, elephants and hyraxes (see Appendix 2). The preferred hosts of *Hy. truncatum* adults are cattle and other domestic ungulates such as goats, sheep, camels and horses as well as wild herbivores such as warthogs, African buffalos, elands, Burchell's zebras (*Equus burchelli*) and giraffes (Table 3.26). Other hosts for adults include dogs, lions, ostriches and humans. Larvae and nymphs feed primarily on domestic livestock and on hares, rodents, hedgehogs and birds. Occasional hosts include elephants, rock hyraxes, aardvarks, elephant shrews and reptiles.

Preferred attachment sites of *Hy. truncatum* adults on cattle are the tail, the legs including around the feet, the inguinal and perineal regions, the scrotum and the udder. On sheep, *Hy. truncatum* is found in the anogenital and inguinal regions, the feet, the brisket and the udder. On wildebeests, it is found on the mane, tail and scrotum.

### Distribution

The geographical distribution of *Hy. truncatum* covers most of Africa and is particularly widespread in sub-Saharan Africa. It has been reported from northern Africa (Egypt), from western Africa (Benin, Burkina Faso, Gambia, Ghana, Guinea, Guinea-Bissau, Ivory Coast, Mali, Mauritania, Niger, Nigeria, Senegal and Togo), from central Africa (Cameroon, Central African Republic, Chad, Congo and Democratic Republic of the Congo), from eastern Africa (Burundi, Djibouti, Eritrea, Ethiopia, Kenya, Rwanda, Somalia, Sudan, Tanzania and Uganda), from southern Africa (Angola, Botswana, Malawi, Mozambique, Namibia, South Africa, Swaziland, Zambia and Zimbabwe) and from the Indian Ocean islands of Madagascar, Seychelles and Zanzibar. Reports have indicated that *Hy. truncatum* has crossed the Red Sea to become established in Saudi Arabia, Yemen and the island of Socotra in the Arabian Sea (Hoogstraal et al., 1981; Al-Khalifa et al., 1987; Ueckermann et al., 2006).

### Life Cycle and Habitat

*Hyalomma truncatum* is adapted to dry habitats but is also found in temperate climates. In southern Africa, the limiting factor in the distribution of *Hy. truncatum* is increasing humidity; in areas of high rainfall, *Hy. truncatum* cannot survive. Adults are most active in the late wet summer months and the nymphs and larvae in the dry months from autumn to spring. The life cycle takes one year to complete.

### Disease Associations

*Hyalomma truncatum* is a tick of much importance to both animal and human health because it is:

- a cause of tick paralysis in humans and sheep (Erasmus, 1952; Swanepoel, 1959; Mans et al., 2004);
- a cause of sweating sickness, a tick toxicosis, in cattle and less commonly in sheep and pigs (Neitz, 1964);
- a cause of tick toxicosis in dogs (Burr, 1983);
- a vector of *Babesia caballi* (de Waal, 1990), a protozoan causing piroplasmosis in horses and other equines;
- a vector of the nairovirus causing Crimean-Congo hemorrhagic fever in humans (Rechav, 1992);
- a cause of foot lameness in sheep (Kok & Fourie, 1995);
- a cause of tissue damage in cattle and sheep that can lead to abscess formation and myiasis due to the blow fly *Chrysomya bezziana* (Fourie & Horak, 2000; Walker et al., 2003);
- a vector of *Rickettsia conorii* (Walker et al., 2003), the bacterium causing tick typhus in humans.

Additionally, *Rickettsia sibirica mongolitimonae*, the bacterium causing lymphangitis-associated rickettsiosis in humans, was isolated from *Hy. truncatum* in Niger (Parola et al., 2001).

### References

Hoogstraal (1956), Theiler (1962), Elbl & Anastos (1966c), Yeoman et al. (1967), Walker (1974), Norval (1982), Matthysse & Colbo (1987), Walker et al. (2003), Apanaskevich & Horak (2008b).

## 3.27. *Ixodes pilosus* Koch

The "sourveld tick" *Ixodes pilosus* was introduced into the continental United States on three occasions, in 1983 on plant material imported into New York from South Africa (USDA, 1985a) and in 1984 and 1985 on cut flowers imported into New York from South Africa (USDA, 1985b, 1987a).

### Hosts

*Ixodes pilosus* is a three-host tick that feeds on ungulates and carnivores (Table 3.27). Common hosts for

Table 3.27. Most commonly reported hosts of the tick *Ixodes pilosus*

| Hosts | Stages[a] | | | No. of reports |
|---|---|---|---|---|
| **CARNIVORES** | | | | |
| *Canis familiaris* (dog) | A | N | L | 16 |
| **UNGULATES** | | | | |
| *Bos taurus* (cattle) | A | | | 14 |
| *Tragelaphus scriptus* (bushbuck) | A | | | 12 |
| *Pelea capreolus* (grey rhebok) | A | N | L | 9 |
| **LAGOMORPHS** | | | | |
| *Lepus saxatilis* (scrub hare) | A | N | L | 8 |

[a] A = adult, N = nymph, L = larva

Table 3.28. Most commonly reported hosts of the tick *Ixodes schillingsi*

| Hosts | Stages[a] | | | No. of reports |
|---|---|---|---|---|
| **PRIMATES** | | | | |
| *Homo sapiens* (human) | A | | | 8 |
| *Colobus polykomos* (king colobus) | A | N | L | 6 |
| *Cercopithecus cephus* (moustached monkey) | A | | | 5 |
| *Colobus guereza* (eastern black-and-white colobus) | A | | | 4 |
| *Otolemur crassicaudatus* (greater bushbaby) | | N | L | 4 |

[a] A = adult, N = nymph, L = larva

adults are domestic livestock such as cattle, goats and sheep, wild ungulates such as bushbucks (*Tragelaphus scriptus*), grey rheboks (*Pelea capreolus*), grey duikers, bonteboks (*Damaliscus pygargus*) and African buffalos, dogs, wild carnivores such as African caracals (*Caracal caracal*) and black-backed jackals, and scrub hares. Many of these hosts also carry larvae and nymphs. Occasional hosts include rodents, elephant shrews, hedgehogs and primates including humans.

Preferred sites of attachment on cattle and sheep are the ears, around the eyes and the neck.

### Distribution

The geographical distribution of *I. pilosus* is limited to South Africa and Swaziland. Older reports of *I. pilosus* in other parts of Africa are now considered in error.

### Disease Associations

None reported.

### References

Theiler (1962), Arthur (1965), Walker et al. (2003).

## 3.28. *Ixodes schillingsi* Neumann

The tick *Ixodes schillingsi* was introduced into the continental United States on one occasion in 1967 on a hyrax imported into Illinois (USDA, 1967b, 1968a).

### Hosts

*Ixodes schillingsi* is a tick of primates on which all stages feed (Table 3.28). Primate hosts are colobus monkeys (*Colobus* spp. and *Piliocolobus badius*), monkeys of the genus *Cercopithecus*, bushbabies (*Galagoides zanzibaricus* and *Otolemur crassicaudatus*) and humans. On monkeys, *I. schillingsi* can be found in the axillae. Occasional hosts include rats and hyraxes.

### Distribution

The geographical distribution of *I. schillingsi* follows that of its primary hosts, colobus monkeys. It has been reported from central Africa (Cameroon and Central African Republic), from eastern Africa (Kenya, Rwanda, Sudan and Tanzania), from southern Africa (Mozambique and Zimbabwe) and from some Indian Ocean islands (Pemba Island and Zanzibar).

**Disease Associations**

None reported.

**References**

Hoogstraal (1956), Arthur (1957, 1965), Theiler (1962), Walker (1974).

### 3.29. *Rhipicephalus appendiculatus* Neumann

The "brown ear tick" *Rhipicephalus appendiculatus* was reported introduced into the continental United States on 10 occasions between 1966 and 1988 on zebras, animal trophies and wool imported into Colorado, Mississippi, New Jersey, South Carolina and Texas from African countries such as Kenya and South Africa (see Appendix 1). The animal trophies included a lion hide. It is worth noting that all stages of *R. appendiculatus* are very similar to those of *R. zambeziensis* (Walker et al., 2000), a tick species first described only in 1981, and the distributions of these two rhipicephalid species overlap in parts of Botswana, South Africa, Tanzania, Zambia and Zimbabwe (Walker et al., 2003). It is possible, therefore, that some ticks introduced into the United States before 1981 were *R. zambeziensis*.

**Hosts**

*Rhipicephalus appendiculatus* is a very important three-host tick with a wide host range. It has been found on 129 species of animals, including 52 species of ungulates, 27 species of carnivores, 18 species of rodents, 13 species of birds, 5 species each of primates and lagomorphs, 4 species of elephant shrews, 2 species of hyraxes, and 1 species each of elephants, pangolins and insectivores (see Appendix 2). The preferred hosts for all stages are ungulates, particularly cattle as well as goats, sheep and horses among domestic animals and impalas, African buffalos, waterbucks (*Kobus ellipsiprymnus*), elands, greater kudus, Burchell's zebras, warthogs, bush pigs (*Potamochoerus larvatus*), bushbucks, grey duikers, blue wildebeests, sable antelopes, giraffes, klipspringers (*Oreotragus oreotragus*) and nyalas (*Tragelaphus angasii*) among wildlife (Table 3.29).

Table 3.29. Most commonly reported hosts of the tick *Rhipicephalus appendiculatus*

| Hosts | Stages[a] | | | No. of reports |
|---|---|---|---|---|
| **Ungulates** | | | | |
| *Bos taurus* (cattle) | A | N | L | 97 |
| *Capra hircus* (goat) | A | N | L | 37 |
| *Ovis aries* (sheep) | A | N | L | 36 |
| *Aepyceros melampus* (impala) | A | N | L | 29 |
| *Syncerus caffer* (African buffalo) | A | N | L | 25 |
| *Taurotragus oryx* (eland) | A | N | L | 20 |
| *Kobus ellipsiprymnus* (waterbuck) | A | N | L | 19 |
| *Equus burchelli* (Burchell's zebra) | A | N | L | 18 |
| *Phacochoerus africanus* (warthog) | A | N | L | 18 |
| *Tragelaphus strepsiceros* (greater kudu) | A | N | L | 18 |
| *Tragelaphus scriptus* (bushbuck) | A | N | L | 16 |
| *Potamochoerus larvatus* (bush pig) | A | N | L | 15 |
| *Hippotragus niger* (sable antelope) | A | N | L | 14 |
| *Sylvicapra grimmia* (grey duiker) | A | N | L | 14 |
| *Connochaetes taurinus* (blue wildebeest) | A | N | L | 13 |
| *Giraffa camelopardalis* (giraffe) | A | N | L | 13 |
| *Equus caballus* (horse) | A | N | L | 12 |
| *Oreotragus oreotragus* (klipspringer) | A | N | | 11 |
| *Tragelaphus angasii* (nyala) | A | N | L | 10 |
| **Carnivores** | | | | |
| *Canis familiaris* (dog) | A | N | L | 27 |
| *Panthera leo* (lion) | A | N | L | 13 |
| **Primates** | | | | |
| *Homo sapiens* (human) | A | N | L | 11 |
| **Lagomorphs** | | | | |
| *Lepus saxatilis* (scrub hare) | A | N | L | 10 |

[a] A = adult, N = nymph, L = larva

Fig. 3.8. *Rhipicephalus appendiculatus* ticks feeding on the ear of a cow.

Other notable hosts for all stages include dogs, lions and scrub hares. All stages of *R. appendiculatus* will feed on humans.

As its common name of brown ear tick implies, the preferred site of attachment of *R. appendiculatus* is the ear (Fig. 3.8). Adults are found attached to the inner surface of the ears. In heavy infestations, adults are also found around the base of the horns, around the eyes and the commissures of the mouth, and on the ventral surface of the body (jowl, intermandibular space, ventral neck, dewlap, axillae, prepuce, scrotum, udder, groin, vulva and escutcheon). Larvae and nymphs attach to cattle mainly on the neck, dewlap, cheeks, eyelids, muzzle and ears.

### Distribution

The geographical distribution of *R. appendiculatus* is limited to the eastern half of Africa from southern Sudan to the southeastern coast of South Africa. It has been reported from central Africa (Central African Republic and Democratic Republic of the Congo), from eastern Africa (Burundi, Kenya, Rwanda, Sudan, Tanzania and Uganda), from southern Africa (Botswana, Malawi, Mozambique, South Africa, Swaziland, Zambia and Zimbabwe) and from some Indian Ocean islands (Comoros, Mauritius and Zanzibar).

### Habitat and Life Cycle

*Rhipicephalus appendiculatus* is found in savanna and temperate climatic regions, ranging from hot coastal areas to cool highland plateaus as long as the climate is humid. In southern Africa the activity of *R. appendiculatus* is strictly seasonal, with a single generation produced each year. Adults are found during the rainy season from December to March, larvae in the cooler late summer to winter period after the rains from April to August and nymphs in the winter to early spring from July to October. In contrast, in hotter regions close to the equator with rainfall more evenly spread throughout the year, several overlapping generations can be completed each year with no clear seasonal pattern. Studies by Lewis (cited by Elbl & Anastos, 1966b) demonstrated that *R. appendiculatus* is able to survive without feeding for prolonged periods of time, with unfed larvae surviving for 9 months, nymphs for 22 months and adults for 2–5 years.

The life cycle of *R. appendiculatus* has been studied in South Africa by Theiler (cited by Elbl & Anastos, 1966b). She reported a preoviposition period of 5–40 days, an egg incubation period of 28 days, a larval feeding period of 3–7 days, a larval premolting period of 10–49 days, a nymphal feeding period of 3–7 days, a nymphal premolting period of 10–61 days and a female feeding period of 4–10 days. Engorged females laid 3,000–5,000 eggs each.

### Disease Associations

*Rhipicephalus appendiculatus* is a very important tick, especially to animal health, because it is:

- the primary vector of *Theileria parva*, the protozoan causing classical East Coast fever (Lounsbury, 1906), Corridor disease (de Vos, 1981) and Zimbabwe theileriosis (Lawrence et al., 2004c), all in cattle;
- the major vector of the nairovirus causing Nairobi sheep disease in sheep and goats (Montgomery, 1917);
- a cause of tick toxicosis in cattle (Neitz, 1964) and probably in wild ungulates (Lightfoot & Norval, 1981);
- a vector of *Theileria taurotragi* (Young et al., 1977), a protozoan causing theileriosis in cattle;
- a vector of *Anaplasma bovis* (de Vos, 1981), a rickettsia causing anaplasmosis in cattle;
- a vector of *Rickettsia conorii* (Walker et al., 2003), the bacterium causing tick typhus in humans.

Experiments have shown that for each female *R. appendiculatus* completing feeding there is a loss of 4.0 g of potential growth of cattle (Walker et al., 2003). In addition, *R. appendiculatus* has been shown to be an experimental vector of *Anaplasma marginale* (Piercy, 1956), another rickettsia causing anaplasmosis in cattle, and of the flavivirus causing louping ill in sheep (de Vos, 1981). Finally, *R. appendiculatus* has been found infected with Thogoto virus (Jongejan & Uilenberg, 2004), which has been reported associated with abortions in sheep and clinical disease in humans (Labuda & Nuttall, 2004).

### References

Theiler (1949b, 1962), Hoogstraal (1956), Elbl & Anastos (1966b), Yeoman et al. (1967), McCulloch et al. (1968), Walker (1974), Matthysse & Colbo (1987), Walker et al. (2000, 2003).

## 3.30. *Rhipicephalus capensis* Koch

The tick *Rhipicephalus capensis* has been confused with two other rhipicephalid ticks, *R. follis* and *R. gertrudae* (Walker et al., 2000). Years ago Theiler (1962) remarked that *R. capensis* appeared to be a catch-all for *R. capensis*-like ticks. Given these identification problems and the belief that *R. capensis* occurs only in South Africa (Walker et al., 2000), data on *R. capensis* relying on records from outside South Africa and data published prior to 2000 have been excluded from the present work.

*Rhipicephalus capensis* was reported introduced into the continental United States on one occasion in 1985 when it was found on cut flowers imported into New York from South Africa (USDA, 1987a).

### Hosts

*Rhipicephalus capensis* is a three-host tick of ungulates (Table 3.30), including elands, bonteboks, gemsboks

Table 3.30. Hosts of the tick *Rhipicephalus capensis*

| Hosts | Stages[a] | No. of reports |
|---|---|---|
| UNGULATES | | |
| *Taurotragus oryx* (eland) | A | 3 |
| *Damaliscus pygargus* (bontebok) | A | 2 |
| *Oryx gazella* (gemsbok) | A | 2 |
| *Bos taurus* (cattle) | A | 1 |
| *Equus caballus* (horse) | A | 1 |
| *Equus zebra* (mountain zebra) | A | 1 |
| CARNIVORES | | |
| *Vulpes chama* (Cape fox) | A | 1 |
| RODENTS | | |
| *Rhabdomys pumilio* (four-striped grass mouse) | - | 1 |

Note: Includes only hosts confirmed in South Africa since 2000
[a] A = adult, - = not recorded

(*Oryx gazella*), cattle, horses and mountain zebras (*Equus zebra*). Foxes and mice have also been found infested.

### Distribution

Walker et al. (2000) considered *R. capensis* to occur only in South Africa despite earlier reports of the tick from several countries in central, eastern and southern Africa.

### Disease Associations

*Rhipicephalus capensis* has been reported a minor vector of *Theileria parva* (Neitz, 1956), the protozoan causing East Coast fever in cattle.

### References

Walker et al. (2000).

## 3.31. *Rhipicephalus compositus* Neumann

The ticks *Rhipicephalus compositus*, *R. longus* and *R. pseudolongus* are difficult to differentiate (Walker et al., 2000). Consequently, data on *R. compositus* relying solely on references published prior to 2000 have been excluded because of the uncertainty of the identity of the ticks in those earlier publications.

*Rhipicephalus compositus* was reported introduced into the continental United States on one occasion in 1983 on plant material imported into New York (USDA, 1985a).

### Hosts

*Rhipicephalus compositus* is a three-host tick of ungulates, in particular cattle, African buffalos and wild pigs (warthogs and bush pigs) (Table 3.31). Other hosts for adults are carnivores, especially dogs, lions and leopards, and humans. Larvae and nymphs are found primarily on murid rodents.

Preferred sites of attachment on cattle are the escutcheon, udder and scrotum. On African buffalos they are found on the ears, axillae, sternum, scrotum, teats and tail.

### Distribution

The geographical distribution of *R. compositus* extends from southern Sudan in the north to Zimbabwe in the south. It has been reported from central Africa (Democratic Republic of the Congo), from eastern Africa (Burundi, Kenya, Rwanda, Sudan, Tanzania and Uganda) and from southern Africa (Angola, Malawi, Zambia and Zimbabwe).

### Life Cycle

The life cycle of *R. compositus* was studied in Zimbabwe by Norval & Tebele (1984). They reported a preoviposition period of 4–10 days, an oviposition period of 14–24 days, an egg incubation period of 29–36 days, a larval feeding period of 3–6 days, a nymphal feeding period of 4–12 days and a female feeding period of 6–7 days. After engorgement and detachment, *R. compositus* females laid up to 6,878 eggs. The duration of the entire life cycle was 87–140 days.

### Disease Associations

Although *R. compositus* has been shown to be an experimental vector of *Theileria parva*, the protozoan causing East Coast fever in cattle, it is unlikely to be an important vector in the field because nymphs and

Table 3.31. Most commonly reported hosts of the tick *Rhipicephalus compositus*

| Hosts | Stages[a] | No. of reports |
| --- | --- | --- |
| UNGULATES | | |
| *Bos taurus* (cattle) | A | 34 |
| *Syncerus caffer* (African buffalo) | A | 21 |
| *Phacochoerus africanus* (warthog) | A | 15 |
| *Potamochoerus larvatus* (bush pig) | A | 12 |
| CARNIVORES | | |
| *Canis familiaris* (dog) | A | 15 |
| *Panthera leo* (lion) | A | 13 |

[a]A = adult

larvae have never been collected from cattle (Norval & Tebele, 1984).

**References**

Norval & Tebele (1984), Matthysse & Colbo (1987), Walker et al. (2000).

### 3.32. *Rhipicephalus evertsi* Neumann

There are two subspecies of *Rhipicephalus evertsi*, *R. e. evertsi* and *R. e. mimeticus*. In this book, ticks of the two subspecies are considered under the species name *R. evertsi* since much of the literature does not differentiate between the subspecies; however, where available, information will be given by subspecies.

The "red-legged tick" *Rhipicephalus evertsi* was introduced into the continental United States on 61 occasions between 1960 and 1990 on domestic animals, wild ungulates and animal trophies imported into Arizona, Colorado, Florida, Georgia, Maine, Mississippi, New Jersey, New York, North Carolina and Ohio from African countries such as Kenya, Namibia, South Africa and Tanzania (see Appendix 1). The domestic animal hosts were camels and a horse, and the wild ungulate hosts zebras, elands, hartebeests, oryxes, Abyssinian asses (*Equus asinus*), gemsboks, a black wildebeest, a giraffe, an impala, a kudu, a sable antelope and a zebra-horse cross, and the animal trophies included a rhinoceros hide and a zebra hide. At least 27 of the importations involved *R. e. evertsi* and five *R. e. mimeticus*.

**Hosts**

*Rhipicephalus evertsi* is a very important two-host tick, particularly to animal health, with a wide host range. It has been found on 114 species of animals, including 54 species of ungulates, 19 species of carnivores, 13 species of rodents, 10 species of birds, 6 species of lagomorphs, 4 species of primates, 4 species of elephant shrews, 2 species of hyraxes, 1 species of elephants and 1 species of reptiles (see Appendix 2). The preferred hosts for all stages are cattle, goats, sheep, horses, dogs and donkeys among domestic animals and impalas, elands, Burchell's zebras, greater kudus and African buffalos among wildlife (Table 3.32). Other hosts infested by all stages include scrub hares, Cape hares and, less commonly, humans.

Attachment sites of *R. evertsi* adults on cattle are under the base of the tail and around the anus; occasionally adults may be found on the neck, fetlocks and heels. In contrast, larvae and nymphs are found deep in the ears in the convolutions of the external auditory meatus, often embedded in wax.

**Distribution**

The geographical distribution of *R. e. evertsi* covers much of sub-Saharan Africa. It has been reported from western Africa (Burkina Faso, Gambia, Ghana, Guinea-Bissau, Ivory Coast, Mali, Mauritania, Niger, Nigeria, Senegal and Togo), from central Africa (Cameroon, Central African Republic, Chad, Congo, Democratic Republic of the Congo and Gabon), from eastern Africa (Burundi, Djibouti, Eritrea, Ethiopia, Kenya, Rwanda, Somalia, Sudan, Tanzania and Uganda), from southern Africa (Angola, Botswana, Lesotho, Malawi, Mozambique, Namibia, South Africa, Swaziland, Zambia and Zimbabwe) and from some Indian Ocean islands (Mauritius, Pemba Island and Zanzibar). It is also reported to have crossed the Red Sea onto the Arabian Peninsula and become established in Yemen and Saudi Arabia (Ueckermann et al., 2006; Kolonin, 2009).

The geographical distribution of *R. e. mimeticus* is limited to central and southern Africa, where it is commonly found in Angola and Namibia and less commonly in Botswana, Democratic Republic of the Congo and South Africa.

*Rhipicephalus e. evertsi* has been found on several occasions outside its established range. It has been detected in Saudi Arabia and Oman on sheep and goats imported from Somalia and in Saudi Arabia on cattle imported from Sudan (Abou-Elela et al., 1981; Hoogstraal et al., 1981; Williams et al., 2000). Al-Khalifa et al. (1987) suggested that, through these movements of infested livestock, *R. e. evertsi* was in the process of becoming established in Saudi Arabia.

**Life Cycle and Habitat**

*Rhipicephalus e. evertsi* occurs in a wide variety of climates, including desert, steppe, savanna, temperate and rain forest regions. It is a two-host tick with both the larval and nymphal stages feeding on the same host, typically in the ear. After engorgement, the nymphs detach and drop off the first host to molt to

Table 3.32. Most commonly reported hosts of the tick Rhipicephalus evertsi

| Hosts | Stages[a] | | | No. of reports |
|---|---|---|---|---|
| **UNGULATES** | | | | |
| *Bos taurus* (cattle) | A | N | L | 131 |
| *Ovis aries* (sheep) | A | N | L | 59 |
| *Capra hircus* (goat) | A | N | L | 56 |
| *Aepyceros melampus* (impala) | A | N | L | 35 |
| *Taurotragus oryx* (eland) | A | N | L | 30 |
| *Equus caballus* (horse) | A | N | L | 26 |
| *Equus burchelli* (Burchell's zebra) | A | N | L | 25 |
| *Equus asinus* (donkey) | A | N | | 23 |
| *Tragelaphus strepsiceros* (greater kudu) | A | N | L | 21 |
| *Syncerus caffer* (African buffalo) | A | N | L | 20 |
| *Giraffa camelopardalis* (giraffe) | A | N | L | 17 |
| *Connochaetes taurinus* (blue wildebeest) | A | N | L | 16 |
| *Alcelaphus buselaphus* (hartebeest) | A | N | L | 15 |
| *Antidorcas marsupialis* (springbok) | A | N | L | 15 |
| *Kobus ellipsiprymnus* (waterbuck) | A | N | L | 13 |
| *Sylvicapra grimmia* (grey duiker) | A | N | L | 13 |
| *Equus zebra* (mountain zebra) | A | N | L | 12 |
| *Hippotragus niger* (sable antelope) | A | N | L | 12 |
| *Tragelaphus scriptus* (bushbuck) | A | N | L | 12 |
| *Raphicerus campestris* (steenbok) | A | N | L | 11 |
| *Redunca arundinum* (southern reedbuck) | A | N | L | 11 |
| *Camelus dromedarius* (camel) | A | | | 10 |
| *Damaliscus lunatus* (korrigum/topi/tsessebe) | A | N | L | 10 |
| *Damaliscus pygargus* (blesbok/bontebok) | A | N | L | 10 |
| **CARNIVORES** | | | | |
| *Canis familiaris* (dog) | A | N | L | 24 |
| **LAGOMORPHS** | | | | |
| *Lepus saxatilis* (scrub hare) | A | N | L | 15 |
| *Lepus capensis* (Cape hare) | A | N | L | 10 |

[a] A = adult, N = nymph, L = larva

adults that in turn attach to the second final host. Its life cycle has been studied in South Africa by Rechav et al. (1977) who reported a preovipostion period of 6 days, an egg incubation period of 25 days, a larval feeding period of 5–8 days, a larval premolting period of 4–5 days, a nymphal feeding period of 5–8 days and a nymphal premolting period of 14 days. After engorgement and detachment, *R. e. evertsi* females laid up to 13,707 eggs. More than one life cycle could be completed in a year.

### Disease Associations

*Rhipicephalus evertsi evertsi* is a very important tick, particularly to animal health, because it is:

- a minor vector of *Theileria parva* (Lounsbury, 1906; Lawrence et al., 2004a), the protozoan causing East Coast fever in cattle;

- a vector of both *Babesia caballi* and *Theileria equi* (Theiler, 1906; Neitz, 1956; de Waal & van Heerden, 2004), two protozoa causing piroplasmosis in horses and other equines;

- a vector of *Babesia bigemina* (Neitz, 1956; de Vos et al., 2004), a protozoan causing babesiosis in cattle;

- a vector of *Borrelia theileri* (Neitz, 1964; Norval & Horak, 2004), the bacterium causing borreliosis in cattle;

- a cause of paralysis in sheep and cattle (Hamel & Gothe, 1978; Norval & Horak, 2004);

- a vector of the nairovirus causing Crimean-Congo hemorrhagic fever in humans (Rechav, 1992);

- a vector of *Ehrlichia ovina* (Norval & Horak, 2004), the rickettsia causing ehrlichiosis in sheep;

- a vector of *Theileria separata* (Lawrence et al., 2004d), a protozoan causing benign theileriosis in sheep;

- a vector of *Theileria taurotragi* (Lawrence & Williamson, 2004a), a protozoan causing theileriosis in cattle.

In addition, *R. e. evertsi* has been shown to be an experimental vector of *Anaplasma marginale* (Potgieter, 1979), the rickettsia causing anaplasmosis in cattle.

*Rhipicephalus evertsi mimeticus* is a tick of some importance to animal and human health because it is:

- a vector of *Theileria ovis* (Neitz, 1972), a protozoan causing benign theileriosis in sheep and goats;
- a cause of paralysis in sheep (Gothe et al., 1986);
- a vector of the nairovirus causing Crimean-Congo hemorrhagic fever in humans (Norval & Horak, 2004);
- a vector of *Theileria equi* (Norval & Horak, 2004), a protozoan causing piroplasmosis in horses;
- a vector of *Theileria separata* (Lawrence, 2004a), a protozoan causing benign theileriosis in sheep.

Finally, *R. evertsi*, subspecies not identified, has been reported to be a vector of *Rickettsia conorii* (Gear, 1992), the bacterium causing tick typhus in humans.

### References

Theiler (1950, 1962), Hoogstraal (1956), Elbl & Anastos (1966b), Yeoman et al. (1967), Walker (1974), Rechav et al. (1977), Mathysse & Colbo (1987), Walker et al. (2000, 2003).

## 3.33. *Rhipicephalus kochi* Donitz

The tick *Rhipicephalus kochi* was introduced into the continental United States on one occasion in 1984 on cut flowers imported into New York (USDA, 1985b).

### Hosts

*Rhipicephalus kochi* is a tick of ungulates, with cattle the preferred host (Table 3.33). Adults are found on domestic and wild ungulates, including cattle, sheep and goats among domestic livestock, and African buffalos, bushbucks, elands, impalas, greater kudus, giraffes and nyalas among wildlife. They are also found on domestic and wild carnivores (especially lions), scrub hares and African savanna hares (*Lepus microtis*). Occasional hosts include rodents, birds, elephant shrews and elephants.

Common sites of attachment on cattle are the udder and flanks.

### Distribution

The geographical distribution of *R. kochi* is limited to the eastern part of sub-Saharan Africa. It has been reported from central Africa (Democratic Republic of the Congo), from eastern Africa (Burundi, Ethiopia, Kenya, Rwanda, Sudan, Tanzania and Uganda), from southern Africa (Malawi, Mozambique, South Africa, Zambia and Zimbabwe) and from the island of Zanzibar in the Indian Ocean.

### Life Cycle

The life cycle of *R. kochi* has been studied in Zambia by Pegram (cited by Clifford et al., 1983). He reported a preoviposition period of 6–9 days, an egg incubation period of 26–29 days, a larval feeding period of 4–7 days, a larval premolting period of 11–15 days, a nymphal feeding period of 3–6 days, a nymphal pre-

Table 3.33. Most commonly reported hosts of the tick *Rhipicephalus kochi*

| Hosts | Stages[a] | | | No. of reports |
|---|---|---|---|---|
| UNGULATES | | | | |
| *Bos taurus* (cattle) | A | | | 22 |
| *Syncerus caffer* (African buffalo) | A | | | 9 |
| *Tragelaphus scriptus* (bushbuck) | A | N | | 8 |
| *Taurotragus oryx* (eland) | A | | | 7 |
| *Aepyceros melampus* (impala) | A | N | L | 6 |
| *Capra hircus* (goat) | A | | | 6 |
| *Ovis aries* (sheep) | A | | | 6 |
| *Tragelaphus strepsiceros* (greater kudu) | A | N | | 6 |
| CARNIVORES | | | | |
| *Panthera leo* (lion) | A | | | 6 |

[a] A = adult, N = nymph, L = larva

molting period of 12–15 days and a female feeding period of 7–8 days.

**Disease Associations**

None reported.

**References**

Hoogstraal (1956), Theiler (1962), Elbl & Anastos (1966b), Walker et al. (2000).

### 3.34. *Rhipicephalus muehlensi* Zumpt

The tick *Rhipicephalus muehlensi* was introduced into the continental United States on two occasions in 1970 on an animal trophy and a nyala imported into California from Africa (USDA, 1970c, 1971).

**Hosts**

*Rhipicephalus muehlensi* is a three-host tick of domestic and wild ungulates, with cattle the preferred domestic host and nyalas (Fig. 3.9), bushbucks, impalas, African buffalos and greater kudus among the commonly infested wildlife hosts (Table 3.34). African elephants and carnivores are also infested. Occasional hosts include rock hyraxes, squirrels, scrub hares and humans.

Attachment sites on cattle include the head and ears and on nyalas the ears.

**Distribution**

The geographical distribution of *R. muehlensi* is restricted to the eastern part of Africa from Somalia in the north to South Africa in the south. It has been reported from central Africa (Democratic Republic of the Congo), from eastern Africa (Kenya, Rwanda, Somalia, Sudan and Tanzania) and from southern Africa (Malawi, Mozambique, South Africa, Swaziland, Zambia and Zimbabwe).

Fig. 3.9. A nyala (*Tragelaphus angasii*), a preferred host of the tick *Rhipicephalus muehlensi*. (Photograph courtesy of iStockphoto.com.)

**Life Cycle**

The life cycle of *R. muehlensi* has been studied in South Africa (Elbl & Anastos, 1966b). They reported a preoviposition period of 9 days, an egg incubation period of 30–40 days, a larval feeding period of 4 days, a larval premolting period of 10–14 days, a nymphal feeding period of 4 days and a nymphal premolting period of 15–19 days.

**Disease Associations**

None reported.

Table 3.34. Most commonly reported hosts of the tick *Rhipicephalus muehlensi*

| Hosts | Stages[a] | | | No. of reports |
|---|---|---|---|---|
| UNGULATES | | | | |
| *Bos taurus* (cattle) | A | | | 14 |
| *Tragelaphus angasii* (nyala) | A | N | L | 14 |
| *Tragelaphus scriptus* (bushbuck) | A | N | | 12 |
| *Aepyceros melampus* (impala) | A | N | L | 11 |
| *Syncerus caffer* (African buffalo) | A | N | L | 10 |
| *Tragelaphus strepsiceros* (greater kudu) | A | N | L | 10 |

[a]A = adult, N = nymph, L = larva

## References

Hoogstraal (1956), Salisbury (1959), Theiler (1962), Elbl & Anastos (1966b), Yeoman et al. (1967), Walker (1974), Walker et al. (2000).

## 3.35. Rhipicephalus pulchellus (Gerstacker)

The "zebra tick" *Rhipicephalus pulchellus* has been introduced into the continental United States on 36 occasions between 1958 and 1976 on wild ungulates, animal trophies and a human imported into or entering Colorado, Florida, Maryland, Michigan, New Jersey, New York, North Carolina, Pennsylvania and Texas from African countries such as Kenya and Tanzania (see Appendix 1). The wild ungulate hosts included giraffes, zebras, a black rhinoceros and an eland, and the animal trophies included zebra hides and lion hides.

### Hosts

*Rhipicephalus pulchellus* is a three-host tick of ungulates. The preferred hosts for adults are domestic animals, including cattle, sheep, goats, camels and horses (Table 3.35). Wild ungulates such as elands, impalas, black rhinoceroses, Thomson's gazelles (*Eudorcas thomsonii*), hartebeests, blue wildebeests, Burchell's zebras, giraffes, warthogs and African buffalos are also commonly infested by adults, as are some carnivores such as dogs and lions. Larvae and nymphs feed on the same hosts as adults as well as on hares. All stages will feed on humans. Occasional hosts include elephants, rodents, birds, elephant shrews and bats.

Preferred attachment sites of adults on cattle are the ears and the underside of the body, including the chest, belly, and the genital and perianal regions. Attachment sites of adults on wild ungulates such as Thomson's gazelles, hartebeests and blue wildebeests are the head, body and tail.

### Distribution

The geographical distribution of *R. pulchellus* is limited to eastern Africa. It has been reported from Djibouti, Eritrea, Ethiopia, Kenya, Somalia, Sudan, Tanzania and Uganda, and from the Indian Ocean island of Zanzibar. Infestations have been reported on animals exported from eastern Africa to Oman, Saudi Arabia and Yemen in the Arabian Peninsula (Hoogstraal et al., 1981; Williams et al., 2000).

### Disease Associations

*Rhipicephalus pulchellus* is a tick of some significance to both animal and human health because it is:

- a poor vector of *Theileria parva* (Brocklesby, 1965), the protozoan causing East Coast fever in cattle;
- a vector of *Theileria taurotragi* (Young et al., 1977), a protozoan causing theileriosis in cattle;
- a vector of *Rickettsia conorii* (Walker et al., 2003), the bacterium causing tick typhus in humans;
- a vector of the nairovirus causing Crimean-Congo hemorrhagic fever in humans (Walker et al., 2003);
- a vector of the nairovirus causing Nairobi sheep disease in sheep and goats (Davies & Terpstra, 2004).

### References

Walker (1955, 1974), Theiler (1962), Yeoman et al. (1967), Matthysse & Colbo (1987), Walker et al. (2000, 2003).

Table 3.35. Most commonly reported hosts of the tick *Rhipicephalus pulchellus*

| Hosts | Stages[a] | No. of reports |
|---|---|---|
| **Ungulates** | | |
| *Bos taurus* (cattle) | A  N  L | 33 |
| *Ovis aries* (sheep) | A | 20 |
| *Capra hircus* (goat) | A  N  L | 18 |
| *Camelus dromedarius* (camel) | A | 14 |
| *Taurotragus oryx* (eland) | A  N | 11 |
| *Equus caballus* (horse) | A | 10 |
| **Primates** | | |
| *Homo sapiens* (human) | A  N  L | 17 |
| **Carnivores** | | |
| *Canis familiaris* (dog) | A  N | 12 |
| *Panthera leo* (lion) | A  N | 10 |

[a]A = adult, N = nymph, L = larva

### 3.36. Rhipicephalus senegalensis Koch

The tick *Rhipicephalus senegalensis* was introduced into the continental United States on one occasion in 1988 on an animal hide imported into Mississippi (USDA, 1994).

**Hosts**

*Rhipicephalus senegalensis* is a three-host tick of ungulates. The preferred hosts for adults are cattle, sheep, pigs and goats among domestic animals and warthogs, African buffalos and bongos (*Tragelaphus eurycerus*) among wildlife (Table 3.36). A few reports have indicated that larvae and nymphs can be found on rodents. Other hosts of adults are wild carnivores, elephants, aardvarks and humans. Occasional hosts include shrews, pangolins, hares, elephant shrews and a duck.

Preferred attachment sites for adults on cattle are the ears, but in heavy infestations they can be found attached anywhere on the head, shoulders, flanks, tail, or around the feet.

**Distribution**

The geographical distribution of *R. senegalensis* occurs in a broad band from Senegal in the west to Kenya and Tanzania in the east. It has been reported from western Africa (Benin, Burkina Faso, Gambia, Ghana, Guinea, Guinea-Bissau, Ivory Coast, Liberia, Mali, Nigeria, Senegal, Sierra Leone and Togo), from central Africa (Cameroon, Central African Republic, Chad, Congo, Democratic Republic of the Congo and Gabon), from eastern Africa (Kenya, Sudan, Tanzania and Uganda) and from southern Africa (Angola and Malawi).

**Disease Associations**

None reported.

**References**

Theiler (1962), Elbl & Anastos (1966b), Mathysse & Colbo (1987), Walker et al. (2000, 2003).

### 3.37. Rhipicephalus simus Koch

There has been some confusion in the past about ticks in the *Rhipicephalus simus* group, with *R. simus* reported from many regions of Africa. This confusion was resolved by Pegram et al. (1987a) and, in this book, *R. simus* refers only to the ticks in this group found in southern Africa.

The "glossy tick" *R. simus* was introduced into the continental United States on two occasions, in 1983 on a warthog hide imported into Mississippi from South Africa (USDA, 1985a) and on a jackal imported into Missouri from South Africa (Keirans & Durden, 2001). The introductions in 1965 on a black rhinoceros imported into Michigan from eastern Africa (USDA, 1966b), in 1969 on a lion hide imported into Texas from Kenya (USDA, 1970a) and in 1979 on a human entering Connecticut from Kenya (Anderson et al., 1981a), were all from eastern Africa where *R. simus* does not occur and thus are likely to be reclassified as *R. praetextatus* in light of the publication of Pegram et al. (1987a). The other two introductions in 1969 on a giraffe hide imported into Texas (USDA, 1970a) and in 1989 on an animal hide imported into Mississippi (USDA, 1994) were both from unknown locations.

Table 3.36. Most commonly reported hosts of the tick *Rhipicephalus senegalensis*

| Hosts | Stages[a] | No. of reports |
| --- | --- | --- |
| **Ungulates** | | |
| *Bos taurus* (cattle) | A | 45 |
| *Phacochoerus africanus* (warthog) | A | 25 |
| *Ovis aries* (sheep) | A | 19 |
| *Syncerus caffer* (African buffalo) | A  N | 18 |
| *Sus scrofa* (pig) | A | 15 |
| *Capra hircus* (goat) | A | 14 |
| *Tragelaphus eurycerus* (bongo) | A | 11 |
| **Carnivores** | | |
| *Canis familiaris* (dog) | A | 14 |

[a]A = adult, N = nymph

Consequently, only the two introductions from South Africa on a warthog hide and on a jackal can be definitively ascribed to *R. simus*.

## Hosts

*Rhipicephalus simus* is a three-host tick of ungulates and carnivores. The preferred hosts for adults are cattle, with other common hosts being dogs, goats and sheep among domestic animals and warthogs, African buffalos, lions, leopards, bush pigs, greater kudus, Burchell's zebras, sable antelopes and elands among wildlife (Table 3.37). Adults are found infrequently on humans. Larvae and nymphs prefer murid rodents (such as red rock rats, Natal multimammate mice and bushveld gerbils), wild carnivores and hares. Occasional hosts include elephants, insectivores, elephant shrews and birds.

Attachment sites vary by species. Adults attach in the tailbrush of cattle, around the feet of cattle and sheep, on the tails of horses and zebras, and on the head and shoulders of dogs and warthogs.

## Distribution

The geographical distribution of *R. simus* is limited to southern Africa. It has been reported from Angola, Botswana, Malawi, Mozambique, Namibia, South Africa, Swaziland, Zambia and Zimbabwe.

## Habitat and Life Cycle

*Rhipicephalus simus* is distributed widely in southern Africa in regions of moderate to high rainfall. Adults are found on ungulates and carnivores in the summer, larvae on rodents in the autumn and winter and nymphs on rodents in the winter and spring. The life cycle of *R. simus* has been studied in Zimbabwe by Norval & Mason (1981). They reported a preoviposition period of 5–6 days, an oviposition period of 21–24 days, an egg incubation period of 21–25 days, a larval feeding period of 3–6 days, a nymphal feeding period of 3–8 days, a nymphal premolting period of 8–13 days and a female feeding period of 6–11 days. After engorgement and detachment, females laid up to 6,415 eggs.

## Disease Associations

*Rhipicephalus simus* is a tick of some significance to both animal and human health because it is:

- a minor vector of *Theileria parva* (Lounsbury, 1906), the protozoan causing East Coast fever in cattle;
- a vector of *Anaplasma marginale* (Theiler, 1912), a rickettsia causing anaplasmosis in cattle;
- a cause of tick paralysis in calves and lambs (Lawrence & Norval, 1979);

Table 3.37. Most commonly reported hosts of the tick *Rhipicephalus simus*

| Hosts | Stages[a] | No. of reports |
|---|---|---|
| **Ungulates** | | |
| *Bos taurus* (cattle) | A | 36 |
| *Phacochoerus africanus* (warthog) | A | 23 |
| *Syncerus caffer* (African buffalo) | A | 19 |
| *Capra hircus* (goat) | A | 14 |
| *Equus burchelli* (Burchell's zebra) | A | 12 |
| *Hippotragus niger* (sable antelope) | A | 12 |
| *Potamochoerus larvatus* (bush pig) | A | 12 |
| *Tragelaphus strepsiceros* (greater kudu) | A | 12 |
| *Ovis aries* (sheep) | A | 11 |
| *Taurotragus oryx* (eland) | A | 11 |
| **Carnivores** | | |
| *Canis familiaris* (dog) | A  N | 27 |
| *Panthera leo* (lion) | A  N  L | 16 |
| *Panthera pardus* (leopard) | A | 13 |
| **Rodents** | | |
| *Hystrix africaeaustralis* (Cape porcupine) | A | 11 |

[a] A = adult, N = nymph, L = larva

- a vector of *Anaplasma centrale* (Potgieter & van Rensburg, 1987), the rickettsia causing benign anaplasmosis in cattle;
- a vector of *Babesia trautmanni* (de Waal et al., 1992), the protozoan causing babesiosis in pigs;
- a vector of *Rickettsia conorii* (Walker et al., 2003), the bacterium causing tick typhus in humans;
- a vector of the nairovirus causing Nairobi sheep disease in sheep and goats (Davies & Terpstra, 2004).

In addition, *R. simus* has been associated with tick paralysis and anaphylactic shock in humans (Zumpt & Glajchen, 1950; Norval & Mason, 1981). Even though *R. simus* has been shown to transmit a number of livestock diseases, its host associations make it unlikely to be important in transstadial disease transmission in the field since only adult *R. simus* are typically found on domestic and wild ungulates.

### References

Norval & Mason (1981), Pegram et al. (1987a), Walker et al. (2000, 2003).

## 3.38. *Rhipicephalus sulcatus* Neumann

The tick *Rhipicephalus sulcatus* has been confused with *R. turanicus* both morphologically and ecologically (Pegram et al., 1987b; Walker et al., 2000), resulting in erroneous conclusions regarding their host preferences and their distributions. Consequently, only data on hosts and distribution confirmed by Pegram et al. (1987b) and/or Walker et al. (2000) have been included herein.

*Rhipicephalus sulcatus* was reported to be introduced into the continental United States on one occasion in 1973 on a hedgehog imported into New York (USDA, 1974a). It is worth noting that *R. sulcatus* has been reported only twice on hedgehogs and, on both occasions, the host was the four-toed hedgehog (Doss et al., 1974b; Morel, 1978), a host not confirmed by either Pegram et al. (1987b) or Walker et al. (2000). This introduction of *R. sulcatus* into the United States in 1973, therefore, must be considered doubtful.

### Hosts

*Rhipicephalus sulcatus* is a three-host tick of mammals. Adults are found most frequently on ungulates and carnivores (Table 3.38). The most common ungulate hosts are cattle, sheep and goats among domestic livestock and hartebeests and oribis (*Ourebia ourebi*) among wildlife. The most common carnivore hosts are dogs, leopards and lions. Adults are also found on hares, particularly scrub and African savanna hares. There are very few records of hosts for immature stages of *R. sulcatus*, with nymphs found only on dogs and Nile grass rats (*Arvicanthis niloticus*). There are a few reports of human infestations. Occasional hosts include birds, rodents and rock hyraxes.

### Distribution

The geographical distribution of *R. sulcatus* covers much of sub-Saharan Africa. It has been reported from western Africa (Benin, Burkina Faso, Gambia, Ghana, Guinea-Bissau, Ivory Coast, Mali, Nigeria, Senegal and Togo), from central Africa (Cameroon, Central African Republic, Chad, Congo, Democratic Republic of the Congo and Gabon), from eastern Africa (Burundi, Ethiopia, Kenya, Rwanda, Sudan, Tanzania and Uganda) and from southern Africa (Angola,

Table 3.38. Most commonly reported hosts of the tick *Rhipicephalus sulcatus*

| Hosts | Stages[a] | No. of reports |
|---|---|---|
| UNGULATES | | |
| *Bos taurus* (cattle) | A | 30 |
| *Ovis aries* (sheep) | A | 11 |
| *Alcelaphus buselaphus* (hartebeest) | A | 10 |
| *Ourebia ourebi* (oribi) | A | 10 |
| CARNIVORES | | |
| *Canis familiaris* (dog) | A  N | 20 |
| *Panthera pardus* (leopard) | A | 13 |
| *Panthera leo* (lion) | A | 10 |

[a] A = adult, N = nymph

Botswana, Malawi, Mozambique, Namibia, South Africa, Zambia and Zimbabwe).

**Disease Associations**

None reported.

**References**

Pegram et al. (1987b), Walker et al. (2000).

## 3.39. *Rhipicephalus (Boophilus) decoloratus* (Koch)

The "blue tick" *Rhipicephalus (Boophilus) decoloratus* was known as *Boophilus decoloratus* prior to the taxonomic revision of ticks by Horak et al. (2002). This tick was introduced into the continental United States on 11 occasions between 1965 and 1989 on wild ungulates and animal trophies imported into Arizona, Colorado, Mississippi, New Jersey and New York from African countries such as Kenya (see Appendix 1). The wild ungulate hosts were giraffes, zebras, a hartebeest and an impala, and the animal trophies included animal hides.

### Hosts

*Rhipicephalus (Boophilus) decoloratus* is an important one-host tick of cattle. Other common hosts are goats, sheep, dogs and horses among domestic animals and impalas, greater kudus, elands, Burchell's zebras, blue wildebeests, African buffalos, bushbucks, giraffes, roan and sable antelopes, hartebeests and nyalas among wildlife (Table 3.39). Occasional hosts include wild carnivores, birds, hares, rats, reptiles and humans.

Sites of attachment are the dewlap for all stages, the brisket, axillae, udder, groin, scrotum and escutcheon for adults and the ears for nymphs and larvae. The entire body may be parasitized in heavy infestations.

### Distribution

The geographical distribution of *R. (B.) decoloratus* covers most countries in Africa south of the Sahara. It has been reported from western Africa (Benin, Burkina Faso, Gambia, Ghana, Guinea, Guinea-Bissau, Ivory Coast, Liberia, Mali, Mauritania, Niger, Nigeria, Senegal, Sierra Leone and Togo), from central Africa (Cameroon, Central African Republic, Chad, Democratic Republic of the Congo and Gabon), from eastern Africa (Burundi, Djibouti, Eritrea, Ethiopia, Kenya, Rwanda, Somalia, Sudan, Tanzania and Uganda), from southern Africa (Angola, Botswana, Lesotho, Malawi, Mozambique, Namibia, South Africa, Swaziland, Zambia and Zimbabwe), and from Cape Verde off the coast of western Africa and Zanzibar off the

Table 3.39. Most commonly reported hosts of the tick *Rhipicephalus (Boophilus) decoloratus*

| Hosts | Stages[a] | No. of reports |
|---|---|---|
| **Ungulates** | | |
| *Bos taurus* (cattle) | A   N   L | 136 |
| *Capra hircus* (goat) | A   N   L | 41 |
| *Aepyceros melampus* (impala) | A   N   L | 37 |
| *Ovis aries* (sheep) | A   N   L | 36 |
| *Equus caballus* (horse) | A   N | 23 |
| *Tragelaphus strepsiceros* (greater kudu) | A   N   L | 20 |
| *Taurotragus oryx* (eland) | A   N   L | 19 |
| *Equus burchelli* (Burchell's zebra) | A   N   L | 18 |
| *Syncerus caffer* (African buffalo) | A   N   L | 18 |
| *Connochaetes taurinus* (blue wildebeest) | A   N   L | 17 |
| *Giraffa camelopardalis* (giraffe) | A   N   L | 14 |
| *Tragelaphus scriptus* (bushbuck) | A   N   L | 14 |
| *Alcelaphus buselaphus* (hartebeest) | A   N   L | 13 |
| *Hippotragus niger* (sable antelope) | A   N   L | 13 |
| *Hippotragus equinus* (roan antelope) | A   N | 11 |
| *Tragelaphus angasii* (nyala) | A   N   L | 10 |
| **Carnivores** | | |
| *Canis familiaris* (dog) | A   N | 25 |

[a] A = adult, N = nymph, L = larva

coast of eastern Africa. *Rhipicephalus (Boophilus) decoloratus* is being replaced in eastern and southeastern Africa by *R. (B.) microplus*, apparently due to the shorter life cycle of *R. (B.) microplus* as well as its tendency to assortative mating and more successful feeding on cattle (Estrada-Pena et al., 2006).

**Habitat and Life Cycle**

*Rhipicephalus (Boophilus) decoloratus* is widespread in sub-Saharan Africa. It is found in savanna and temperate climates in grasslands and wooded areas used as cattle pasture, but it is absent from deserts and the most humid parts of central Africa. It is present at most altitudes and at most levels of rainfall, being most common in those areas receiving 500–1,000 mm of rain a year. It can survive moderate levels of frost.

All three stages spend a total of about three weeks on the same host. The engorged females detach and lay up to 2,500 eggs within a week which hatch in 3–6 weeks. The entire life cycle can be completed in about two months, allowing for several generations a year.

**Disease Associations**

*Rhipicephalus (Boophilus) decoloratus* is an important tick to bovine health because it is :

- a vector of *Borrelia theileri* (Neitz, 1964), the bacterium causing borreliosis in cattle, sheep, goats and horses;
- a vector of *Babesia bigemina* (Gray & Potgieter, 1982), a protozoan causing babesiosis in cattle.

In addition, *R. (B.) decoloratus* is an experimental vector of *Anaplasma marginale* (Potgieter & Stoltsz, 2004), a rickettsia causing anaplasmosis in cattle.

**References**

Theiler (1949a, 1962), Hoogstraal (1956), Arthur (1960), Elbl & Anastos (1966c), Yeoman et al. (1967), Walker (1974), Matthysse & Colbo (1987), Walker et al. (2003), Horak et al. (2006a).

# 4

# Ticks from the Afro-Caribbean Region

The only invasive tick species whose distribution includes both Africa (see Map 3.1) and the Caribbean (Map 4.1), *Amblyomma variegatum*, is discussed in this chapter.

## 4.1. *Amblyomma variegatum* (Fabricius)

The "tropical bont tick" *Amblyomma variegatum* has been introduced into the continental United States on 17 occasions between 1965 and 1997 on domestic animals, African wildlife, animal trophies, ostriches, humans and an African savanna monitor imported into or entering Arizona, California, Colorado, Florida, Michigan, New York, North Carolina, Pennsylvania and Texas from African countries such as Ghana, Kenya and Tanzania (see Appendix 1). The domestic hosts were cattle, sheep and a horse, the wildlife hosts were a black rhinoceros, an eland, a giraffe, a kudu and a zebra, and the animal trophies included an African buffalo hide.

### Hosts

*Amblyomma variegatum* (Fig. 4.1) is a very important three-host tick with a wide host range. It has been found on 157 species of animals, including 54 species of ungulates, 47 species of birds, 23 species of carnivores, 10 species of rodents, 8 species of reptiles, 6 species of primates, 4 species of insectivores, 2 species of lagomorphs, and one species each of elephants, aardvarks and pangolins (see Appendix 2). The preferred hosts for adults are ungulates, in particular cattle, sheep and goats (Table 4.1). Other important hosts include dogs, horses, donkeys and pigs among domestic animals, and African buffalos, warthogs, elands, hartebeests and waterbucks among wildlife. Larvae and nymphs feed on the same hosts as adults and also on birds, wild carnivores and rodents. All stages of *A. variegatum* will feed on humans. Less common hosts include non-human primates, African elephants, hares, hedgehogs and reptiles (monitors, chameleons and snakes).

Attachment sites of *A. variegatum* adults on domestic stock are the dewlap, brisket, axillae, abdomen, udder, groin and escutcheon, including the male genitalia, and sometimes the heels, elbows, hocks, interdigital cleft, vulva, head and tip of the tail. Adults have been found attached to the buccal mucous membrane in the mouth as well as in the nostrils of dogs. Nymphs are found attached especially to the heels of cattle.

### Distribution

The geographical distribution of *A. variegatum* covers all but the southernmost part of sub-Saharan Africa, associated islands in the Atlantic and Indian oceans, parts of the Arabian Peninsula, and islands in the eastern Caribbean. In Africa, it has been reported from western Africa (Benin, Burkina Faso, Gambia, Ghana, Guinea, Guinea-Bissau, Ivory Coast, Liberia, Mali, Mauritania, Niger, Nigeria, Senegal, Sierra Leone and Togo), from central Africa (Cameroon, Central African Republic, Chad, Congo, Democratic Republic of the Congo, Equatorial Guinea and Gabon), from eastern Africa (Burundi, Djibouti, Eritrea, Ethiopia, Kenya, Rwanda, Somalia, Sudan, Tanzania and Uganda), from southern Africa (Angola, Botswana, Malawi,

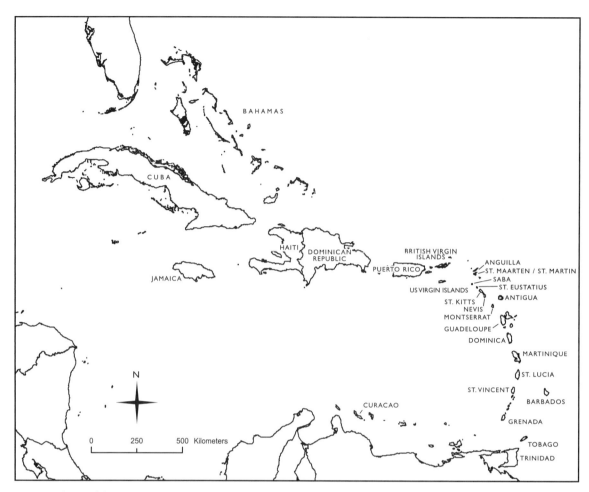

Map 4.1. The Caribbean

Fig. 4.1. Engorged *Amblyomma variegatum* ticks taken from cattle in Antigua. (Photograph courtesy of the U.S. Department of Agriculture.)

Table 4.1. Most commonly reported hosts of the tick *Amblyomma variegatum*

| Hosts | Stages[a] | | | No. of reports |
|---|---|---|---|---|
| **UNGULATES** | | | | |
| *Bos taurus* (cattle) | A | N | L | 188 |
| *Ovis aries* (sheep) | A | N | L | 68 |
| *Capra hircus* (goat) | A | N | L | 66 |
| *Equus caballus* (horse) | A | N | L | 33 |
| *Syncerus caffer* (African buffalo) | A | N | | 23 |
| *Taurotragus oryx* (eland) | A | N | | 23 |
| *Phacochoerus africanus* (warthog) | A | N | L | 20 |
| *Equus asinus* (donkey) | A | N | L | 19 |
| *Alcelaphus buselaphus* (hartebeest) | A | N | L | 18 |
| *Sus scrofa* (pig) | A | N | L | 17 |
| *Kobus ellipsiprymnus* (waterbuck) | A | N | L | 15 |
| *Tragelaphus scriptus* (bushbuck) | A | N | L | 14 |
| *Camelus dromedarius* (camel) | A | | | 12 |
| *Equus burchelli* (Burchell's zebra) | A | N | | 12 |
| *Aepyceros melampus* (impala) | A | N | L | 11 |
| *Hippotragus equinus* (roan antelope) | A | N | | 11 |
| *Ourebia ourebi* (oribi) | A | N | L | 11 |
| *Hippotragus niger* (sable antelope) | A | N | | 10 |
| **CARNIVORES** | | | | |
| *Canis familiaris* (dog) | A | N | L | 47 |
| *Felis catus* (cat) | A | N | L | 13 |
| **PRIMATES** | | | | |
| *Homo sapiens* (human) | A | N | L | 21 |
| **BIRDS** | | | | |
| *Numida meleagris* (helmeted guineafowl) | | N | L | 14 |
| *Gallus gallus* (chicken) | A | N | L | 11 |

[a] A = adult, N = nymph, L = larva

Mozambique, Namibia, Zambia and Zimbabwe), from Cape Verde in the Atlantic Ocean and from Comoros, Madagascar, Mauritius, Mayotte, Pemba Island, Reunion and Zanzibar in the Indian Ocean. In the Arabian Peninsula, it has been reported from Oman, Saudi Arabia and Yemen. In the eastern Caribbean, it has been reported from Anguilla, Antigua, Barbados, Dominica, Guadeloupe including La Desirade and Marie Galante, Martinique, Montserrat, Puerto Rico including Culebra and Vieques, St. Kitts and Nevis, St. Lucia, St. Maarten/St. Martin, St. Vincent and the Grenadines, and St. Croix in the U.S. Virgin Islands.

*Amblyomma variegatum* is an African tick that has spread outside Africa to the eastern Caribbean and southwestern Asia. French colonists transported infested cattle from Senegal in French West Africa to Guadeloupe in the French West Indies around 1828 and thus introduced *A. variegatum* into the eastern Caribbean (Barre et al., 1995). Cattle from Guadeloupe spread *A. variegatum* to the French island of Marie Galante around 1835. Antigua became infested around 1895. Martinique became infested in 1948 when oxen were imported from Guadeloupe. One focus of infestation was found in St. Croix in the U.S. Virgin Islands in 1967, and, over the next 21 years, *A. variegatum* rapidly spread throughout much of the eastern Caribbean as a result of increased intra-island movement of animals and probably migratory cattle egrets (*Bubulcus ibis*). Spread of *A. variegatum* to the Arabian Peninsula was probably through importation of infested livestock from Africa; for example, *A. variegatum* has been found on cattle in Saudi Arabia imported from eastern Africa (Hoogstraal et al., 1981). Its spread to Madagascar, and probably other African islands, was through transportation of infested cattle (Walker et al., 2003).

*Amblyomma variegatum* has been found on several occasions outside its established range. Nymphs were found on tree pipits in Cyprus during their spring migration (Kaiser et al., 1974), demonstrating the ability of birds to carry *A. variegatum* from Africa to other continents. Adult *A. variegatum* have been found on a dog in France (Hoogstraal, 1967), on animals in Sicily (Albanese et al., 1971) and on cattle in the Krasnodar region of Russia (Voltzit & Keirans, 2003), and, in each case, the authors speculated that the infestations had been introduced by migrating birds carrying *A. variegatum* nymphs. Adult *A. variegatum* were also found on a dog in Bangladesh (Rahman & Mondal, 1985), and the authors believed the tick had been introduced on imported cattle or dogs.

**Life Cycle and Habitat**

*Amblyomma variegatum* occurs in a wide variety of climates, including rain forests, temperate highlands and savanna, with its lack of tolerance for aridity a determining factor in its distribution. It has been collected from areas with a very wide range of rainfall intensity (650–1,900 mm/year) and from areas with 6–7 months of continuous dry season to those with no severe dry season. In Africa, adults are most abundant in the wet season (October to February), nymphs in May to September and larvae in March to May, with one generation completed in a year in areas with a single annual rainy season. However, in regions with two rainy seasons such as Kenya and Uganda, two or three generations may be produced in one year. Detachment of engorged females occurs throughout the rainy season, but large-scale larval emergence does not take place until the beginning of the dry season.

The life cycle of *A. variegatum* has been studied in Kenya in Africa by Walker (cited by Elbl & Anastos, 1966a). She reported a preoviposition period of 12 days, an egg incubation period of 53 days, a larval prefeeding period of 7 days, a larval feeding period of 5 days, a larval premolting period of 14 days, a nymphal prefeeding period of 7 days, a nymphal feeding period of 5 days, a nymphal premolting period of 19 days, a female prefeeding period of 7 days and a female feeding period of 12 days.

The life cycle of *A. variegatum* has been studied also in Puerto Rico in the Caribbean by Garris (1984). He reported a preoviposition period of 9–21 days, an oviposition period of 17–50 days, an egg incubation period of 43–62 days, a larval prefeeding period of 1–10 days, a larval feeding period of 6–13 days, a larval premolting period of 15–22 days, a nymphal prefeeding period of 1–5 days, a nymphal feeding period of 5–10 days, a nymphal premolting period of 19–28 days, a female prefeeding period of 1–7 days and a female feeding period of 11–16 days.

**Disease Associations**

*Amblyomma variegatum* is a very important tick to both human and animal health because it is:

- a major vector of *Ehrlichia ruminantium* (Daubney, 1930), the rickettsia causing heartwater in domestic and wild ruminants, being the most widespread vector in Africa and the only vector in the Caribbean;
- a vector of *Theileria mutans* (Uilenberg et al., 1974), a protozoan causing theileriosis in cattle;
- a vector of *Theileria velifera* (Uilenberg & Schreuder, 1976), a protozoan causing benign theileriosis in cattle;
- a vector of *Anaplasma bovis* (Stewart, 1992), a rickettsia causing anaplasmosis in cattle;
- a vector of the nairovirus causing Crimean-Congo hemorrhagic fever in humans (Faye et al., 1999);
- a vector of the nairovirus causing Nairobi sheep disease in sheep and goats (Davies & Terpstra, 2004);
- a vector of *Rickettsia conorii* (Brown et al., 2005), the rickettsia causing tick typhus in humans;
- a putative vector of *Rickettsia africae* (Kelly, 2006), the rickettsia causing African tick-bite fever in humans.

In addition, *A. variegatum* is closely associated with the development of acute dermatophilosis in cattle (Burridge et al., 1984), a horrific skin condition caused by the bacterium *Dermatophilus congolensis*. Furthermore, heavy infestations on cattle reduce liveweight gain (Pegram & Oosterwijk, 1990), damage teats (Walker et al., 2003) and lead to loss of udder quarters and to lameness (Jongejan & Uilenberg, 2004).

*Amblyomma variegatum* has been found infected with Dugbe, Jos and Thogoto viruses (Labuda & Nuttall, 2004); one of these viruses, the Thogoto virus, has been reported associated with abortions in sheep and clinical disease in humans (Labuda & Nuttall, 2004). There have been several reports of infection of *A. variegatum* with *Coxiella burnetii*, the rickettsia causing Q fever in humans (e.g., Berge & Lennette, 1953; Babudieri, 1959), but its role as a vector of *C. burnetii* remains unclear. Finally, *A. variegatum* has been implicated in tick paralysis in sheep although no conclusive experimental data exist (Mans et al., 2004).

**References**

Robinson (1926), Hoogstraal (1956), Floch & Fauran (1958), Theiler (1962), Elbl & Anastos (1966a), Yeoman et al. (1967), Walker (1974), Uilenberg et al. (1979), Matthysse & Colbo (1987), Petney et al. (1987), Barre et al. (1988, 1991, 1995), Barre & Garris (1990), Yonow (1995), Voltzit & Keirans (2003), Walker et al. (2003).

# 5

# Ticks from the Afro-Asian Region

The only invasive tick species whose distribution includes only Africa (see Map 3.1) and Asia (see Map 10.1), *Hyalomma dromedarii*, is discussed in this chapter.

## 5.1. *Hyalomma dromedarii* Koch

The tick *Hyalomma dromedarii*, known colloquially as the camel hyalomma, was introduced into the continental United States on one occasion on a camel imported into New York (Becklund, 1968).

### Hosts

*Hyalomma dromedarii* is a one-, two- or three-host tick, with the two-host life cycle the most common, whose preferred hosts are camels (Fig. 5.1). Common hosts of adults are other domestic ungulates, such as cattle, sheep, goats, horses, water buffalos (*Bubalus bubalis*) and donkeys, and dogs (Table 5.1). Nymphs are found on camels, cattle, sheep and horses, and both larvae and nymphs feed on a variety of rodents, especially gerbils and jirds, and on dogs. All stages will attach to humans. Occasional hosts include insectivores, lagomorphs, lizards and birds.

Reports from Asia suggest that the preferred sites of attachment for adults on camels are the groin, udder, perianal region and tail, with nymphs found on the back, around the hump and in the intermandibular space. In contrast, one report from Africa found adults predominantly attached to the nose, nostrils and ears of camels. Adults are found on the ears and in the perianal region of sheep.

### Distribution

*Hyalomma dromedarii* is found in Africa and Asia wherever camels occur, namely from Africa north of the Equator through the Arabian Peninsula to Asia as far east as Tibet and Mongolia. On the African continent, *Hy. dromedarii* has been reported from northern Africa (Algeria, Egypt, Libya, Morocco and Tunisia), from western Africa (Burkina Faso, Mali, Mauritania, Niger, Nigeria, Senegal and Western Sahara), from central Africa (Cameroon, Central African Republic, Chad and Democratic Republic of the Congo), from eastern Africa (Djibouti, Eritrea, Ethiopia, Kenya, Somalia, Sudan, Tanzania and Uganda) and from the Canary Islands in the Atlantic Ocean off the coast of Morocco; it has also been introduced into Namibia in southern Africa. In Asia, it has been reported from Afghanistan, Armenia, Azerbaijan, Bahrain, China, India, Iran, Iraq, Israel, Jordan, Kazakhstan, Kuwait, Kyrgyzstan, Lebanon, Mongolia, Oman, Pakistan, the Palestinian Territories, Qatar, Russia, Saudi Arabia, Syria, Tajikistan, Turkey, Turkmenistan, United Arab Emirates, Uzbekistan and Yemen.

### Life Cycle

*Hyalomma dromedarii* is well adapted to extreme dryness and to its camel hosts. Its life cycle is continuous throughout the year. The life cycle of *Hy. dromedarii* has been studied in Morocco by Ouhelli (1994). He reported a preovipostion period of 4–6 days, an oviposition period of 24–36 days, an egg incubation period of 30–37 days, a larval feeding period of 6–11 days, a larval premolting period of 6–8 days, a nymphal feeding period of 9–12 days, a nymphal premolting period of 20–23 days and a female feeding period of 10–14 days. Following engorgement and detachment, females laid up to 9,111 eggs.

## Disease Associations

*Hy. dromedarii* is a tick of some significance to animal health because it is:

- a vector of both *Babesia caballi* (Sippel et al., 1962) and *Theileria equi* (Mehlhorn & Schein, 1998), two protozoa causing piroplasmosis in horses and other equines;
- a vector of *Babesia merionis* (Friedhoff, 1988), a protozoan causing babesiosis in jirds;
- a vector of *Theileria annulata* (Pipano & Shkap, 2004), the protozoan causing tropical theileriosis in cattle.

In addition, it has been found to be an experimental vector of the orbivirus causing African horse sickness (Awad et al., 1981).

## References

Delpy & Gouchey (1937), Hoogstraal (1956), Theiler (1962), Kaiser & Hoogstraal (1964), Singh & Dhanda (1965), Das & Subramarian (1972b), Walker (1974), Hoogstraal et al. (1981), Matthysse & Colbo (1987), Khan & Srivastava (1994), Ouhelli (1994), Walker et al. (2003), Estrada-Pena et al. (2004a), Apanaskevich et al. (2008b).

Fig. 5.1. A camel (or dromedary) (*Camelus dromedarius*), the preferred host of the tick *Hyalomma dromedarii* and a common host of many other African and Asian ticks. (Photograph courtesy of iStockphoto.com.)

Table 5.1. Most commonly reported hosts of the tick *Hyalomma dromedarii*

| Hosts | Stages[a] | | | No. of reports |
|---|---|---|---|---|
| **Ungulates** | | | | |
| *Camelus dromedarius* (camel) | A | N | L | 108 |
| *Bos taurus* (cattle) | A | N | | 68 |
| *Ovis aries* (sheep) | A | N | | 36 |
| *Capra hircus* (goat) | A | | | 33 |
| *Equus caballus* (horse) | A | N | | 27 |
| *Bubalus bubalis* (water buffalo) | A | | | 12 |
| *Equus asinus* (donkey) | A | | | 12 |
| **Carnivores** | | | | |
| *Canis familiaris* (dog) | A | N | L | 14 |

[a] A = adult, N = nymph, L = larva

# 6

# Ticks from the Afro-European Region

The only invasive tick species whose distribution includes only Africa (see Map 3.1) and Europe (see Map 8.1), *Hyalomma lusitanicum*, is discussed in this chapter.

## 6.1. *Hyalomma lusitanicum* Koch

The tick *Hyalomma lusitanicum* was introduced into the continental United States on two occasions in 1989 on an animal hide imported into Mississippi (USDA, 1994) and again on ostriches imported into New York from Portugal (Mertins & Schlater, 1991; USDA, 1994).

### Hosts

*Hyalomma lusitanicum* is a three-host tick whose preferred hosts are cattle (Table 6.1). Other hosts for adults are sheep, goats, pigs and humans. Nymphs and larvae are found on rabbits (*Oryctolagus cuniculus*). Occasional hosts include dogs, foxes, polecats, weasels, hares, rodents, hedgehogs, birds, camels and deer.

### Distribution

The geographical distribution of *Hy. lusitanicum* is limited to northern Africa and southwestern Europe. It has been reported from Algeria, Egypt and Morocco in northern Africa and from France, Italy, Portugal and Spain in continental Europe. It has also been found on islands in the Atlantic Ocean (Azores, Canary Islands and Madeira Islands) and on islands in the Mediterranean Sea (Balearic Islands, Sicily and Sardinia).

### Life Cycle

Adult *Hy. lusitanicum* are active throughout the year, with peaks of activity from May to July and again from October to November. Larvae are active from May

Table 6.1. Most commonly reported hosts of the tick *Hyalomma lusitanicum*

| Hosts | Stages[a] | No. of reports |
|---|---|---|
| **Ungulates** | | |
| *Bos taurus* (cattle) | A | 29 |
| *Ovis aries* (sheep/Mediterranean mouflon) | A | 14 |
| *Capra hircus* (goat) | A | 11 |
| *Sus scrofa* (domestic and wild pigs) | A | 8 |
| **Primates** | | |
| *Homo sapiens* (human) | A | 7 |
| **Lagomorphs** | | |
| *Oryctolagus cuniculus* (rabbit) | A  N  L | 7 |

[a] A = adult, N = nymph, L = larva

to September and nymphs from July to September. The life cycle of *Hy. lusitanicum* has been studied by Ouhelli & Pandey (1984) and by Ouhelli (1994) in Morocco. They reported a preoviposition period of 40–60 days, an oviposition period of 20–30 days, an egg incubation period of 32–40 days, a larval feeding period of 4–6 days, a larval premolting period of 12–14 days, a nymphal feeding period of 6–12 days, a nymphal premolting period of 16–21 days and a female feeding period of 8–13 days.

Following engorgement and detachment, females laid up to 7,124 eggs. From these data it was estimated that the life cycle of *Hy. lusitanicum* takes 138–196 days to complete.

**Disease Associations**

*Hyalomma lusitanicum* is a tick of some significance to animal and human health because it is:

- a vector of *Anaplasma marginale* (Sergent et al., 1928), a rickettsia causing anaplasmosis in cattle;
- a vector of *Theileria annulata* (Viseras et al., 1999), the protozoan causing tropical theileriosis in cattle;
- a vector of *Coxiella burnetii* (Walker et al., 2003), the rickettsia causing Q fever in humans.

**References**

Ouhelli & Pandey (1984), Ouhelli (1994), Perez-Eid & Cabrita (2003), Walker et al. (2003), Estrada-Pena et al. (2004a), Apanaskevich et al. (2008a).

# 7

# Ticks from the Afro-Eurasian Region

The eight invasive tick species whose distributions include Africa (see Map 3.1), Europe (see Map 8.1) and Asia (see Map 10.1) are discussed in this chapter.

## 7.1. *Haemaphysalis punctata* Canestrini & Fanzago

The "red sheep tick" *Haemaphysalis punctata* was introduced into the continental United States on three occasions on ostriches imported into California and New York from Portugal in 1989 (Mertins & Schlater, 1991; USDA, 1994) and on a pig imported into Florida (Wilson & Bram, 1998).

### Hosts

*Haemaphysalis punctata* is a three-host tick with a wide host range. It has been collected from 183 species including 121 species of birds, 17 species of rodents, 14 species of ungulates, 13 species of reptiles, 9 species of carnivores, 5 species of lagomorphs, 3 species of insectivores and one species of primates (see Appendix 2). The preferred hosts for adults are ungulates, particularly cattle, sheep, goats and horses (Table 7.1). Other hosts for adults include donkeys, pigs, dogs, brown hares (*Lepus europaeus*), rabbits, northern hedgehogs (*Erinaceus europaeus*) and humans. Nymphs and larvae will feed on the same hosts as adults as well as on a large variety of birds and rodents. Occasional hosts include lizards and other reptiles.

Attachment sites on sheep include the head, axillae and groin for adults and nymphs and the upper rim of the hoof and the muzzle for larvae.

### Distribution

The geographical distribution of *Ha. punctata* extends from western Europe and northern Africa eastward into central Asia. It has been reported from most European countries (Albania, Belarus, Bosnia & Herzegovina, Bulgaria, Croatia, Denmark, Estonia, France, Germany, Greece, Hungary, Italy, Macedonia, Moldova, Montenegro, Netherlands, Poland, Portugal, Romania, Russia, Serbia, Slovakia, Spain, Sweden, Switzerland, Ukraine and United Kingdom), from several Mediterranean islands (Corsica, Crete, Sardinia and Sicily), from northern Africa (Algeria, Egypt, Libya, Morocco and Tunisia), from the Madeira Islands off the coast of northwestern Africa and from eastern and central Asia (Armenia, Azerbaijan, western China, Georgia, Iran, Iraq, Israel, Jordan, Kazakhstan, Kyrgyzstan, Lebanon, Russia, Syria, Tajikistan, Turkey, Turkmenistan and Uzbekistan).

### Life Cycle and Habitat

*Haemaphysalis punctata* is found in a wide variety of habitats, provided humidity is high. Its life cycle can be completed in one year, but it usually takes three years.

### Disease Associations

*Haemaphysalis punctata* is an important tick, especially to animal health, because it is:

- a cause of tick paralysis in chickens, goats and sheep (Pavlov, 1963; Mans et al., 2004);

- a vector of *Rickettsia sibirica sibirica* (Hoogstraal, 1967), the bacterium causing North Asian tick typhus in humans;
- the vector of *Babesia major* (Morzaria et al., 1977), a protozoan causing babesiosis in cattle;
- the vector in Europe of *Babesia motasi* (Uilenberg et al., 1980), a protozoan causing babesiosis in sheep and goats;
- a vector of *Theileria buffeli/Theileria orientalis* (Uilenberg & Hashemi-Fesharki, 1984; Walker et al., 2003), a protozoal group causing theileriosis in cattle;
- a vector of *Theileria ovis* (Alani & Herbert, 1988), a protozoan causing benign theileriosis in sheep.

In addition, *Ha. punctata* has been shown to be an experimental vector of *Babesia bigemina* (Neitz, 1956), a protozoan causing babesiosis in cattle.

### References

Arthur (1963), Nosek (1971), Garben et al. (1981), Walker et al. (2003), Estrada-Pena et al. (2004a).

### 7.2. *Hyalomma aegyptium* (Linnaeus)

Many earlier records of the tick *Hyalomma aegyptium* are considered erroneous since in those days the name *Hy. aegyptium* was used as a catch-all for a number of different tick species (Hoogstraal, 1956). Consequently, all references to *Hy. aegyptium* prior to 1957 have been excluded because of the uncertainty of the identity of the ticks in those earlier publications.

*Hyalomma aegyptium*, known colloquially as the "tortoise hyalomma," was introduced into the continental United States on seven occasions between 1970 and 2002 on tortoises imported into Florida, New York and North Carolina from Egypt and Italy (see Appendix 1). The chelonian hosts included Egyptian tortoises (*Testudo kleinmanni*) and Greek tortoises (*Testudo graeca*).

### Hosts

*Hyalomma aegyptium* is a three-host tick of tortoises and in particular Greek tortoises (Table 7.2). All stages have been found on Greek tortoises, Hermann's tortoises (*Testudo hermanni*), Afghan tortoises (*Testudo horsfieldii*) and marginated tortoises (*Testudo marginata*), with nymphs and larvae also on a variety of avian species. Occasional hosts include lizards, domestic ungulates, rodents, hedgehogs, hares, dogs and humans.

Studies on Greek tortoises have found *Hy. aegyptium* attached to the forelegs, hindlegs and tail and, in damaged specimens, to fractures of the carapace and plastron.

Table 7.1. Most commonly reported hosts of the tick *Haemaphysalis punctata*

| Hosts | Stages[a] | | | No. of reports |
|---|---|---|---|---|
| UNGULATES | | | | |
| *Bos taurus* (cattle) | A | N | | 45 |
| *Ovis aries* (sheep/Mediterranean mouflon) | A | N | L | 44 |
| *Capra hircus* (goat) | A | N | | 27 |
| *Equus caballus* (horse) | A | N | L | 18 |
| *Equus asinus* (donkey) | A | | | 10 |
| *Sus scrofa* (domestic & wild pigs) | A | | | 10 |
| PRIMATES | | | | |
| *Homo sapiens* (human) | A | | | 19 |
| BIRDS | | | | |
| *Turdus merula* (blackbird) | | N | L | 17 |
| *Phoenicurus phoenicurus* (redstart) | A | N | L | 10 |
| CARNIVORES | | | | |
| *Canis familiaris* (dog) | A | N | L | 16 |
| LAGOMORPHS | | | | |
| *Lepus europaeus* (brown hare) | A | N | L | 11 |

[a] A = adult, N = nymph, L = larva

Table 7.2. Most commonly reported hosts of the tick *Hyalomma aegyptium*

| Hosts | Stages[a] | | | No. of reports |
|---|---|---|---|---|
| **Tortoises** | | | | |
| *Testudo graeca* (Greek tortoise) | A | N | L | 19 |
| *Testudo hermanni* (Hermann's tortoise) | A | N | L | 7 |
| *Testudo horsfieldii* (Afghan tortoise) | A | N | L | 5 |
| **Birds** | | | | |
| *Coturnix coturnix* (common quail) | - | | | 6 |
| **Reptiles** | | | | |
| *Laudakia stellio* (rough-tailed rock agama) | A | N | | 5 |
| **Ungulates** | | | | |
| *Ovis aries* (sheep) | A | | | 5 |

[a] A = adult, N = nymph, L = larva, - = not recorded

## Distribution

The geographical distribution of *Hy. aegyptium* extends from eastern Europe to central Asia. It has been reported from Albania, Bulgaria, Greece, Hungary, Italy, Romania, Russia, Ukraine and the Mediterranean island of Corsica in Europe and from Afghanistan, Armenia, Azerbaijan, China, Georgia, India, Iran, Iraq, Israel, Jordan, Kazakhstan, Lebanon, Pakistan, Russia, Saudi Arabia, Syria, Tajikistan, Turkey, Turkmenistan and Uzbekistan in Asia. Hoogstraal (1956) and Hoogstraal & Kaiser (1958) stated that *Hy. aegyptium* was introduced into Egypt but did not become established. However, Clark & Doten (1995) reported finding *Hy. aegyptium* in Florida on tortoises imported from Egypt, raising the possibility that *Hy. aegyptium* does indeed now occur in Egypt. There are isolated reports of *Hy. aegyptium* occurring in other African countries such as Tunisia (Hoogstraal & Kaiser, 1960; Kolonin, 2009), Libya and Senegal (Arthur, 1963), Morocco (Bailly-Choumara et al., 1976; Kolonin, 2009; Siroky et al., 2009), Nigeria (Okaeme, 1986), Ghana (Ntiamoa-Baidu et al., 2004) and Algeria (Kolonin, 2009; Siroky et al., 2009), so it appears that the natural distribution of *Hy. aegyptium* now extends into Africa.

## Life Cycle

The life cycle of *Hy. aegyptium* has been studied in Lebanon by Sweatman (1968). He reported that the preoviposition and oviposition periods for females were 1–22 and 6–32 days, respectively, with up to 16,427 eggs laid. The optimal temperature for egg laying was between 20 and 35 °C, with no eggs laid at 40 °C.

## Disease Associations

*Hyalomma aegyptium* is a vector of two parasites of reptiles, *Hemolivia mauritanicum* (Michel, 1973; Landau & Paperna, 1997), a hemogregarine parasite of Greek tortoises, and *Hepatozoon kisrae* (Paperna et al., 2002), a hemogregarine parasite of rough-tailed rock agamas (*Laudakia stellio*).

## References

Arthur (1963), Kaiser & Hoogstraal (1964), Petney & Al-Yaman (1985), Burridge & Simmons (2003), Siroky et al. (2006).

## 7.3. *Hyalomma anatolicum* Koch spp. (*Hy. anatolicum* & *Hy. excavatum*)

Historically there has been much confusion about the taxonomy of *Hyalomma* species. In their recent publication of valid tick names, Horak et al. (2002) considered *Hyalomma anatolicum* to consist of two subspecies, *Hy. anatolicum anatolicum* and *Hy. anatolicum excavatum*. More recently, Apanaskevich & Horak (2005) have suggested that these two subspecies should be considered separate species, *Hy. anatolicum* and *Hy. excavatum*. In this book, ticks of the two species will be designated by the name *Hy. anatolicum* spp. since much of the literature does not differentiate between the species; however, where available, information will be given by species.

*Hyalomma anatolicum* spp., known colloquially as the "small hyalommas," were introduced into the continental United States on four occasions before 1985 on horses imported into New York from Jordan and

Table 7.3. Most commonly reported hosts of the *Hyalomma anatolicum* spp. ticks

| Hosts | Stages[a] | | | No. of reports |
|---|---|---|---|---|
| **UNGULATES** | | | | |
| *Bos taurus* (cattle) | A | N | L | 111 |
| *Ovis aries* (sheep) | A | N | L | 68 |
| *Capra hircus* (goat) | A | N | L | 59 |
| *Camelus dromedarius* (camel) | A | N | | 42 |
| *Equus caballus* (horse) | A | N | | 40 |
| *Bubalus bubalis* (water buffalo) | A | N | | 29 |
| *Equus asinus* (donkey) | A | N | | 22 |
| *Sus scrofa* (domestic & wild pigs) | A | | | 17 |
| **CARNIVORES** | | | | |
| *Canis familiaris* (dog) | A | | | 16 |
| **PRIMATES** | | | | |
| *Homo sapiens* (human) | A | N | L | 15 |

*Note*: Tick species include *Hy. anatolicum* and *Hy. excavatum*
[a] A = adult, N = nymph, L = larva

into New Jersey from Spain (Becklund, 1968; USDA, 1968b), on an onager (*Equus hemionus*) imported into New York from Iran (Becklund, 1968) and on baggage arriving in Delaware (USDA, 1985b).

## Hosts

*Hyalomma anatolicum* spp. are two- or three-host ticks of great importance, especially to animal health. The preferred hosts of *Hy. anatolicum* spp. adults are domestic livestock, in particular cattle, sheep, goats, camels, horses, water buffalos, donkeys and pigs (Table 7.3). The preferred hosts of *Hy. anatolicum* spp. immatures differ by species; nymphs and larvae of *Hy. anatolicum* typically parasitize medium-sized to large ungulates whereas nymphs and larvae of *Hy. excavatum* parasitize small mammals such as hares and rodents, especially gerbils and jirds. These differing preferences for immatures reflect the fact that *Hy. anatolicum* is primarily a domestic tick, being closely associated with livestock housing, whereas *Hy. excavatum* is not a domestic tick. All stages of *Hy. anatolicum* spp. will feed on humans. Occasional hosts include insectivores, birds, lizards and tortoises.

Sites of attachment on cattle include the axillae, groin, genitalia, perineum and udder. On sheep and goats, adults are found in the inguinal region, with nymphs and larvae on the ears, eyelids and lips. Adults have been found attached to the scrotum of humans.

## Distribution

The geographical distribution of *Hy. anatolicum* spp. extends from Africa north of the equator through southern Europe into Asia as far east as Bangladesh and China (Estrada-Pena et al., 2004a). The distribution of the two species has been compared by Apanaskevich & Horak (2005). Both *Hy. anatolicum* and *Hy. excavatum* have been reported in Africa from Algeria, Djibouti, Egypt, Ethiopia, Libya, Somalia, Sudan and Tunisia; in addition, *Hy. excavatum* has been found in Mauritania, Morocco and Western Sahara. Only *Hy. excavatum* has been reported in Europe, from Albania, Cyprus, Greece and Italy. Both *Hy. anatolicum* and *Hy. excavatum* have been reported in Asia from Afghanistan, Iran, Iraq, Israel, Jordan, Kazakhstan, Lebanon, Saudi Arabia, Syria, Tajikistan, Turkey, Turkmenistan, Uzbekistan and Yemen; in addition, *Hy. anatolicum* has been found in Armenia, Azerbaijan, Bangladesh, India, Kyrgyzstan, Nepal, Oman, Pakistan, Russia and United Arab Emirates.

## Life Cycle and Habitat

*Hyalomma anatolicum* spp. are ticks adapted to dry conditions. The species *Hy. anatolicum* is closely associated with livestock housing, and, when it feeds on domestic livestock, it has a strict three-host life cycle. Its activity is restricted to the summer in areas with a distinct winter season but it is active year-round where seasonal changes are minimal, such as in Sudan.

The life cycle of *Hy. excavatum* has been studied in Israel by Rubina et al. (1982) and in Morocco by Ouhelli (1994). They reported a preoviposition period of 7–24 days, an oviposition period of 28–38 days, an egg incubation period of 21–54 days, a larval prefeeding period of 10–20 days, a larval feeding period of 3–9 days, a larval premolting period of 10–22 days, a nymphal prefeeding period of 6–16 days, a nymphal feeding period of 6–14 days, a nymphal premolting period of 18–29 days, a female prefeeding period of 21–25 days and a female feeding period of 8–18 days. Following engorgement and detachment, females laid up to 10,114 eggs. From these data it was estimated that the life cycle of *Hy. excavatum* takes 138–269 days to complete. Rubina et al. (1982) studied the longevity of unfed *Hy. excavatum* and found that unfed adults could successfully overwinter, sometimes surviving two winter seasons with a longevity of 665–706 days.

**Disease Associations**

*Hyalomma anatolicum* is an important tick both to animal and human health because it is:

- a vector of both *Babesia caballi* and *Theileria equi* (Neitz, 1956; Friedhoff, 1988), two protozoa causing piroplasmosis in horses and other equines;
- the vector of *Theileria lestoquardi* (Hooshmand-Rad & Hawa, 1973; Uilenberg, 1997), a protozoan causing malignant theileriosis in sheep and goats;
- a vector of *Theileria annulata* (Bhattacharyulu et al., 1975), a protozoan causing tropical theileriosis in cattle;
- a vector of *Babesia merionis* (Friedhoff, 1988), a protozoan causing babesiosis in rodents;
- a vector of *Trypanosoma theileri* (Walker et al., 2003), a protozoan causing benign trypanosomiasis in cattle;
- a vector of the nairovirus causing Crimean-Congo hemorrhagic fever in humans (Labuda & Nuttall, 2004);
- a vector of *Anaplasma bovis* (Sumption & Scott, 2004), a rickettsia causing anaplasmosis in cattle.

The disease relationships of *Hy. excavatum* are poorly understood. Under natural conditions the immature stages characteristically feed on small mammals and not larger domestic livestock; thus, the ability of this subspecies to act as a vector of pathogens of livestock is uncertain. However, *Hy. excavatum* has been associated with a number of diseases of livestock, having been reported to be:

- a vector of *Theileria equi* (Mehlhorn & Schein, 1998), a protozoan causing piroplasmosis in horses and other equines;
- a vector of *Theileria annulata* (Pipano & Shkap, 2004), the protozoan causing tropical theileriosis in cattle;
- a vector of *Anaplasma bovis* (Sumption & Scott, 2004), a rickettsia causing anaplasmosis in cattle.

Additionally, *Hy. excavatum* has been shown to be an experimented vector of *Babesia ovis* (Friedhoff, 1997), a protozoan causing babesiosis in sheep, and it has been found to harbor *Rickettsia sibirica mongolitimonae*, the bacterium causing lymphangitis-associated rickettsiosis in humans, in Greece (Psaroulaki et al., 2005).

**References**

Hoogstraal (1956), Hoogstraal & Kaiser (1959a), Theiler (1962), Kaiser & Hoogstraal (1964), Singh & Dhanda (1965), Chaudhuri et al. (1969), Hoogstraal et al. (1981), Rubina et al. (1982), Ouhelli (1994), Walker et al. (2003), Estrada-Pena et al. (2004a), Apanaskevich & Horak (2005).

## 7.4. *Hyalomma detritum* Schulze (= *Hy. scupense*)

In a recent publication Guglielmone et al. (2009) pointed out that *Hyalomma detritum* is, in fact, a junior synonym of *Hyalomma scupense* Schulze, and the authors proposed using "*Hyalomma scupense* (= *Hyalomma detritum*)" when referring to this tick species; however, the name *Hy. detritum* will be retained in this book because that is the name by which the tick has been known in virtually all the literature in the West.

*Hyalomma detritum* was introduced into the continental United States on one occasion in 1969 on a horse imported into Florida from Portugal (USDA, 1970a, b).

**Hosts**

*Hyalomma detritum* is a two-host tick whose preferred hosts for all stages are domestic ungulates, particularly

Table 7.4. Most commonly reported hosts of the tick *Hyalomma detritum*

| Hosts | Stages[a] | | | No. of reports |
|---|---|---|---|---|
| **UNGULATES** | | | | |
| *Bos taurus* (cattle) | A | N | L | 57 |
| *Ovis aries* (sheep) | A | N | | 26 |
| *Capra hircus* (goat) | A | N | | 19 |
| *Camelus dromedarius* (camel) | A | | | 18 |
| *Equus caballus* (horse) | A | N | L | 17 |
| *Bubalus bubalis* (water buffalo) | A | | | 11 |
| **PRIMATES** | | | | |
| *Homo sapiens* (human) | A | | | 12 |

[a] A = adult, N = nymph, L = larva

cattle, but also sheep, goats, horses, camels and water buffalos (Table 7.4). Adults will feed on dogs and humans. Occasional hosts include birds, hares, rodents and reptiles.

Sites of attachment of adults to cattle include the inner thigh, udder, scrotum and perineum. Nymphs and larvae attach to the neck and shoulders.

### Distribution

The geographical distribution of *Hy. detritum* extends from north Africa through southern Europe into Asia. It has been reported from Algeria, Egypt, Morocco, Sudan and Tunisia in Africa, from Albania, Bulgaria, France, Greece, Italy, Moldova, Romania, Russia, Spain, Ukraine and the Mediterranean island of Sicily in Europe, and from Afghanistan, Armenia, Azerbaijan, China, Georgia, India, Iran, Iraq, Israel, Jordan, Kazakhstan, Nepal, Pakistan, Syria, Tajikistan, Turkey, Turkmenistan and Uzbekistan in Asia.

### Life Cycle

The life cycle of *Hy. detritum* has been studied in Morocco by Ouhelli (1994). He reported a preoviposition period of 3–16 days, an oviposition period of 10–35 days, an egg incubation period of 32–40 days, a larval premolting period of 15–22 days, a nymphal premolting period of 23–25 days and a female feeding period of 15–16 days. Following engorgement and detachment, females laid up to 7,181 eggs. Adult *Hy. detritum* are found on cattle from late spring to late summer and larvae and nymphs in the autumn. The nymphs then enter diapause and molt to adults the following summer, with the life cycle taking one year to complete.

### Disease Associations

*Hyalomma detritum* is a tick of significance to both animal and human health because it is:

- a vector of *Anaplasma marginale* (Neitz, 1964), a rickettsia causing anaplasmosis in cattle;
- a vector of *Theileria annulata* (Neitz, 1964), the protozoan causing tropical theileriosis in cattle;
- a vector of *Theileria equi* (Friedhoff, 1988), a protozoan causing piroplasmosis in horses and donkeys;
- a vector of *Coxiella burnetii* (Walker et al., 2003), the rickettsia causing Q fever in livestock, small domestic animals and humans.

### References

Hoogstraal (1956), Theiler (1962), Kaiser & Hoogstraal (1964), Ouhelli (1994), Walker et al. (2003), Estrada-Pena et al. (2004a).

### 7.5. *Hyalomma marginatum* Koch group (incl. *Hy. marginatum* & *Hy. rufipes*)

In their recent publication of valid tick names, Horak et al. (2002) considered *Hyalomma marginatum* to consist of three subspecies: *Hy. marginatum isaaci*, *Hy. m. marginatum* and *Hy. m. rufipes*. More recently, Apanaskevich & Horak (2008a) established these subspecies as three separate species, namely *Hy. isaaci*, *Hy. marginatum* and *Hy. rufipes*. In this book, ticks of the three newly created species will be considered the *Hy. marginatum* group since much of the literature does not differentiate between species; however, where available, information will be given by species.

The *Hy. marginatum* group was introduced into the continental United States on 14 occasions between 1966 and 1989 on horses, wild animals, ostriches, an animal hide, a human and a shipment of virgin cork imported into or entering California, Indiana, Maryland, New Jersey, New York and Texas from European countries such as Germany, Greece and Spain (see Appendix 1). The wild animal hosts were zebras and a rhinoceros. These data on the *Hy. marginatum* group demonstrate clearly the variety of means by which invasive ticks can be introduced into the United States, whether on a human or animal host or on animal hides or on plant material.

**Hosts**

The *Hy. marginatum* group contains two-host ticks of great importance to both animal and human health. This group of ticks has a very wide host range, having been found on 351 species of animals, including 262 species of birds, 42 species of ungulates, 16 species of rodents, 9 species of lagomorphs, 8 species of carnivores, 5 species of insectivores, 3 species of reptiles, 2 species each of primates and elephant shrews, and one species each of elephants and hyraxes (see Appendix 2). The preferred hosts of adult ticks in the *Hy. marginatum* group are cattle, with other domestic animals such as sheep, goats, horses, camels, water buffalos, pigs and dogs also commonly infested (Table 7.5). In Africa, *Hy. rufipes* adults are also commonly found on wild ungulates such as African buffalos, elands, Burchell's zebras and giraffes. Nymphs and larvae of ticks in the *Hy. marginatum* group are found on birds, hares, rodents and insectivores. All stages infest humans.

Preferred attachment sites of adult ticks in the *Hy. marginatum* group are the hairless parts of the hindquarters, including the perineum, udder and genitalia. Adults can also be found on the heels, dewlap, brisket and tail tip of ungulates. Nymphs and larvae are found on the ears of rodents and lagomorphs and on the head of birds and hedgehogs.

**Distribution**

The geographical distribution of ticks in the *Hy. marginatum* group extends from Africa through southern Europe into central Asia. The species found in Africa are *Hy. marginatum* and *Hy. rufipes*; *Hy. rufipes* is widely distributed in Africa, particularly in sub-Saharan regions, whereas *Hy. marginatum* is limited to northern Africa. The species found in Europe are also *Hy. marginatum* and *Hy. rufipes*; *Hy. marginatum* is widely distributed in southern Europe, whereas *Hy. rufipes* is confined to southeastern Europe. All species are found in Asia.

*Hyalomma marginatum* has been reported from Africa (Algeria, Egypt, Libya, Morocco and Tunisia), from Europe (Albania, Bosnia & Herzegovina, Bulgaria, Croatia, Cyprus, France, Greece, Italy, Macedonia, Moldova, Montenegro, Portugal, Romania, Russia, Serbia, Spain and Ukraine) and from Asia (Armenia, Azerbaijan, Georgia, Iran, Iraq, Israel, Syria, Turkey and Turkmenistan). Records of *Hy. marginatum* from sub-Saharan Africa and northern Europe are most probably the result of dissemination of the immature stages by migratory birds.

*Hyalomma rufipes* has been reported from Africa (Angola, Benin, Botswana, Burkina Faso, Cameroon, Central African Republic, Chad, Democratic Republic of the Congo, Djibouti, Egypt, Ethiopia, Ghana, Guinea, Ivory Coast, Kenya, Lesotho, Libya, Malawi, Mali, Mauritania, Namibia, Niger, Nigeria, Senegal, Sierra Leone, Somalia, South Africa, Sudan, Swaziland, Tanzania, Togo, Tunisia, Uganda, Zambia and Zimbabwe), from Europe (Macedonia, Malta, Russia and Ukraine) and from Asia (Iran, Israel, Kazakhstan, Oman, Russia, Saudi Arabia, Syria, Tajikistan, Turkmenistan, Uzbekistan and Yemen).

*Hyalomma isaaci* has been reported only from Asia (Afghanistan, China, India, Myanmar, Nepal, Pakistan, Sri Lanka and Vietnam).

**Life Cycle and Habitat**

Adults of *Hy. marginatum* are active from March to November, with immatures active in the summer from May to September. Whereas *Hy. marginatum* cannot survive under desert conditions, *Hy. rufipes* can exist in a wide variety of climatic environments from desert to rain forest, but it is more common in drier areas. Adults of *Hy. rufipes* are active in the wet season in Africa, with immatures active in the dry season.

The life cycle of *Hy. rufipes* has been studied by Theiler (cited by Elbl & Anastos, 1966c). She reported a preoviposition period of 4–19 days, an oviposition period of 37–59 days, an egg incubation period of

Table 7.5. Most commonly reported hosts of ticks of the *Hyalomma marginatum* group

| Hosts | Stages[a] | | | No. of reports |
|---|---|---|---|---|
| **Ungulates** | | | | |
| *Bos taurus* (cattle) | A | N | L | 210 |
| *Ovis aries* (sheep) | A | N | L | 81 |
| *Capra hircus* (goat) | A | | | 57 |
| *Camelus dromedarius* (camel) | A | | | 47 |
| *Equus caballus* (horse) | A | N | | 47 |
| *Syncerus caffer* (African buffalo) | A | | | 24 |
| *Bubalus bubalis* (water buffalo) | A | | L | 22 |
| *Taurotragus oryx* (eland) | A | | | 20 |
| *Giraffa camelopardalis* (giraffe) | A | | | 17 |
| *Sus scrofa* (domestic & wild pigs) | A | N | | 17 |
| *Equus burchellii* (Burchell's zebra) | A | | | 16 |
| *Equus asinus* (donkey) | A | N | | 12 |
| *Equus zebra* (mountain zebra) | A | | | 11 |
| *Antidorcas marsupialis* (springbok) | A | | | 10 |
| *Diceros bicornis* (black rhinoceros) | A | | | 10 |
| *Oryx gazella* (gemsbok) | A | | | 10 |
| *Phacochoerus africanus* (warthog) | A | | | 10 |
| **Primates** | | | | |
| *Homo sapiens* (human) | A | N | L | 32 |
| **Carnivores** | | | | |
| *Canis familiaris* (dog) | A | | | 20 |
| **Birds** | | | | |
| *Phoenicurus phoenicurus* (redstart) | | N | L | 18 |
| *Struthio camelus* (ostrich) | A | | | 18 |
| *Oenanthe oenanthe* (northern wheatear) | | N | L | 16 |
| *Anthus trivialis* (tree pipit) | | N | L | 14 |
| *Motacilla alba* (white wagtail) | | N | L | 13 |
| *Hirundo rustica* (barn swallow) | A | N | L | 12 |
| *Numida meleagris* (helmeted guineafowl) | | N | L | 12 |
| *Motacilla flava* (yellow wagtail) | | N | L | 11 |
| *Muscicapa striata* (spotted flycatcher) | | | L | 11 |
| *Oenanthe hispanica* (black-eared wheatear) | | N | L | 11 |
| *Galerida cristata* (crested lark) | | N | L | 10 |
| *Lanius collurio* (red-backed shrike) | | N | L | 10 |
| *Monticola saxatilis* (rock thrush) | | N | L | 10 |
| **Lagomorphs** | | | | |
| *Lepus capensis* (Cape hare) | | N | L | 11 |
| *Lepus saxatilis* (scrub hare) | A | N | L | 11 |

*Note*: Tick group includes *Hy. isaaci*, *Hy. marginatum*, and *Hy. rufipes*
[a] A = adult, N = nymph, L = larva

28–66 days, a larval feeding period of 5–7 days, a larval premolting period of 2–15 days, a nymphal feeding period of 7–10 days, a nymphal premolting period of 14–95 days and a female feeding period of 5–12 days. Elbl & Anastos (1966c) estimated that the life cycle of *Hy. rufipes* takes a minimum of about 4–5 months to complete which may be doubled under unfavorable conditions.

The life cycle of *Hy. isaaci* has been studied in India by Das & Subramanian (1972a). They reported a preoviposition period of 8–10 days, an oviposition period of 16–19 days, an egg incubation period of 7–15 days, a larval prefeeding period of 7–10 days, a larval feeding period of 6–8 days, a larval premolting period of 3–16 days, a nymphal prefeeding period of 1–2 days, a nymphal feeding period of 5–9 days, a nymphal premolting period of 19–27 days, a female prefeeding period of 7 days and a female feeding period of 14–19 days. Following engorgement and detachment, females laid up to 12,727 eggs. From these data it was estimated that the life cycle of *Hy. isaaci* takes 101–131 days to complete.

### Disease Associations

*Hyalomma marginatum* is an important tick to both animal and human health because it is:

- a vector of both *Theileria equi* (Mehlhorn & Schein, 1998) and *Babesia cabulli* (Walker et al., 2003), two protozoa causing piroplasmosis in horses and other equines;
- a vector of *Theileria annulata* (Jongejan & Uilenberg, 2004), the protozoan causing tropical theileriosis in cattle;
- a vector of *Rickettsia aeschlimannii* (Parola et al., 2005), a bacterium causing rickettsiosis in humans;
- a major vector in Europe of the nairovirus causing Crimean-Congo hemorrhagic fever in humans (Ergonul, 2006).

*Hyalomma rufipes* is a very important tick to both animal and human health because it is:

- a cause of lesions at attachment sites on cattle, leading to the formation of abscesses and sloughing of host skin (Knight et al., 1978);
- a vector of *Babesia occultans* (Gray & de Vos, 1981), a protozoan causing babesiosis in cattle;
- the main vector in southern Africa of the nairovirus causing Crimean-Congo hemorrhagic fever in humans (Shepherd et al., 1991);
- a vector of *Anaplasma marginale* (Walker et al., 2003), a rickettsia causing anaplasmosis in cattle;
- a vector of *Rickettsia conorii* (Walker et al., 2003), the bacterium causing tick typhus in humans.

The disease relationships of *Hy. isaaci* are not well documented. It has been reported to be a vector of *Ehrlichia ovina* (Sumption & Scott, 2004), the rickettsia causing ehrlichiosis in sheep. Experimentally, it has been shown to be able to transmit the flavivirus causing Kyasanur Forest disease in humans (Singh & Bhatt, 1968) and *Theileria annulata* (Bhattacharyulu et al., 1975), the protozoan causing tropical theileriosis in cattle.

### References

Hoogstraal (1956), Theiler (1962), Kaiser & Hoogstraal (1964), Singh & Dhanda (1965), Elbl & Anastos (1966c), Yeoman et al. (1967), Das & Subramanian (1972a), Walker (1974), Knight et al. (1978), Norval (1982), Matthysse & Colbo (1987), Khan (1993), Ouhelli (1994), Walker et al. (2003), Estrada-Pena et al. (2004a), Apanaskevich & Horak (2008a).

## 7.6. *Ixodes ricinus* (Linnaeus)

The "castor bean or sheep tick" *Ixodes ricinus* has been introduced into the continental United States on seven occasions between 1973 and 1989 on horses, a donkey, a lizard and a human imported into or entering California, Georgia, New York and Pennsylvania from European countries such as Germany, Switzerland and United Kingdom (see Appendix 1).

### Hosts

*Ixodes ricinus* is a very important three-host tick with a wide host range. It has been found on 297 species of animals, including 161 species of birds, 49 species of rodents, 23 species each of ungulates and carnivores, 17 species of reptiles, 16 species of insectivores, 5 species of lagomorphs, 2 species of bats and one species of primates (see Appendix 2). The preferred hosts for adults are ungulates including cattle, sheep, roe deer (*Capreolus capreolus*), goats, horses, red deer (*Cervus elaphus*), pigs and fallow deer (*Dama dama*), and carnivores including dogs, cats, red foxes (*Vulpes vulpes*), badgers (*Meles meles*), stoats (*Mustela erminea*) and weasels (*Mustela nivalis*) (Table 7.6). Adults are also found on rabbits and hares. Nymphs and larvae feed on the same hosts as adults and also on a wide variety of birds, rodents, insectivores and lizards. Humans are commonly infested by all stages of *I. ricinus*. Occasional hosts include tortoises, snakes and bats.

Attachment sites of adults include the axillae, forelegs, hindlegs, udder, flank and neck for cattle, the head, axillae, forelegs and inguinal region for sheep, and the inguinal area and the underside of the body for deer. Attachment sites of larvae and nymphs are primarily the head of birds, the ears of rodents such as squirrels and the foreleg region for reptiles such as lizards. Attachment sites on humans include the eyelids.

Table 7.6. Most commonly reported hosts of the tick *Ixodes ricinus*

| Hosts | Stages[a] | | | No. of reports |
|---|---|---|---|---|
| **Ungulates** | | | | |
| *Bos taurus* (cattle) | A | N | L | 74 |
| *Ovis aries* (sheep/Mediterranean mouflon) | A | N | L | 62 |
| *Capreolus capreolus* (roe deer) | A | N | L | 53 |
| *Capra hircus* (goat) | A | N | L | 34 |
| *Equus caballus* (horse) | A | N | L | 32 |
| *Cervus elaphus* (red deer) | A | N | L | 30 |
| *Dama dama* (fallow deer) | A | N | L | 22 |
| *Sus scrofa* (domestic & wild pigs) | A | | | 22 |
| **Carnivores** | | | | |
| *Canis familiaris* (dog) | A | N | L | 64 |
| *Felis catus* (cat) | A | N | | 38 |
| *Vulpes vulpes* (red fox) | A | N | L | 29 |
| *Meles meles* (badger) | A | N | L | 16 |
| *Mustela erminea* (stoat) | A | N | L | 13 |
| *Mustela nivalis* (weasel) | | N | L | 10 |
| **Primates** | | | | |
| *Homo sapiens* (human) | A | N | L | 60 |
| **Birds** | | | | |
| *Turdus merula* (blackbird) | A | N | L | 47 |
| *Erithacus rubecula* (European robin) | | N | L | 41 |
| *Turdus philomelos* (song thrush) | | N | L | 33 |
| *Parus major* (great tit) | | N | L | 29 |
| *Fringilla coelebs* (chaffinch) | | N | L | 28 |
| *Prunella modularis* (dunnock) | | N | L | 27 |
| *Phoenicurus phoenicurus* (redstart) | | N | L | 26 |
| *Sylvia communis* (greater whitethroat) | A | N | L | 25 |
| *Anthus trivialis* (tree pipit) | A | N | L | 24 |
| *Phylloscopus trochilus* (willow warbler) | | N | L | 24 |
| *Sylvia atricapilla* (blackcap) | | N | L | 23 |
| *Troglodytes troglodytes* (winter wren) | | N | L | 22 |
| *Sturnus vulgaris* (common starling) | A | N | L | 21 |
| *Phylloscopus collybita* (chiffchaff) | | N | L | 18 |
| *Parus caeruleus* (blue tit) | | N | L | 17 |
| *Sylvia borin* (garden warbler) | | N | L | 17 |
| *Parus palustris* (marsh tit) | | N | L | 16 |
| *Acrocephalus scirpaceus* (reed warbler) | A | N | L | 15 |
| *Coccothraustes coccothraustes* (hawfinch) | | N | L | 15 |
| *Pyrrhula pyrrhula* (bullfinch) | | N | L | 15 |
| *Sitta europaea* (nuthatch) | | N | L | 15 |
| *Carduelis spinus* (siskin) | | N | L | 14 |
| *Garrulus glandarius* (Eurasian jay) | | N | L | 14 |
| *Luscinia luscinia* (thrush nightingale) | | N | L | 14 |
| *Muscicapa striata* (spotted flycatcher) | | N | L | 14 |
| *Phasianus colchicus* (pheasant) | A | N | L | 14 |
| *Turdus iliacus* (redwing) | | N | L | 14 |
| *Turdus pilaris* (fieldfare) | | N | L | 14 |
| *Accipiter nisus* (sparrowhawk) | | N | L | 13 |
| *Lagopus lagopus* (red grouse) | A | N | L | 13 |
| *Luscinia megarhynchos* (nightingale) | | N | L | 13 |
| *Sylvia curruca* (lesser whitethroat) | | N | L | 13 |
| *Saxicola rubetra* (whinchat) | | | L | 12 |
| *Anthus pratensis* (meadow pipit) | A | N | L | 11 |
| *Asio otus* (long-eared owl) | | N | L | 11 |

*continued*

Table 7.6.—Continued

| Hosts | Stages[a] | | | No. of reports |
|---|---|---|---|---|
| *Emberiza citrinella* (yellowhammer) | | N | | 11 |
| *Ficedula hypoleuca* (pied flycatcher) | A | N | L | 11 |
| *Lanius collurio* (red-backed shrike) | | N | L | 11 |
| *Luscinia svecica* (bluethroat) | | N | L | 11 |
| *Oenanthe oenanthe* (northern wheatear) | | N | L | 11 |
| *Passer domesticus* (house sparrow) | | N | L | 11 |
| *Acrocephalus palustris* (marsh warbler) | | N | L | 10 |
| *Falco tinnunculus* (kestrel) | | N | L | 10 |
| *Fringilla montifringilla* (brambling) | | N | L | 10 |
| *Pica pica* (magpie) | A | N | L | 10 |
| *Turdus viscivorus* (mistle thrush) | | N | L | 10 |
| **RODENTS** | | | | |
| *Myodes glareolus* (bank vole) | | N | L | 44 |
| *Apodemus sylvaticus* (wood mouse) | | N | L | 41 |
| *Apodemus flavicollis* (yellow-necked mouse) | | N | L | 40 |
| *Sciurus vulgaris* (red squirrel) | A | N | L | 22 |
| *Microtus agrestis* (field vole) | | N | L | 21 |
| *Microtus arvalis* (common vole) | | N | L | 15 |
| *Apodemus agrarius* (striped field mouse) | A | N | L | 12 |
| **INSECTIVORES** | | | | |
| *Erinaceus europaeus* (northern hedgehog) | A | N | L | 29 |
| *Sorex araneus* (Eurasian shrew) | | N | L | 20 |
| *Talpa europaea* (European mole) | | N | L | 10 |
| **LAGOMORPHS** | | | | |
| *Oryctolagus cuniculus* (rabbit) | A | N | L | 21 |
| *Lepus europaeus* (brown hare) | A | N | L | 18 |
| *Lepus timidus* (mountain hare) | A | N | L | 16 |
| **REPTILES** | | | | |
| *Lacerta agilis* (sand lizard) | A | N | L | 17 |
| *Lacerta viridis* (green lizard) | | N | L | 13 |
| *Lacerta vivipara* (viviparous lizard) | | N | L | 12 |

[a] A = adult, N = nymph, L = larva

## Distribution

The geographical distribution of *I. ricinus* covers virtually all of Europe and extends into northern Africa and western Asia. In Europe, *I. ricinus* has been reported from Albania, Austria, Belgium, Bosnia & Herzegovina, Bulgaria, Croatia, Czech Republic, Denmark, Estonia, Finland, France, Germany, Greece, Hungary, Iceland, Ireland, Italy, Kosovo, Latvia, Lithuania, Luxembourg, Macedonia, Moldova, Netherlands, Norway, Poland, Portugal, Romania, Russia, Serbia, Slovakia, Slovenia, Spain, Sweden, Switzerland, Ukraine, United Kingdom, several Mediterranean islands (Crete, Cyprus, Rhodes, Sardinia and Sicily), and the Faroe Islands between the Norwegian Sea and the North Atlantic Ocean. In northern Africa, it is established in Algeria, Morocco, Tunisia and the Madeira Islands off the coast of northwestern Africa. In Asia, it has been reported from Armenia, Azerbaijan, Iran, Israel, Russia and Turkey.

Studies in Sweden have suggested that *I. ricinus* is increasing its geographical range in that country, probably related to climate change (Talleklint & Jaenson, 1998; Lindgren et al., 2000).

## Life Cycle and Habitat

The life cycle of *I. ricinus* has been studied by Gray and others (Gray, 1991). They reported a preoviposition period of 7–28 days, an oviposition period of 30–224 days, an egg incubation period of 14–378 days, a larval feeding period of 2–6 days, a larval premolting period of 28–392 days, a nymphal feeding period of 3–7 days, a nymphal premolting period of 56–322 days and a female feeding period of 5–14 days. Each stage in the life cycle takes approximately one year to develop to the next stage, so the life cycle usually takes three years to complete, but this can vary from 2 to 6 years throughout the full geographical range of *I. ricinus*.

The non-parasitic phases of *I. ricinus* require a good cover of vegetation and a mat of decaying material where the relative humidity of their microclimate remains above 80% to survive. Furthermore, they will not survive in areas prone to winter flooding. In consequence, *I. ricinus* is found mainly in deciduous woodlands containing small mammals and deer or in open habitats with sufficient rainfall such as meadows and moorlands containing livestock. In North Africa, *I. ricinus* is restricted mainly to the cooler and more humid areas associated with the Atlas Mountains.

*Ixodes ricinus* has distinct patterns of seasonal activity that start in March in the spring and last into early summer. In some areas, a second but less intense phase of activity occurs in the autumn. Diapause is the most important regulator of the seasonal activity of *I. ricinus*, allowing the tick to avoid having to quest for hosts during unfavorable times of year, such as in midsummer when temperatures are high and relative humidity low, and in winter when temperatures are too low. Diapause, which is a neurohormonally mediated dynamic state of low metabolic activity, usually occurs before the onset of unfavorable conditions and extends well after they end.

**Disease Associations**

*Ixodes ricinus* is a very important tick to both human and animal health because it is:

- the vector of the flavivirus causing louping ill in sheep (MacLeod & Gordon, 1932);
- the vector of *Babesia divergens* (Donnelly & Peirce, 1975; Lewis & Young, 1980), a protozoan causing babesiosis in cattle and humans;
- the major vector in Europe of *Borrelia burgdorferi* (Granstrom, 1997), the bacterium causing Lyme borreliosis in humans and animals;
- a vector of the flavivirus causing tickborne encephalitis in humans (Gresikova & Kaluzova, 1997);
- the vector in Europe of *Anaplasma phagocytophilum* (Uilenberg, 1997; Goodman, 2005), the rickettsia causing tickborne fever in sheep, pasture fever in cattle and granulocytic anaplasmosis in humans;
- a vector of *Francisella tularensis* (Parola & Raoult, 2001), the bacterium causing tularemia in humans;
- a vector of *Babesia microti* (Gray et al., 2002), a protozoan causing babesiosis in rodents and humans;
- the vector of the coltivirus causing Eyach fever in humans (Labuda & Nuttall, 2004);
- the vector of *Rickettsia helvetica* (Parola et al., 2005), a bacterium causing rickettsiosis in humans;
- a vector of *Babesia* sp. EU1 (Becker et al., 2009), a newly identified protozoan causing babesiosis in splenectomized humans in Europe (Austria, Germany and Italy);
- able to induce anemia in northern hedgehogs (Pfafle et al., 2009).

In addition, *I. ricinus* has been found to be an experimental vector of *Bartonella henselae* (Cotte et al., 2008), the bacterium causing cat-scratch disease in humans. Finally, *I. ricinus* has been implicated in tick paralysis in humans and sheep (Mans et al., 2004).

**References**

MacLeod (1932, 1940), Milne (1945, 1947, 1949a, b), Edwards & Arthur (1947), Evans (1951a, b, c), Ohman (1961), Arthur (1963, 1965), Aeschlimann (1972), Gilot et al. (1975), Gray et al. (1978), Bauwens et al. (1983), Gray (1984, 1991, 1998), Matuschka et al. (1991), L'Hostis et al. (1996), Ogden et al. (1998), Walker et al. (2003), Estrada-Pena et al. (2004a).

### 7.7. *Rhipicephalus bursa* Canestrini & Fanzago

The "Anatolian brown tick" *Rhipicephalus bursa* has been introduced into the continental United States on two occasions, in 1966 on a horse imported into New Jersey from Spain (USDA, 1966b) and on a Cretan goat (*Capra aegagrus*) imported into New York from Italy (Becklund, 1968).

**Hosts**

*Rhipicephalus bursa* is a two-host tick of ungulates. Its preferred hosts for all stages are cattle, sheep, goats, horses and donkeys (Table 7.7). Dogs and humans are also common hosts of adults. Occasional hosts include birds, lagomorphs, rodents, hedgehogs and lizards.

Table 7.7. Most commonly reported hosts of the tick *Rhipicephalus bursa*

| Hosts | Stages[a] | | | No. of reports |
|---|---|---|---|---|
| **UNGULATES** | | | | |
| *Bos taurus* (cattle) | A | N | L | 64 |
| *Ovis aries* (sheep/Mediterranean mouflon) | A | N | L | 64 |
| *Capra hircus* (goat) | A | N | L | 54 |
| *Equus caballus* (horse) | A | N | | 20 |
| *Equus asinus* (donkey) | A | N | L | 11 |
| **CARNIVORES** | | | | |
| *Canis familiaris* (dog) | A | | | 19 |
| **PRIMATES** | | | | |
| *Homo sapiens* (human) | A | | | 19 |

[a] A = adult, N = nymph, L = larva

Sites of attachment for adults are the tail, groin, udder, scrotum and belly, whereas nymphs and larvae attach preferentially to the ears and also to the legs.

## Distribution

The geographical distribution of *R. bursa* extends from northern Africa through southern Europe into western Asia. In Africa, it is confined to coastal areas of Algeria, Libya, Morocco and Tunisia, and to the Madeira Islands off the coast of northwestern Africa. In Europe, it has been reported from Albania, Bosnia & Herzegovina, Bulgaria, Croatia, France, Greece, Italy, Macedonia, Montenegro, Portugal, Romania, Russia, Serbia, Slovenia, Spain, Switzerland, Ukraine and several Mediterranean islands (Balearic Islands, Corsica, Cyprus, Sardinia and Sicily). In Asia, it has been reported from Armenia, Azerbaijan, China, Georgia, Iran, Iraq, Israel, Kazakhstan, Syria, Turkey, Turkmenistan and Uzbekistan.

## Habitat and Life Cycle

The habitat of *R. bursa* is characterized by humid winters and long dry summers, with the tick absent from areas with average annual rainfall of less than 300 mm and temperature below 10 °C. The life cycle of *R. bursa* is strictly seasonal with one generation per year. In the Northern Hemisphere, larvae are found on the pasture and both larvae and nymphs on animals during the winter (October to February), with peak numbers in November and December. Adults appear in April, reaching peak numbers in May, and persist in the field and on animals until July.

## Disease Associations

*Rhipicephalus bursa* is a very important tick to animal health because it is:

- a vector of *Anaplasma marginale* (Sergent et al., 1928), a rickettsia causing anaplasmosis in cattle;
- a vector of both *Babesia caballi* and *Theileria equi* (Neitz, 1956), two protozoa causing piroplasmosis in horses and other equines;
- the vector of both *Babesia ovis* (Neitz, 1956) and *Babesia motasi* (Bock et al., 2004), two protozoa causing babesiosis in sheep and goats;
- a vector of *Anaplasma ovis* (Jongejan & Uilenberg, 2004), the rickettsia causing anaplasmosis in sheep;
- a vector of *Ehrlichia ovina* (Norval & Horak, 2004), the rickettsia causing ehrlichiosis in sheep;
- a vector of both *Theileria ovis* (Lawrence, 2004a) and *Theileria separata* (Norval & Horak, 2004), two protozoa causing benign theileriosis in sheep;
- a cause of tick paralysis in sheep (Norval & Horak, 2004).

## References

Feldman-Muhsam (1953), Yeruham et al. (1985, 1989), Walker et al. (2003), Estrada-Pena et al. (2004a).

## 7.8. *Rhipicephalus turanicus* Pomerantsev

Strong evidence has emerged of a close relationship between the ticks *Rhipicephalus turanicus* and *R. sanguineus*. Differences in these two species from various parts of the world have caused considerable confusion in precise identification of either species (Szabo et al., 2005), and sequences of specimens identified morphologically as *R. turanicus* have been characterized by high levels of variability, indicating that *R. turanicus*-like morphology may cover a spectrum of species (Beati & Keirans, 2001). Consequently, there is still uncertainty about the ticks identified as *R. turanicus* in the literature, and all such records of *R. turanicus* should be considered tentative until they are confirmed.

The tick *R. turanicus* was introduced into the continental United States on three occasions in 1989 on ostriches imported into New York from Botswana, Portugal and Tanzania (Mertins & Schlater, 1991; USDA, 1994).

### Hosts

*Rhipicephalus turanicus* is a three-host tick with an apparent wide host range. It has been reported on 138 species of animals, including 35 species of rodents, 31 species of carnivores, 29 species of ungulates, 19 species of birds, 10 species of insectivores, 8 species of lagomorphs, 3 species of reptiles, 2 species of primates and one species of antbears (see Appendix 2). The preferred hosts for adults are domestic ungulates and carnivores, particularly sheep, cattle, goats, donkeys, horses, dogs and cats (Table 7.8). Adults are also commonly found on ostriches and humans. Hosts of larvae and nymphs are not well documented, but they appear to include rodents, hedgehogs, shrews and hares. Occasional hosts include birds (other than ostriches), reptiles, aardvarks and monkeys.

### Distribution

The geographical distribution of *R. turanicus* extends from Africa through southern Europe into Asia as far east as China. In Africa, *R. turanicus* has been reported from northern Africa (Algeria, Egypt, Libya, Morocco and Tunisia), from western Africa (Guinea, Nigeria and Senegal), from central Africa (Cameroon), from eastern Africa (Burundi, Djibouti, Ethiopia, Kenya, Somalia, Sudan, Tanzania and Uganda) and from southern Africa (Angola, Botswana, Malawi, Mozambique, Namibia, South Africa, Swaziland, Zambia and Zimbabwe). In Europe, it has been reported from Albania, Bulgaria, Croatia, France, Greece, Italy, Macedonia, Portugal, Russia, Spain, Switzerland, Ukraine and several Mediterranean islands (Balearic Islands, Corsica, Crete, Cyprus, Sardinia and Sicily). In Asia, it has been reported from Afghanistan, Armenia, Azerbaijan, Bahrain, China, Georgia, India, Iran, Iraq, Israel, Jordan, Kazakhstan, Kyrgyzstan, Lebanon, Nepal, Oman, Pakistan, Russia, Saudi Arabia, Sri Lanka, Syria, Tajikistan, Turkey, Turkmenistan, United Arab Emirates, Uzbekistan and Yemen.

Table 7.8. Most commonly reported hosts of the tick *Rhipicephalus turanicus*

| Hosts | Stages[a] | No. of reports |
|---|---|---|
| **Ungulates** | | |
| *Ovis aries* (sheep) | A  N | 58 |
| *Bos taurus* (cattle) | A  N | 53 |
| *Capra hircus* (goat) | A | 43 |
| *Equus caballus* (horse) | A  N | 13 |
| *Equus asinus* (donkey) | A | 12 |
| **Carnivores** | | |
| *Canis familiaris* (dog) | A  N | 46 |
| *Felis catus* (cat) | A | 13 |
| **Birds** | | |
| *Struthio camelus* (ostrich) | A | 13 |
| **Primates** | | |
| *Homo sapiens* (human) | A | 13 |
| **Rodents** | | |
| *Rattus rattus* (black rat) | N  L | 11 |

[a] A = adult, N = nymph, L = larva

## Life Cycle

Adult *R. turanicus* are most numerous during the late rainy season until the early dry season. The life cycle of *R. turanicus* has been studied in India by Mahadev (1977) and in Israel by Ioffe-Uspensky et al. (1997). They reported a preovipostion period of 2–9 days, an oviposition period of 17–44 days, an egg incubation period of 16–25 days, a larval prefeeding period of 1–2 days, a larval feeding period of 2–6 days, a larval premolting period of 6–11 days, a nymphal prefeeding period of 1–2 days, a nymphal feeding period of 3–13 days, a nymphal premolting period of 12–21 days and a female feeding period of 8–12 days. After engorgement and detachment, females laid up to 12,276 eggs.

## Disease Associations

*Rhipicephalus turanicus* is not considered a major vector of any pathogen, but it is still a tick of some significance to animal and human health because it is:

- a vector of *Theileria equi* (de Waal & van Heerden, 2004), a protozoan causing piroplasmosis in horses;
- a vector of *Coxiella burnetii* (Brown et al., 2005), the rickettsia causing Q fever in humans and animals;
- a vector of *Rickettsia conorii* (Brown et al., 2005), the bacterium causing Israel tick typhus in humans.

In addition, *R. turanicus* is a vector of *Rickettsia massiliae* (Matsumoto et al., 2005), a bacterium of unknown pathogenicity to humans, and it also is an experimental vector of *Anaplasma ovis* (Stoltsz, 2004), a rickettsia causing anaplasmosis in sheep and goats, of *Babesia trautmanni* (de Waal, 2004), the protozoan causing babesiosis in pigs, and of *Babesia ovis* (Friedhoff, 1997), a protozoan causing babesiosis in sheep.

## References

Mitchell & Spillett (1968), Saratsiotis & Battelli (1975), Mahadev (1977), Hoogstraal et al. (1981), Ouhelli et al. (1982), Matthysse & Colbo (1987), Pegram et al. (1987b), Papadopoulos et al. (1992), Ioffe-Uspensky et al. (1997), Walker et al. (2000, 2003), Estrada-Pena et al. (2004a).

# 8

# Ticks from Europe

A total of 12 invasive tick species whose distribution includes Europe (Map 8.1) have been introduced into the continental United States. The only species whose distribution is limited to Europe, *Ixodes hexagonus*, is discussed in this chapter. The other 11 species with more extensive distributions are discussed in chapter 6 (*Hyalomma lusitanicum*), chapter 7 (*Haemaphysalis punctata*, *Hy. aegyptium*, *Hy. anatolicum* spp., *Hy. detritum*, *Hy. marginatum* group, *Ixodes ricinus*, *Rhipicephalus bursa* and *R. turanicus*), chapter 9 (*Dermacentor reticulatus*) and chapter 14 (*R. (Boophilus) annulatus*).

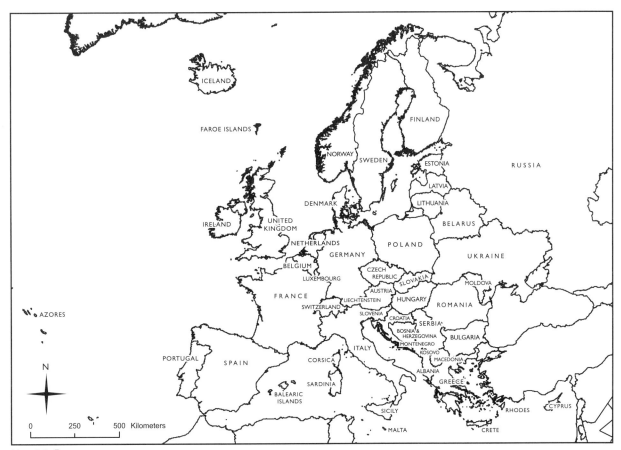

Map 8.1. Europe

## 8.1. *Ixodes hexagonus* Leach

The "hedgehog tick" *Ixodes hexagonus*, known in the past as *Pholeoixodes hexagonus*, was introduced into the continental United States on four occasions between 1976 and 1985 on hedgehogs and a shipment of virgin cork imported into Maryland and New York from European countries such as France, Portugal and United Kingdom (see Appendix 1).

### Hosts

*Ixodes hexagonus* is a three-host tick of carnivores and hedgehogs. The preferred hosts for all stages are dogs, cats, wild carnivores such as red foxes, European polecats (*Mustela putorius*), badgers, stoats and weasels, and northern hedgehogs (Table 8.1). It is also commonly found on humans. Less common hosts are cattle and other domestic livestock and rabbits. Occasional hosts include rodents, hares, birds and bats.

Attachment sites for adults include the axillae, the inguinal area and the genito-anal region, whereas larvae prefer to attach to the head and neck.

### Distribution

The geographical distribution of *I. hexagonus* covers much of Europe from the Mediterranean countries in the south to Iceland and Scandinavia in the north. It has been reported from Austria, Belgium, Bosnia & Herzegovina, Bulgaria, Croatia, Czech Republic, Denmark, France, Germany, Greece, Hungary, Iceland, Ireland, Italy, Luxembourg, Macedonia, Netherlands, Norway, Poland, Portugal, Romania, Russia, Serbia, Slovakia, Slovenia, Spain, Sweden, Switzerland, Ukraine, United Kingdom and the Mediterranean island of Sicily.

### Habitat

*Ixodes hexagonus* is found in a variety of habitats including the nests and burrows of hosts, forests and caves, and urban and suburban areas.

### Disease Associations

*Ixodes hexagonus* is a tick of minor importance to animal and human health since it is:

- a minor vector of *Borrelia burgdorferi* (Gray, 1998), the bacterium causing Lyme borreliosis in humans and animals;
- a vector of *Rickettsia conorii* (Brown et al., 2005), the bacterium causing tick typhus in humans;
- able to induce anemia in northern hedgehogs (Pfafle et al., 2009).

Additionally, *I. hexagonus* has been implicated in tick paralysis in humans (Mans et al., 2004).

### References

Arthur (1947, 1953, 1963, 1965), Estrada-Pena et al. (2004a).

Table 8.1. Most commonly reported hosts of the tick *Ixodes hexagonus*

| Hosts | Stages[a] | No. of reports |
|---|---|---|
| **Carnivores** | | |
| *Canis familiaris* (dog) | A  N  L | 35 |
| *Vulpes vulpes* (red fox) | A  N  L | 23 |
| *Mustela putorius* (European polecat) | A  N  L | 23 |
| *Felis catus* (cat) | A  N  L | 19 |
| *Meles meles* (badger) | A  N  L | 15 |
| *Mustela erminea* (stoat) | A  N  L | 13 |
| *Mustela nivalis* (weasel) | A  N  L | 11 |
| **Insectivores** | | |
| *Erinaceus europaeus* (northern hedgehog) | A  N  L | 30 |
| **Primates** | | |
| *Homo sapiens* (human) | A  N  L | 22 |
| **Ungulates** | | |
| *Bos taurus* (cattle) | A | 12 |

[a] A = adult, N = nymph, L = larva

# 9

# Ticks from Eurasia

The only invasive tick species whose distribution includes only Europe (see Map 8.1) and Asia (see Map 10.1), *Dermacentor reticulatus*, is discussed in this chapter.

## 9.1. *Dermacentor reticulatus* (Fabricius)

The "ornate dog tick" *Dermacentor reticulatus* was introduced into the continental United States on four occasions, including on an oryx in 1961, on a horse imported into New York from France in 1985 and on a hedgehog imported into New York from France (see Appendix 1).

### Hosts

*Dermacentor reticulatus* is a three-host tick of carnivores, particularly dogs and red foxes (Table 9.1). Other common hosts are ungulates, including cattle, horses, sheep, pigs and red deer. Adults prefer to feed on carnivores and ungulates, whereas larvae and nymphs favor rodents and insectivores. Adults also feed on humans. Occasional hosts include birds, lagomorphs and lizards. It has been found attached to the ears of European bison (*Bison bonasus*).

### Distribution

The geographical distribution of *D. reticulatus* ranges over much of Europe and into Asia. In Europe, it has been reported from Austria, Belarus, Belgium, Bulgaria, Croatia, Czech Republic, France, Germany, Hungary, Italy, Lithuania, Macedonia, Moldova, Netherlands, Poland, Portugal, Romania, Russia, Serbia, Slovakia, Slovenia, Spain, Switzerland, Ukraine, United Kingdom and the Mediterranean island of Corsica, but it is absent from Ireland and the Scandinavian countries. In Asia, it occurs in Russia including

Table 9.1. Most commonly reported hosts of the tick *Dermacentor reticulatus*

| Hosts | Stages[a] | | No. of reports |
|---|---|---|---|
| **Carnivores** | | | |
| *Canis familiaris* (dog) | A | N | 23 |
| *Vulpes vulpes* (red fox) | A | | 15 |
| **Primates** | | | |
| *Homo sapiens* (human) | A | | 20 |
| **Ungulates** | | | |
| *Bos taurus* (cattle) | A | | 16 |
| *Equus caballus* (horse) | A | N | 13 |
| *Cervus elaphus* (red deer/elk) | A | N | 12 |
| *Ovis aries* (sheep) | A | | 12 |
| *Sus scrofa* (domestic pig/wild boar) | A | | 11 |

[a] A = adult, N = nymph

Siberia, with isolated reports from Armenia, Azerbaijan, China, India and Kazakhstan.

### Life Cycle and Habitat

*Dermacentor reticulatus* occurs in woodlands, grasslands and pastures with high humidity and mild winters. It is found feeding from October to March in the Mediterranean region. Its life cycle can be completed in one year, but it usually takes three years.

### Disease Associations

*Dermacentor reticulatus* is an important tick to both human and animal health because it is:

- a vector of both *Babesia caballi* and *Theileria equi* (Friedhoff, 1988), two protozoa causing piroplasmosis in horses and other equines;
- a vector of *Francisella tularensis* (Hubalek et al., 1997), the bacterium causing tularemia in humans;
- a vector of *Coxiella burnetii* (Parola & Raoult, 2001), the rickettsia causing Q fever in humans and animals;
- a vector of *Babesia canis* (Shaw et al., 2001), a protozoan causing babesiosis in dogs;
- a vector of *Rickettsia sibirica sibirica* (Jongejan & Uilenberg, 2004), the bacterium causing North Asian tick typhus in humans;
- the major vector of the flavivirus causing Omsk hemorrhagic fever virus in humans (Labuda & Nuttall, 2004);
- a vector of *Rickettsia slovaca* (Brouqui et al., 2007), the bacterium causing tickborne lymphadenopathy and *Dermacentor*-borne necrosis-erythema-lymphadenopathy in humans.

In addition, *D. reticulatus* has been found to be an experimental vector of the flavivirus causing tickborne encephalitis virus in humans (Kozuch & Nosek, 1971).

### References

Olenev (1927), Arthur (1960, 1963), Nosek (1972), Gilot et al. (1974), Szymanski (1987), Estrada-Pena et al. (2004a).

# 10

# Ticks from Asia

A total of 28 invasive tick species whose distribution includes Asia (Map 10.1) have been introduced into the continental United States. The 12 species whose distribution is limited to Asia are discussed in this chapter. The other 16 species with more extensive distributions are discussed in chapter 5 (*Hyalomma dromedarii*), in chapter 7 (*Haemaphysalis punctata*, *Hy. aegyptium*, *Hy. anatolicum* spp., *Hy. detritum*, *Hy. marginatum* group, *Ixodes ricinus*, *Rhipicephalus bursa* and *R. turanicus*), in chapter 9 (*Dermacentor reticulatus*), in chapter 11

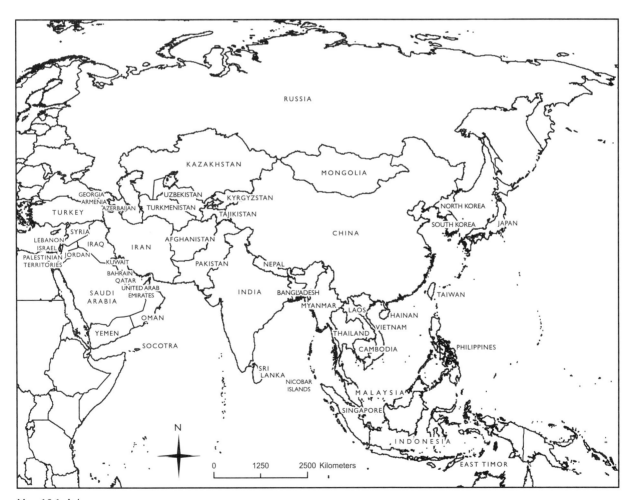

Map 10.1. Asia

Fig. 10.1. An Indian star tortoise (*Geochelone elegans*), the preferred host of the tick *Amblyomma clypeolatum*. (Photograph courtesy of Ray Ashton.)

(*Amblyomma fimbriatum*, *A. helvolum*, *A. trimaculatum* and *Haemaphysalis longicornis*) and in chapter 14 (*R. (Boophilus) annulatus* and *R. (B) microplus*).

## 10.1. *Amblyomma clypeolatum* Neumann

The tick *Amblyomma clypeolatum* has been introduced into the continental United States on two occasions, in 1987 on a star tortoise (*Geochelone elegans*) imported into Texas (USDA, 1988) and again when it was found on an unidentified tortoise imported into New York (Keirans & Durden, 2001).

### Hosts

*Amblyomma clypeolatum* is a tick of Asian tortoises. Its preferred hosts are Indian star tortoises (Fig. 10.1) on which all stages feed, attached to the forearms, bases of the hindlegs and tail, and the midplastral and carapacal sutures. Other tortoises infrequently infested are Burmese starred tortoises (*Geochelone platynota*), forest hinged tortoises (*Kinixys erosa*) and elongated tortoises (*Indotestudo elongata*) (Table 10.1). The only non-reptilian hosts on which *A. clypeolatum* have been found are water buffalos and wild boars.

### Distribution

The geographical distribution of *A. clypeolatum* is limited to India, Myanmar and Sri Lanka.

### Disease Associations

None reported.

### References

Robinson (1926), Sharif (1928), Voltzit & Keirans (2002), Burridge & Simmons (2003).

Table 10.1. Hosts of the tick *Amblyomma clypeolatum*

| Hosts | Stages[a] | | | No. of reports |
|---|---|---|---|---|
| **Tortoises** | | | | |
| *Geochelone elegans* (Indian star tortoise) | A | N | L | 14 |
| *Geochelone platynota* (Burmese starred tortoise) | A | N | | 3 |
| *Indotestudo elongata* (elongated tortoise) | A | | | 2 |
| *Kinixys erosa* (forest hinged tortoise) | A | | | 2 |
| **Ungulates** | | | | |
| *Bubalus bubalis* (water buffalo) | - | | | 1 |
| *Sus* sp. (wild boar) | - | | | 1 |

[a] A = adult, N = nymph, L = larva, - = not recorded

## 10.2. *Amblyomma crassipes* Neumann

The tick *Amblyomma crassipes*, previously known as *Aponomma crassipes*, has been introduced into the continental United States on two occasions, on a Bengal monitor (*Varanus bengalensis*) imported into Florida from Thailand in 1988 (Burridge & Simmons, 2003) and on an unidentified lizard imported into Florida (Keirans & Durden, 2001).

### Hosts

*Amblyomma crassipes* is a rarely reported tick of monitor lizards and snakes. It has been collected from three species of lizards, grey monitors (*Varanus griseus*), water monitors and Bengal monitors, and from four species of snakes, Burmese pythons, reticulated pythons, banded kraits (*Bungarus fasciatus*) and Indian cobras (*Naja naja*) (Table 10.2). All stages have been found only on water monitors. Occasional hosts include carnivores, rodents, serows (*capricornis sumatraensis*), and tree shrews.

### Distribution

The geographical distribution of *A. crassipes* is limited to Indochina, with collections reported from Cambodia, Laos, Thailand and Vietnam.

### Disease Associations

None reported.

### References

Kaufman (1972), Burridge & Simmons (2003).

## 10.3. *Amblyomma fuscolineatum* (Lucas)

The tick *Amblyomma fuscolineatum*, previously known as *Aponomma fuscolineatum*, was introduced into the continental United States on one occasion, when it was found on a Burmese python in the New York Zoological Park (Kaufman, 1972).

### Hosts

*Amblyomma fuscolineatum* is a rarely reported tick of snakes and lizards. It has been collected from only one species, the Burmese python, on more than one occasion (Table 10.3). In addition to the three hosts listed in Table 10.3, other species of snakes and lizards, and several species of ungulates, carnivores, rodents, pangolins and turtles have been reported as hosts by Santos Dias (1993), but these latter hosts have been attributed to *A. varanense*, which Santos Dias considers synonymous with *A. fuscolineatum*.

Table 10.2. Most commonly reported hosts of the tick *Amblyomma crassipes*

| Hosts | Stages[a] | No. of reports |
|---|---|---|
| **Lizards** | | |
| *Varanus griseus* (grey monitor) | A | 7 |
| *Varanus salvator* (water monitor) | A N L | 6 |
| **Snakes** | | |
| *Python molurus* (Burmese python) | A | 5 |
| *Python reticulatus* (reticulated python) | A | 5 |
| *Bungarus fasciatus* (banded krait) | A | 4 |

[a] A = adult, N = nymph, L = larva

Table 10.3. Hosts of the tick *Amblyomma fuscolineatum*

| Hosts | Stages[a] | No. of reports |
|---|---|---|
| **Snakes** | | |
| *Python molurus* (Burmese python) | - | 5 |
| **Lizards** | | |
| *Varanus salvadorii* (crocodile monitor) | A | 1 |
| *Varanus salvator* (water monitor) | A | 1 |

[a] A = adult, - = not recorded

### Distribution

*Amblyomma fuscolineatum* has been reported from China, India, Indonesia and New Guinea, and from the island of Hainan off the southern coast of China in the South China Sea and from Pescador Island off Taiwan.

### Disease Associations

None reported.

### References

Kaufman (1972).

## 10.4. *Amblyomma geoemydae* (Cantor)

The tick *Amblyomma geoemydae* was introduced into the continental United States on six occasions between 2000 and 2003 on turtles imported into California, Florida, Massachusetts and New York from Asian countries such as China (see Appendix 1). The chelonian hosts were jagged-shelled turtles (*Cuora mouhotii*), black-breasted leaf turtles (*Geoemyda splengleri*), giant Asian pond turtles (*Heosemys grandis*) and a spiny terrapin (*Heosemys spinosa*).

### Hosts

*Amblyomma geoemydae* is a tick of reptiles, especially turtles and tortoises (Table 10.4). The preferred hosts are spiny terrapins on which all stages feed. Both adults and immatures feed also on water monitors. Larvae and nymphs have been found feeding on 15 species of birds. Occasional hosts include snakes, carnivores, pigs, pangolins and humans.

Sites of attachment of *A. geoemydae* on tortoises, turtles and terrapins include the lower jaw, neck, forelegs and axillae, hindlegs, base of the tail and the shell (both carapace and plastron).

### Distribution

The geographical distribution of *A. geoemydae* covers much of southern Asia. It has been reported from Cambodia, China, India, Indonesia, Japan, Malaysia, Myanmar, Philippines, Singapore, Sri Lanka, Taiwan, Thailand and Vietnam. It has also been found on the island of Hainan off the southern coast of China in the South China Sea.

### Life Cycle

*Amblyomma geoemydae* can complete its life cycle in 185–271 days on the Asian giant tortoise (*Manouria emys*) with larvae taking 8–20 days, nymphs 12–28 days and adults 18–48 days to engorge (Nadchatram, 1960). Engorgement time for females is dependent upon the site of attachment, with those attaching to the skin feeding more rapidly than those attaching along the sutures of the shell. Unfed *A. geoemydae* can survive without a blood meal for extended periods, with maximum survival times of 190, 240 and 315 days reported for larvae, nymphs and adults, respectively.

### Disease Associations

None reported.

### References

Robinson (1926), Simmons & Burridge (2000), Voltzit & Keirans (2002).

Table 10.4. Most commonly reported hosts of the tick *Amblyomma geoemydae*

| Hosts | Stages[a] | | | No. of reports |
|---|---|---|---|---|
| **Chelonians** | | | | |
| *Heosemys spinosa* (spiny terrapin) | A | N | L | 12 |
| *Cuora mouhotii* (jagged-shelled turtle) | A | | | 5 |
| *Geoemyda spengleri* (black-breasted leaf turtle) | A | N | L | 5 |
| *Manouria emys* (Asian giant tortoise) | - | | | 5 |
| *Indotestudo forstenii* (Forsten's tortoise) | - | | | 4 |
| **Lizards** | | | | |
| *Varanus salvator* (water monitor) | A | N | | 5 |
| **Birds** | | | | |
| *Megalaima mystacophanos* (red-throated barbet) | | N | | 4 |

[a] A = adult, N = nymph, L = larva, - = not recorded

## 10.5. Amblyomma javanense (Supino)

The tick *Amblyomma javanense* (syn. *Amblyomma subleave*) was introduced into the continental United States on six occasions between 1968 and 1973 on pangolins imported into Illinois, New York and Oklahoma from Asian countries such as India and Thailand (see Appendix 1). The hosts included an Indian pangolin (*Manis crassicaudata*).

### Hosts

*Amblyomma javanense* is a tick of pangolins, with the Malayan pangolin (*Manis javanica*) the preferred host (Table 10.5). Other hosts include water monitors, striped hyenas (*Hyaena hyaena*) and three-keeled land tortoises (*Melanochelys tricarinata*). Occasional hosts include other reptiles, ungulates, monkeys and porcupines.

### Distribution

The geographical distribution of *A. javanense* ranges over much of southern Asia from Pakistan in the west to Indonesia in the east. It has been reported from Cambodia, China, India, Indonesia, Malaysia, Myanmar, Pakistan, Philippines, Singapore, Sri Lanka, Thailand and Vietnam.

### Disease Associations

None reported.

### References

Voltzit & Keirans (2002).

## 10.6. Amblyomma komodoense (Oudemans)

The tick *Amblyomma komodoense*, previously known as *Aponomma komodoense*, has been introduced into the continental United States on two occasions, once in 1994 when a zoo in Florida acquired a pair of infested Komodo dragons (*Varanus komodoensis*) from the Indonesian island of Flores (Burridge & Simmons, 2003; Burridge et al., 2004b) and again on a Komodo dragon imported into the District of Columbia from Indonesia (Keirans & Durden, 2001).

### Hosts

*Amblyomma komodoense* is almost exclusively a tick of Komodo dragons (Fig. 10.2), with the only report of this tick on another host being of an infestation on a water monitor in Jakarta zoo in Indonesia (Table 10.6). The sites of attachment of *A. komodoense* on Komodo dragons are primarily on the head and in the ear canals (Fig. 10.2).

### Distribution

The geographical distribution of *A. komodoense* is limited to just two islands in south central Indonesia, the islands of Flores and Komodo.

### Disease Associations

None reported.

### References

Kaufman (1972), Burridge & Simmons (2003).

Table 10.5. Most commonly reported hosts of the tick *Amblyomma javanense*

| Hosts | Stages[a] | No. of reports |
|---|---|---|
| **Pangolins** | | |
| *Manis javanica* (Malayan pangolin) | A  N  L | 12 |
| *Manis crassicaudata* (Indian pangolin) | A  N  L | 3 |
| *Manis pentadactyla* (Chinese pangolin) | A | 3 |
| **Lizards** | | |
| *Varanus salvator* (water monitor) | A | 7 |
| **Carnivores** | | |
| *Hyaena hyaena* (striped hyena) | A  N | 5 |
| **Tortoises** | | |
| *Melanochelys tricarinata* (three-keeled land tortoise) | - | 4 |

[a] A = adult, N = nymph, L = larva, - = not recorded

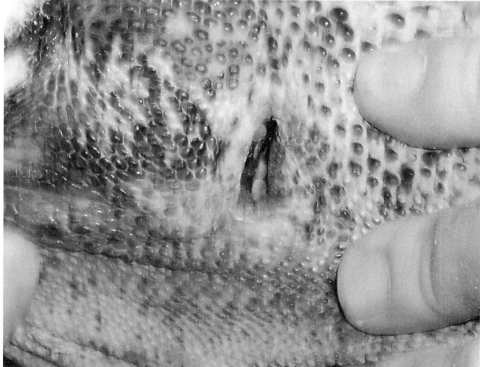

Fig. 10.2. An adult Komodo dragon (*Varanus komodoensis*) (*top*), with *Amblyomma komodoense* ticks feeding in its ear canal (*bottom*).

Table 10.6. Hosts of the tick *Amblyomma komodoense*

| Hosts | Stages[a] | No. of reports |
|---|---|---|
| **Lizards** | | |
| *Varanus komodoensis* (Komodo dragon) | A  N  L | 18 |
| *Varanus salvator* (water monitor) | A | 4 |

[a]A = adult, N = nymph, L = larva

## 10.7. *Amblyomma kraneveldi* (Anastos)

The tick *Amblyomma kraneveldi*, previously known as *Aponomma kraneveldi*, was introduced into the continental United States on one occasion on a reticulated python imported into Florida (Burridge et al., 2006).

### Hosts

*Amblyomma kraneveldi* is a rarely reported tick of pythons. It has been found on only three species of snakes, the reticulated python, Macklot's python (*Liasis mackloti*) and the Timor python (*Python timoriensis*) (Table 10.7). Occasional hosts include water monitors.

### Distribution

The geographical distribution of *A. kraneveldi* is limited to just three islands in south central Indonesia, the islands of Flores, Sulawesi and Sumbawa.

### Disease Associations

None reported.

### References

Kaufman (1972).

## 10.8. *Amblyomma testudinarium* Koch

The tick *Amblyomma testudinarium* was introduced into the continental United States on two occasions on rhinoceroses imported into California and District of Columbia (USDA, 1988; Keirans & Durden, 2001).

### Hosts

*Amblyomma testudinarium* is a tick of mammals. The principal hosts for adults are ungulates, in particular cattle, pigs, water buffalos and sambars (*Rusa unicolor*), and carnivores, especially tigers (*Panthera tigris*) (Table 10.8). Nymphs and larvae feed on the same hosts as adults and also on tree shrews (*Tupaia* spp.) and rodents. All stages feed commonly on humans. Occasional hosts include Asian elephants (*Elephas maximus*), pangolins, rabbits, reptiles, amphibians and birds.

Sites of attachment of adults on humans include the inguinal region and the glans penis, whereas larvae have been found over the whole body.

### Distribution

The geographical distribution of *A. testudinarium* extends over much of southern Asia from India in the

Table 10.7. Hosts of the tick *Amblyomma kraneveldi*

| Hosts | Stages[a] | No. of reports |
|---|---|---|
| SNAKES | | |
| *Python reticulatus* (reticulated python) | A | 5 |
| *Liasis mackloti* (Macklot's python) | - | 3 |
| *Python timoriensis* (Timor python) | A | 1 |
| LIZARDS | | |
| *Varanus salvator* (water monitor) | A | 1 |

[a] A = adult, - = not recorded

Table 10.8. Most commonly reported hosts of the tick *Amblyomma testudinarium*

| Hosts | Stages[a] | No. of reports |
|---|---|---|
| UNGULATES | | |
| *Bos taurus* (cattle) | A  N | 26 |
| *Sus scrofa* (domestic & wild pigs) | A  N  L | 18 |
| *Bubalus bubalis* (water buffalo) | A  N | 16 |
| *Rusa unicolor* (sambar) | A  N | 11 |
| PRIMATES | | |
| *Homo sapiens* (human) | A  N  L | 25 |
| CARNIVORES | | |
| *Panthera tigris* (tiger) | A  N | 14 |

[a] A = adult, N = nymph, L = larva

west to Indonesia in the east. It has been reported from Bangladesh, Cambodia, China, India, Indonesia, Japan (southern islands), Laos, Malaysia, Myanmar, Nepal, Philippines, Sri Lanka, Taiwan, Thailand and Vietnam.

**Disease Associations**

The only relationship between *A. testudinarium* and any disease condition is the skin eruptions reported in humans at the site of attachment of larvae (Nakamura-Uchiyama et al., 2000).

**References**

Robinson (1926), Sharif (1928), Voltzit & Keirans (2002).

## 10.9. *Amblyomma varanense* (Supino)

The tick *Amblyomma varanense*, previously known as *Aponomma varanense* and *Aponomma varanensis* (syn. *Aponomma lucasi*), was introduced into the continental United States on at least 20 occasions between 1969 and 2005 on lizards and snakes imported into California, Florida, Georgia and Maryland from Asian countries such as India, Indonesia and Thailand and from New Guinea (see Appendix 1). The saurian hosts were water monitors, Bengal monitors and harlequin monitors (*Varanus rudicollis*), and the ophidian hosts a green tree python (*Morelia viridis*), an Indian cobra, a white-lipped python (*Leiopython albertisii*) and a Sumatran short-tailed python (*Python curtus*).

**Hosts**

*Amblyomma varanense* is a tick of reptiles, especially lizards and snakes. The most commonly infested hosts are water monitors (Fig 10.3), Bengal monitors and harlequin monitors among lizards, and king cobras (*Ophiophagus hannah*), reticulated pythons, Indian cobras and Chinese cobras (*Naja atra*) among snakes (Table 10.9). All stages have been found on Indian and Chinese cobras and on Burmese pythons. Occasional hosts include chelonians, carnivores, ungulates, rodents and Malayan pangolins.

In experimental studies, all stages of *A. varanense* fed successfully on North American rat snakes (*Elaphe obsoleta*), and larvae also fed on eastern box turtles (*Terrapene carolina*), reptiles native to the United States (Oliver et al., 1988). In addition, although *A. varanense* is an Asian tick, it was found infesting a host native to Africa, the ball python, while it was in captivity (Burridge et al., 2000a). Both these experimental and field findings demonstrate the potential for *A. varanense*

Fig. 10.3. A water monitor (*Varanus salvator*), the preferred host of the ticks *Amblyomma varanense*, *A. fimbriatum* and *A. helvolum* and a common host of several other Asian ticks. (Photograph courtesy of iStockphoto.com.)

Table 10.9. Most commonly reported hosts of the tick *Amblyomma varanense*

| Hosts | Stages[a] | | | No. of reports |
|---|---|---|---|---|
| LIZARDS | | | | |
| *Varanus salvator* (water monitor) | A | N | | 15 |
| *Varanus bengalensis* (Bengal monitor) | A | N | | 9 |
| SNAKES | | | | |
| *Ophiophagus hannah* (king cobra) | A | | | 8 |
| *Python reticulatus* (reticulated python) | A | N | | 8 |
| *Naja naja* (Indian cobra) | A | N | L | 8 |

[a] A = adult, N = nymph, L = larva

Table 10.10. Most commonly reported hosts of the tick *Dermacentor auratus*

| Hosts | Stages[a] | | | No. of reports |
|---|---|---|---|---|
| PRIMATES | | | | |
| *Homo sapiens* (human) | A | N | L | 22 |
| UNGULATES | | | | |
| *Sus scrofa* (domestic & wild pigs) | A | N | | 17 |
| *Bos taurus* (cattle) | A | N | L | 7 |
| RODENTS | | | | |
| *Rattus rattus* (black rat) | A | N | L | 9 |
| CARNIVORES | | | | |
| *Panthera pardus* (leopard) | A | N | | 7 |
| TREE SHREWS | | | | |
| *Tupaia glis* (Malaysian tree shrew) | A | N | L | 7 |

[a] A = adult, N = nymph, L = larva

to spread outside its native geographical range to new reptilian hosts.

## Distribution

The geographical distribution of *A. varanense* ranges over much of southern Asia from India in the west to Indonesia in the east. It has been reported from Bangladesh, Cambodia, China, India, Indonesia, Malaysia, Myanmar, Philippines, Singapore, Sri Lanka, Taiwan, Thailand and Vietnam. There have been isolated reports of *A. varanense* in Australia (Keirans, 1984) and New Guinea (Kolonin, 1995; Kenny et al., 2004), suggesting that the tick has spread to Australasia.

## Life Cycle

Female *A. varanense* take 22–82 days to mate and feed, and they can lay up to 1,285 eggs (Oliver et al., 1988). Unfed larvae are exceptionally long lived, with some remaining alive and active for more than 22 months.

## Disease Associations

None reported.

## References

Kaufman (1972), Oliver et al. (1988), Burridge (2001).

## 10.10. *Dermacentor auratus* Supino

Several *Dermacentor* spp. ticks from Asia were confused in the literature before *Dermacentor auratus* was redescribed by Hoogstraal & Wassef (1985). Consequently, data on *D. auratus* relying solely on references published prior to 1985 have been excluded because of the uncertainty of the identity of the ticks in those earlier publications.

The tick *Dermacentor auratus* has been introduced into the continental United States on three occasions, once in 1986 on a plant imported into Illinois (USDA, 1987b) and twice on humans entering Arizona and Maine from Nepal (Keirans & Durden, 2001).

## Hosts

*Dermacentor auratus* is a three-host tick of mammals. The principal hosts for adults are wild pigs (Table 10.10). Other ungulate hosts include domestic pigs, cattle and sambars, whereas carnivore hosts include leopards, sloth bears (*Melursus ursinus*) and Asian black bears (*Ursus thibetanus*). Larvae and nymphs feed on the same hosts as adults and also on a wide variety of rodents, especially black rats and Indian crested porcupines (*Hystrix indica*), and on Malaysian tree shrews (*Tupaia glis*). All stages feed frequently on humans. Occasional hosts include birds and hares. Reported infestations of reticulated pythons require verification (Hoogstraal & Wassef, 1985).

Sites of attachment of nymphs on humans include the scrotum and the ear canal.

### Distribution

The geographical distribution of *D. auratus* ranges over much of southern Asia from India in the west to Indonesia in the east. It has been reported from Bangladesh, China, India, Indonesia, Laos, Malaysia, Myanmar, Nepal, Sri Lanka, Thailand and Vietnam. It has also been found on the island of Hainan off the southern coast of China in the South China Sea.

### Disease Associations

*Dermacentor auratus* has recently been shown to be of significance to human health as a vector of *Rickettsia sibirica sibirica* (Fournier et al., 2006), the bacterium causing North Asian tick typhus in humans. Additionally, experimental studies in India have demonstrated that *D. auratus* can transmit Kyasanur forest disease virus transstadially between rodents (Sreenivasan et al., 1979). *D. auratus* has also been implicated in tick paralysis in humans (Mans et al., 2004).

### References

Wassef & Hoogstraal (1984), Hoogstraal & Wassef (1985).

## 10.11. *Dermacentor nuttalli* Olenev

The tick *Dermacentor nuttalli* was introduced into the continental United States on one occasion in 1982 on an animal hide imported into Illinois (USDA, 1983).

### Hosts

*Dermacentor nuttalli* is a rarely reported tick of ungulates including horses, sheep, cattle, camels, goats, Asian wild asses (*Equus hemionus*), Mongolian wild horses (*Equus przewalskii*) and pigs (Table 10.11). It has also been found on carnivores, rodents, lagomorphs and humans.

### Distribution

The geographical distribution of *D. nuttalli* is limited to eastern Asia, having been reported only from China, Kazakhstan, Mongolia and Russia including Siberia.

### Habitat

*Dermacentor nuttalli* is a tick of the steppes and high grasslands.

### Disease Associations

*Dermacentor nuttalli* is a tick of significance to both animal and human health because it is:

- a vector of *Rickettsia sibirica sibirica* (Hoogstraal, 1967), the bacterium causing North Asian tick typhus in humans;
- a vector of *Babesia caballi* (Friedhoff, 1988), a protozoan causing piroplasmosis in horses and other equines;
- a vector of *Anaplasma ovis* (Yin & Luo, 2007), a bacterium causing anaplasmosis in sheep.

In addition, *D. nuttalli* has been implicated in tick paralysis in sheep (Gothe & Neitz, 1991; Mans et al., 2004).

### References

Arthur (1960).

## 10.12. *Haemaphysalis hystricis* Supino

The tick *Haemaphysalis hystricis* was confused with several other *Haemaphysalis* spp. in Asia before it was redescribed by Hoogstraal et al. (1965). Consequently, data on *Ha. hystricis* relying only on references published prior to 1965 have been excluded because of the uncertainty of the identity of the ticks in those earlier publications.

*Haemaphysalis hystricis* was introduced into the continental United States on one occasion in 1971 on a tapir imported into Massachusetts (USDA, 1972).

Table 10.11. Most commonly reported hosts of the tick *Dermacentor nuttalli*

| Hosts | Stages[a] | No. of reports |
|---|---|---|
| **Ungulates** | | |
| *Equus caballus* (horse) | A | 6 |
| *Bos taurus* (cattle) | A | 5 |
| *Ovis aries* (sheep) | A | 5 |
| *Camelus* sp. (camel) | A | 3 |
| **Primates** | | |
| *Homo sapiens* (human) | A | 5 |
| **Carnivores** | | |
| *Canis familiaris* (dog) | A  N | 3 |
| *Felis catus* (cat) | N | 3 |

[a] A = adult, N = nymph

Table 10.12. Most commonly reported hosts of the tick *Haemaphysalis hystricis*

| Hosts | Stages[a] | No. of reports |
|---|---|---|
| **Carnivores** | | |
| *Canis familiaris* (dog) | A  N | 15 |
| *Panthera tigris* (tiger) | A | 11 |
| *Arctonyx collaris* (hog badger) | A | 6 |
| **Primates** | | |
| *Homo sapiens* (human) | A  N | 13 |
| **Ungulates** | | |
| *Sus scrofa* (domestic & wild pigs) | A  N  L | 12 |

[a] A = adult, N = nymph, L = larva

## Hosts

*Haemaphysalis hystricis* is a tick of mammals. The principal hosts of adults are carnivores, in particular dogs and tigers, and ungulates, especially domestic and wild pigs (Table 10.12). Humans are commonly bitten by both adults and nymphs. Immatures are found on the same hosts as adults as well as on rodents and Malaysian tree shrews. Occasional hosts include rabbits and hares. The often-quoted record of a single female *Ha. hystricis* on a spiny terrapin is considered highly questionable by Hoogstraal et al. (1965).

## Distribution

The geographical distribution of *Ha. hystricis* ranges over much of southern Asia from India in the west to Japan and Indonesia in the east. It has been reported from China (including Hong Kong), India, Indonesia, the Ryukyu Islands of Japan, Laos, Malaysia, Myanmar, Taiwan, Thailand and Vietnam. It has also been found on the island of Hainan off the southern coast of China in the South China Sea.

## Disease Associations

Recently *Ha. hystricis* was identified as a vector of *Rickettsia japonica* (Fournier & Raoult, 2005), the rickettsia causing Japanese spotted fever in humans.

## References

Hoogstraal et al. (1965).

# 11

# Ticks from the Australasian-Asiatic Region

The four invasive tick species whose distribution includes Asia (see Map 10.1) and Australasia (Map 11.1) are discussed in this chapter.

## 11.1. *Amblyomma fimbriatum* Koch

The tick *Amblyomma fimbriatum*, previously known as *Aponomma fimbriatum*, was introduced into the continental United States on three occasions between 1988 and 2002 on monitors, including a harlequin monitor, and an amethystine python (*Morelia amethistina*) imported into Florida and New Mexico from Asian countries such as Indonesia (see Appendix 1).

### Hosts

*Amblyomma fimbriatum* is a reptilian tick whose preferred hosts are a variety of snakes and monitor lizards (Table 11.1). Ophidian hosts include black-headed

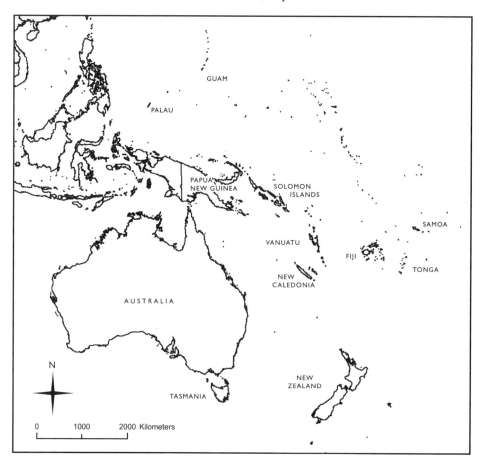

Map 11.1. Australasia and the Pacific

Table 11.1. Most commonly reported hosts of the tick *Amblyomma fimbriatum*

| Hosts | Stages[a] | | No. of reports |
|---|---|---|---|
| **LIZARDS** | | | |
| *Varanus salvator* (water monitor) | A | N | 13 |
| *Varanus gouldii* (Gould's monitor) | A | N | 9 |
| *Varanus varius* (lace monitor) | A | | 8 |
| **SNAKES** | | | |
| *Aspidites melanocephalus* (black-headed python) | - | | 8 |
| *Notechis scutatus* (mainland tiger snake) | A | | 8 |
| *Pseudechis porphyriacus* (red-bellied black snake) | A | | 8 |
| *Python reticulatus* (reticulated python) | A | N | 8 |

[a] A = adult, N = nymph, - = not recorded

Table 11.2. Most commonly reported hosts of the tick *Amblyomma helvolum*

| Hosts | Stages[a] | | | No. of reports |
|---|---|---|---|---|
| **LIZARDS** | | | | |
| *Varanus salvator* (water monitor) | A | N | | 15 |
| *Mabuya multifasciata* (Indonesian brown mabuya) | A | N | L | 6 |
| *Varanus komodoensis* (Komodo dragon) | A | | | 6 |
| **SNAKES** | | | | |
| *Python reticulatus* (reticulated python) | A | | | 9 |
| *Naja naja* (Indian cobra) | A | | | 6 |
| *Ptyas korros* (Chinese rat snake) | A | | | 6 |

[a] A = adult, N = nymph, L = larva

pythons (*Aspidites melanocephalus*), mainland tiger snakes (*Notechis scutatus*), red-bellied black snakes (*Pseudechis porphyriacus*) and reticulated pythons, while saurian hosts include Gould's monitors (*Varanus gouldii*), water monitors (see Fig. 10.3) and lace monitors (*Varanus varius*). Occasional hosts include ungulates (cattle and horses) and birds.

### Distribution

The geographical distribution of *A. fimbriatum* includes parts of Asia, Australasia and the Pacific, ranging from Malaysia and Philippines in the north to Australia in the south. It has been reported from Indonesia, Malaysia and Philippines in Asia, from Australia and Papua New Guinea in Australasia and from the Solomon Islands in the Pacific.

### Disease Associations

None reported.

### References

Roberts (1964a, 1970), Kaufman (1972), Burridge & Simmons (2003).

## 11.2. *Amblyomma helvolum* Koch

The tick *Amblyomma helvolum* was introduced into the continental United States on six occasions between 1970 and 2001 on water monitors and snakes imported into Florida, Maryland and Michigan from Asian countries such as Indonesia and Thailand (see Appendix 1). The ophidian hosts were king cobras and a reticulated python.

### Hosts

*Amblyomma helvolum* is a tick of reptiles. The preferred hosts are lizards and snakes, with most reports from water monitors (see Fig. 10.3) and reticulated pythons (Table 11.2). Occasional hosts include turtles and rats. The most common attachment sites for *A. helvolum* on female monitors are between the toes of the front feet and in the axillae, and on male monitors, the ventral surface and the base of the claws.

### Distribution

The geographical distribution of *A. helvolum* includes parts of Asia and Australasia, ranging from India in

Table 11.3. Most commonly reported hosts of the tick *Amblyomma trimaculatum*

| Hosts | Stages[a] | | | No. of reports |
|---|---|---|---|---|
| **Lizards** | | | | |
| *Varanus indicus* (mangrove monitor) | A | N | L | 12 |
| *Varanus salvator* (water monitor) | A | | | 8 |
| *Corucia zebrata* (Solomon Islands giant skink) | A | N | | 6 |
| *Varanus tristis* (freckled monitor) | - | | | 6 |
| **Snakes** | | | | |
| *Morelia amethistina* (amethystine python) | - | | | 9 |

[a] A = adult, N = nymph, L = larva, - = not recorded

Table 11.4. Most commonly reported hosts of the tick *Haemaphysalis longicornis*

| Hosts | Stages[a] | | | No. of reports |
|---|---|---|---|---|
| **Ungulates** | | | | |
| *Bos taurus* (cattle) | A | N | L | 29 |
| *Equus caballus* (horse) | A | N | L | 18 |
| *Ovis aries* (sheep) | A | N | | 14 |
| *Capra hircus* (goat) | A | N | L | 12 |
| **Carnivores** | | | | |
| *Canis familiaris* (dog) | A | N | | 13 |
| **Primates** | | | | |
| *Homo sapiens* (human) | A | N | L | 10 |

[a] A = adult, N = nymph, L = larva

the north and west to Papua New Guinea and Australia in the south and east. It has been reported from India (including the Nicobar Islands), Indonesia, Laos, Malaysia, Philippines, Singapore, Sri Lanka, Taiwan, Thailand and Vietnam and the island of Hainan off the southern coast of China in Asia and from Australia and Papua New Guinea in Australasia.

### Disease Associations

None reported.

### References

Robinson (1926), Sharif (1928), Auffenberg (1988), Simmons et al. (2002), Voltzit & Keirans (2002).

## 11.3. *Amblyomma trimaculatum* (Lucas)

The tick *Amblyomma trimaculatum*, previously known as *Aponomma trimaculatum*, was introduced into the continental United States on six occasions between 1976 and 2005 on monitor lizards imported into Florida and South Carolina from Asian countries such as Indonesia, Philippines and Thailand and from New Guinea (see Appendix 1). The hosts were a Bengal monitor, a blue-tailed monitor (*Varanus doreanus*), a blue tree monitor (*Varanus macraei*), a Gray's monitor, a peachthroat monitor (*Varanus jobiensis*) and a water monitor.

### Hosts

*Amblyomma trimaculatum* is a reptilian tick whose preferred hosts are lizards, especially mangrove monitors (*Varanus indicus*), and snakes, particularly amethystine pythons (Table 11.3). Occasional hosts include cattle, horses and deer.

### Distribution

The geographical distribution of *A. trimaculatum* includes parts of Asia, Australasia and the Pacific. It has been reported from Indonesia, Philippines, Sri Lanka and Thailand in Asia, from Australia and Papua New Guinea (including the islands of New Britain and New Ireland) in Australasia, and from the Solomon Islands in the Pacific.

### Disease Associations

None reported.

## References

Roberts (1953, 1964a, 1970), Kaufman (1972), Burridge & Simmons (2003).

## 11.4. *Haemaphysalis longicornis* Neumann

The tick *Haemaphysalis longicornis* was introduced into the continental United States on six occasions between 1969 and 1997 on horses imported into California and New Jersey (see Appendix 1).

### Hosts

*Haemaphysalis longicornis* is a three-host tick of ungulates, with cattle, horses, sheep and goats the preferred hosts of adults (Table 11.4). It is also a significant pest of red deer in areas where they are farmed such as in New Zealand. Other hosts of adults include dogs, cats, rabbits and hares. All stages will feed on humans. Larvae and nymphs will feed on the same hosts as adults as well as on a variety of avian species. Occasional hosts include marsupials, rodents and hedgehogs.

The preferred site of attachment is the ears. On deer, they are found also on the submandibular region, on the face, neck and shoulders and along the ventral abdomen.

### Distribution

The geographical distribution of *Ha. longicornis* includes parts of Asia, Australasia and the Pacific, ranging from Russia in the north to New Zealand in the south. It has been reported from China, Japan, North and South Korea, and Russia in Asia, from Australia and New Zealand in Australasia, and from Fiji, New Caledonia, Samoa, Tonga and Vanuatu in the Pacific.

The original distribution of *Ha. longicornis* was in eastern Asia, from where it was introduced by movement of infested cattle and other animals to Australia and then to New Zealand and islands in the Pacific (Hoogstraal et al., 1968, 1981). In Australasia and the Pacific, *Ha. longicornis* found ecologically favorable conditions where it has established thriving populations. This extension of the distribution of *Ha. longicornis* beyond Asia through animal movements demonstrates how invasive ticks can establish breeding populations on new continents where they become problems for domestic animals, wildlife and humans.

### Life Cycle and Habitat

*Haemaphysalis longicornis* populations are parthenogenetic (i.e., reproduce without involvement of males) in the northern colder areas of its range in continental Asia and Japan, whereas those in the southern warmer areas of southern Japan and South Korea are bisexual (Hoogstraal et al., 1981). All *Ha. longicornis* populations from northern areas introduced into Australia, New Zealand and Pacific islands remain parthenogenetic.

*Haemaphysalis longicornis* is a tick of forests and pastures. Its life cycle has been summarized by Hoogstraal et al. (1968). They estimated a preoviposition period of 3–7 days, an oviposition period of 11–27 days, an egg incubation period of 24–39 days, a larval prefeeding period of 3–5 days, a larval feeding period of 4–7 days, a larval premolting period of 14–22 days, a nymphal prefeeding period of 2–3 days, a nymphal feeding period of 4–7 days, a nymphal premolting period of 12–16 days, a female prefeeding period of 2–6 days and a female feeding period of 11–19 days. After engorgement and detachment, *Ha. longicornis* females laid up to 2,740 eggs.

### Disease Associations

*Haemaphysalis longicornis* is a tick of some significance to animal health because it is:

- a vector of *Babesia gibsoni* (Otsuka, 1974), a protozoan causing babesiosis in dogs;
- a cause of fatal anemia in heavy infestations on deer fawns (Neilson & Mossman, 1982);
- a vector of *Theileria buffeli*/*Theileria orientalis* (Uilenberg et al., 1985; Jongejan & Uilenberg, 2004), a protozoal group causing theileriosis in cattle;
- the vector of *Theileria uilenbergi* (Li et al., 2009), a newly discovered protozoal parasite causing theileriosis in sheep and goats in northern China.

In addition, *Ha. longicornis* is an experimental vector of *Babesia ovata* (Minami & Ishihara, 1980), a protozoan causing babesiosis in cattle in Japan, and of a *Theileria* sp. causing theileriosis in goats and sheep in China (Li et al., 2007).

### References

Hoogstraal et al. (1968), Roberts (1969, 1970).

# 12

# Ticks from Australia

A total of nine invasive tick species whose distribution includes Australia (see Map 11.1) have been introduced into the continental United States. The four species whose distribution is limited to Australia are discussed in this chapter. The other five species with more extensive distributions are discussed in chapter 11 (*Amblyomma fimbriatum*, *A. helvolum*, *A. trimaculatum* and *Haemaphysalis longicornis*) and in chapter 14 (*Rhipicephalus (Boophilus) microplus*).

## 12.1. *Amblyomma moreliae* (Koch)

The tick *Amblyomma moreliae* was introduced into the continental United States on two occasions, in 1973 and 1988 on a python and a skink imported into Texas and New Mexico, respectively (USDA, 1974a,b, 1994). The tick found on the skink was misidentified as *Amblyomma echidnae* (USDA, 1994) but on reexamination was found to be *A. moreliae* (Burridge & Simmons, 2003).

### Hosts

*Amblyomma moreliae* is a tick of snakes and lizards (Table 12.1). It has been collected from 10 species of snakes and 8 species of lizards. Other hosts include cattle, horses, kangaroos/wallabies and humans.

A Balkan whip snake (*Coluber gemonensis*), a native of Europe, was found infested in 1911 at a zoo in India (Keirans, 1984), demonstrating the potential for *A. moreliae* to spread outside its native geographical range to new reptilian hosts.

Table 12.1. Most commonly reported hosts of the tick *Amblyomma moreliae*

| Hosts | Stages[a] | No. of reports |
|---|---|---|
| **Lizards** | | |
| *Varanus varius* (lace monitor) | A | 8 |
| **Snakes** | | |
| *Morelia spilota* (carpet python) | - | 6 |
| *Notechis scutatus* (mainland tiger snake) | A | 6 |
| *Pseudechis porphyriacus* (red-bellied black snake) | L | 6 |
| **Ungulates** | | |
| *Bos taurus* (cattle) | - | 6 |
| *Equus caballus* (horse) | - | 6 |
| **Marsupials** | | |
| *Macropus* sp. (kangaroo/wallaby) | - | 6 |

[a]A = adult, L = larva, - = not recorded

## Distribution

The geographical distribution of *A. moreliae* is limited to Australia.

## Disease Associations

None reported.

## References

Robinson (1926), Roberts (1964a, 1969, 1970), Burridge & Simmons (2003).

## 12.2. *Amblyomma triguttatum* Koch

The "ornate kangaroo tick" *Amblyomma triguttatum* was introduced into the continental United States on one occasion on a human entering Georgia from Australia (Keirans & Durden, 2001).

## Hosts

*Amblyomma triguttatum* is a three-host tick associated with Australian marsupials, in particular kangaroos and wallabies, on which all stages feed (Table 12.2). Adults, nymphs and larvae are also found on domestic livestock (cattle, horses, sheep and pigs) and humans. Other hosts include dingoes (*Canis familiaris*), dogs, duck-billed platypuses (*Ornithorhynchus anatinus*) and skinks.

Preferred sites of attachment of *A. triguttatum* on the eastern grey kangaroo (*Macropus giganteus*) are the ears for all stages, with the hind legs also infested. On ungulates, *A. triguttatum* is also found on the ears of feral pigs, on the ears of cattle, on the ears and axillae of sheep, and in the pelvic and preputial regions of the horse. On humans, *A. triguttatum* has been found over the whole body of soldiers bivouacked in Australia, with the highest density on the trunk, especially around the waist and on the frontal aspect.

## Distribution

The geographical distribution of *A. triguttatum* is limited to Australia.

## Disease Associations

*Amblyomma triguttatum* has been reported to be a vector of *Coxiella burnetii* (McDiarmid et al., 2000), the rickettsia causing Q fever in humans and animals, and *A. triguttatum* larvae have been found to cause allergic dermatitis in humans (Guglielmone, 1994).

## References

Robinson (1926), Roberts (1953, 1962, 1969, 1970), McCarthy (1960), Guglielmone & Moorhouse (1986), Guglielmone (1990, 1994), Waudby & Petit (2007), Waudby et al. (2007).

## 12.3. *Bothriocroton concolor* (Neumann)

The tick *Bothriocroton concolor*, previously known as *Aponomma concolor*, was introduced into the continental United States on one occasion in 1973 on an

Table 12.2. Most commonly reported hosts of the tick *Amblyomma triguttatum*

| Hosts | Stages[a] | | | No. of reports |
|---|---|---|---|---|
| **Primates** | | | | |
| *Homo sapiens* (human) | A | N | L | 14 |
| **Ungulates** | | | | |
| *Bos taurus* (cattle) | A | N | L | 12 |
| *Equus caballus* (horse) | A | N | L | 12 |
| *Ovis aries* (sheep) | A | N | L | 9 |
| **Carnivores** | | | | |
| *Canis familiaris* (dingo/dog) | A | N | | 11 |
| **Monotremes** | | | | |
| *Ornithorhynchus anatinus* (duck-billed platypus) | | - | | 9 |
| **Marsupials** | | | | |
| *Macropus rufus* (red kangaroo) | A | N | L | 8 |

[a] A = adult, N = nymph, L = larva, - = not recorded

echidna imported into Illinois from Australia (USDA, 1974a,b).

### Hosts

*Bothriocroton concolor* is a tick of echidnas, especially short-beaked echidnas (*Tachyglossus aculeatus*) (Fig. 12.1), on which it can be found in the ear canals (Table 12.3). Occasional hosts include lace monitors and western grey kangaroos (*Macropus fuliginosus*).

### Distribution

The geographical distribution of *B. concolor* is limited to Australasia. All reports but one are from Australia, with the other report from New Guinea.

### Disease Associations

None reported.

### References

Roberts (1969, 1970), Kaufman (1972).

## 12.4. *Bothriocroton hydrosauri* (Denny)

The tick *Bothriocroton hydrosauri*, previously known as *Aponomma hydrosauri*, was introduced into the continental United States on one occasion on an echidna imported into Illinois (Keirans & Durden, 2001).

### Hosts

*Bothriocroton hydrosauri* is a three-host tick of lizards, snakes and turtles, with shingleback skinks or sleepy lizards (*Trachydosaurus rugosus*) (Fig. 12.2) the most common host and the host on which all three parasitic stages feed. Other reptilian hosts include Gould's mon-

Table 12.3. Hosts of the tick *Bothriocroton concolor*

| Hosts | Stages[a] | | No. of reports |
|---|---|---|---|
| **MONOTREMES** | | | |
| *Tachyglossus aculeatus* (short-beaked echidna) | A | | 7 |
| *Zaglossus bruijni* (long-beaked echidna) | A | | 1 |
| **MARSUPIALS** | | | |
| *Macropus fuliginosus* (western grey kangaroo) | A | N | 1 |
| **LIZARDS** | | | |
| *Varanus varius* (lace monitor) | - | | 1 |

[a]A = adult, N = nymph, - = not recorded

Fig. 12.1. A short-beaked echidna (*Tachyglossus aculeatus*), the preferred host of the tick *Bothriocroton concolor*. (Photograph courtesy of iStockphoto.com.)

itors, eastern blue-tongued skinks (*Tiliqua scincoides*), blotched blue-tongued skinks (*Tiliqua nigrolutea*) and mainland tiger snakes (Table 12.4). Occasional hosts include echidnas, horses, cattle, bandicoots and humans.

Sites of attachment of *B. hydrosauri* on lizards are the ears, neck, forelegs, back and tail. Attachment on shingleback skinks is under posteriorly raised scales, particularly in the axillae and midback for female ticks and in the midback, lower back and tail for male ticks. The tick has also been reported in clusters in the ear canals of reptiles.

### Distribution

The geographical distribution of *B. hydrosauri* is limited to southeastern and southwestern Australia, including Tasmania and Flinders Island.

Table 12.4. Most commonly reported hosts of the tick *Bothriocroton hydrosauri*

| Hosts | Stages[a] | | | No. of reports |
|---|---|---|---|---|
| **Lizards** | | | | |
| *Trachydosaurus rugosus* (shingleback skink) | A | N | L | 25 |
| *Tiliqua nigrolutea* (blotched blue-tongued skink) | A | | | 11 |
| *Varanus gouldii* (Gould's monitor) | A | | | 11 |
| *Tiliqua scincoides* (eastern blue-tongued skink) | A | | | 10 |
| **Snakes** | | | | |
| *Notechis scutatus* (mainland tiger snake) | A | N | | 11 |

[a]A = adult, N = nymph, L = larva

Fig. 12.2. A shingleback skink (*Trachydosaurus rugosus*), the preferred host of the tick *Bothriocroton hydrosauri*. (Photograph courtesy of iStockphoto.com.)

### Life Cycle

Studies on the life cycle of *B. hydrosauri* have been conducted by Bull, Chilton and their colleagues (Bull et al., 1977; Chilton & Bull, 1991, 1993, 1994; Chilton, 1994; Chilton et al., 2000). They reported that unfed larvae preferred cooler and wetter conditions, with a mean survival time of 139 days at 4 °C and 80–85% relative humidity reduced to only one day at 37 °C and 0% relative humidity, indicating that *B. hydrosauri* is susceptible to dessication. Unfed nymphs survived for 5–17 days at 30 °C. The premolt period for engorged larvae varied from 10–72 days and for engorged nymphs from 13–124 days, depending on temperature and relative humidity. The period of female engorgement was 6–25 days and the mean preoviposition period 24 days, with 1,000–2,000 eggs laid by the females.

### Disease Associations

*Bothriocroton hydrosauri* is a tick of some significance to animal and human health because it is:
- a vector of *Hemolivia mariae* (Smallridge & Bull, 1999), a hemogregarine parasite of skinks;
- the reservoir and a putative vector of *Rickettsia honei* (Stenos et al., 2003), the bacterium causing Flinders Island spotted fever in humans.

### References

Roberts (1953, 1969, 1970), Kaufman (1972), Smyth (1973), Burridge & Simmons (2003).

# 13

# Ticks from the Americas

A total of 29 invasive tick species whose distribution includes the Americas (Map 13.1) have been introduced into the continental United States. The 26 species whose distribution is limited to the Americas are discussed in this chapter. The other three species with more extensive distributions are discussed in chapter 4 (*Amblyomma variegatum*) and in chapter 14 (*Rhipicephalus (Boophilus) annulatus* and *R. (B.) microplus*).

## 13.1. *Amblyomma albopictum* Neumann

The tick *Amblyomma albopictum* has been introduced into the continental United States on one occasion on an iguana imported into Florida from Hispaniola (Keirans & Durden, 2001).

### Hosts

*Amblyomma albopictum* is a rarely reported tick of iguanid lizards. Its preferred hosts are iguanas of the genus *Cyclura*, notably Cayman Islands ground iguanas (*C. nubila*) (Fig. 13.1), Bahamas rock iguanas (*C. carinata*) and rhinoceros iguanas (*C. cornuta*) (Table 13.1). Occasional hosts include Cuban boas (*Epicrates angulifer*) and northern curly-tailed lizards (*Leiocephalus carinatus*).

### Distribution

The geographical distribution of *A. albopictum* includes parts of the Caribbean, Central America and South America. It has been reported from Cuba, Dominican Republic and Haiti in the Caribbean, from Costa Rica and Honduras in Central America, and

Map 13.1. The Americas

from French Guiana in South America. Reports of *A. albopictum* from Brazil require verification.

**Disease Associations**

None reported.

**References**

Robinson (1926), Guglielmone et al. (2003a).

## 13.2. *Amblyomma argentinae* Neumann

The tick *Amblyomma argentinae* (Fig. 13.2), previously known as *A. testudinis*, has been introduced into the continental United States on one occasion on an unidentified reptile (Becklund, 1968). Another tick found on a boa constrictor in Oregon in 1972 was misidentified as *A. testudinis* (USDA, 1973a) but upon reexamination was found to be *Amblyomma dissimile* (Burridge & Simmons, 2003).

**Hosts**

*Amblyomma argentinae* is a three-host tick of Argentine tortoises (*Chelonoidis chilensis*) (Fig. 13.3) (Table 13.2). It appears so strongly dependent on the presence of this host that Guglielmone et al. (2001) suggested that should the Argentine tortoise "become extinct, *A. argentinae* would almost certainly share its fate." Other reptilian hosts include snakes and turtles. Occasional hosts include giant toads (*Bufo marinus*) and birds.

Table 13.1. Hosts of the tick *Amblyomma albopictum*

| Hosts | Stages[a] | No. of reports |
|---|---|---|
| **LIZARDS** | | |
| *Cyclura nubila* (Cayman Islands ground iguana) | A | 7 |
| *Cyclura cornuta* (rhinoceros iguana) | A | 4 |
| *Cyclura carinata* (Bahamas rock iguana) | - | 3 |
| *Leiocephalus carinatus* (northern curly-tailed lizard) | - | 1 |
| **SNAKES** | | |
| *Epicrates angulifer* (Cuban boa) | - | 2 |

[a]A = adult, - = not recorded

Fig. 13.1. A Cayman Islands ground iguana (*Cyclura nubila*), the preferred host of the tick *Amblyomma albopictum*. (Photograph courtesy of Darryl Heard.)

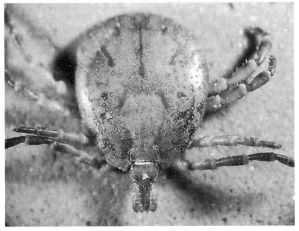

Fig. 13.2. Male (*left*) and female (*right*) *Amblyomma argentinae* ticks. (Photographs courtesy of Santiago Nava.)

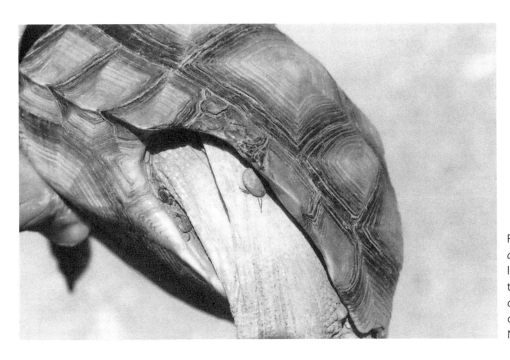

Fig. 13.3. *Amblyomma argentinae* ticks on the leg of an Argentine tortoise (*Chelonoidis chilensis*). (Photograph courtesy of Santiago Nava.)

Table 13.2. Most commonly reported hosts of the tick *Amblyomma argentinae*

| Hosts | Stages[a] | No. of reports |
|---|---|---|
| **TORTOISES** | | |
| *Chelonoidis chilensis* (Argentine tortoise) | A | 13 |
| **SNAKES** | | |
| *Boa constrictor* (boa constrictor) | A | 5 |
| *Crotalus durissus* (cascabel rattlesnake) | A  N | 3 |
| *Eunectes notaeus* (yellow anaconda) | A | 2 |
| **AMPHIBIANS** | | |
| *Bufo marinus* (giant toad) | A  N | 2 |

[a] A = adult, N = nymph

## Distribution

The natural geographic distribution of *A. argentinae* appears to be limited to Argentina. Reports of the tick from Chile are considered attributable to importations of infested hosts from Argentina since no tortoises are native to Chile (Gonzalez-Acuna & Guglielmone, 2005).

## Disease Associations

No associations between *A. argentinae* and any disease have been confirmed. Mans et al. (2004) reported in a review article that *A. argentinae* caused paralysis in reptiles in Argentina and Peru, but Guglielmone (personal communication) indicated that there are no records of natural cases of tick paralysis due to *A. argentinae*. Furthermore, Hanson et al. (2007) made a convincing argument that the studies of Lehmann et al. (1969), which were quoted by Mans et al. (2004), may have been conducted using *A. dissimile* rather than *A. argentinae*.

## References

Robinson (1926), Estrada-Pena et al. (1993), Guglielmone et al. (2001, 2003a).

## 13.3. *Amblyomma auricularium* (Conil)

The tick *Amblyomma auricularium* (Fig. 13.4) has been reported in the continental United States on five occasions: in 1981 on a bovine in Texas (USDA, 1982), in 1989 on a dog in Florida (USDA, 1994), in 1991 on a nine-banded armadillo (*Dasypus novemcinctus*) in Florida (Lord & Day, 2000), in 1998 on a feral pig in Florida (Allan et al., 2001) and in 2000 on a Virgin-

Fig. 13.4. A female *Amblyomma auricularium* tick. (Photograph courtesy of Santiago Nava.)

Fig. 13.5. A six-banded armadillo (*Euphractus sexcinctus*), a common host of the ticks *Amblyomma auricularium* and *A. pseudoconcolor*. (Photograph courtesy of Mariano Mastropaolo.)

Table 13.3. Most commonly reported hosts of the tick *Amblyomma auricularium*

| Hosts | Stages[a] | | | No. of reports |
|---|---|---|---|---|
| **XENARTHRANS** | | | | |
| *Dasypus novemcinctus* (nine-banded armadillo) | A | N | L | 16 |
| *Euphractus sexcinctus* (six-banded armadillo) | A | N | | 10 |
| *Zaedyus pichiy* (pichi) | A | N | | 8 |
| *Tolypeutes matacus* (southern three-banded armadillo) | A | N | | 7 |
| **CARNIVORES** | | | | |
| *Canis familiaris* (dog) | A | N | | 8 |

[a] A = adult, N = nymph, L = larva

ian opossum (*Didelphis virginiana*) in Florida (USDA, unpublished data).

## Hosts

*Amblyomma auricularium* is a tick of armadillos (Fig. 13.5), in particular nine-banded armadillos, on which all stages of the tick feed (Table 13.3). Other less common hosts are dogs and anteaters. Occasional hosts include cats, wild carnivores, rodents, opossums, domestic ungulates and monkeys.

## Distribution

The geographical distribution of *A. auricularium* ranges from northern Patagonia in Argentina in the south to Mexico in the north. It has been reported from Argentina, Bolivia, Brazil, Colombia, French Guiana, Paraguay, Uruguay and Venezuela in South America, from Belize, Costa Rica, Guatemala, Honduras, Nicaragua and Panama in Central America, from Mexico and from Trinidad & Tobago in the Caribbean. The distribution of *A. auricularium* mirrors that of its principal host, the nine-banded armadillo (McBee & Baker, 1982).

*Amblyomma auricularium* has already been found on three native wildlife species in Florida, suggesting it has become established in that state. The current distribution of established populations of the nine-banded armadillo in the United States extends from the Gulf Coast states of Texas, Louisiana, Mississippi, Alabama and Florida northward to Oklahoma, Kansas, Arkansas, Missouri, Tennessee, Georgia and South Carolina (Taulman & Robbins, 1996), providing adequate populations of preferred hosts for the spread of *A. auricularium* in the United States.

## Disease Associations

None reported.

## References

Jones et al. (1972), Amorim & Serra-Freire (2000), Guglielmone et al. (2003a,b).

## 13.4. *Amblyomma cajennense* (Fabricius)

The "Cayenne tick" *Amblyomma cajennense* (Figs. 2.4, 13.6) was introduced into the continental United States on at least 62 occasions between 1913 and 1989, on cattle, horses, a three-toed sloth, a capybara, humans, a shipment of frozen beef, plants and inanimate objects imported into or entering Arizona, Colorado, Connecticut, Delaware, Florida, Georgia, Illinois, Louisiana, Maryland, Nevada, Oklahoma, Pennsylvania and Texas from Mexico, Jamaica and Central American countries such as Costa Rica, Guatemala, Honduras, Nicaragua and Panama (see Appendix 1). The plants included orchids, palm fronds, ferns, annasas, limes and plantains, and the inanimate objects included baggage. These data on *A. cajennense* demonstrate clearly the variety of means by which an invasive tick can be introduced into the United States, whether on a human or animal host or on plants or inanimate objects.

## Hosts

*Amblyomma cajennense* is an important three-host tick with a wide host range. It has been found on 117 species, including 41 species of birds, 20 species of ungulates, 17 species of carnivores, 13 species of rodents,

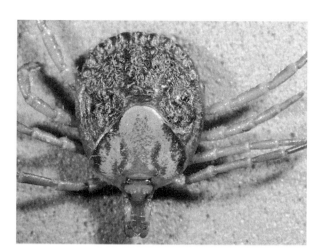

Fig. 13.6. Male (*left*) and female (*right*) Cayenne ticks (*Amblyomma cajennense*). (Photographs courtesy of Santiago Nava.)

7 species of reptiles, 5 species each of marsupials and primates, 4 species of xenarthrans, 3 species of bats, and one species each of lagomorphs and amphibians (see Appendix 2). The preferred hosts are mammals, in particular cattle, horses, dogs and pigs among domestic animals and capybaras (*Hydrochaerus hydrochaeris*) and collared peccaries (*Pecari tajacu*) among wildlife (Table 13.4). However, Estrada-Pena et al. (2004b) believe that records of *A. cajennense* from capybaras should be treated with caution because some of these ticks may have been *Amblyomma dubitatum*, which utilizes the capybara as its principal host. *Amblyomma cajennense* is an anthropophilic tick species of which all stages frequently bite humans. In addition to mammals, *A. cajennense* has been reported infrequently from reptiles, toads and birds. Although *A. cajennense* has been found on 42 species of birds, they are considered secondary hosts, helping to disperse immature stages of the tick.

The ease with which *A. cajennense* will attach to new hosts can be seen from observations in Texas. African Barbary sheep (*Ammotragus lervia*), Asian

Table 13.4. Most commonly reported hosts of the tick *Amblyomma cajennense*

| Hosts | Stages[a] | | | No. of reports |
|---|---|---|---|---|
| **Ungulates** | | | | |
| *Bos taurus* (cattle) | A | N | L | 88 |
| *Equus caballus* (horse) | A | N | L | 84 |
| *Sus scrofa* (domestic & feral pigs) | A | N | | 38 |
| *Pecari tajacu* (collared peccary/javelina) | A | N | L | 23 |
| *Capra hircus* (goat) | A | N | | 18 |
| *Ovis aries* (sheep) | A | | | 18 |
| *Tapirus terrestris* (South American tapir) | A | | | 13 |
| *Odocoileus virginianus* (white-tailed deer) | A | | | 12 |
| **Primates** | | | | |
| *Homo sapiens* (human) | A | N | L | 57 |
| **Carnivores** | | | | |
| *Canis familiaris* (dog) | A | N | L | 53 |
| *Nasua nasua* (South American coati) | A | | | 10 |
| **Rodents** | | | | |
| *Hydrochaerus hydrochaeris* (capybara) | A | N | L | 28 |
| **Xenarthrans** | | | | |
| *Tamandua tetradactyla* (collared anteater) | A | N | | 16 |

[a] A = adult, N = nymph, L = larva

nilgais (*Boselaphus tragocamelus*) and African black rhinoceroses have all been kept on game ranches in Texas and each has become infested with *A. cajennense* acquired from the pasture (Wilson & Richard, 1984; USDA, 1985b, 1987a, 1994). These findings demonstrate the potential for *A. cajennense* to become established on mammals in the United States in areas where climatic conditions are favorable for the tick.

An interesting relationship between capuchin monkeys (*Cebus apella*) and *A. cajennense* has been proposed by Verderane et al. (2007). They described how capuchins rub their fur with carpenter ants (*Camponotus rufipes*) in Brazil so the fur is covered in formic acid from the ants. They noted that this behavior, termed anting, occurs significantly more often during the months when *A. cajennense* nymphs are active and hypothesized that this anting behavior was developed by capuchins to protect them against *A. cajennense* infestations. A subsequent experimental study (Falotico et al., 2007) demonstrated that both formic acid and abdominal secretions from carpenter ants do indeed have significant repellent effects against *A. cajennense* nymphs, adding weight to their hypothesis of an association between anting by capuchins and control of *A. cajennense* infestations.

## Distribution

The geographical distribution of *A. cajennense* ranges from northern Argentina in the south to southern Texas in the north, that is, from about latitude 29°S to 27°N. It has been reported from all countries in South America except Chile and Uruguay (i.e., Argentina, Bolivia, Brazil, Colombia, Ecuador, French Guiana, Guyana, Paraguay, Peru, Suriname and Venezuela), from all countries in Central America (Belize, Costa Rica, El Salvador, Guatemala, Honduras, Nicaragua and Panama), from three Caribbean islands (Cuba, Jamaica and Trinidad), from Mexico and from an adjoining part of Texas.

Within the United States, *A. cajennense* is firmly established in southern Texas. Adult, nymphal and larval stages have been found on cattle, dogs, feral pigs, collared peccaries, white-tailed deer (*Odocoileus virginianus*), bobwhite quail (*Colinus virginianus*), meadowlarks (*Sturnella* sp.) and captive black rhinoceroses in Aransas, Brooks, Cameron, Hidalgo, Jim Wells, Kennedy, Kleberg, Live Oak, Nueces, Sam Patricio, Starr, Travis, Uvalde and Willacy counties (Cooley & Kohls, 1944; Bishopp & Trembley, 1945; Hightower et al., 1953; Eads et al., 1956; Samuel & Trainer, 1970; Coombs & Springer, 1974; Wilson & Richard, 1984; Billings et al., 1998; Teel et al., 1998). Its apparent inability to extend its distribution northward in the United States is possibly a result of its lack of cold tolerance (Estrada-Pena et al., 2004b). In addition, *A. cajennense* was reported on occasion in Florida on native cattle in 1968 in Lee County in the southwestern part of the state, and again in 1969, 1972 and 1979 (USDA, 1969, 1970a, 1973a, 1981), but no evidence has yet been presented to suggest that it has become established in Florida.

## Life Cycle and Habitat

Studies in Trinidad have indicated that factors that promote establishment of *A. cajennense* populations include annual rainfall of <175 cm, soil with good drainage as found in deep sandy soils, adequate grass cover, well-distributed tree shade and available large animals as hosts (Smith, 1974, 1975). Other authors have noted that *A. cajennense* is not tolerant of either low or high temperatures but is tolerant of semiarid environments such as those seen in southern Texas (Estrada-Pena et al., 2004b).

The life cycle of *A. cajennense* was studied in Trinidad by Smith (1975). He reported a preoviposition period of 7–13 days, an egg incubation period of 32–43 days, a larval feeding period of 3–6 days, a larval premolting period of 8–19 days, a nymphal feeding period of 4–7 days, a nymphal premolting period of 14–18 days and a female feeding period of 12–14 days.

In Argentina, there is one generation of *A. cajennense* each year, with larvae and nymphs found during the cooler drier season and adults during the hotter rainier season; larvae are most abundant in the winter from June to August, nymphs in the spring from August to November and adults in the summer from November to January.

## Disease Associations

*Amblyomma cajennense* is a tick of significance to both human and animal health because it is:

- a vector of *Rickettsia rickettsii* (Dias & Martins, 1939), the rickettsia causing Brazilian spotted fever in humans;

- a common parasite of humans, with bites by all stages of the tick producing intense itching and uncomfortable indurated areas that are slow to heal (Aitken et al., 1958);
- a cause of paralysis in cattle, sheep and goats (Serra-Freire, 1983).

Additionally, *A. cajennense* is an experimental vector of *Ehrlichia ruminantium* (Uilenberg, 1983a), the rickettsia causing heartwater in domestic and wild ruminants; Uilenberg (1983a) was only able to transmit heartwater by *A. cajennense* in one trial, whereas other transmission attempts by Uilenberg (1982, 1983a) and Mahan et al. (2000) were unsuccessful, suggesting that *A. cajennense* may be at best a poor vector of heartwater.

### References

Robinson (1926), Cooley & Kohls (1944), Bishopp & Trembley (1945), Aitken et al. (1958), Floch & Fauran (1958), Jones et al. (1972), Prieto (1974), Smith (1975), Alvarez et al. (2000), Oliveira et al. (2000, 2003), Estrada-Pena et al. (2004b).

### 13.5. *Amblyomma calcaratum* Neumann

The tick *Amblyomma calcaratum* was found in the continental United States on one occasion in 1985 when it was identified off-host during a flannel drag in Kentucky (Bloemer et al., 1987).

### Hosts

*Amblyomma calcaratum* is a tick of anteaters, with collared anteaters (*Tamandua tetradactyla*) and giant anteaters (*Myrmecophaga tridactyla*) (Fig. 13.7) the preferred hosts of adults (Table 13.5). Other less common hosts of adults are dogs. Occasional hosts include sloths, deer, raccoons, tapirs, bats, birds and, rarely, humans.

### Distribution

The geographical distribution of *A. calcaratum* ranges from Argentina in the south to Mexico in the north. It has been reported from Argentina, Bolivia, Brazil, Colombia, Ecuador, French Guiana, Paraguay, Peru, Suriname and Venezuela in South America, from Belize, Costa Rica, Honduras, Nicaragua and Panama in

Fig. 13.7. A giant anteater (*Myrmecophaga tridactyla*), a common host of the ticks *Amblyomma calcaratum*, *A. nodosum* and *A. pictum*. (Photograph courtesy of Andres Pautasso.)

Central America, from Mexico and from Trinidad & Tobago in the Caribbean.

### Disease Associations

None reported.

### References

Robinson (1926), Jones et al. (1972), Guglielmone et al. (2003a).

## 13.6. Amblyomma coelebs Neumann

The tick *Amblyomma coelebs* has been introduced into the continental United States on one occasion on seeds imported into Louisiana from Costa Rica (Keirans & Durden, 2001).

### Hosts

*Amblyomma coelebs* is a tick of perissodactyls. Its preferred hosts are South American tapirs (*Tapirus terrestris*) (Fig. 13.8), Central American tapirs (*Tapirus bairdii*) and horses (Table 13.6). Humans are also infested by adults and nymphs of *A. coelebs*. Occasional hosts include rodents, armadillos, anteaters, wild carnivores, opossums and birds.

### Distribution

The geographical distribution of *A. coelebs* ranges from northern Argentina in the south to Mexico in the north. It has been reported from Argentina, Bolivia, Brazil, Colombia, French Guiana, Guyana, Paraguay, Peru, Suriname and Venezuela in South America,

Table 13.5. Most commonly reported hosts of the tick *Amblyomma calcaratum*

| Hosts | Stages[a] | No. of reports |
|---|---|---|
| **Xenarthrans** | | |
| *Tamandua tetradactyla* (collared anteater) | A | 21 |
| *Myrmecophaga tridactyla* (giant anteater) | A  N | 11 |
| *Choloepus hoffmanni* (Hoffmann's two-toed sloth) | A | 3 |
| **Carnivores** | | |
| *Canis familiaris* (dog) | A | 8 |
| *Procyon cancrivorus* (crab-eating raccoon) | A | 3 |
| **Ungulates** | | |
| *Mazama americana* (red brocket deer) | A | 3 |

[a]A = adult, N = nymph

Fig. 13.8. A South American tapir (*Tapirus terrestris*) and newborn calf, the preferred host of the ticks *Amblyomma coelebs* and *A. incisum* and a common host of the ticks *A. multipunctum*, *A. oblongoguttatum* and *Haemaphysalis juxtakochi*. (Photograph courtesy of Scott Citino.)

Table 13.6. Most commonly reported hosts of the tick *Amblyomma coelebs*

| Hosts | Stages[a] | No. of reports |
|---|---|---|
| **UNGULATES** | | |
| *Tapirus terrestris* (South American tapir) | A | 12 |
| *Equus caballus* (horse) | A | 7 |
| *Tapirus bairdii* (Central American tapir) | A | 7 |
| *Mazama americana* (red brocket deer) | - | 4 |
| **PRIMATES** | | |
| *Homo sapiens* (human) | A  N | 9 |
| **RODENTS** | | |
| *Cuniculus paca* (paca) | A | 5 |

[a]A = adult, N = nymph, - = not recorded

from Belize, Costa Rica, Guatemala, Honduras, Nicaragua and Panama in Central America, and from Mexico. The distribution of *A. coelebs* mirrors that of its two principal hosts, the South American tapir in South America and the Central American tapir in Central America and southern Mexico (Baker, 1963; Padilla & Dowler, 1994).

**Disease Associations**

None reported.

**References**

Robinson (1926), Floch & Fauran (1958), Jones et al. (1972), Guglielmone et al. (2003a).

## 13.7. *Amblyomma dissimile* Koch

The "iguana tick" *Amblyomma dissimile*, which became established in southern Florida by early last century (Bequaert, 1932), was introduced into the continental United States on at least 106 occasions between 1954 and 2002 on snakes, lizards, turtles, alligators and a giant toad imported into Alabama, Alaska, Arizona, California, Colorado, Connecticut, District of Columbia, Florida, Georgia, Idaho, Illinois, Indiana, Kansas, Louisiana, Maryland, Michigan, Minnesota, Montana, New Jersey, New Mexico, New York, North Dakota, Ohio, Oklahoma, Oregon, Pennsylvania, South Carolina, Tennessee, Texas, Utah, Virginia and Washington from Mexico, Central American countries such as Costa Rica, El Salvador, Guatemala, Honduras and Nicaragua, and South American countries such as Argentina, Colombia, Guyana, Paraguay, Peru, Suriname and Venezuela (see Appendix 1). The ophidian hosts included boa constrictors, rainbow boas (*Epicrates cenchria*), a black racer (*Alsophis ater*), a green pampas rat snake (*Philodryas baroni*), a Neuwied's false boa (*Pseudoboa neuwiedii*), a red-tailed boa (*Boa constrictor*) and a yellow anaconda (*Eunectes notaeus*). The saurian hosts included green iguanas, Mexican spiny-tailed iguanas (*Ctenosaura pectinata*), a black iguana (*Ctenosaura similis*), an Ecuadorean desert tegu (*Dicrodon heterolepis*), a green basilisk (*Basiliscus plumifrons*), a green spiny lizard (*Sceloporus malachitinus*), an Oaxacan spiny-tailed iguana (*Ctenosaura quinquecarinatus*) and a smooth-helmeted iguana (*Corytophanes cristatus*). The chelonian hosts included a Central American wood turtle (*Rhinoclemmys pulcherrima*), a narrow-bridged mud turtle (*Kinosternon angustipons*) and a scorpion mud turtle (*Kinosternon scorpioides*).

**Hosts**

*Amblyomma dissimile* is a common three-host tick of reptiles and amphibians, especially of boas, iguanas and toads (Table 13.7). It has been collected from 34 species of snakes, 15 species of lizards, 12 species of chelonians, 3 species of amphibians and 2 species of crocodilians (see Appendix 2). The most common hosts are boa constrictors (Fig. 13.9), giant toads (Fig. 13.13) and green iguanas (Fig. 2.2). Mammalian hosts include capybaras, cattle and sheep. Occasional hosts include smaller rodents, carnivores and, rarely, humans.

Although *A. dissimile* is a tick of the Western Hemisphere, it has infested hosts native to other geographi-

Fig. 13.9. A boa constrictor (*Boa constrictor*), a common host of the tick *Amblyomma dissimile*. (Photograph courtesy of Kevin Enge.)

Table 13.7. Most commonly reported hosts of the tick *Amblyomma dissimile*

| Hosts | Stages[a] | | No. of reports |
|---|---|---|---|
| **SNAKES** | | | |
| *Boa contrictor* (boa constictor) | A | N | 45 |
| *Epicrates cenchria* (rainbow boa) | A | N | 11 |
| **AMPHIBIANS** | | | |
| *Bufo marinus* (giant toad) | A | | 29 |
| **LIZARDS** | | | |
| *Iguana iguana* (green iguana) | A | | 23 |
| **RODENTS** | | | |
| *Hydrochaerus hydrochaeris* (capybara) | - | | 10 |

[a]A = adult, N = nymph, - = not recorded

cal regions while they were in captivity, including a Mediterranean turtle (*Mauremys leprosa*) from Africa and Europe (Brumpt, 1934), and a king cobra (USDA, 1970a,c) and a Burmese python (USDA, unpublished data) from Asia. These findings demonstrate the potential for *A. dissimile* to spread outside its native geographical range to new reptilian hosts.

### Distribution

The geographical distribution of *A. dissimile* ranges from Argentina in the south to southern Mexico in the north. It has been reported from Argentina, Brazil, Colombia, Ecuador, French Guiana, Guyana, Paraguay, Peru, Suriname and Venezuela in South America, from all countries in Central America (Belize, Costa Rica, El Salvador, Guatemala, Honduras, Nicaragua and Panama), from Mexico and from many Caribbean nations or territories (Antigua & Barbuda, Bahamas, Barbados, Cuba, Dominican Republic, Grenada, Guadeloupe, Haiti, Jamaica, Puerto Rico, St. Lucia and Trinidad & Tobago). In addition, *A. dissimile* has been established in southern Florida since early last century, where it has been collected from water moccasins (*Agkistrodon piscivorus*), eastern diamondback rattlesnakes (*Crotalus adamanteus*), southern indigo snakes (*Drymarchon corais*), corn snakes (*Elaphe gut-

*tata*), North American rat snakes, common kingsnakes (*Lampropeltis getula*), pine snakes (*Pituophis melanoleucus*), pygmy rattlesnakes (*Sistrurus miliarus*), green iguanas, giant toads, gopher tortoises, cotton mice (*Peromyscus gossypinus*) and cattle (Bequaert, 1932, 1945; Cooley & Kohls, 1944; Bishopp & Trembley, 1945; Wilson & Kale, 1972; Durden et al., 1993; Foster et al., 2000; USDA, unpublished data). Finally, there is one report of *A. dissimile* on a native snake in Texas (USDA, 1974a).

### Life Cycle

The life cycle of *A. dissimile* was studied in Panama by Dunn (1918). He reported a preoviposition period of 4–7 days, an oviposition period of 6–25 days, an egg incubation period of 38–47 days, a larval feeding period of 11–18 days, a larval premolting period of 11–15 days, a nymphal feeding period of 11–22 days, a nymphal premolting period of 8–16 days and a female feeding period of 15–36 days. After engorgement and detachment, *A. dissimile* females laid up to 9,254 eggs. All stages are frequently found on an individual host, with up to 1,800 ticks reported from a single snake.

### Disease Associations

*Amblyomma dissimile* is a tick of minor significance to reptilian health since it is:

- a cause of ulcers in rainbow boas (Dunn, 1918) and giant toads (Jakowska, 1972) and of pustular dermatitis in indigo snakes (Foster et al., 2000);
- a vector of *Hepatozoon fusifex* (Ball et al., 1969), a hemogregarine parasite of boa constrictors.

Additionally, *A. dissimile* is an experimental vector of *Ehrlichia ruminantium* (Jongejan, 1992), the rickettsia causing heartwater; however, since heartwater is a disease of ruminants and *A. dissimile* is primarily a tick of reptiles and amphibians, *A. dissimile* is unlikely to play any significant role in *E. ruminantium* transmission.

### References

Bodkin (1918–19), Dunn (1918), Robinson (1926), Cooley & Kohls (1944), Floch & Fauran (1958), Jones et al. (1972), Amorim & Serra-Freire (1994b), Schumaker & Barros (1994), Burridge & Simmons (2003), Guglielmone et al. (2003a).

## 13.8. *Amblyomma geayi* Neumann

The tick *Amblyomma geayi* was introduced into the continental United States on two occasions in 1969 and 1973 on sloths imported into Pennsylvania and Illinois from Central American countries such as Panama (USDA, 1970a,b, 1974a).

### Hosts

*Amblyomma geayi* is a tick of sloths and, in particular, pale-throated sloths (*Bradypus tridactylus*) (Table 13.8). Occasional hosts include porcupines and opossums.

### Distribution

The geographical distribution of *A. geayi* is limited to southern Central America and northern South America. It has been reported from Costa Rica and Panama in Central America, and from Brazil, Colombia,

Table 13.8. Most commonly reported hosts of the tick *Amblyomma geayi*

| Hosts | Stages[a] | No. of reports |
|---|---|---|
| **Xenarthrans** | | |
| *Bradypus tridactylus* (pale-throated sloth) | A | 11 |
| *Choloepus hoffmanni* (Hoffmann's two-toed sloth) | A | 5 |
| *Bradypus variegatus* (brown-throated sloth) | A  N | 3 |
| **Rodents** | | |
| *Coendou prehensilis* (Brazilian tree porcupine) | A | 5 |
| *Sphiggurus spinosus* (spiny tree porcupine) | A | 2 |
| **Marsupials** | | |
| *Caluromys derbianus* (Central American woolly opossum) | N | 2 |
| *Philander opossum* (grey four-eyed opossum) | - | 2 |

[a]A = adult, N = nymph, - = not recorded

French Guiana, Guyana, Peru and Suriname in South America. The distribution of *A. geayi* is similar to that of its sloth hosts, which are restricted to forested areas of Central and South America (de Moraes-Barros et al., 2006).

**Disease Associations**

None reported.

**References**

Robinson (1926), Floch & Fauran (1958), Guglielmone et al. (2003a).

## 13.9. *Amblyomma humerale* Koch

The tick *Amblyomma humerale* was introduced into the continental United States on two occasions in 1999 and 2001 on yellow-footed tortoises imported into Florida from South American countries such as Guyana (Simmons & Burridge, 2000).

**Hosts**

*Amblyomma humerale* is a reptilian tick parasitizing primarily tortoises and turtles (Table 13.9). Its preferred hosts are yellow-footed tortoises. Other reptilian hosts include lizards and caimans. Occasional hosts include toads, armadillos, anteaters, rats, opossums, ungulates, dogs and birds.

A grey-cheeked thrush was found infested with an *A. humerale* nymph in Manitoba, Canada, following its migration from South America (Morshed et al., 2005), showing the potential for birds to transport ticks over great distances.

The sites of attachment of *A. humerale* adults on yellow-footed tortoises differ by sex, with females attached to the skin and males typically found attached in clusters on the ventral side of the carapace.

**Distribution**

The geographical distribution of *A. humerale* ranges over much of South America. It has been reported from Brazil, Bolivia, Colombia, Ecuador, French Guiana, Guyana, Peru, Suriname, Uruguay and Venezuela. In addition to South America, *A. humerale* is also present in the southern Caribbean in Trinidad & Tobago.

**Disease Associations**

None reported.

**References**

Robinson (1926), Floch & Fauran (1958), Simmons & Burridge (2000), Labruna et al. (2002), Guglielmone et al. (2003a).

## 13.10. *Amblyomma incisum* Neumann

A recent study by Labruna et al. (2005) reexamined ticks identified as *Amblyomma incisum* and found many to be either *Amblyomma latepunctatum* or *Amblyomma scalpturatum*. The primary host of the adult stage of all three tick species is the South American tapir. Consequently, data on *A. incisum* relying solely on references published prior to 2005 have been excluded because of the uncertainty of the identity of the ticks in those earlier publications.

Table 13.9. Most commonly reported hosts of the tick *Amblyomma humerale*

| Hosts | Stages[a] | No. of reports |
|---|---|---|
| **Chelonians** | | |
| *Chelonoidis denticulata* (yellow-footed tortoise) | A | 19 |
| *Chelonoidis carbonaria* (red-footed tortoise) | A | 4 |
| *Podocnemis* sp. (South American river turtle) | - | 4 |
| **Xenarthrans** | | |
| *Cyclopes didactylus* (pygmy anteater) | N | 4 |
| *Dasypus novemcinctus* (nine-banded armadillo) | - | 4 |
| *Kentropyx calcarata* (Spix's kentropyx) | N | 3 |
| **Marsupials** | | |
| *Didelphis marsupialis* (southern opossum) | N | 4 |

[a]A = adult, N = nymph, - = not recorded

Table 13.10. Hosts of the tick *Amblyomma incisum*

| Hosts | Stages[a] | | No. of reports |
|---|---|---|---|
| **UNGULATES** | | | |
| *Tapirus terrestris* (South American tapir) | A | N | 19 |
| *Mazama bororo* (Sao Paulo bororo) | | N | 4 |
| *Mazama gouazoubira* (brown brocket deer) | | N | 1 |
| **PRIMATES** | | | |
| *Homo sapiens* (human) | A | N | 10 |
| **RODENTS** | | | |
| *Hydrochaerus hydrochaeris* (capybara) | A | | 2 |

[a] A = adult, N = nymph

*Amblyomma incisum* was introduced into the continental United States on one occasion in 1964 when it was found on a South American tapir in a New York zoo (USDA, 1964b, 1965a); however, given the fact that many specimens of *A. latepunctatum* and *A. scalpturatum* were misidentified as *A. incisum* in the past (Labruna et al., 2005), there is some uncertainty about the true identity of the tick found on the tapir in New York in 1964.

### Hosts

*Amblyomma incisum* is a tick of South American tapirs (Fig. 13.8) (Table 13.10). It is also found feeding on humans. Occasional hosts include deer and capybaras.

### Distribution

The geographical distribution of *A. incisum* is limited to South America. It has been confirmed in Argentina, Bolivia, Brazil, Paraguay and Peru. In addition, it has been reported from Ecuador, French Guiana, Guyana and Venezuela, but its presence in these countries is uncertain (Nava et al., 2007).

### Life Cycle

The life cycle of *A. incisum* has been studied by Szabo et al. (2009) in Brazil using rabbits for immatures and horses for adults. They reported a preoviposition period of 71–148 days, an egg incubation period of 152–256 days, a larval feeding period of 5–11 days, a larval premolting period of 28–36 days, a nymphal feeding period of 5–15 days, a nymphal premolting period of 49–51 days and a female feeding period of 7–15 days. From these data it was estimated that *A. incisum* produces one generation a year.

### Disease Associations

None reported.

### References

Robinson (1926), Floch & Fauran (1958), Jones et al. (1972), Guglielmone et al. (2003a).

## 13.11. *Amblyomma longirostre* (Koch)

The tick *Amblyomma longirostre* was introduced into the continental United States on six occasions between 1929 and 1967 on migrating birds and imported porcupines found in Michigan, New Jersey, Pennsylvania, Tennessee and Texas (see Appendix 1). The migrating birds were a white-eyed vireo, a northern cardinal and a summer tanager, and one of the porcupines was a Brazilian tree porcupine (*Coendou prehensilis*).

### Hosts

*Amblyomma longirostre* is a tick of rodents, with porcupines the preferred hosts for adults (Table 13.11). In contrast, nymphs and larvae parasitize birds, with 99 avian species known to have been infested with immature *A. longirostre*. Adults and nymphs will feed on humans. Occasional hosts include capuchin monkeys, anteaters, sloths, wild carnivores, bats and deer.

### Distribution

The geographical distribution of *A. longirostre* ranges from Argentina in the south to Mexico in the north. It has been reported from Argentina, Bolivia, Brazil, Colombia, French Guiana, Paraguay, Uruguay and Venezuela in South America, from Belize, Costa Rica and

Table 13.11. Most commonly reported hosts of the tick *Amblyomma longirostre*

| Hosts | Stages[a] | No. of reports |
|---|---|---|
| **RODENTS** | | |
| *Coendou prehensilis* (Brazilian tree porcupine) | A | 13 |
| *Coendou rothschildi* (Rothschild's porcupine) | A | 7 |
| **PRIMATES** | | |
| *Homo sapiens* (human) | A N | 8 |
| *Cebus* sp. (capuchin) | A | 5 |
| **BIRDS** | | |
| *Saltator maximus* (buff-throated saltator) | N L | 6 |
| *Lepidocolaptes souleyetii* (streak-headed woodcreeper) | N L | 5 |

[a] A = adult, N = nymph, L = larva

Table 13.12. Hosts of the tick *Amblyomma multipunctum*

| Hosts | Stages[a] | No. of reports |
|---|---|---|
| **UNGULATES** | | |
| *Tapirus terrestris* (South American tapir) | - | 2 |
| *Tapirus pinchaque* (Andean tapir) | A | 1 |

[a] A = adult, - = not recorded

Panama in Central America, from Mexico and from Trinidad & Tobago in the Caribbean.

Nymphal and larval *A. longirostre* are found commonly on a wide variety of birds and, as the birds migrate, immature ticks are disseminated outside the normal range of *A. longirostre*. In this manner, *A. longirostre* has been introduced by migrating birds not only to the United States, as mentioned above, but also to Canada. Scott et al. (2001) found immature *A. longirostre* on two willow flycatchers, a red-eyed vireo and a Canada warbler that had migrated to Manitoba and Ontario from South or Central America. Similarly, Morshed et al. (2005) found an *A. longirostre* nymph on a yellow-bellied flycatcher that had migrated to Manitoba from Central America.

### Disease Associations

None reported.

### References

Robinson (1926), Floch & Fauran (1958), Jones et al. (1972), Guglielmone et al. (2003a).

## 13.12. *Amblyomma multipunctum* Neumann

The tick *Amblyomma multipunctum* was introduced into the continental United States on one occasion in 1968 on a tapir imported into California from Ecuador (USDA, 1968b, 1969). Additional *A. multipunctum* ticks were found in the cargo pit of the plane carrying the tapir (USDA, 1968b).

### Hosts

*Amblyomma multipunctum* is a rarely reported and poorly understood tick of tapirs (Table 13.12). The only other host report is of the tick on an unidentified tortoise. The sparse literature on *A. multipunctum* is confusing, and this controversy has been summarized by Guglielmone et al. (2003a).

### Distribution

The geographical distribution of *A. multipunctum* is not well documented, but it has been reported from five South American countries: Bolivia, Brazil, Colombia, Ecuador and Venezuela.

### Disease Associations

None reported.

### References

Robinson (1926), Guglielmone et al. (2003a).

Table 13.13. Most commonly reported hosts of the tick *Amblyomma nodosum*

| Hosts | Stages[a] | No. of reports |
|---|---|---|
| **XENARTHRANS** | | |
| *Tamandua tetradactyla* (collared anteater) | A | 24 |
| *Myrmecophaga tridactyla* (giant anteater) | A | 16 |
| *Bradypus* sp. (three-toed sloth) | - | 4 |

[a]A = adult, - = not recorded

Table 13.14. Most commonly reported hosts of the tick *Amblyomma oblongoguttatum*

| Hosts | Stages[a] | No. of reports |
|---|---|---|
| **PRIMATES** | | |
| *Homo sapiens* (human) | A  N | 17 |
| **CARNIVORES** | | |
| *Canis familiaris* (dog) | A | 12 |
| **UNGULATES** | | |
| *Tapirus terrestris* (South American tapir) | A | 9 |
| *Bos taurus* (cattle) | A | 8 |
| *Sus scrofa* (pig) | A  N | 8 |
| *Tayassu pecari* (white-lipped peccary) | A | 8 |

[a]A = adult, N = nymph

## 13.13. *Amblyomma nodosum* Neumann

The tick *Amblyomma nodosum* was introduced into the continental United States on six occasions between 1969 and 2002 on anteaters, a three-toed sloth (*Bradypus* sp.) and pineapples imported into Florida, Illinois, North Carolina and Texas from Central American countries such as Panama (see Appendix 1).

### Hosts

*Amblyomma nodosum* is a tick of anteaters, in particular giant anteaters (Fig. 13.7) and collared anteaters (Table 13.13). Other xenarthran hosts are three-toed sloths and six-banded armadillos (*Euphractus sexcinctus*). Occasional hosts include cattle, tapirs, pacas (*Cuniculus paca*) and birds.

### Distribution

The geographical distribution of *A. nodosum* ranges from Argentina in the south to Mexico in the north. It has been reported from Argentina, Bolivia, Brazil, Colombia, Paraguay and Venezuela in South America, from Costa Rica, Guatemala, Nicaragua and Panama in Central America, from Mexico and from Trinidad & Tobago in the Caribbean.

### Disease Associations

None reported.

### References

Robinson (1926), Jones et al. (1972), Amorim & Serra-Freire (1994a), Guglielmone et al. (2003a).

## 13.14. *Amblyomma oblongoguttatum* Koch

The tick *Amblyomma oblongoguttatum* has been introduced into the continental United States on one occasion on a human entering Michigan from Venezuela (Walker et al., 1998).

### Hosts

*Amblyomma oblonguttatum* is a tick with no particular host preference. It has been reported most commonly from humans, dogs, tapirs, cattle, pigs and peccaries (Table 13.14). The tick has also been collected from rodents, other ungulates (goats, horses, water buffalos and deer), birds, armadillos, anteaters, sloths, wild carnivores, opossums, capuchin monkeys and bats.

## Distribution

The geographical distribution of *A. oblonguttatum* ranges from Bolivia and Brazil in the south to Mexico in the north. It has been reported from Bolivia, Brazil, Colombia, French Guiana, Guyana, Peru, Suriname and Venezuela in South America, from Belize, Costa Rica, Guatemala, Nicaragua and Panama in Central America, and from Mexico.

## Disease Associations

None reported.

## References

Robinson (1926), Floch & Fauran (1958), Jones et al. (1972), Guglielmone et al. (2003a).

## 13.15. *Amblyomma ovale* Koch

The tick *Amblyomma ovale* was introduced into the continental United States on three occasions between 1968 and 1988, on an ocelot (*Leopardus pardalis*), a cougar (*Puma concolor*) and a dog imported into California, New York and Washington from South American countries such as Ecuador (see Appendix 1). In addition to these importations, *A. ovale* was found on a dog in Iowa in 1941 (Eddy & Joyce, 1942), on a hispid cotton rat (*Sigmodon hispidus*) in Tennessee in 1964 (Durden & Kollars, 1992), on a dog in Texas in 1971 (USDA, 1973b) and on a coyote (*Canis latrans*) in Texas in 1973 (USDA, 1974b).

## Hosts

*Amblyomma ovale* is a tick of carnivores. The most common hosts are dogs, on which all stages feed (Table 13.15). Wild carnivores are frequently infested, including South American coatis (*Nasua nasua*), jaguars (*Panthera onca*) and crab-eating raccoons (*Procyon cancrivorus*). Other mammalian hosts include ungulates (in particular tapirs), rodents, anteaters, opossums and humans. The most common hosts for adults are carnivores, ungulates and humans, whereas larvae and nymphs feed primarily on rodents. Occasional hosts include monkeys, birds, armadillos and toads.

## Distribution

The geographical distribution of *A. ovale* ranges from north-central Argentina in the south to southern Mexico in the north. It has been reported from Argentina, Bolivia, Brazil, Colombia, Ecuador, French Guiana, Guyana, Paraguay, Peru, Suriname and Venezuela in South America, from Belize, Costa Rica, Guatemala, Nicaragua and Panama in Central America, from Mexico and from Trinidad & Tobago in the Caribbean. Despite the finding of *A. ovale* on dogs in Iowa and Texas, on a hispid cotton rat in Tennessee and on a coyote in Texas, there is no published evidence that *A. ovale* has become established in the United States.

## Disease Associations

*Amblyomma ovale* can cause tick paralysis in humans (Baeza, 1979). Additionally, *A. ovale* has been shown to be an experimental vector of *Hepatozoon* spp. (Forlano et al., 2005), protozoal parasites that cause canine hepatozoonosis in dogs that ingest infected ticks.

## References

Robinson (1926), Jones et al. (1972), Guglielmone et al. (2003a,c).

Table 13.15. Most commonly reported hosts of the tick *Amblyomma ovale*

| Hosts | Stages[a] | | | No. of reports |
|---|---|---|---|---|
| **Carnivores** | | | | |
| *Canis familiaris* (dog) | A | N | L | 43 |
| *Nasua nasua* (South American coati) | A | N | | 11 |
| *Panthera onca* (jaguar) | A | | | 11 |
| *Procyon cancrivorus* (crab-eating raccoon) | A | | | 10 |
| **Primates** | | | | |
| *Homo sapiens* (human) | A | | | 22 |

[a] A = adult, N = nymph, L = larva

## 13.16. *Amblyomma parvum* Aragao

The tick *Amblyomma parvum* (Fig. 13.10) has been introduced into the continental United States on two occasions, once on an armadillo imported into Ohio from Paraguay (Keirans & Durden, 2001) and later on a human in Florida (Berube, 2006).

### Hosts

*Amblyomma parvum* is a tick with no particular host preference. The most common hosts for adults are cattle, horses, dogs and armadillos (Table 13.16), whereas larvae and nymphs appear to prefer rodents. Adults often infest humans. Occasional hosts include capuchin monkeys, anteaters, rabbits, bats, birds and tortoises. Adults feed on the head of cattle and goats, predominantly on the ears and in the periocular region, whereas larvae and nymphs feed on the ears of rodents such as common yellow-toothed cavies (*Galea musteloides*).

### Distribution

The geographical distribution of *A. parvum* ranges from Argentina in the south to southern Mexico in the north. It has been reported from Argentina, Bolivia, Brazil, Colombia, French Guiana, Paraguay and Venezuela in South America, from Costa Rica, El Salvador, Guatemala, Nicaragua and Panama in Central America, and from Mexico.

### Life Cycle

The life cycle of *A. parvum* has been studied in Argentina by Guglielmone et al. (1991) and Nava et al. (2008b). They reported a preoviposition period of

Fig. 13.10. A female *Amblyomma parvum* tick. (Photograph courtesy of Santiago Nava.)

4–9 days, an oviposition period of 13–23 days, an egg incubation period of 27–37 days, a larval feeding period of 1–6 days, a larval premolting period of 8–17 days, a nymphal feeding period of 3–7 days, a nymphal premolting period of 15–23 days and a female feeding period of 7–9 days. After engorgement and detachment, *A. parvum* females laid a mean of 1,500 eggs. One generation of ticks was produced each year, with the peak activity for larvae in the autumn, for nymphs in midwinter and for females in early to middle summer.

### Disease Associations

None reported.

### References

Robinson (1926), Floch & Fauran (1958), Jones et al. (1972), Guglielmone et al. (1990, 2003a), Nava et al. (2006, 2008a).

Table 13.16. Most commonly reported hosts of the tick *Amblyomma parvum*

| Hosts | Stages[a] | No. of reports |
|---|---|---|
| **UNGULATES** | | |
| *Bos taurus* (cattle) | A  N  L | 22 |
| *Equus caballus* (horse) | A | 17 |
| **PRIMATES** | | |
| *Homo sapiens* (human) | A | 20 |
| **CARNIVORES** | | |
| *Canis familiaris* (dog) | A | 15 |
| **XENARTHRANS** | | |
| *Dasypus novemcinctus* (nine-banded armadillo) | A | 10 |

[a] A = adult, N = nymph, L = larva

## 13.17. Amblyomma pictum Neumann

The tick *Amblyomma pictum* was introduced into the continental United States on one occasion in 1972 on an anteater imported into New York (USDA, 1973a).

### Hosts

*Amblyomma pictum* is a rarely reported tick of giant anteaters (Fig. 13.7) and collared anteaters (Table 13.17). The only other known hosts are dogs.

### Distribution

The geographical distribution of *A. pictum* includes three countries in South America (Brazil, French Guiana and Guyana) and one in Central America (Panama).

### Disease Associations

None reported.

### References

Robinson (1926), Guglielmone et al. (2003a).

## 13.18. Amblyomma pseudoconcolor Aragao

The tick *Amblyomma pseudoconcolor* (Fig. 13.11) has been introduced into the continental United States on two occasions since 1973 on armadillos imported into Illinois and Ohio from Bolivia and Paraguay (USDA, 1974a,b; Keirans & Durden, 2001). One of the hosts was a six-banded armadillo (Fig. 13.5).

### Hosts

*Amblyomma pseudoconcolor* is a tick of armadillos, having been found on 10 different armadillo species (Figs. 13.5, 13.12). Other hosts are ungulates (cattle, tapirs and water buffalos), birds and dogs (Table 13.18). Occasional hosts include opossums and anteaters. From the limited data available, it would appear that

Fig. 13.11. A female *Amblyomma pseudoconcolor* tick. (Photograph courtesy of Santiago Nava.)

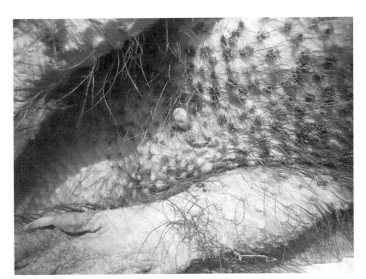

Fig. 13.12. An engorged female and larvae of the tick *Amblyomma pseudoconcolor* on the body and a limb, respectively, of a larger hairy armadillo (*Chaetophractus villosus*). (Photograph courtesy of Santiago Nava.)

Table 13.17. Hosts of the tick *Amblyomma pictum*

| Hosts | Stages[a] | No. of reports |
|---|---|---|
| **Xenarthrans** | | |
| *Myrmecophaga tridactyla* (giant anteater) | A | 8 |
| *Tamandua tetradactyla* (collared anteater) | A | 4 |
| **Carnivores** | | |
| *Canis familiaris* (dog) | A | 7 |

[a] A = adult

Table 13.18. Most commonly reported hosts of the tick *Amblyomma pseudoconcolor*

| Hosts | Stages[a] | No. of reports |
|---|---|---|
| **XENARTHRANS** | | |
| *Euphractus sexcinctus* (six-banded armadillo) | A | 6 |
| *Chaetophractus vellerosus* (lesser hairy armadillo) | - | 4 |
| *Dasypus novemcinctus* (nine-banded armadillo) | A | 3 |
| **UNGULATES** | | |
| *Bos taurus* (cattle) | A | 6 |
| *Bubalus bubalis* (water buffalo) | A | 3 |
| *Tapirus terrestris* (South American tapir) | A | 3 |
| **BIRDS** | | |
| *Nothura maculosa* (spotted nothura) | A N L | 5 |
| *Rhynchotus rufescens* (red-winged tinamou) | N L | 3 |
| **CARNIVORES** | | |
| *Canis familiaris* (dog) | A | 3 |

[a] A = adult, N = nymph, L = larva, - = not recorded

Table 13.19. Most commonly reported hosts of the tick *Amblyomma quadricavum*

| Hosts | Stages[a] | No. of reports |
|---|---|---|
| **SNAKES** | | |
| *Epicrates striatus* (Haitian boa) | A | 8 |
| *Alsophis cantherigerus* (Cuban racer) | - | 6 |
| *Epicrates angulifer* (Cuban boa) | - | 6 |
| *Alsophis portoricensis* (Puerto Rican racer) | - | 3 |
| *Boa constrictor* (boa constrictor) | - | 2 |
| *Epicrates subflavus* (Jamaican boa) | - | 2 |

[a] A = adult, - = not recorded

*A. pseudoconcolor* adults prefer to feed on armadillos, whereas nymphs and larvae prefer birds.

### Distribution

The geographical distribution of *A. pseudoconcolor* is limited to South America, where it has been reported from Argentina, Bolivia, Brazil, French Guiana, Paraguay, Suriname and Uruguay.

### Disease Associations

None reported.

### References

Robinson (1926), Floch & Fauran (1958), Guglielmone et al. (2003b).

## 13.19. *Amblyomma quadricavum* (Schulze)

The tick *Amblyomma quadricavum*, previously known as *Aponomma quadricavum*, was introduced into the continental United States on one occasion in 1979 on a Haitian boa (*Epicrates striatus*) imported into Connecticut (Anderson et al., 1981b).

### Hosts

*Amblyomma quadricavum* is a rarely reported tick of snakes, in particular boas and racers (Table 13.19). The only other known hosts are green iguanas.

### Distribution

The geographical distribution of *A. quadricavum* is limited to Cuba, Haiti, Jamaica and Puerto Rico in the Caribbean.

### Disease Associations

None reported.

### References

Keirans & Klompen (1996), Burridge & Simmons (2003), Guglielmone et al. (2003a).

Fig. 13.13. A giant toad (*Bufo marinus*), the preferred host of the tick *Amblyomma rotundatum* and a common host of the tick *A. dissimile*. (Photograph courtesy of Kenneth Krysko.)

## 13.20. *Amblyomma rotundatum* Koch

The "rotund toad tick" *Amblyomma rotundatum*, which became established in Florida by 1979 or earlier (Oliver et al., 1993), was introduced into the continental United States on 14 occasions between 1964 and 2005 on snakes, yellow-footed tortoises, a box turtle, an iguana and a Suriname toad (*Pipa pipa*) imported into Florida, Illinois and Maryland from South American countries such as Colombia, Guyana and Suriname (see Appendix 1). The ophidian hosts were boa constrictors, an anaconda, an emerald tree boa (*Corallus caninus*), a rainbow boa and a red-tailed boa.

### Hosts

*Amblyomma rotundatum* is a three-host tick of amphibians and reptiles. Its preferred hosts are toads, in particular giant toads (Fig. 13.13), and snakes, especially boa constrictors (Table 13.20). Occasional hosts include armadillos, anteaters, tortoises, turtles, lizards, coatis, capybaras and, rarely, humans.

Although *A. rotundatum* is an American tick, it has infested hosts native to Eurasia and Africa, such as Greek tortoises, and hosts native to Asia, such as yellow-headed temple turtles (*Heosemys annandalii*), while in captivity (Doss et al., 1974a; UF, unpublished data), demonstrating the potential for *A. rotundatum* to spread outside its native geographical range to new reptilian hosts.

### Distribution

The geographical distribution of *A. rotundatum* ranges from Argentina in the south to Mexico and the Caribbean in the north. It has been reported from Argen-

Table 13.20. Most commonly reported hosts of the tick *Amblyomma rotundatum*

| Hosts | Stages[a] | | | No. of reports |
|---|---|---|---|---|
| **AMPHIBIANS** | | | | |
| *Bufo marinus* (giant toad) | A | N | L | 23 |
| *Bufo granulosus* (common lesser toad) | | - | | 5 |
| **SNAKES** | | | | |
| *Boa constrictor* (boa constrictor) | A | N | L | 16 |
| **XENARTHRANS** | | | | |
| *Dasypus novemcinctus* (nine-banded armadillo) | | - | | 5 |
| **TORTOISES** | | | | |
| *Chelonoidis chilensis* (Argentine tortoise) | | - | | 4 |
| *Chelonoidis denticulata* (yellow-footed tortoise) | A | | | 4 |

[a] A = adult, N = nymph, L = larva, - = not recorded

Table 13.21. Most commonly reported hosts of the tick *Amblyomma sabanerae*

| Hosts | Stages[a] | No. of reports |
|---|---|---|
| **TURTLES** | | |
| *Rhinoclemmys funerea* (black wood turtle) | A N L | 6 |
| *Rhinoclemmys pulcherrima* (Central American wood turtle) | A N L | 6 |
| *Terrapene carolina* (eastern box turtle) | A N | 6 |
| *Rhinoclemmys annulata* (brown wood turtle) | A N L | 5 |
| *Rhinoclemmys areolata* (furrowed wood turtle) | A N | 5 |
| *Trachemys scripta* (common slider) | A | 5 |

[a]A = adult, N = nymph, L = larva

tina, Bolivia, Brazil, Colombia, French Guiana, Guyana, Paraguay, Peru, Suriname and Venezuela in South America, from Costa Rica, Guatemala, Honduras and Panama in Central America, from Mexico and from several Caribbean nations and territories (Antigua, Barbados, Cuba, Grenada, Guadeloupe, Jamaica, Martinique, St. Lucia and Trinidad & Tobago). In addition, *A. rotundatum* has been established in southern Florida since 1979, where it has been collected from a racer (*Coluber constrictor*) (Hanson et al., 2007).

### Life Cycle

*Amblyomma rotundatum* is a parthenogenetic species in which the female reproduces without involvement of the male (Oliver et al., 1993). The life cycle of *A. rotundatum* was studied in Brazil by Oba & Schumaker (1983). They reported a preoviposition period of 15–44 days, an egg incubation period of 39–68 days, a larval prefeeding period of 4 days, a larval feeding period of 27–38 days, a larval premolting period of 11–31 days, a nymphal prefeeding period of 4 days, a nymphal feeding period of 18–48 days, a nymphal premolting period of 10–28 days, a female prefeeding period of 4 days and a female feeding period of 3–17 days.

### Disease Associations

*Amblyomma rotundatum* is associated with two disease conditions, being a vector of *Hemolivia stellata* (Petit et al., 1990; Boulard et al., 2001), a hemogregarine parasite of giant toads, and causing tick paralysis in snakes (Hanson et al., 2007).

### References

Robinson (1926), Jones et al. (1972), Keirans & Oliver (1993), Amorim & Serra-Freire (1995), Burridge & Simmons (2003), Guglielmone et al. (2003a).

## 13.21. *Amblyomma sabanerae* Stoll

The tick *Amblyomma sabanerae* was introduced into the continental United States on six occasions between 1995 and 1998 on turtles, tortoises and an emerald tree boa imported into Florida and Indiana from Mexico and Central American countries such as Nicaragua (see Appendix 1). The chelonian hosts were Central American wood turtles, narrow-bridged mud turtles, yellow-footed tortoises and a red-footed tortoise (*Chelonoidis carbonaria*).

### Hosts

*Amblyomma sabanerae* is a tick of turtles, especially of the genus *Rhinoclemmys*, on which all stages feed (Table 13.21). Occasional hosts include iguanas, boas, opossums and birds.

The preferred site of attachment of *A. sabanerae* adults on turtles is the shell, whereas immatures almost always attach to the skin. Most adults are found on the carapace of the shell attached in pits dissolved between the scutes. Adults have also been found attached to the head and forelegs of tortoises.

### Distribution

The geographical distribution of *A. sabanerae* ranges from northern South America in the south to Mexico in the north. It has been reported from Colombia and Suriname in South America, from all countries in Central America (Belize, Costa Rica, El Salvador, Guatemala, Honduras, Nicaragua and Panama), and from Mexico.

Apart from its introduction into the United States, *A. sabanerae* has been introduced into Canada on a migrating bird. Scott et al. (2001) found two *A. sabanerae* nymphs on a veery that had migrated to Ontario, demonstrating that immature ticks can be disseminated outside the normal range of *A. sabanerae* on birds.

**Disease Associations**

None reported.

**References**

Robinson (1926), Burridge (2001), Guglielmone et al. (2003a).

## 13.22. *Amblyomma scutatum* Neumann

The tick *Ambloyomma scutatum* was introduced into the continental United States on five occasions between 1964 and 2005 on iguanas imported into Florida, Illinois and New York from Central American countries such as Honduras and Nicaragua (see Appendix 1). The hosts included smooth-helmeted iguanas and a black iguana. The tick found on an iguana in Illinois in 1964 was originally misidentified as *A. cyprium* (USDA, 1965b), but upon reexamination was found to be *A. scutatum* (Burridge & Simmons, 2003).

**Hosts**

*Amblyomma scutatum* is a tick of lizards and snakes (Table 13.22). Occasional hosts include opossums, bats, anteaters, agoutis and birds.

**Distribution**

The geographical distribution of *A. scutatum* extends from Brazil and Paraguay in the south to Mexico in the north. It has been reported from Brazil, Paraguay and Venezuela in South America, from Costa Rica, El Salvador, Guatemala, Honduras and Nicaragua in Central America, and from Mexico.

**Disease Associations**

None reported.

**References**

Robinson (1926), Burridge & Simmons (2003), Guglielmone et al. (2003a).

## 13.23. *Amblyomma varium* Koch

The tick *Amblyomma varium* was introduced into the continental United States on three occasions between 1966 and 1972 on sloths imported into California and Illinois from Central and South American countries such as Panama and Peru (see Appendix 1). The hosts included southern two-toed sloths (*Choloepus didactylus*) (Fig. 13.14). *Amblyomma varium* is a very large tick, with females weighing up to 8.6 g when engorged and laying up to 15,000 eggs (Marques et al., 2002); hence *A. varium* is commonly known in Brazil as the *carrapato-gigante-da-preguica* or "giant tick of sloths."

**Hosts**

*Amblyomma varium* is a tick of sloths, in particular pale-throated sloths and Hoffmann's two-toed sloths (*Choloepus hoffmanni*) (Table 13.23). Occasional hosts include dogs, iguanas, toads, peccaries and opossums.

The site of attachment on maned sloths (*Bradypus torquatus*) is underneath the thighs.

Table 13.22. Most commonly reported hosts of the tick *Amblyomma scutatum*

| Hosts | Stages[a] | No. of reports |
|---|---|---|
| **Lizards** | | |
| *Iguana iguana* (green iguana) | A | 8 |
| *Ameiva ameiva* (giant ameiva) | - | 5 |
| **Bats** | | |
| *Noctilio albiventris* (lesser bulldog bat) | N | 6 |
| **Snakes** | | |
| *Boa constrictor* (boa constrictor) | - | 5 |
| *Bothrops lanceolatus* (Martinique lancehead) | A | 5 |
| *Drymarchon corais* (southern indigo snake) | - | 5 |

[a]A = adult, N = nymph, - = not recorded

Fig. 13.14. A southern two-toed sloth (*Choloepus didactylus*), a host of the tick *Amblyomma varium*. (Photograph courtesy of iStockphoto.com)

Table 13.23. Most commonly reported hosts of the tick *Amblyomma varium*

| Hosts | Stages[a] | | No. of reports |
|---|---|---|---|
| **XENARTHRANS** | | | |
| *Bradypus tridactylus* (pale-throated sloth) | A | N | 20 |
| *Choloepus hoffmanni* (Hoffmann's two-toed sloth) | A | | 11 |
| *Bradypus variegatus* (brown-throated sloth) | A | | 7 |
| *Choloepus didactylus* (southern two-toed sloth) | A | | 5 |
| **CARNIVORES** | | | |
| *Canis familiaris* (dog) | - | | 4 |

[a]A = adult, N = nymph, - = not recorded

## Distribution

The geographical distribution of *A. varium* extends from Argentina in the south to Mexico in the north. It has been reported from Argentina, Brazil, Colombia, French Guiana, Guyana, Peru, Suriname and Venezuela in South America, from Costa Rica, Guatemala, Nicaragua and Panama in Central America, and from Mexico. Reports of *A. varium* in Chile are considered questionable since the sloth hosts of this tick do not occur in that country (Onofrio et al., 2008).

## Disease Associations

None reported.

## References

Robinson (1926), Floch & Fauran (1958), Guglielmone et al. (2003a).

## 13.24. *Dermacentor nitens* Neumann

The "tropical horse tick" *Dermacentor nitens* (syn. *Anocentor nitens*) was introduced into the continental United States on at least 46 occasions between 1941 and 1999 on horses, cattle, a donkey, a deer and an ocelot imported into Alabama, California, Florida, Illinois, Louisiana, Maryland, Mississippi, New Jersey, Tennessee and Texas from Mexico, Central American countries such as Belize, Costa Rica, Guatemala, Honduras and Nicaragua, South American countries such as Colombia, Ecuador, Peru and Venezuela, and Caribbean islands such as Cuba, Dominican Republic/Haiti, Jamaica and Puerto Rico (see Appendix 1).

## Hosts

*Dermacentor nitens* is an important one-host tick of equines (horses, donkeys and mules) (Table 13.24).

Table 13.24. Most commonly reported hosts of the tick *Dermacentor nitens*

| Hosts | Stages[a] | | | No. of reports |
|---|---|---|---|---|
| UNGULATES | | | | |
| *Equus caballus* (horse) | A | N | L | 63 |
| *Bos taurus* (cattle) | A | N | | 25 |
| *Equus asinus* (donkey) | A | N | L | 16 |
| *Equus asinus* × *Equus caballus* (mule) | A | N | L | 15 |
| *Odocoileus virginianus* (white-tailed deer) | A | N | | 11 |
| *Capra hircus* (goat) | - | | | 10 |

[a] A = adult, N = nymph, L = larva, - = not recorded

Other mammalian hosts include cattle, white-tailed deer, goats, sheep, dogs and humans. Occasional hosts are wild carnivores, snakes, rodents, rabbits and toads.

The preferred site of attachment on horses and mules is the ears for all stages, but in heavy infestations the ticks may also be found on the muzzle and face, in the nasal diverticulum, in the mane and the perineal region and along the ventral midline of the body. On cattle, all stages are found most frequently on the ears and neck.

## Distribution

The geographical distribution of *D. nitens* ranges from northern Argentina in the south to Mexico in the north. It has been reported from Argentina, Bolivia, Brazil, Colombia, Ecuador including the Galapagos Islands, Guyana, Paraguay, Peru, Suriname and Venezuela in South America, from Belize, Costa Rica, Guatemala, Honduras, Nicaragua and Panama in Central America, from Antigua, Bahamas, Barbados, British Virgin Islands, Cuba, Dominica, Dominican Republic, Grenada, Guadeloupe, Haiti, Jamaica, Martinique, Montserrat, Puerto Rico including Culebra, St. Kitts & Nevis, St. Maarten/St. Martin, St. Vincent, Trinidad & Tobago and the U.S. Virgin Islands in the Caribbean, and from Mexico.

Within the continental United States, *D. nitens* was found in the ears of horses in Texas as early as 1907 and identified in Florida in 1958 (USDA, 1963). By 1963, *D. nitens* was established in southern Texas and in southern Florida (Strickland & Gerrish, 1964). USDA reports for the years 1962 through 1989 indicate that *D. nitens* was present on equines in Florida and Texas throughout that period (Table 13.25). *Dermacentor nitens* is now considered eradicated from Florida,

Table 13.25. *Dermacentor nitens* infestations found on horses in Florida and Texas by the U.S. Department of Agriculture, 1958–1989

| Year | No. of equine infestations in | | Reference |
|---|---|---|---|
| | Florida | Texas | |
| 1958 | 1[a] | 0 | USDA (1963) |
| 1962 | 148 | 1 | USDA (1962a) |
| 1963 | 374 | 1 | USDA (1964a) |
| 1964 | 213 | 15 | USDA (1965a) |
| 1965 | 253 | 12 | USDA (1966a) |
| 1966 | 330 | 4 | USDA (1967a) |
| 1967 | 257 | 5 | USDA (1968a) |
| 1968 | 240 | 4 | USDA (1969) |
| 1969 | 207 | 5 | USDA (1970a) |
| 1970 | 110 | 9 | USDA (1971) |
| 1971 | 149 | 20 | USDA (1972) |
| 1972 | 79 | 14 | USDA (1973a) |
| 1973 | 125 | 3 | USDA (1974a) |
| 1974 | 100 | 0 | USDA (1975) |
| 1975 | 147 | 0 | USDA (1977a) |
| 1976 | 85 | 4 | USDA (1977b) |
| 1977 | 77 | 1 | USDA (1978) |
| 1978 | 49 | 2 | USDA (1980) |
| 1979 | 55 | 0 | USDA (1981) |
| 1980 | 38 | 0 | USDA (1982) |
| 1981 | 10 | 1 | USDA (1982) |
| 1982 | 12 | 4 | USDA (1983) |
| 1983 | 15 | 5 | USDA (1985a) |
| 1984 | 11 | 3 | USDA (1985b) |
| 1985 | 18 | 6 | USDA (1987a) |
| 1986 | 5 | 13 | USDA (1987b) |
| 1987 | 9 | 5 | USDA (1988) |
| 1988 | 2 | 4 | USDA (1994) |
| 1989 | 2 | 5 | USDA (1994) |

[a] First identification of *D. nitens* in Florida

with the last infestation found on a native horse in 1991 (USDA, unpublished data).

## Life Cycle

Being a one-host tick, the larval, nymphal and adult stages of *D. nitens* remain on the same host during their development, with the mean time required for larvae to develop into engorged adults being only 36 days while feeding on horses (Roby et al., 1964). The life cycle of *D. nitens* has been studied in Panama by Dunn (1915, 1923). He reported a preoviposition period of 5–7 days, an oviposition period of 8–16 days and an egg incubation period of 25–27 days. Following engorgement and detachment, females laid up to 2,401 eggs.

### Disease Associations

*Dermacentor nitens* is an important tick for the equine industry because it is:

- a vector of *Babesia caballi* (Roby & Anthony, 1963: Roby et al., 1964), a protozoan causing piroplasmosis in horses and other equines;
- a cause of damage to infested ears of horses, with the ears becoming crusty and greatly thickened, making them susceptible to secondary blowfly or screwworm infestation (Strickland & Gerrish, 1964).

### References

Dunn (1915), Bishopp & Trembley (1945), Floch & Fauran (1958, 1959), Arthur (1960), Jones et al. (1972), Amorim et al. (1997), Borges et al. (2000), Guglielmone et al. (2003a).

## 13.25. *Haemaphysalis juxtakochi* Cooley

The tick *Haemaphysalis juxtakochi* (Fig. 13.15) was introduced into the continental United States on one occasion in 1978 on a palm frond imported into Louisiana (USDA, 1980). In 1992 it was also found on a white-tailed deer in Ohio (Keirans & Restifo, 1993).

### Hosts

*Haemaphysalis juxtakochi* is a tick of ungulates such as South American tapirs and white-tailed deer (Table 13.26). Adults are found on ungulates, dogs and humans. Nymphs and larvae are found on the same hosts as adults, but also on rodents and birds. Occasional hosts include rabbits, armadillos, capuchin monkeys and bats.

### Distribution

The geographical distribution of *Ha. juxtakochi* ranges from Argentina in the south to Mexico in the north.

Fig. 13.15. A female *Haemaphysalis juxtakochi* tick. (Photograph courtesy of Santiago Nava.)

It has been reported from Argentina, Brazil, Colombia, Ecuador, French Guiana, Guyana, Paraguay, Peru, Suriname, Uruguay and Venezuela in South America, from Belize, Costa Rica and Panama in Central America, from Trinidad & Tobago in the Caribbean, and from Mexico.

### Disease Associations

None reported.

### References

Cooley (1946), Kohls (1960), Keirans & Restifo (1993), Guglielmone et al. (2003a).

## 13.26. *Ixodes luciae* Senevet

The tick *Ixodes luciae* was introduced into the continental United States on one occasion in 1965 on a greyish mouse-opossum (*Marmosa canescens*) imported into Texas from Mexico in a shipment of bananas (Eads et al., 1966).

### Hosts

*Ixodes luciae* is a tick of opossums, especially southern opossums (*Didelphis marsupialis*) and grey four-eyed opossums (*Philander opossum*) (Table 13.27). While all stages of the tick are found on opossums, larvae and nymphs are also found on rodents, especially rice rats. Occasional hosts include dogs, deer and humans.

Table 13.26. Most commonly reported hosts of the tick *Haemaphysalis juxtakochi*

| Hosts | Stages[a] | | | No. of reports |
|---|---|---|---|---|
| **PRIMATES** | | | | |
| *Homo sapiens* (human) | A | N | | 7 |
| **UNGULATES** | | | | |
| *Tapirus terrestris* (South American tapir) | A | | | 6 |
| *Odocoileus virginianus* (white-tailed deer) | A | | | 5 |
| *Mazama gouazoubira* (brown brocket deer) | A | N | L | 4 |
| **CARNIVORES** | | | | |
| *Canis familiaris* (dog) | A | N | | 6 |

[a] A = adult, N = nymph, L = larva

Table 13.27. Most commonly reported hosts of the tick *Ixodes luciae*

| Hosts | Stages[a] | | | No. of reports |
|---|---|---|---|---|
| **MARSUPIALS** | | | | |
| *Didelphis marsupialis* (southern opossum) | A | N | | 12 |
| *Philander opossum* (grey four-eyed opossum) | A | | | 6 |
| *Marmosa robinsoni* (Robinson's mouse-opossum) | | N | L | 3 |
| **RODENTS** | | | | |
| *Oryzomys megacephalus* (large-headed rice rat) | | N | L | 4 |
| *Oecomys concolor* (unicolored arboreal rice rat) | | N | L | 3 |
| *Oryzomys xantheolus* (yellowish rice rat) | A | | | 3 |

[a] A = adult, N = nymph, L = larva

## Distribution

The geographical distribution of *I. luciae* extends from Argentina in the south to southern Mexico in the north. It has been reported from Argentina, Bolivia, Brazil, Colombia, Ecuador, French Guiana, Peru, Suriname and Venezuela in South America, from Belize, Costa Rica, Guatemala, Honduras, Nicaragua and Panama in Central America, from Mexico and from Trinidad & Tobago in the Caribbean.

## Disease Associations

None reported.

## References

Floch & Fauran (1958), Guglielmone et al. (2003a).

# 14

# Ticks with Widespread Distributions

The two invasive tick species whose distribution extends to more than three continents are discussed in this chapter.

## 14.1. *Rhipicephalus (Boophilus) annulatus* (Say)

The "cattle tick" *Rhipicephalus (Boophilus) annulatus* was known as *Boophilus annulatus* prior to the taxonomic revision of ticks by Horak et al. (2002). This tick was introduced into the continental United States on at least 50 occasions between 1962 and 1989 on cattle, horses, deer and deer hides imported into Arizona, California, New York and Texas, with the vast majority of the introductions involving Texas (see Appendix 1). The origin of the infested animals was Mexico in the few instances where these data were recorded.

### Hosts

*Rhipicephalus (Boophilus) annulatus* is a very important one-host tick of cattle. Other common hosts are sheep, goats, horses and dogs (Table 14.1). Occasional hosts include lagomorphs, rodents, hedgehogs, birds and, rarely, humans.

The preferred sites of attachment are the legs, belly, neck and dewlap, with the back and shoulders parasitized in heavy infestations.

### Distribution

The geographical distribution of *R. (B.) annulatus* extends from Africa north of the equator into southern Europe and western Asia. In Africa, *R. (B.) annulatus* has been reported from northern Africa (Algeria, Egypt, Libya, Morocco and Tunisia), from western Africa (Benin, Burkina Faso, Ghana, Guinea, Guinea-Bissau, Ivory Coast, Liberia, Mali, Niger, Nigeria, Sierra Leone and Togo), from central Africa (Cameroon, Central African Republic, Chad, Congo and Democratic Republic of the Congo), from eastern Africa (Eritrea, Ethiopia and Sudan) and from the Madeira Islands off the coast of northwestern Africa. In Europe, it has been reported from Albania, Bulgaria, Greece, Italy, Portugal, Romania, Russia, Spain and the Mediterranean islands of Corsica, Cyprus, Sardinia, Sicily and the Balearic Islands. In Asia, it has been reported from Afghanistan, Armenia, Azerbaijan, Georgia, India, Iran, Iraq, Israel, Kazakhstan, Lebanon, Oman, Saudi Arabia, Syria, Tajikistan, Turkey, Turkmenistan, Uzbekistan, Yemen and the island of Socotra in the Arabian Sea.

Several centuries ago *R. (B.) annulatus* was introduced into northeastern Mexico by Spanish colonialists on their livestock where it is presently found. While it has not spread to Central America, it did spread north into the United States. By the end of the eighteenth century, populations of *R. (B.) annulatus* seem to have become established throughout the southern United States (George et al., 2002). At the beginning of the twentieth century, *R. (B.) annulatus* was widely distributed in the southern United States, being present in Alabama, Arkansas, California, Florida, Georgia, Kentucky, Louisiana, Mississippi, Missouri, North Carolina, Oklahoma, South Carolina, Ten-

Table 14.1. Most commonly reported hosts of the tick *Rhipicephalus (Boophilus) annulatus*

| Hosts | Stages[a] | | | No. of reports |
|---|---|---|---|---|
| UNGULATES | | | | |
| *Bos taurus* (cattle) | A | N | L | 109 |
| *Ovis aries* (sheep) | A | N | L | 31 |
| *Capra hircus* (goat) | A | N | L | 25 |
| *Equus caballus* (horse) | A | N | L | 23 |
| CARNIVORES | | | | |
| *Canis familiaris* (dog) | A | | | 14 |

[a] A = adult, N = nymph, L = larva

nessee, Texas and Virginia (Cooley, 1946). It was the vector of bovine babesiosis (or Texas cattle fever), the most important impediment to the development of a strong cattle industry in the southern United States. Consequently, a cooperative state-federal eradication program was initiated in 1907 and successfully completed in 1960 (Graham & Hourrigan, 1977). Initially, the eradication program targeted only *R. (B.) annulatus*, but soon included *R. (B.) microplus* after it was discovered in Florida in 1912. During the 53-year-long eradication program these two ticks were eliminated from an area of more than 1.8 million km$^2$, primarily by dipping cattle and other livestock in an arsenical solution every 14–18 days. Bovine babesiosis disappeared from the United States after these two tick vectors were eradicated, with resultant savings to the livestock industry of well over $1 billion a year.

Despite the successful eradication of *R. (B.) annulatus* from the southern United States, this tick still remains a constant threat to the U.S. cattle industry because of its presence in Mexico. Reintroductions of *R. (B.) annulatus* occur regularly in the southern Texas counties where the United States is separated from Mexico by the Rio Grande River (George, 1990). From time to time the limited flow of water in the Rio Grande River makes it easy for tick-infested domestic livestock to stray or to be brought across from Mexico. Furthermore, large populations of wildlife, including white-tailed deer and exotic animals such as nilgai, are abundant in southern Texas and northern Mexico and can transport ticks across the river. This area of southern Texas, which is adjacent to the parts of northern Mexico populated with *R. (B.) annulatus* and *R. (B.) microplus*, is the most likely to become infested with these invasive ticks and has been declared a quarantine or "buffer" zone. It extends from the border community of Del Rio southeast for about 800 km to the Gulf of Mexico. USDA reports for the years 1962 through 1989 show that *R. (B.) annulatus* was found each year on native cattle in Texas, with infestations some years also on equines and wildlife such as deer, nilgais, coyotes and wapitis (*Cervus elaphus*) (Table 14.2).

### Life Cycle

The life cycle of *R. (B.) annulatus* can be completed in two months, allowing for six generations in a year in conditions of high temperature and humidity.

### Disease Associations

*Rhipicephalus (Boophilus) annulatus* is a very important tick to bovine health because it is:

- a major vector of both *Babesia bigemina* (Neitz, 1956) and *Babesia bovis* (Estrada-Pena et al., 2006), two protozoa causing babesiosis in cattle;
- a vector of *Borrelia theileri* (Trees, 1978), the bacterium causing borreliosis in cattle;
- a cause of damage to the hides of cattle with heavy infestations (Walker et al., 2003);
- a major vector of *Anaplasma marginale* (Estrada-Pena et al., 2006), a rickettsia causing anaplasmosis in cattle.

### References

Cooley (1946), Hoogstraal (1956), Arthur (1960), Elbl & Anastos (1966c), Walker et al. (2003), Estrada-Pena et al. (2004a).

Table 14.2. *Rhipicephalus (Boophilus) annulatus* infestations found on native animals in Texas by the U.S. Department of Agriculture, 1962–1989

| Year | No. of infestations by species | | | Reference |
|---|---|---|---|---|
| | cattle | horses & mules | other species | |
| 1962 | 20 | 4 | 0 | USDA (1962a) |
| 1964 | 10 | 0 | 0 | USDA (1965a) |
| 1965 | 13 | 2 | 0 | USDA (1966a) |
| 1966 | 12 | 3 | 0 | USDA (1967a) |
| 1967 | 12 | 3 | 0 | USDA (1968a) |
| 1968 | 27 | 2 | 2 deer | USDA (1969) |
| 1969 | 36 | 6 | 0 | USDA (1971) |
| 1971 | 26 | 28 | 0 | USDA (1972) |
| 1972 | 25 | 6 | 0 | USDA (1973a) |
| 1973 | 40 | 2 | 0 | USDA (1974a) |
| 1974 | 16 | 1 | 0 | USDA (1975) |
| 1975 | 51 | 1 | 0 | USDA (1977a) |
| 1976 | 50 | 0 | 16 deer | USDA (1977b) |
| 1977 | 26 | 0 | 8 deer | USDA (1978) |
| 1978 | 39 | 2 | 7 deer<br>1 nilgai | USDA (1980) |
| 1979 | 40 | 0 | 61 deer<br>1 coyote | USDA (1981) |
| 1980 | 37 | 0 | 24 deer<br>1 nilgai | USDA (1981) |
| 1981 | 21 | 0 | 7 deer | USDA (1982) |
| 1982 | 18 | 0 | 0 | USDA (1983) |
| 1983 | 11 | 2 | 2 deer | USDA (1985a) |
| 1984 | 9 | 6 | 0 | USDA (1985b) |
| 1985 | 12 | 0 | 0 | USDA (1987a) |
| 1986 | 15 | 0 | 0 | USDA (1987b) |
| 1987 | 5 | 0 | 0 | USDA (1988) |
| 1988 | 12 | 2 | 3 deer | USDA (1994) |
| 1989 | 35 | 0 | 7 deer<br>1 wapiti | USDA (1994) |

## 14.2. *Rhipicephalus (Boophilus) microplus* (Canestrini)

The "southern cattle tick" *Rhipicephalus (Boophilus) microplus* was known as *Boophilus microplus* prior to the taxonomic revision of ticks by Horak et al. (2002). Recent studies by Labruna et al. (2009) provide evidence that *R. (B.) microplus* may in fact be two different species, with populations from Australia not being conspecific with populations from Africa and the Americas.

*Rhipicephalus (Boophilus) microplus* was introduced into the continental United States on at least 44 occasions between 1962 and 1989 on cattle (including beef hindquarters and refrigerated beef), horses, a southern two-toed sloth, cattle and deer hides, hair and a plant imported into Arizona, Florida, Illinois, New York and in particular Texas from Mexico, Cuba, Brazil and Central American countries such as Belize, Costa Rica, Guatemala and Honduras (see Appendix 1).

### Hosts

*Rhipicephalus (Boophilus) microplus* is a one-host tick of cattle and is considered the most important parasite of livestock in the world (Estrada-Pena et al., 2006). Other common hosts are goats, horses, sheep, water buffalos, dogs, white-tailed deer and pigs (Table 14.3). Several human infestations have also been reported. Occasional hosts include wild carnivores, rodents, lagomorphs, sloths, nonhuman primates, shrews, wallabies, kangaroos, birds, reptiles and toads.

Attachment sites on cattle include the escutcheon, tail butt, shoulders, dewlap and ears in light infestations, but in heavy infestations the ticks can be found anywhere on the body. Adults have been found

Table 14.3. Most commonly reported hosts of the tick *Rhipicephalus (Boophilus) microplus*

| Hosts | Stages[a] | | | No. of reports |
|---|---|---|---|---|
| **UNGULATES** | | | | |
| *Bos taurus* (cattle) | A | N | L | 197 |
| *Capra hircus* (goat) | A | N | L | 57 |
| *Equus caballus* (horse) | A | N | | 52 |
| *Ovis aries* (sheep) | A | N | | 41 |
| *Bubalus bubalis* (water buffalo/carabao) | A | N | L | 39 |
| *Odocoileus virginianus* (white-tailed deer) | - | | | 20 |
| *Sus scrofa* (domestic & wild pigs) | A | | | 12 |
| **CARNIVORES** | | | | |
| *Canis familiaris* (dog) | A | N | L | 40 |
| **PRIMATES** | | | | |
| *Homo sapiens* (human) | A | | L | 21 |

[a]A = adult, N = nymph, L = larva, - = not recorded

attached to the ears of humans and to the head of snakes.

## Distribution

The geographical distribution of *R. (B.) microplus* covers much of the tropics including Australasia and parts of the Pacific and Asia, Mexico and parts of Central and South America and the Caribbean, and several countries in Africa.

In Australasia, *R. (B.) microplus* is common in Australia and Papua-New Guinea. In the Pacific, it is common in New Caledonia and has been reported from Guam, Northern Mariana Islands, Palau and Solomon Islands. In Asia, it occurs from Afghanistan eastward and has been reported from Afghanistan, Bangladesh, Cambodia, China, India, Indonesia, Japan, Laos, Malaysia, Myanmar, Nepal, Pakistan, Philippines, Singapore, South Korea, Taiwan, Thailand, Timor-Leste (formerly East Timor), and Vietnam. It is common in Mexico and has been reported from Belize, Costa Rica, Guatemala, Honduras and Panama in Central America. In the Caribbean, it has been reported from Antigua & Barbuda, Bahamas, Barbados, Cuba, Curacao, Dominica, Dominican Republic, Grenada, Guadeloupe, Haiti, Jamaica, Martinique, Montserrat, Puerto Rico, Saba, St. Eustatius, St. Kitts & Nevis, St. Lucia, St. Martin, St. Vincent & the Grenadines, Trinidad & Tobago, and U.S. Virgin Islands. In South America, it has been reported from all countries except Chile, namely Argentina, Bolivia, Brazil, Colombia, Ecuador, French Guiana, Guyana, Paraguay, Peru, Suriname, Uruguay and Venezuela. In Africa, the distribution of *R. (B.) microplus* is more restricted; it is found from the southern coastal strip of Kenya south to the southern coastal region of South Africa. It has been reported in Africa from eastern Africa (Kenya and Tanzania), from southern Africa (Malawi, Mozambique, South Africa, Swaziland, Zambia and Zimbabwe) and from the Indian Ocean islands of Comoros, Madagascar, Mauritius, Pemba Island, Reunion, Rodriguez, Seychelles and Zanzibar. Recently an infestation of *R. (B.) microplus* was confirmed in western Africa on cattle in the Ivory Coast (Madder et al., 2007).

As mentioned in section 14.1 of this chapter, both *R. (B.) microplus* and *R. (B.) annulatus* are present in northern Mexico and remain constant threats to the U.S. cattle industry. USDA reports for the years 1962 through 1989 show that *R. (B.) microplus* was found each year on native cattle in Texas, with infestations some years also on equines and wildlife such as deer (Table 14.4). As recently as July 2008, the preventive cattle fever tick quarantine area in south Texas was enlarged to cover over 1 million acres (c. 405,000 hectares) in Dimmit, Jim Hogg, Maverick, Starr, Webb and Zapata counties in addition to the 500,000 acres (c. 202,500 hectares) in the permanent cattle fever tick quarantine zone that runs alongside the Rio Grande from Del Rio to Brownsville. During the 12-month period prior to July 2008, *R. (B.) microplus* and *R. (B.) annulatus* were detected on livestock or wildlife on 139 Texas premises, demonstrating the need for continuous vigilance to prevent these ticks from becoming re-established in the continental United States.

Table 14.4. *Rhipicephalus (Boophilus) microplus* infestations found on native animals in Texas by the U.S. Department of Agriculture, 1962–1989

| Year | No. of infestations by species | | Reference |
|---|---|---|---|
| | cattle | other species | |
| 1962 | 12 | 2 equines[a] | USDA (1962a) |
| 1964 | 8 | 0 | USDA (1965a) |
| 1965 | 9 | 0 | USDA (1966a) |
| 1966 | 10 | 0 | USDA (1967a) |
| 1967 | 4 | 2 equines | USDA (1968a) |
| 1968 | 12 | 0 | USDA (1969) |
| 1969 | 9 | 1 equine | USDA (1970a) |
| 1970 | 10 | 0 | USDA (1971) |
| 1971 | 99 | 5 equines | USDA (1972) |
| 1972 | 152 | 2 equines 2 deer | USDA (1973a) |
| 1973 | 46 | 1 equine | USDA (1974a) |
| 1974 | 52 | 1 equine | USDA (1975) |
| 1975 | 57 | 0 | USDA (1977a) |
| 1976 | 162 | 12 deer | USDA (1977b) |
| 1977 | 23 | 2 deer | USDA (1978) |
| 1978 | 11 | 1 equine | USDA (1980) |
| 1979 | 13 | 0 | USDA (1981) |
| 1980 | 17 | 0 | USDA (1982) |
| 1981 | 20 | 0 | USDA (1982) |
| 1982 | 39 | 1 equine | USDA (1983) |
| 1983 | 11 | 1 equine | USDA (1985a) |
| 1984 | 7 | 1 equine | USDA (1985b) |
| 1985 | 20 | 0 | USDA (1987a) |
| 1986 | 36 | 2 equines | USDA (1987b) |
| 1987 | 12 | 1 equine 1 deer | USDA (1988) |
| 1988 | 15 | 0 | USDA (1994) |
| 1989 | 9 | 0 | USDA (1994) |

[a]Horses or mules

## Habitat and Life Cycle

*Rhipicephalus (B.) microplus* occurs in savanna climatic regions where there are habitats of wooded grassland used as cattle pasture (Walker et al., 2003). As a one-host tick, all three stages spend a total of only three weeks on the same host, with laying of up to 3,000 eggs by the females completed in four weeks, allowing the life cycle to be completed in seven weeks. This high reproductive potential has enabled *R. (B.) microplus* to compete successfully with *R. (B.) decoloratus* in Africa where they occur together in climates favorable to *R. (B.) microplus*.

## Disease Associations

*Rhipicephalus (B.) microplus* is a very important tick to animal health because it is:

- a major vector of *Babesia bigemina* (Callow & Hoyte, 1961) and *Babesia bovis* (Riek, 1966), two protozoa causing babesiosis in cattle;
- a major vector of *Anaplasma marginale* (Connell & Hall, 1972), a rickettsia causing anaplasmosis in cattle;
- a vector of *Theileria equi* (de Waal & van Heerden, 2004), a protozoan causing piroplasmosis in horses;
- a vector of *Borrelia theileri* (Norval & Horak, 2004), the bacterium causing borreliosis in cattle;
- a cause of anemia and decreased productivity in cattle (Jonsson, 2006), with heavy infestations producing damage to hides by formation of scars at the feeding sites (Walker et al., 2003); experiments in Australia have shown that, for each female tick that completes feeding, there is a loss of 0.6 g of potential growth by cattle (Walker et al., 2003).

A recent incident in New Caledonia graphically demonstrates the risks of introductions of diseases when ticks such as *R. (B.) microplus* are established in a territory. In November 2007, a total of 43 bulls were exported from Australia to the nearby French Pacific territory of New Caledonia. While New Caledonia is infested with *R. (B.) microplus*, it is free of *Babesia* parasites. Unfortunately, the Australian bulls had been vaccinated with a live attenuated vaccine containing *Babesia bovis* organisms prior to shipment in contravention of Australian law, apparently as a result of a paperwork error. The imported bulls infected *R. (B.) microplus* ticks in New Caledonia with babesiosis which then passed the disease on to local cattle, causing four farm outbreaks of babesiosis with 11 deaths by March 2008. The Australian government immediately agreed to treat all cattle in the affected areas of New Caledonia with imidocarb to kill the *Babesia* parasites, and the New Caledonian government closed all pastures in the affected areas to grazing for up to 10 months.

## References

Cooley (1946), Hoogstraal (1956), Floch & Fauran (1958, 1959), Arthur (1960), Yeoman et al. (1967), Roberts (1970), Uilenberg et al. (1979), Guglielmone et al. (2003a), Walker et al. (2003).

# 15

# Tickborne Diseases

Only those tickborne diseases associated with the invasive tick species discussed in this book will be reviewed in this chapter. The diseases are arranged alphabetically for ease of reference.

## 15.1. Anaplasmosis

There are several forms of anaplasmosis caused by different species of the rickettsial bacterium *Anaplasma*; the forms discussed below are caused by *A. marginale*, *A. centrale*, *A. bovis*, *A. ovis* and *A. phagocytophilum*.

### 15.1.1. *Anaplasma marginale* infection

Bovine anaplasmosis caused by *A. marginale* (or gallsickness) is the classical pathogenic form of the disease found in cattle throughout much of the tropical and subtropical regions of the world, including the United States, Central and South America, southern Europe, Africa, Asia and Australia. *Anaplasma marginale* is an obligate intraerythrocytic rickettsia appearing as dense inclusions usually located toward the margin of the infected erythrocyte. The principal biological vectors of *A. marginale* are ixodid ticks such as *Dermacentor albipictus*, *D. andersoni*, *D. variabilis*, *Hyalomma detritum*, *Hy. lusitanicum*, *Hy. rufipes*, *Rhipicephalus bursa*, *R. simus*, *R. (B.) annulatus* and *R. (B.) microplus*. Transmission can also occur mechanically via biting flies or contaminated needles or equipment such as that used for dehorning.

*Anaplasma marginale* is pathogenic for cattle, but it usually produces only inapparent or mild infections in other ruminants such as sheep, goats, camels, water buffalos, white-tailed deer, mule deer (*Odocoileus hemionus*), black-tailed deer (*Odocoileus hemionus*), pronghorns (*Antilocapra americana*), elk (*Cervus elaphus*), bighorn sheep (*Ovis canadensis*), black wildebeests, blesboks and grey duikers. In cattle less than one year of age anaplasmosis is usually subclinical, in those 1–2 years old it is moderately severe and in older cattle it is severe and can be fatal. Anaplasmosis is characterized by progressive anemia due to extravascular destruction of infected and uninfected erythrocytes, with usually 10–30% of erythrocytes infected at peak rickettsemia. Acutely infected cattle lose condition rapidly, with reduced milk production, anorexia and loss of coordination. A transient febrile response occurs at about the time of peak rickettsemia. Mucous membranes appear pale, and pregnant cows may abort.

Clinical anaplasmosis can be treated with tetracyclines and imidocarb. Vaccination with a live *A. centrale* vaccine will provide cattle with partial protection against anaplasmosis caused by *A. marginale*, and this method is used to protect cattle in South Africa, Israel, South America and Australia. Tick and fly control are also helpful preventive measures.

Anaplasmosis caused by *A. marginale* has been reviewed by Kocan et al. (2003) and Potgieter & Stoltsz (2004).

### 15.1.2. *Anaplasma centrale* infection

In contrast to *A. marginale*, *A. centrale* produces only a mild disease in cattle, with subclinical infections found in Africa in African buffalos and blesboks. Recently, *A. centrale* infections were found in cattle in southern Italy (Carelli et al., 2008). *Anaplasma cen-*

*trale* is also an obligate intraerythrocytic rickettsia appearing as dense inclusions, but, unlike *A. marginale*, the inclusions are located more centrally within the infected erythrocyte. It has been transmitted in southern Africa by the tick *R. simus*. In the early twentieth century it was found that cattle recovered from *A. centrale* infection developed an incomplete cross-immunity to *A. marginale* infection. As a result, infection with a live *A. centrale* strain is used as a vaccine to provide cattle with partial protection against the more virulent *A. marginale*.

### 15.1.3. *Anaplasma bovis* infection

Bovine anaplasmosis caused by *A. bovis* (formerly known as *Ehrlichia bovis*) is a disease typically of low pathogenicity to cattle and reported from various African countries (Chad, Congo, Niger, South Africa, Sudan and Zimbabwe). *Anaplasma bovis* is a rickettsia appearing as compact inclusions in a tightly packed morula within the cytoplasm of monocytes. It has been transmitted by the ticks *Amblyomma variegatum*, *Hy. anatolicum*, *Hy. excavatum* and *R. appendiculatus*. Infection with *A. bovis* can cause peracute, acute and chronic forms of anaplasmosis in cattle. The peracute form is rare and results in sudden death. The acute form is characterized by head shaking with drooping of an ear and enlargement of lymph nodes, followed by circling, recumbency and death. The chronic form is the most common and is subclinical.

Bovine anaplasmosis caused by *A. bovis* has been reviewed by Stewart (1992).

### 15.1.4. *Anaplasma ovis* infection

Ovine and caprine anaplasmosis caused by *A. ovis* is a disease typically of low pathogenicity to sheep and goats and has been reported from most tropical and subtropical parts of the world, including countries in Africa, North and South America, Asia, southern and central Europe, and the former USSR. *Anaplasma ovis* is an intraerythrocytic rickettsia morphologically indistinguishable from *A. marginale*, with most inclusions located marginally within the infected erythrocyte. It is transmitted by ticks, including *Dermacentor nuttalli*, *Hy. asiaticum*, *R. bursa* and *R. pumilio*.

Infection with *A. ovis* is more pathogenic for goats than for sheep; consequently, clinical signs of infection are more frequently observed in goats. Generally only slight listlessness, weakness or loss of condition is noticeable in infected animals, but in more severe cases anemia due to erythrophagocytosis is evident, with signs of fatigue and severe respiratory distress seen on exertion. Higher morbidity and mortality rates of 40–50% and 17%, respectively, have been reported from China. Some wild ruminants such as blesboks, elands and white-tailed deer are susceptible to infection with *A. ovis*, developing only mild or subclinical disease.

Ovine and caprine anaplasmosis have been reviewed by Stoltsz (2004).

### 15.1.5. *Anaplasma phagocytophilum* infection

Anaplasmosis caused by *A. phagocytophilum* (formerly *Ehrlichia phagocytophila*) is a disease recorded in sheep, goats, cattle, horses, dogs, cats, roe deer, reindeer (*Rangifer tarandus*) and humans in Europe, the United States (northern California, the upper midwest and the northeast) and Canada. In Europe, *A. phagocytophilum* infects livestock, rodents and humans, with some species such as cattle experiencing severe disease, whereas in the United States cattle infection is rare. *Anaplasma phagocytophilum* is a rickettsia that infects mainly neutrophils as intracytoplasmic inclusions. It is transmitted by ticks of the genus *Ixodes*: *I. ricinus* in Europe, *I. scapularis* in the eastern United States and *I. pacificus* in the western United States.

The disease is known as tickborne fever in sheep and pasture fever in cattle. Infection of sheep can cause high fever, severe neutropenia and reduced weight gain. Disease in cattle occurs as an annual event in Europe when dairy cows are turned out from winter housing onto tick-infested pastures in the spring or early summer. The cows become depressed, with marked loss of appetite and milk yield, and often exhibit respiratory distress. Complications of infection in both sheep and cattle include abortions and impaired spermatogenesis. Perhaps the most significant effect of *A. phagocytophilum* infection in sheep and cattle is impairment of humoral and cellular immunity, resulting in increased susceptibility to secondary infections such as tick pyemia, pneumonic pasteurellosis, louping ill and listeriosis. Infection in dogs can manifest as acute lethargy, anorexia and fever. Infection in horses produces lethargy and fever, sometimes with ataxia and distal limb edema.

The disease is known as granulocytic anaplasmosis in humans. Clinical manifestations in humans range from a mild self-limiting flulike illness to a life-threatening infection. Most patients have a fever with headache, myalgia and malaise, and with leukopenia and thrombocytopenia present. Although the majority of human infections are either mild or subclinical, some can be severe, with prolonged fever, shock, confusion, seizures, pneumonitis, renal failure and death.

Anaplasmosis caused by *A. phagocytophilum* has been reviewed by Dumler & Bakken (1998) and Dumler et al. (2005) for human infections and by Stuen (2007) for animal infections.

## 15.2. Babesiosis

The babesioses are tickborne diseases caused by intraerythrocytic protozoal parasites of the genus *Babesia*. The diseases affect a wide range of domestic and wild animals and occasionally humans. The diseases discussed below are bovine babesiosis (or redwater) caused by *B. bigemina*, *B. bovis*, *B. divergens* and *B. major*, caprine and ovine babesiosis caused by *B. motasi* and *B. ovis*, porcine babesiosis caused by *B. trautmanni*, canine babesiosis caused by *B. canis*, *B. gibsoni* and *B. rossi*, and human babesiosis caused by *B. divergens* and *B. microti*. Equine babesiosis caused by *B. caballi* is more commonly referred to as equine piroplasmosis and thus will be discussed under piroplasmosis.

Economic losses from bovine babesiosis result not only from mortality, poor condition, abortions, loss of milk or meat production and draft power, and from control measures such as vaccination, therapeutics and acaricidal treatment, but also through its impact on international trade in cattle. In Africa *B. bigemina* is the parasite most often associated with bovine babesiosis, whereas in Australia *B. bovis* is by far the more economically important cause of bovine babesiosis.

### 15.2.1. *Babesia bigemina* infection

Bovine babesiosis caused by *B. bigemina* is a very economically important tickborne disease of cattle throughout much of the tropical and subtropical world, including Africa, Asia, Australia, Central and South America, Mexico and southern Europe. The principal vectors of *B. bigemina* are the ticks *R. (B.) annulatus*, *R. (B.) decoloratus*, *R. (B.) geigyi*, *R. (B.) microplus* and *R. e. evertsi*. Transmission by *Rhipicephalus (Boophilus)* ticks is transovarial, with engorging adult ticks ingesting the *B. bigemina* parasites and the nymphal and adult stages of the next generation transmitting the infection to cattle. In addition, wandering male *Rhipicephalus (Boophilus)* ticks are known to migrate to uninfested cattle and may play a role in transmission of *B. bigemina*. Transmission by *R. e. evertsi* is transovarial, with only the nymphal stage infecting the bovine host.

*Babesia bigemina* is highly pathogenic for cattle. In addition, African buffalos and water buffalos can develop latent infections, and *R. e. evertsi* infected with *B. bigemina* have been collected from a sable antelope with clinical babesiosis. In cattle, calves less than two months old born to previously unexposed cows are susceptible to *B. bigemina* infection, while offspring of immune dams are resistant to infection due to the passive transfer of immunity via colostrum. After the age of two months, a natural nonspecific innate immunity protects calves for a further 4–6 months which is not dependent on the immune status of the cow. The disease in older cattle can develop very rapidly, with a sudden and severe hemolytic anemia, jaundice and death. Hemoglobinuria is present. Recovery in nonfatal cases is usually rapid and complete. Cattle recovered from *B. bigemina* infection show some resistance to *B. bovis* challenge, but the reverse is not true.

Cattle develop a durable immunity after a single infection with *B. bigemina*; consequently, live attenuated vaccines containing bovine erythrocytes infected with selected strains are used to immunize cattle against *B. bigemina* infection. Data from Australia on the efficacy, degree and duration of immunity provided by these live vaccines have demonstrated that they provide greater than 95% protection for the life of the animal (Bock et al., 2004).

Bovine babesiosis caused by *B. bigemina* has been reviewed by Bock et al. (2004) and de Vos et al. (2004).

### 15.2.2. *Babesia bovis* infection

Bovine babesiosis caused by *B. bovis*, formerly known as *B. argentina*, is another very economically important tickborne disease of cattle throughout much of the tropical and subtropical world, including Africa,

Asia, Australia, Central and South America, Mexico and southern Europe. It is less widespread in Africa than *B. bigemina*. The principal vectors of *B. bovis* are the ticks *R. (B.) annulatus*, *R. (B.) geigyi* and *R. (B.) microplus*. Transmission is transovarial, with engorging adult ticks ingesting the *B. bovis* parasites and the larval stages of the next generation transmitting the infection to cattle.

*Babesia bovis*, like *B. bigemina*, is highly pathogenic for cattle. Resistance to infection in calves less than nine months of age is similar to that for *B. bigemina*. In acute infections in older cattle, a fever is usually present for several days, followed by anorexia, depression, weakness and a reluctance to move. Hemoglobinuria is often present, and hemolytic anemia and jaundice develop, especially in protracted cases. Diarrhea is common, and pregnant cows may abort. In nonfatal cases recovery may take several weeks but is usually complete. Unlike with *B. bigemina* infection, cerebral babesiosis develops in some *B. bovis* infections, manifested by hyperesthesia, nystagmus, circling, head pressing, aggression, convulsions and paralysis. The course of cerebral babesiosis is usually short and the outcome almost invariably fatal.

Cattle develop a durable immunity after a single infection with *B. bovis;* consequently, live attenuated vaccines containing bovine erythrocytes infected with selected strains are used to immunize cattle against *B. bovis* infection. As with *B. bigemina*, data from Australia have demonstrated that these live *B. bovis* vaccines provide greater than 95% protection for the life of the animal.

Bovine babesiosis caused by *B. bovis* has been reviewed by Bock et al. (2004) and de Vos et al. (2004).

### 15.2.3. *Babesia divergens* infection

Babesiosis caused by *B. divergens* is a tickborne disease of cattle and humans in Europe. Clinical disease has been produced experimentally in reindeer. It is transmitted by the tick *Ixodes ricinus* both transstadially and transovarially.

Acute disease in cattle is manifested by fever, anemia, anorexia, depression, weakness and pipestem diarrhea due to spasms of the anal sphincter. As the parasitemia rises, extensive erythrocyte destruction occurs, with hemoglobinuria evident. Case fatality rates are influenced by the speed of diagnosis and treatment, and in Ireland they have been estimated at 10%. Imidocarb is an effective babesiacide for *B. divergens* that has also proved to be an effective prophylactic at twice the therapeutic dose.

Calves up to one year of age are fully susceptible to *B. divergens* infection but are resistant to disease, whereas adult cattle are more likely to respond to infection with clinical symptoms. This inverse age resistance is due to innate resistance in calves and is independent of the maternal immune status.

Human infection with *B. divergens* is rare but invariably fatal unless treated immediately. All cases have occurred in splenectomized individuals. Acute illness appears suddenly, with hemoglobinuria as the initial symptom, followed by jaundice due to severe hemolysis, a persistent high fever, shaking chills, intense sweats, headaches, myalgia, and lumbar and abdominal pain. Most patients develop shocklike symptoms with renal failure induced by intravascular hemolysis and pulmonary edema. Fast and aggressive treatment using massive blood exchange transfusion followed by intravenous clindamycin and oral quinine has reduced the mortality rate from *B. divergens* infection to 40%.

Babesiosis caused by *B. divergens* has been reviewed by Zintl et al. (2003).

### 15.2.4. *Babesia major* infection

Babesiosis caused by *B. major* is a tickborne disease of cattle in Europe. Natural clinical disease has also been observed in American bison (*Bison bison*) maintained in England. *Babesia major* is transmitted by the tick *Haemaphysalis punctata*.

*Babesia major* is, like *B. divergens*, a temperate-zone species found in Europe, but it is of lower pathogenicity to cattle. Clinical disease is manifested by anorexia, depression, respiratory distress and hemoglobinuria, as seen in the redwater outbreak in the American bison herd in England in which several animals died.

### 15.2.5. *Babesia motasi* infection

Babesiosis caused by *B. motasi* affects all breeds of sheep and, more rarely, goats. It appears that some isolates of *B. motasi* are infective for both sheep and goats, while others are infective only for sheep. *Babesia motasi* causes severe disease and mortality in sheep in Asia and Africa, whereas it is regarded as a benign parasite in western Europe, occasionally causing anemia in lambs. It is transmitted by *Haemaphysalis punctata* in Europe, by *Ha. bispinosa* and *Ha. intermedia* in India, and by *R. bursa*. Transmission by *Ha. punctata* is both transstadial and transovarial.

Clinical signs of *B. motasi* infection include fever, followed by anorexia and depression. As the disease progresses and anemia develops, visible mucous membranes become pale and jaundiced, but hemoglobinuria is uncommon. Clinical babesiosis can be treated with babesiacidal drugs such as diminazene and imidocarb. No ovine or caprine babesiosis vaccine is available commercially.

Ovine babesiosis caused by *B. motasi* has been reviewed by Yeruham & Hadani (2004).

### 15.2.6. *Babesia ovis* infection

*Babesia ovis* infects sheep, goats, mouflons, and Asian wild sheep (*Ovis ammon*), but in the field it causes disease primarily in domestic sheep, where it is of considerable economic importance in the Mediterranean basin, the Balkans and western Asia. It is transmitted by the tick *Rhipicephalus bursa* both transstadially and transovarially.

*Babesia ovis* is highly pathogenic for sheep, with mortality rates of up to 50% reported in susceptible animals. Clinical signs include fever, jaundice, hemoglobinuria and severe anemia, with histopathological findings in lambs that died of acute babesiosis indicative of liver and kidney damage. As with *B. motasi* infection, clinical babesiosis caused by *B. ovis* can be treated with drugs such as diminazene and imidocarb. No vaccine is available commercially.

Ovine babesiosis caused by *B. ovis* has been reviewed by Friedhoff (1997), Yeruham et al. (1998) and Yeruham & Hadani (2004).

### 15.2.7. *Babesia trautmanni* infection

Porcine babesiosis is a tickborne disease of pigs in Africa and Europe caused by *B. trautmanni*. A vector of the disease is the tick *Rhipicephalus simus*. Clinical signs are milder in younger pigs than in adults and include fever, listlessness, anorexia and anemia. Jaundice and hemoglobinuria may appear during the later stages of the disease. Pregnant sows may abort. The disease can be successfully treated with a variety of drugs, including diminazene, phenamidine and trypan blue. No vaccine is currently available.

Porcine babesiosis has been reviewed by de Waal (2004).

### 15.2.8. *Babesia canis* and *Babesia rossi* infections

The large *Babesia* species of dogs, formerly known as *B. canis*, has been divided into three new species, *B. canis*, *B. vogeli* and *B. rossi* (Schetters, 2005). The pathogenicity of the *B. canis* group for dogs differs by species: *B. rossi* is highly pathogenic, causing severe clinical disease in South Africa; *B. canis* is moderately pathogenic in Europe; and *B. vogeli* causes relatively mild disease worldwide. Also, the tick vectors of the *B. canis* group differ by species: *B. canis* is transmitted in tropical and subtropical regions by *Dermacentor marginatus*, *D. reticulatus* and *Rhipicephalus sanguineus*, *B. vogeli* in tropical and subtropical regions by *R. sanguineus*, and *B. rossi* in southern Africa by *Haemaphysalis leachi*.

Clinically *B. canis* infection is characteristic of a progressive hemolytic anemia, with infected dogs showing sudden fever, anorexia, depression, anemia and bilirubinuria. In contrast, *B. rossi* infection is more severe, involving hypoxic hypotensive shock with disseminated intravascular coagulation, systemic inflammatory response and multiple organ dysfunction. Some cases of *B. rossi* infection are so severe they present as clinical collapse, with the dog unable to walk unassisted.

Canine babesiosis can be treated using the drug imidocarb. A canine babesiosis vaccine is available in Europe based on exoantigens of *B. canis* obtained from cell culture. The vaccine does not prevent infection, but it does reduce parasitemia and limits the severity of anemia and splenomegaly; however, the vaccine will protect only against homologous strains of *B. canis*.

### 15.2.9. *Babesia gibsoni* infection

*Babesia gibsoni* causes babesiosis in domestic and wild canids in Asia, Africa, Europe, the Middle East and North America. It is transmitted by the ticks *Haemaphysalis bispinosa* and *Ha. longicornis* in Asia.

Clinical signs of *B. gibsoni* infection include fever, anorexia, depression, anemia and splenomegaly. Dogs that do not receive prompt effective treatment generally die with severe hemolytic anemia. The case-fatality rate in one study of dogs in California was 40%.

Various drugs have been used to treat *B. gibsoni* infection, with diminazene, phenamidine, pentamidine and parvaquone the most efficacious.

Canine babesiosis caused by *B. gibsoni* has been reviewed by Yamane et al. (1993).

### 15.2.10. *Babesia microti* infection

*Babesia microti* is a parasite of rodents in the United States (northeastern and upper midwestern states), Europe and Japan. In the United States the rodent hosts include the white-footed mouse (*Peromyscus leucopus*) and the meadow vole (*Microtus pennsylvanicus*), and they form reservoirs of infection for ticks. Tick vectors in the United States include *Ixodes angustus*, *I. muris*, *I. spinipalpis* and *I. scapularis*, with *I. scapularis* the primary vector for humans. Tick vectors in Europe are *I. trianguliceps*, a rodent-specific tick, and *I. ricinus*. The tick vector in Japan is suspected to be *I. ovatus*. Human babesiosis occurs when ticks that have acquired infection from mice feed on humans or when humans receive blood transfusions contaminated with *B. microti* organisms. In the case of *I. scapularis*, transmission is transstadial, with infection acquired from mice by larval or nymphal ticks and transmitted to humans by nymphal or adult ticks.

Cases of human babesiosis caused by *B. microti* in the United States have resulted both from bites of the tick *I. scapularis* and from blood transfusions from asymptomatic donors, while those in Japan have occurred through blood transfusions only. Although *B. microti* is present throughout Europe, no cases of human disease have been confirmed, probably because the major vector in Europe, the rodent tick *I. trianguliceps*, does not bite humans.

Human infection with *B. microti* presents a wide spectrum of clinical manifestations, from asymptomatic infections to acute and fatal disease. In most persons infection is asymptomatic or seen as a transient flulike illness. In some, infection may become chronic and manifest itself only when the patient becomes immunocompromised or is splenectomized. In others, infection may produce an acute and fatal disease, particularly in the elderly, the immunocompromised and the splenectomized. Human babesiosis caused by *B. microti* is typically less acute than the disease caused by *B. divergens*. It has a gradual onset with malaise, anorexia, fatigue, shaking chills, sweating, headaches, myalgia and arthralgia, often followed by fever and nausea. In severe cases hemoglobinuria and jaundice may occur. The parasites produce a hemolytic anemia and thrombocytopenia. Complications from severe infections include acute respiratory distress, congestive heart failure, disseminated intravascular coagulation and renal failure. In the United States about 25% of acute cases require hospitalization, and the mortality rate is about 6%.

Treatment of symptomatic cases has utilized a combination of clindamycin and quinine historically, but more recently a combination of atovaquone and azithromycin has been favored to reduce side effects. In very serious cases where drug therapy is not adequate, procedures such as erythrocyte exchange transfusion can be beneficial or even life-saving.

Human babesiosis caused by *B. microti* has been reviewed by Gorenflot et al. (1998) and Kjemtrup & Conrad (2000).

## 15.3. Borreliosis

The borrelioses are tickborne diseases caused by spirochete bacteria of the genus *Borrelia*. The two diseases discussed below are Lyme borreliosis and bovine borreliosis.

### 15.3.1. Lyme borreliosis

Lyme borreliosis (or Lyme disease) is considered the most prevalent arthropod-borne human disease in Europe and the United States and, although rarely fatal, can cause severe debilitation in untreated chronic cases. The name of this form of borreliosis goes back to 1975, when a cluster of cases of the disease was recognized in children in the town of Old Lyme in Connecticut. Lyme borreliosis occurs in the United States (the northeast from Maine to Maryland, the upper midwest in Minnesota and Wisconsin, and on the west coast in northern California and Oregon), in Europe (where it is particularly prevalent in Austria, Germany, Slovenia and Sweden) and in Russia, China, and Japan. The cause of Lyme borreliosis is *Borrelia burgdorferi* sensu lato, which consists of several genospecies, including the pathogens *B. burgdorferi* sensu stricto, *B. afzelii* and *B. garinii*. *Borrelia burgdorferi* s.s. is the only genospecies causing disease in the United States and is also found in Europe, whereas *B. afzelii* and *B. garinii* are the primary pathogens in Europe and Asia. The vectors of *B. burgdorferi* are ticks of the genus *Ixodes*. The principal vectors for humans are *I. scapularis* in the east and upper midwest and *I. pacificus* in the west of the United States, *I. ricinus* in Europe, and *I. persulcatus* in Asia. The role of *I. uriae*, a tick of seabirds in both the Northern and Southern Hemispheres, is uncertain; while *I. uriae* bites humans, no disease has so far been associated with this tick. In addition to the vectors that transmit infection to humans, there are other tick vectors involved in enclosed enzootic cycles that maintain the infection in nature, and they include *I. hexagonus* in Europe and *I. dentatus*, *I. neotomae* and *I. spinipalpis* in the United States.

Rodents and birds form the primary reservoir of *B. burgdorferi* for infection of *Ixodes* ticks. In the northeastern and upper midwestern United States, there is a highly efficient cycle of *B. burgdorferi* transmission between larval and nymphal *I. scapularis* ticks and white-footed mice that results in high rates of infection in nymphal ticks and a high frequency of Lyme borreliosis in humans during the late spring and summer months. The epidemiology of the disease is somewhat more complex in the western United States, where *B. burgdorferi* is maintained in a cycle between the dusky-footed woodrat (*Neotoma fuscipes*) and *I. neotomae* ticks, which, unlike *I. scapularis*, do not bite humans. The vector of human Lyme borreliosis in the western United States is *I. pacificus*, whose immature stages prefer to feed on lizards, particularly the western fence lizard (*Sceloporus occidentalis*), which are not susceptible to *B. burgdorferi* infection. Only the relatively few nymphal *I. pacificus* that fed on infected woodrats as larvae are responsible for transmission of *B. burgdorferi* to humans. Similarly, in the southeastern United States nymphal *I. scapularis* feed primarily on lizards rather than rodents, and human Lyme borreliosis occurs rarely in that region.

Human Lyme borreliosis is a multisystem infection, with inflammatory complications that commonly affect the skin, central nervous system and joints. It typically begins with a slowly expanding skin lesion called erythema migrans at the site of the tick bite. This skin lesion is often accompanied by flulike symptoms such as malaise, fatigue, headache, arthralgia, myalgia, fever and regional lymphadenopathy. Two other skin lesions not seen in the United States and associated with *B. afzelii* infection in Europe are borrelial lymphocytoma (a nodule appearing on the ear lobe, nipple or scrotum) and acrodermatitis chronica atrophicans (an atrophic lesion on sun-exposed acral surfaces). Within weeks of disease onset, signs of acute neurological involvement develop in about 15% of untreated patients in the United States; these include aseptic meningitis, episodic headaches and mild neck stiffness. Also within weeks of disease onset, signs of acute cardiac involvement develop in about 5% of untreated patients, with fluctuating degrees of atrioventricular block most commonly seen. Months after disease onset, signs of joint involvement develop in about 60% of untreated patients, with intermittent attacks of joint swelling and pain, primarily in the larger joints such as the knee.

Treatment with antibiotics is beneficial for all clinical manifestations of human Lyme borreliosis, but such therapy is most effective if used early in the course of the disease. Treatment with doxycycline is used for persons older than eight years of age except for pregnant women; amoxicillin is used for young children and pregnant women.

Protective measures for the prevention of Lyme borreliosis include avoidance of tick-infested areas, the use of protective clothing, repellents and acaricides, bodily tick checks, and modifications of landscapes in or near residential areas. Other preventive measures include quick removal of attached ticks and a single dose of doxycycline given within 72 hours of a tick bite. A recombinant vaccine against Lyme borreliosis was developed and licensed for use in humans in the United States in 1998, but it was withdrawn from the market in 2002.

Apart from humans, Lyme borreliosis has been reported in dogs, cats, horses, sheep and cattle. The majority of infections with *B. burgdorferi* in these domestic animal species are asymptomatic. When clinical disease is evident, the predominant signs are lameness with or without joint swelling and fever.

Human Lyme borreliosis has been reviewed by Steere (2001), Hengge et al. (2003) and Stanek & Strle (2003), and Lyme borreliosis in domestic animals has been reviewed by Bushmich (1994).

### 15.3.2. Bovine borreliosis

Bovine borreliosis (or spirochetosis) is a mild tickborne disease of cattle caused by *Borrelia theileri*. Infection by *B. theileri* is often associated with infection by *Babesia* spp. Bovine borreliosis has been reported from several countries in Africa, Europe, and South America and from Australia, Madagascar and Mauritius. *Borrelia theileri* infection has also been found in sheep and horses. Vectors of *B. theileri* are the ticks *R. e. evertsi*, *R. (B.) annulatus*, *R. (B.) decoloratus* and *R. (B.) microplus*. The clinical signs of bovine borreliosis are fever, followed by anorexia and depression. The disease is of little economic importance, and few deaths have been reported.

Bovine borreliosis has been reviewed by Bishop (2004).

## 15.4. Crimean-Congo Hemorrhagic Fever

In 1944–45 Soviet military personnel were infected by a disease while assisting peasants in Crimea (present-day Ukraine) in the wake of World War II, and the disease was called Crimean hemorrhagic fever. The causative virus was eventually isolated in 1967 and later shown to be antigenically indistinguishable from the Congo virus, which had been isolated in 1956 from a patient with Congo hemorrhagic fever in the Belgian Congo (present-day Democratic Republic of the Congo). Consequently the disease caused by the virus isolated in Crimea and the Belgian Congo was renamed Crimean-Congo hemorrhagic fever (CCHF).

Crimean-Congo hemorrhagic fever is an often fatal tickborne infection of humans caused by a nairovirus of the family Bunyaviridae. Infection of livestock such as cattle and sheep is typically inapparent or causes only mild fever. CCHF virus has been isolated from many countries in Africa, eastern Europe and southwestern Asia. It is transmissible transstadially by members of three genera of ixodid ticks, *Hyalomma*, *Dermacentor* and *Rhipicephalus*, but the similarity in the distributions of CCHF virus and *Hyalomma* ticks strongly suggests that members of this genus are the most important vectors of the virus. Tick vectors of the CCHF virus include *Hy. marginatum* in Europe, *Hy. anatolicum* in Asia, and *Hy. rufipes*, *Hy. truncatum*, *R. evertsi evertsi*, *R. evertsi mimeticus*, *R. pulchellus* and *A. variegatum* in Africa.

CCHF virus circulates in nature unnoticed in an enzootic tick-vertebrate-tick cycle. The virus has been isolated from numerous species of domestic and wild vertebrates such as cattle, goats, sheep, hares, hedgehogs and rodents. Although most species of birds tested appear to be refractory to CCHF virus infection, ostriches appear to be more susceptible, developing a viremia and strong antibody response. However, birds may play a significant role in the dissemination of CCHF virus-infected ticks between countries; for example, birds migrating from the Balkans have been suggested to be the transport hosts of the virus responsible for the 2002 CCHF outbreak in Turkey.

Humans acquire CCHF through tick bites, by crushing infected ticks, by contact with patients with CCHF during the acute phase of the infection, or by contact with blood or tissues from viremic livestock. Consequently those at highest risk of infection include (1) people such as livestock farmers, hunters, hikers and campers who have the greatest likelihood of contacting the tick vectors; (2) veterinarians, abattoir workers and others who contact viremic blood while performing procedures such as castrations, vaccinations, insertion of ear tags and slaughtering of livestock; and (3) health-care workers who care for CCHF patients. There have been numerous nosocomial outbreaks of CCHF with high mortality, and the available evidence suggests that human infections were acquired through contact of viremic blood with broken skin such as when medical personnel accidentally prick themselves with needles contaminated with the blood of patients or similar mishaps.

Onset of CCHF is usually very sudden, with fever, chills, severe headaches, dizziness, photophobia, back pains and myalgia. Hemorrhagic manifestations develop 3–6 days after onset of disease and range from petechiae to large hematomas appearing on the mucous membranes and skin. Bleeding from other sites can occur, including from the nose, gastrointestinal system, uterus, urinary tract, respiratory tract, vagina, gums and even the brain. Thrombocytopenia is a consistent feature. Severely ill patients may enter a state of hepatorenal and pulmonary failure, becoming progressively drowsy and comatose. The mortality rate is about 30%, but rates as high as 80% have been reported from Asia. Mortality rates of nosocomial infections are often higher than those acquired naturally through tick bites. For those who do not succumb to the disease, the convalescent period is generally characterized by prolonged and pronounced generalized

Fig. 15.1. Cow in Puerto Rico suffering from acute dermatophilosis (*left*); cow in Nevis that had recently died from the disease (*above*). (Photograph on left courtesy of U.S. Department of Agriculture.)

weakness, a weak pulse and sometimes complete loss of hair.

Treatment of CCHF is based upon supportive therapy, including the administration of thrombocytes, fresh frozen plasma and erythrocyte preparations. Ribavirin is the recommended antiviral agent for infected patients. Treatment of CCHF patients must include the use of barrier-nursing techniques to protect medical staff from infection with CCHF virus.

The highly pathogenic nature of the CCHF virus to humans has led to the fear that it might be used as an agent of bioterrorism and/or biowarfare. This has resulted in its inclusion by the U.S. Centers for Disease Control and Prevention as a category C priority pathogen for bioterrorism.

CCHF has been reviewed by Swanepoel & Burt (2004), Whitehouse (2004) and Ergonul (2006).

## 15.5. Dermatophilosis

Dermatophilosis (syn. streptothricosis, lumpy wool disease, strawberry footrot, Senkobo disease) is an acute, subacute, or chronic tick-associated skin disease that affects a wide range of domestic and wild animal species as well as humans. It is characterized by the development of an exudative dermatitis followed by scab formation, alopecia and thickening of the skin, with the lesions either localized or generalized. Dermatophilosis is caused by the bacterium *Dermatophilus congolensis*, an actinomycete that produces motile zoospores that invade the skin. It has a worldwide distribution but is most prevalent in humid, tropical and subtropical regions. It is seen most commonly in cattle, sheep, goats and horses, but also in mules, donkeys, water buffalos and pigs. It is seen rarely in dogs, cats, camels and wildlife such as monkeys, antelope, deer, chamois (*Rupicapra rupicapra*), giraffes, zebras, polar bears (*Ursus maritimus*), raccoons, skunks, rabbits, rodents and lizards. Dermatophilosis is transmitted by direct or indirect contact; however, infestations of cattle with the tropical bont tick *Amblyomma variegatum* have long been associated with an increase in prevalence and severity of dermatophilosis.

Clinically the disease is characterized by an exudative to proliferative dermatitis with subsequent formation of scabs and crusts under which the hair or fleece tends to break or matte together. The matted hair or wool may become detached, leaving raw red exudative areas (Fig. 15.1). In cattle, the lesions may be localized or become extensive, covering a large part of the back and sides, the perineal region and the legs (Fig. 15.1). In sheep, the lesions often appear first on the lips, ears and legs from the coronet to the knee. In equines, the lesions are frequently localized on the muzzle and ears, with eventual progression to the withers, back, croup, rump and tail. In humans, the lesions are usually localized as pustules on the hands or occasionally the feet, and they heal spontaneously or with minimal treatment.

The mortality rate from dermatophilosis is low in animals except in Africa and the Caribbean, where it is associated with the tick *A. variegatum*. The association between *A. variegatum* and acute bovine dermatophilosis is particularly strong in the Caribbean. The severe form of this disease has been seen in cattle on the Caribbean islands of Puerto Rico, Vieques, St. Maarten/St. Martin, St. Kitts, Nevis, Antigua, Montserrat, Guadeloupe, La Desirade, Marie Galante, Dominica, Martinique and St. Lucia only after the introduction of the tick *A. variegatum* (Burridge, 1985) and then only on infested farms. Furthermore, acute bovine dermatophilosis was controlled only by strict control of tropical bont ticks. The author visited a Caribbean island in the early 1980s and was told by the veterinary authorities that *A. variegatum* had not been introduced onto that island; upon learning that a farm on the island had an outbreak of acute bovine dermatophilosis, the author visited the farm and promptly found affected cattle infested with *A. variegatum*! The cattle populations of some Caribbean islands heavily infested with *A. variegatum* have been decimated by acute dermatophilosis, with herds of cattle becoming emaciated with high mortality rates. For example, acute dermatophilosis exacerbated by *A. variegatum* infestations has been considered responsible for the dramatic reduction in the cattle population on the island of Nevis from 2,740 head in 1975 to 720 head in 1988 (Hadrill et al., 1990). The exact nature of the association between *A. variegatum* and acute bovine dermatophilosis is not clearly understood, but it has been postulated that the feeding of *A. variegatum* ticks has a systemic effect on the host that allows dermatophilosis to become aggravated via an effect on immunological reactions in the skin (Walker & Lloyd, 1993).

Dermatophilosis has been reviewed by Hyslop (1980), Zaria (1993) and Zaria & Amin (2004).

## 15.6. Heartwater

Heartwater (syn. cowdriosis) is an acute tickborne disease of ruminants caused by the rickettsia *Ehrlichia ruminantium* (formerly *Cowdria ruminantium*).

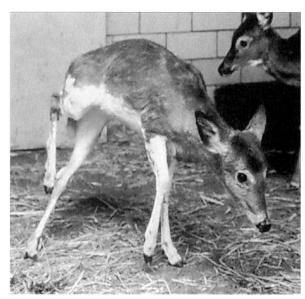

Fig. 15.2. White-tailed deer (*Odocoileus virginianus*) exhibiting incoordination while in the terminal stages of heartwater. (Photograph courtesy of U.S. Department of Agriculture.)

It occurs in most countries in sub-Saharan Africa and was introduced into the Western Hemisphere by infected ticks infesting cattle imported from Africa to the French West Indies. Thus far heartwater has been confirmed on three islands in the eastern Caribbean, Guadeloupe, Marie Galante and Antigua. Heartwater is primarily a disease of cattle, sheep and goats; however, at least 15 wild ruminant species have also been found to be susceptible to *E. ruminantium* infection; they are 12 species from Africa, African buffalos, black wildebeests, blesboks, blue wildebeests, elands, giraffes, greater kudus, lechwes (*Kobus leche*), sable antelopes, sitatungas (*Tragelaphus spekii*), springboks and steenboks, two species from Asia, spotted deer (*Axis axis*) and Timor deer (*Rusa timorensis*), and one species from the Americas, white-tailed deer (Fig. 15.2). Heartwater is transmitted transstadially by *Amblyomma* spp. ticks; *A. variegatum* is the most important and widely distributed vector in Africa and the only vector in the Caribbean, *A. hebraeum* is the predominant field vector in southern Africa, *A. lepidum* has been involved in field outbreaks in Sudan, *A. pomposum* is a vector in Angola, *A. astrion* is a vector on the islands of Sao Tome and Principe off the coast of western Africa, and eight other *Amblyomma* spp. (*A. cohaerens*, *A. gemma*, *A. marmoreum*, *A. sparsum* and *A. tholloni* from Africa and *A. cajennense*, *A. dissim-*

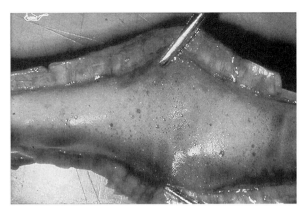

Fig. 15.3. Froth in trachea of a cow that died from heartwater.

*ile* and *A. maculatum* from the Americas) have been shown to be experimental vectors.

The first clinical sign seen in acute cases of heartwater is a sudden rise in body temperature with loss of appetite and respiratory distress, followed by nervous signs such as circling motions, incoordination and recumbency with paddling movements of the limbs. In peracute cases seen in exotic breeds of cattle and goats, the animals collapse with fever and die in convulsions without premonitory symptoms. Mortality rates in susceptible domestic ruminants can range from 40% to approaching 100%, with losses of 90% common in highly susceptible Angora goats. On postmortem examination, the most constant feature is accumulation of fluid in various organs and cavities, resulting in pulmonary edema with froth in the trachea and bronchi (Fig. 15.3), hydropericardium (hence the name heartwater for the disease), hydrothorax and ascites.

Control of heartwater is dependent on control of its tick vectors. No practical method of treatment is available. The only vaccine commercially available at the present time for prevention for heartwater is an infection-and-treatment vaccine produced in South Africa whereby livestock are inoculated intravenously with blood infected with virulent *E. ruminantium*, followed by treatment with tetracyclines at the beginning of the febrile response.

Heartwater has been reviewed by Uilenberg (1983b), Camus et al. (1996) and Allsopp et al. (2004).

## 15.7. Hemogregarine Infections

Hemogregarines are coccidial parasites and the most common blood parasites affecting reptiles. Their life cycle involves sporogonic development in the definitive invertebrate vector and asexual stages in the intermediate vertebrate host. Despite the high prevalence of hemogregarine infections in many reptiles, their pathogenicity for reptilian hosts remains obscure, with hemogregarines frequently appearing innocuous to vertebrate hosts. The vectors for five hemogregarine parasites are invasive ticks: *Amblyomma dissimile* is a vector of *Hepatozoon fusifex* to boa constrictors, *A. rotundatum* is a vector of *Hemolivia stellata* to giant toads, *Bothriocroton hydrosauri* is a vector of *Hemolivia mariae* to skinks, and *Hyalomma aegyptium* is a vector of *Hemolivia mauritanicum* to Greek tortoises and of *Hepatozoon kisrae* to rough-tailed rock agamas.

Hemogregarine infections have been reviewed by Telford (1984).

## 15.8. Louping Ill

Louping ill (louping in Scottish dialect = leaping) is an acute tickborne encephalomyelitis primarily of sheep caused by a flavivirus of the family Flaviviridae. It occurs in the hill farming areas of Scotland, northern England, Wales, and Ireland. Closely related viruses have been reported to cause encephalomyelitis of sheep in Bulgaria, Norway, Spain and Turkey. While louping-ill virus is primarily a pathogen of sheep, it also produces clinical disease in cattle, horses, pigs, humans and two avian species, red grouse (*Lagopus lagopus*) and ptarmigans (*Lagopus mutus*). In addition, infection with the virus has been demonstrated in feral goats, red deer, roe deer, dogs, mountain hares (*Lepus timidus*), badgers, wood mice (*Apodemus sylvaticus*), field voles (*Microtus agrestis*) and common shrews (*Sorex araneus*). Louping ill is transmitted transstadially by the tick *Ixodes ricinus*. Other ixodid ticks, including *Hyalomma anatolicum* spp., *Ixodes persulcatus* and *Rhipicephalus appendiculatus*, have been shown to be experimental vectors. Apart from bites of *I. ricinus*, the louping-ill virus can be transmitted by ingestion of *I. ricinus* ticks by red-grouse chicks, and by penetration of virus through skin wounds, by aerosol exposure in the laboratory and by ingestion of infected milk in the case of humans.

Initial clinical signs of louping ill in sheep include fever, depression and anorexia. The fever is biphasic, with the second rise occurring when the virus invades the central nervous system. The signs of nervous involvement include muscular tremors, incoordination, hyperesthesia and development of a high-stepping gait. At this stage of the disease, sheep are often hypersensitive to noise and touch and will go into convulsive spasms if disturbed. The disease progresses with head pressing, paraplegia, convulsions, opisthotonus and coma. Mortality rates as high as 60% can occur in sheep, especially in lambs whose passively acquired immunity has declined when they are introduced onto heavily tick-infested pastures for the first time. Red grouse and ptarmigan are also highly susceptible to infection with the louping-ill virus, and they develop high viremias and succumb rapidly to the infection. Mortality rates of 50% have been reported in red grouse.

Humans at risk of infection with the louping-ill virus are mainly those with occupational exposure such as sheep farmers, veterinarians, abattoir workers and butchers with direct contact with infected animals. Like sheep, humans also experience a biphasic illness, with headache, retro-orbital pain, photophobia, malaise, drowsiness, nausea, myalgia and arthralgia characterizing the first febrile response, and severe headaches, double vision and confusion occurring during the second fever.

There is no specific treatment for encephalitic cases of louping ill. The single most important method of control in endemic areas is vaccination of sheep using a formalin-inactivated commercial vaccine.

Louping ill has been reviewed by Timoney (1998) and Swanepoel & Laurenson (2004).

### 15.9. Nairobi Sheep Disease

Nairobi sheep disease (NSD) is an acute tickborne gastroenteritis of sheep and goats caused by a nairovirus of the family Bunyaviridae. It occurs in eastern Africa, having been reported from Ethiopia, Kenya, Rwanda, Somalia, Tanzania and Uganda. In addition to domestic small ruminants, fatal cases of Nairobi sheep disease have been reported in blue duikers (*Philantomba monticola*), and rare laboratory-acquired infections have occurred in humans. The NSD virus is transmitted transovarially and transstadially by the brown ear tick *Rhipicephalus appendiculatus*; other less efficient vectors are the ticks *R. pulchellus*, *R. simus* and *Amblyomma variegatum*.

Clinical signs of Nairobi sheep disease in sheep and goats begin with a sudden rise in temperature. The animal may become depressed, anorexic and disinclined to move. There is usually a serosanguineous nasal discharge. Diarrhea develops within 1–3 days of the onset of fever; initially it is profuse and watery, becoming mucoid and bloody later. Pregnant animals frequently abort. The mortality rate can be as high as 90% in indigenous breeds of sheep and goats.

Accidental human infections with the NSD virus have been acquired in the laboratory, but they appear to be very rare. When they do occur, they result in fever, joint pains and general malaise.

Two types of vaccines have been developed for Nairobi sheep disease, a modified-live-virus vaccine and an inactivated oil-adjuvant vaccine; however, neither has been used extensively in the field.

Nairobi sheep disease has been reviewed by Davies & Terpstra (2004).

### 15.10. Omsk Hemorrhagic Fever

Omsk hemorrhagic fever (OHF) is a hemorrhagic tickborne disease of humans caused by a flavivirus of the family Flaviviridae. It is endemic in western Siberia, where the virus is maintained in a sylvatic cycle involving water voles (*Arvicola amphibius*) and muskrats (*Ondatra zibethicus*). Muskrats were first introduced into Siberia from Canada in 1928 and are highly susceptible to infection with the OHF virus. The principal vector of the virus is the tick *Dermacentor reticulatus*. The virus can also be transmitted to humans by ingestion of infected milk from sheep and goats or by contact with bodily fluids or carcasses of infected animals, as seen in the frequent infections of muskrat trappers.

Human infection with the OHF virus does not produce symptoms of neurological involvement, but rather vascular and circulatory disorders, with capillary damage responsible for the hemorrhagic manifestations. Clinical signs begin with a sudden fever followed by a second more severe fever with headaches, myalgia, cough and a distinct hemorrhagic syndrome with bleeding in the nose, gums, uterus and lungs. The case fatality rate is up to 3%, and recovery is usually slow. In recent years, most human cases of Omsk hemorrhagic fever have been related to direct contact by trappers with infected muskrats.

Recently the OHF virus has been listed as a virus that poses a serious risk as a biological weapon (Borio et al., 2008) and that has a high potential for use in bioterrorism (Davis, 2004a).

Omsk hemorrhagic fever has been reviewed by Gritsun et al. (2003).

## 15.11. Piroplasmosis

Equine piroplasmoses (or biliary fever) are tickborne diseases of horses, donkeys, mules and zebras caused by the protozoan parasites *Babesia caballi* and *Theileria equi* (formerly *Babesia equi*) and characterized by fever, progressive anemia, jaundice, hepatomegaly and splenomegaly. Equine piroplasmoses present a global problem for the equine trade and for equine sporting events since horses with antibodies to either *B. caballi* or *T. equi* are restricted from entering a number of countries, including the United States.

*Babesia caballi* was introduced into southern Florida in 1961 following the importation of ponies from Cuba (Sippel et al., 1962; Bruning, 1996). A state-federal control program was initiated in south Florida in 1962. Affected equines were quarantined, and infected and exposed animals were sprayed with acaricide every three weeks, with ears and false nostrils treated. In addition, animals were treated with drugs to kill *B. caballi* parasites and thus eliminate the carrier state. After 26 years, equine piroplasmosis was finally eradicated from southern Florida in 1988. The United States was thereafter considered free of equine piroplasmosis until August 2008, when a horse was found to be infected with *T. equi* in Manatee County in southwestern Florida. Initial investigations traced the source of the outbreak to two horses imported from Mexico, with transmission of infection in Florida apparently due to management practices such as reuse of needles and "blood packing" (i.e., transfusion of erythrocytes to horses prior to racing). No evidence of tick transmission was detected in the quarantine area. A total of 20 horses on seven premises were found to be infected during the outbreak. Florida was declared free of *T. equi* infection in February 2009.

Later in October 2009 another outbreak of equine piroplasmosis due to *T. equi* was discovered in southern Texas. Initial investigations of this outbreak found 288 infected horses on the index ranch in Kleberg County and other *T. equi* infections in "epidemiologically linked" horses on other premises in Texas and on premises in Florida, Louisiana, New Jersey, California, North Carolina, Alabama, Georgia, Minnesota, Tennessee and Wisconsin (http://www.oie.int/wahis/public.php?page=single_report&pop=1&reportid=8647). All infected horses were either living on the index ranch or had previously lived on that ranch. This ongoing outbreak of equine piroplasmosis has had two worrying features. Two of the horses found infected in New Jersey were bought from the index ranch in 2008, suggesting that the disease might have been present but undetected on the Texas property for many months. Furthermore, *D. nitens* ticks were collected from horses on the index ranch. These two features beg the question: how well established are the disease agent *T. equi* and the invasive tick *D. nitens* in southern Texas?

### 15.11.1. *Babesia caballi* infection

Equine piroplasmosis caused by *B. caballi* has a widespread distribution, occurring in Africa, Asia, Europe, and South and Central America. *Babesia caballi* develops in the erythrocytes of equines. Ticks of three genera, *Dermacentor*, *Hyalomma* and *Rhipicephalus*, have been identified as vectors of *B. caballi*, and they include *D. nitens*, *D. nuttalli*, *D. reticulatus*, *Hy. anatolicum*, *Hy. dromedarii*, *Hy. marginatum* group, *Hy. truncatum*, *R. bursa* and *R. e. evertsi*. The parasite can be transmitted by ticks both transstadially and transovarially.

Clinical signs include fever, depression, anorexia, swelling of the eyelids and jaundice. In acute cases, hemoglobinuria may become evident. Pregnant mares may abort. Affected animals lose weight, may become emaciated and exhibit weakness in the hindquarters. Mortality rates are about 10%. Chronically affected horses may be ill for several weeks or months, with recovery slow.

A number of drugs are effective in treating *B. ca-*

*balli* infection, including diminazene, amicarbalide and imidocarb. No vaccine is currently available.

Equine piroplasmosis caused by *B. caballi* has been reviewed by Sippel et al. (1962) and de Waal & van Heerden (2004).

### 15.11.2. *Theileria equi* infection

Equine piroplasmosis caused by *T. equi* has the same widespread distribution as the disease caused by *B. caballi*. However, unlike *B. caballi*, *T. equi* develops initially in lymphocytes before its intraerythrocytic cycle in equines. Ticks of three genera, *Dermacentor*, *Hyalomma*, and *Rhipicephalus*, have been identified as vectors of *T. equi*, and they include *D. reticulatus*, *Hy. anatolicum*, *Hy. excavatum*, *Hy. detritum*, *Hy. dromedarii*, *Hy. marginatum* group, *R. bursa*, *R. e. evertsi*, *R. e. mimeticus*, *R. turanicus* and *R. (B.) microplus*. *Theileria equi* is only transmitted transstadially through the bites of nymphal or adult ticks.

The clinical signs of *T. equi* infection are similar to those of *B. caballi* infection although *T. equi* is more pathogenic to equines. There is no cross-immunity between *B. caballi* and *T. equi*.

Since *T. equi* develops in both lymphocytes as schizonts and erythrocytes as piroplasms, both antitheilerial drugs (e.g., buparvoquone, halofuginone, parvaquone) and babesiacidal drugs (e.g., amicarbalide, diminazene, imidocarb) have been used to treat *T. equi* infections with some success. It has been suggested that the combined use of buparvoquone and imidocarb might eliminate *T. equi* infections.

Equine piroplasmosis caused by *T. equi* has been reviewed by Mehlhorn and Schein (1998) and de Waal & van Heerden (2004).

### 15.12. Q Fever

Q fever is a tick-associated zoonotic disease caused by the rickettsia *Coxiella burnetii*. In 1935 an outbreak of a febrile illness occurred among abattoir workers in Australia, and the disease was given the name Q (for query) fever. *Coxiella burnetii* is found worldwide with the notable exception of New Zealand. It can infect a broad range of hosts, including livestock, domestic pets, wild mammals, reptiles, fishes, birds and ticks. The most commonly identified sources of human infections are cattle, sheep and goats, whereas cats, dogs and rabbits have been the source of urban outbreaks.

Human infection with *C. burnetii* results primarily from aerosols, generated typically by infected livestock when they give birth or abort since *C. burnetii* multiplies to reach high concentrations in the placenta of mammals. Infected parturient pets, especially cats but also rabbits and dogs, have been similarly involved as sources of human infection in urban outbreaks. Other less common modes of transmission of *C. burnetii* to humans include ingestion of unpasteurized milk, infection via the percutaneous route and vertical transmission from mother to child. *Coxiella burnetii* can also be transported by wind. Finally, more than 40 species of ticks have been found to be naturally infected with *C. burnetii*, and transmission of the organism may occur during tick feeding. Although ticks appear to play a role in transmission of *C. burnetii* among wild vertebrates, particularly rodents, lagomorphs and birds, there is little to indicate that ticks are important in the maintenance of infections in livestock or humans and, in fact, transmission of *C. burnetii* to humans by a tick bite has seldom been reported. Nevertheless, a few tick species have been reported to be vectors of Q fever, including *A. triguttatum*, *D. reticulatus*, *Hy. detritum*, *Hy. lusitanicum* and *R. turanicus*, and several have been found to be naturally infected with *C. burnetii*, including *A. nuttalli*, *A. splendidum*, *A. variegatum*, *Ha. leachi*, *Ha. punctata*, *Hy. anatolicum* spp., *Hy. dromedarii*, *Hy. marginatum* group, *I. ricinus*, *R. bursa*, *R. senegalensis*, *R. simus* and *R. (B.) decoloratus*. However, since *C. burnetii* does multiply in the gut cells of ticks and large numbers of the organisms are shed in tick feces, hides and wool contaminated with infected tick feces may be a source of human infection, either by direct contact or indirectly after the feces have dried and been inhaled as airborne dust particles.

Q fever in humans is usually either a mild or asymptomatic infection, but it can be a serious and even life-threatening disease. Recent reviews indicate that about 60% of cases are asymptomatic, about 38% of patients experience a mild disease without the need for hospitalization and about 2% of cases result in acute or chronic disease requiring hospitalization. The onset of acute Q fever is sudden, with a high fever, fatigue, chills and severe headaches. Patients also frequently exhibit atypical pneumonia and hepatitis. Other rarer clinical manifestations of acute Q fever are myocarditis, pericarditis and meningoencephalitis. The major clinical presentation in chronic Q fever is endocarditis. Cases of mild Q fever resemble a flulike illness with fever, headaches and myalgia, and they resolve spontaneously within a week.

Infection of animals with *C. burnetii* seldom results in clinical disease, although abortion in sheep and goats has been reported; however, the rickettsiae are excreted in the milk, urine and feces of infected animals and, most significantly for human exposure, they are shed in high numbers within the amniotic fluid and the placenta during parturition.

Acute Q fever can be treated with doxycycline, but the drug is contraindicated in children under eight years of age and in pregnant women. Chronic Q fever is treated with a combination of doxycycline and hydroxychloroquine. Since Q fever in humans is often an occupational hazard, vaccination should be considered in high-risk groups such as livestock handlers, abattoir workers, veterinarians and laboratory personnel working with *C. burnetii*-infected animals, especially pregnant sheep. A formalin-inactivated whole-cell vaccine is available and has proved highly effective in reducing Q fever infections in abattoir workers in Australia.

Two outbreaks of Q fever were reported in the Netherlands in 2007 and 2008, with the ongoing outbreak in the southeastern part of the country becoming the largest community outbreak of Q fever ever recorded (Delsing & Kullberg, 2008). *Coxiella burnetii* was first identified as the cause of abortion on two dairy-goat farms in 2005, from where it spread to other goat and sheep premises. Q fever was rarely known among humans in the Netherlands until 2007, when 168 cases were reported, increasing to 3,461 cases through November 25, 2009, including at least six deaths. The outbreaks on goat and sheep farms are believed to be associated with the human infections. Thus the Netherlands has become the country with the highest rate of Q fever infection in Europe. As a consequence of this public health problem, the Dutch government designated Q fever a reportable disease in goats and sheep in 2008, and special preventive and control measures in animals were implemented, with compulsory vaccination of "at risk" goat and sheep populations during the 2009–10 seasons.

Recently *C. burnetii* has been listed as a category B bioterrorism agent of public health importance by the U.S. Centers for Disease Control and Prevention (Davis, 2004b). The World Health Organization has estimated that 50 kg of dried *C. burnetii* organisms, released near a developed city of 5 million inhabitants, would result in 250,000 human illnesses and 250 human deaths (Davis, 2004b). Furthermore, *C. burnetii* has been evaluated as a potential biological warfare agent by the United States, but stocks (except for those required for vaccine research) were destroyed in 1971–72 (Byrne, 1997).

Q fever has been reviewed by Maurin and Raoult (1999) and Kelly (2004).

## 15.13. Rickettsiosis

The rickettsioses discussed in this section are the tickborne diseases primarily of humans caused by obligate intracellular rickettsia of the genus *Rickettsia*. These diseases are African tick-bite fever caused by *R. africae*, tick typhus caused by *R. conorii*, Flinders Island spotted fever caused by *R. honei*, Japanese spotted fever caused by *R. japonica*, Rocky Mountain spotted fever and Brazilian spotted fever caused by *R. rickettsii*, North Asian tick typhus caused by *R. sibirica sibirica*, lymphangitis-associated rickettsiosis caused by *R. sibirica mongolitimonae* and tickborne lymphadenopathy caused by *R. slovaca*.

### 15.13.1. *Rickettsia africae* infection

African tick-bite fever is a mild tickborne disease of humans in sub-Saharan Africa and the eastern Caribbean caused by the bacterium *R. africae*. It is transmitted in southern Africa by the South African bont tick *A. hebraeum* and in the rest of sub-Saharan Africa and in the Caribbean by the tropical bont tick *A. variegatum*. Transmission by *A. hebraeum* is both transovarial and transstadial.

There are numerous reports of cases of African tick-bite fever in tourists returning from trips to rural Africa. In most instances the disease has a benign or self-limiting course, but prolonged fever, reactive arthritis, and subacute neuropathy are occasionally seen. Infection with *R. africae* is characterized by fever, headaches, myalgia, multiple inoculation eschars (necrotic lesions covered by a black crust) at the site of tick bites, regional lymphadenopathy and the frequent absence of a cutaneous rash. The drug of choice for treatment of *R. africae* infection is doxycycline. African tick-bite fever has a mortality rate approaching zero.

African tick-bite fever has been reviewed by Jensenius et al. (2003).

### 15.13.2. *Rickettsia conorii* infection

Tick typhus (or Mediterranean spotted fever or boutonneuse fever) is a mild tickborne disease of humans in southern Europe, India, and northern, central, eastern, and southern Africa caused by the bacterium *R. conorii*. It is generally accepted that the main vector in the Mediterranean countries of Europe and northern Africa is the brown dog tick of the *Rhipicephalus sanguineus* group. In other regions the vector situation is not so clear, although several tick species (*A. variegatum*, *Ha. leachi*, *Hy. albiparmatum*, *Hy. rufipes*, *Hy. truncatum*, *I. hexagonus*, *R. appendiculatus*, *R. evertsi*, *R. pulchellus*, *R. simus* and *R. turanicus*) have been reported to be vectors of *R. conorii*.

The clinical signs of tick typhus are fever, headache, an eschar and a diffuse maculopapular rash. Most cases are mild, but complications are not uncommon and include neurological involvement, peripheral gangrene, respiratory distress syndrome and circulatory problems such as deep-vein thrombosis and development of a hemorrhagic crisis. The mortality rate is 3–6%, but a rate as high as 32% has been reported in hospitalized patients in Portugal. The drug of choice for treatment of *R. conorii* infection is doxycycline.

Tick typhus has been reviewed by Gear (1992).

### 15.13.3. *Rickettsia honei* infection

Flinders Island spotted fever is a tickborne disease of humans in southeastern Australia caused by the bacterium *R. honei*. The first cases of the disease were discovered on Flinders Island in the Bass Strait between the Australian mainland and Tasmania, with subsequent cases found on Schouten Island off the east coast of Tasmania and in the southeastern corner of the state of South Australia. The putative vector of *R. honei* is the reptilian tick *Bothriocroton hydrosauri*, in which the rickettsia is maintained by vertical transmission. In Australia, *R. honei* has been isolated from its arthropod reservoir, *B. hydrosauri*, collected from blotched blue-tongued skinks, mainland tiger snakes and Australian copperheads (*Austrelaps superbus*). Additionally, *R. honei* has been detected in an *Ixodes granulatus* tick collected from a black rat in Thailand and in a Cayenne tick *A. cajennense* collected from cattle in Texas. The only human infection with *R. honei* outside Australia was reported from Thailand in 2005 (Jiang et al., 2005).

Flinders Island spotted fever is characterized by fever, headache, myalgia, arthralgia with joint swelling, a slight cough and a diffuse maculopapular rash on the trunk and limbs. Some patients also exhibit inoculation eschars. The drug of choice for treatment is doxycycline. No deaths have been reported from the disease.

### 15.13.4. *Rickettsia japonica* infection

Japanese spotted fever (or oriental spotted fever) is a tickborne disease of humans in Japan caused by the bacterium *R. japonica*. Recent studies have indicated that the tick vectors of *R. japonica* are *Ha. flava* and *Ha. hystricis*. The clinical signs of Japanese spotted fever are fever, headache, malaise, shaking chills, skin eruptions and an eschar. Fatal cases are rare. Treatment with doxycycline has been successful.

Japanese spotted fever has been reviewed by Mahara (1997, 2006).

### 15.13.5. *Rickettsia rickettsii* infection

*Rickettsia rickettsii* causes an acute tickborne disease of humans and dogs in the Americas, with cases seen from Canada in the north to Argentina in the south. The disease has a number of names depending on geographical location, the most common being Rocky Mountain spotted fever in North America and Brazilian spotted fever in Brazil. The major vectors of *R. rickettsii* are the American dog tick *Dermacentor variabilis* in most of the United States, the wood tick *D. andersoni* in the Rocky Mountain region and in Canada, the brown dog tick of the *R. sanguineus* group in Mexico, and the Cayenne tick *A. cajennense* in Central and South America. Transmission in ticks is both transovarial and transstadial.

Most humans have a moderate to severe illness. Initial clinical signs include fever, headache, myalgia and malaise. A skin rash develops around the ankles and wrists, spreading to the trunk, palms and soles, becoming maculopapular with central petechiae. Complications can occur in *R. rickettsii* infection since the rickettsiae induce substantial multisystem vasculitis, and they include encephalitis, pulmonary edema, respiratory distress syndrome, gastrointestinal bleeding and skin necrosis. The antibiotic of choice is doxycycline (or chloramphenicol for pregnant women). Case fatality rates have been reported to be 15–30% in untreated cases and about 5% in cases with delayed treatment. No vaccine is available for *R. rickettsii* infection.

Clinical signs in dogs include fever, anorexia, polyarthritis, coughing or dyspnea, abdominal pain and edema of the face or extremities. Focal retinal hemorrhages may be observed, and petechial hemorrhages may be seen in the conjunctiva and oral mucosa in severe cases. Antibiotics of choice for treating *R. rickettsii* infection in dogs are tetracycline, doxycycline and chloramphenicol.

Recently *R. rickettsii* has been listed as a category B bioterrorism agent of public health importance by the U.S. Centers for Disease Control and Prevention (Azad and Radulovic, 2003).

Rocky Mountain spotted fever has been reviewed by Thorner et al. (1998), Warner & Marsh (2002) and Dantas-Torres (2007), and Brazilian spotted fever by Greca et al. (2008).

### 15.13.6. *Rickettsia sibirica sibirica* infection

North Asian tick typhus (or Siberian tick typhus) is a tickborne disease of humans in Siberia and western China caused by the bacterium *R. sibirica sibirica*. It is transmitted primarily by *Dermacentor* spp. (including *D. auratus*, *D. nuttalli* and *D. reticulatus*), but also by *Haemaphysalis* spp. (including *Ha. punctata*). The clinical signs of North Asian tick typhus are fever, headache, myalgia, an inoculation eschar, a rash and lymphadenopathy. Treatment with doxycycline has been successful.

### 15.13.7. *Rickettsia sibirica mongolitimonae* infection

Lymphangitis-associated rickettsiosis is a mild tickborne disease of humans in Europe (France, Greece, Portugal), Africa (Algeria, Niger, South Africa) and Asia (China) caused by the bacterium *R. sibirica mongolitimonae*. Its tick vector is not known, but the causative rickettsia has been isolated from *Hy. asiaticum* in China, *Hy. truncatum* in Niger and *Hy. excavatum* in Greece. The clinical signs of lymphangitis-associated rickettsiosis are fever, inoculation eschars, maculopapular rash, enlarged regional lymph nodes and lymphangitis.

### 15.13.8. *Rickettsia slovaca* infection

*Rickettsia slovaca* causes a mild tickborne disease of humans in Europe. The disease is known as tickborne lymphadenopathy (TIBOLA) in France and Hungary and as *Dermacentor*-borne necrosis-erythema-lymphadenopathy (DEBONEL) in Spain. It is transmitted by the ticks *D. marginatus* and *D. reticulatus*, and the majority of human cases are seen in the winter when these tick vectors are most active. The clinical signs of the disease are a necrotic black lesion (eschar) surrounded by erythema at the site of the tick bite, which is typically on the scalp, headache and painful enlarged regional lymph nodes. These signs are often followed by persistent fatigue and residual alopecia at the site of the tick bite. Treatment with doxycycline has been successful.

## 15.14. Theileriosis

The theilerioses are a group of tickborne diseases of domestic livestock caused by protozoal parasites of the genus *Theileria*. The infective sporozoite stage of the parasites is transmitted to its mammalian host in the saliva of the vector tick as it feeds. The sporozoites invade lymphocytes where they develop into intracytoplasmic schizonts that in turn develop into merozoites that infect erythrocytes as piroplasms. The life cycle is completed when the vector tick ingests erythrocytes from an infected mammalian host. The theilerial diseases discussed below fall into two broad categories: (1) those which are highly pathogenic, namely East Coast fever, Corridor disease, Zimbabwe theileriosis and turning sickness, all caused by *T. parva*, tropical theileriosis caused by *T. annulata*, and malignant theileriosis caused by *T. lestoquardi*; and (2) those which produce mild infections that may on occasion become pathogenic, namely bovine theileriosis caused by the *T. buffeli/T. orientalis* group, by *T. mutans* and by

*T. taurotragi*. In addition, there are those that produce benign infections, namely benign caprine theileriosis caused by *T. ovis*, benign ovine theileriosis caused by *T. ovis* and by *T. separata*, and benign bovine theileriosis caused by *T. velifera*. Equine theileriosis caused by *T. equi* is more commonly referred to as equine piroplasmosis and is thus discussed under piroplasmosis.

### 15.14.1. *Theileria parva* infection

*Theileria parva* probably originated from African buffalo populations in eastern Africa and became adapted to cattle following their introduction and dissemination in the region. Cattle derived *T. parva*, formerly known as *T. parva parva*, causes East Coast fever, a highly pathogenic tickborne disease of cattle in eastern, central and southern Africa north of the Zambezi River. Buffalo-derived *T. parva*, formerly known as *T. lawrencei* and *T. parva lawrencei*, causes Corridor disease, an acute and usually fatal tickborne disease of cattle in eastern, central and southern Africa wherever contact occurs between cattle and infected African buffalos in the presence of vector ticks. Other strains of *T. parva*, formerly known as *T. parva bovis*, cause Zimbabwe theileriosis (syn. January disease), an acute and frequently fatal tickborne disease of cattle in the highveld of Zimbabwe. Finally, *T. parva* also causes turning sickness (syn. cerebral theileriosis), a highly fatal tickborne disease of cattle in eastern Africa.

*Theileria parva* is transmitted transstadially by the brown ear tick *R. appendiculatus*. Other rhipicephalid ticks may play a minor role as vectors of East Coast fever, including *R. capensis*, *R. evertsi evertsi*, *R. pulchellus* and *R. simus*. *Theileria parva* is highly pathogenic to cattle and experimentally to water buffalos, with natural subclinical or mild infections found in African buffalos and waterbucks.

East Coast fever is characterized by fever, enlargement of superficial lymph nodes, severe pulmonary edema and emaciation, usually with a fatal outcome. The initial clinical signs are fever and enlargement of the parotid lymph nodes that drain the ears, to which the tick vectors attach. There may be a sharp decline in milk production. The animal becomes depressed and lethargic with enlargement of other superficial lymph nodes, including the prescapular and precrural nodes. It progresses to anorexia and a severe loss of body condition, with ataxia and frequent recumbency.

Fig. 15.4. Frothy fluid exuding from the nostrils of a cow that died from East Coast fever.

Pregnant cows may abort. In the terminal stages, dyspnea develops with a discharge of frothy fluid from the nostrils (Fig. 15.4). The disease is characterized by a severe panleukopenia and, at necropsy, the most prominent feature typically is severe pulmonary edema with copious amounts of froth in the bronchi and trachea and often also in the nasal passages. The mortality rate from East Coast fever in susceptible cattle can be greater than 90%.

Corridor disease occurs only in cattle that graze pastures on which African buffalos are or have recently been present. Cattle with Corridor disease exhibit the same clinical signs as those with East Coast fever except that the course is usually shorter, with death sometimes occurring only 3–4 days after onset of first signs. Severe pulmonary edema precedes death. The mortality rate from Corridor disease is about 80%.

Most cases of Zimbabwe theileriosis occur between January and March, coinciding with the seasonal distribution of the adult tick vectors in the highveld of Zimbabwe. The disease usually affects cattle more than one year of age, being rarely seen in calves. As with Corridor disease, cattle with Zimbabwe theileriosis exhibit the same clinical signs as those with East Coast fever except that the course is often shorter, with death sometimes occurring only 3–4 days after onset of first signs. One of the presenting signs can be blindness associated with corneal opacity. The mortality rate from Zimbabwe theileriosis may reach 90% in severe field outbreaks.

Turning sickness is an aberrant form of theileriosis characterized by the accumulation of parasitized lymphoblasts in the cerebral blood vessels, leading to thrombosis and infarction. In eastern Africa it is caused by *T. parva*, whereas in South Africa it is caused by *T. taurotragi*. The condition is usually afebrile, occurring with the sudden onset of nervous signs including circling, head pressing, blindness, ataxia, opisthotonus and paralysis. Death usually occurs 2–21 days after onset of nervous signs, but occasionally cases may become chronic, with cattle living up to six months with blindness and muscular incoordination predominating.

Currently three effective drugs are available for the treatment of *T. parva* infection; they are parvaquone and buparvaquone, effective against both the schizont and piroplasm stages, and halofuginone lactate, effective against only the schizont stage. However, treatment of cattle with Corridor disease is not permitted in South Africa by law.

An infection-and-treatment method of vaccination has been developed for East Coast fever. A stabilate of live *T. parva* organisms is inoculated subcutaneously into cattle with the simultaneous administration intramuscularly of a therapeutic drug such as long-acting oxytetracycline. This method of vaccination provides good immunity against homologous challenge. Although this technique is relatively expensive and cumbersome, it is being implemented on an increasing scale in a number of African countries with success. A similar infection-and-treatment method of vaccination has been developed for Zimbabwe theileriosis using a mild strain of cattle-derived *T. parva*; this mild strain has declined in virulence on passage to the extent that the vaccine can now be used without simultaneous treatment with antitheilerial drugs. There is no vaccine currently available for Corridor disease.

East Coast fever has been reviewed by Lawrence et al. (2004a), Corridor disease by Lawrence et al. (2004b), Zimbabwe theileriosis by Lawrence et al. (2004c) and turning sickness by Lawrence & Williamson (2004a).

### 15.14.2. *Theileria annulata* infection

Tropical theileriosis is a serious tickborne disease of cattle caused by the protozoal parasite *T. annulata*. It occurs in northern Africa, Sudan and Eritrea, southern Europe, the Near and Middle East, central Asia, India and northern China. It is transmitted transstadially by *Hyalomma* ticks, including *Hy. anatolicum*, *Hy. excavatum*, *Hy. detritum*, *Hy. dromedarii*, *Hy. lusitanicum* and *Hy. marginatum* group. In addition to cattle, water buffalos and yaks (*Bos mutus*) are susceptible to *T. annulata* infection; infections of water buffalos are inapparent whereas those of yaks are acute with mortalities. Tropical theileriosis has a seasonal occurrence related to the biology of the *Hyalomma* tick vectors, with the majority of cases occurring between June and September in most endemic areas.

Invasion of macrophages and lymphoid cells by *T. annulata* results in their replication or destruction, and invasion of erythrocytes results in anemia. Tropical theileriosis may present as a peracute, acute, subacute, mild or chronic disease. In a typical acute case, the initial clinical signs are fever, swelling of superficial lymph nodes and anorexia. Emaciation is rapid, and anemia becomes progressively more severe. Petechiae on the visible mucous membranes usually indicate an unfavorable outcome. Dairy cattle exhibit a sudden and marked drop in milk production, and pregnant cows often abort. Mortality rates of 20–90% have been reported.

Menoctone, parvaquone, buparvaquone and halofuginone are all effective in the treatment of tropical theileriosis. A cell-culture-derived vaccine was developed in the 1960s and is currently used for control of *T. annulata* infection.

Tropical theileriosis has been reviewed by Pipano & Shkap (2004).

### 15.14.3. *Theileria lestoquardi* infection

Malignant theileriosis is a serious tickborne disease of sheep and goats caused by the protozoal parasite *T. lestoquardi* (syn. *T. hirci*). It occurs in the Mediterranean basin, Sudan, western and central Asia, India, and western China. It is transmitted transstadially by *Hy. anatolicum*.

Malignant theileriosis is widespread in some important sheep-breeding areas, where it causes heavy losses in both indigenous and exotic breeds and their crosses. It is seen in acute, subacute and chronic forms. In acute cases fever, anemia, swelling of superficial lymph nodes and emaciation are the most common signs of the disease. Mortality rates may approach 100%.

Buparvaquone is effective in the treatment of malignant theileriosis. An attenuated schizont vaccine prepared in lymphoid cell culture has been developed and used successfully to control the disease in Iran.

Malignant theileriosis has been reviewed by Lawrence (2004a).

### 15.14.4. *Theileria buffeli/Theileria orientalis* infection

The *T. buffeli/T. orientalis* group of parasites is widely distributed in cattle around the world, where it usually causes benign or mild bovine theileriosis, but in eastern Asia it poses an economically significant threat, causing clinical bovine theileriosis. The name *T. buffeli* was given to a parasite recovered from a water buffalo in Indochina in 1912 and the name *T. orientalis* to a parasite isolated from cattle in Siberia in 1931. Parasites from the *T. buffeli/T. orientalis* group have been reported from central and eastern Africa, South Africa, Asia, Australia, the Mediterranean basin, western Europe and North America. The primary vectors are *Ha. punctata* in Europe, the Mediterranean basin and western Asia, *Ha. concinna*, *Ha. longicornis* and *Ha. japonica* in eastern Asia, and *Ha. bancrofti* and *Ha. humerosa* in Australia; the vectors in North America and sub-Saharan Africa have not been identified. In addition to cattle, *T. buffeli/T. orientalis* also infects water buffalos and African buffalos.

The *T. buffeli/T. orientalis* group usually causes clinical bovine theileriosis in cattle subjected to stress such as intercurrent infection with other parasites or viruses. The clinical signs include fever, anorexia, anemia, reduced milk production and enlarged superficial lymph nodes. The condition is occasionally fatal.

Bovine theileriosis caused by the *T. buffeli/T. orientalis* group has been treated successfully with antimalarials such as primaquine and pamaquine and with imidocarb. In eastern Asia it is a common practice to stimulate immunity to *T. buffeli/T. orientalis* by deliberately exposing cattle to tick challenge and treating them prophylactically.

*Theileria buffeli/T. orientalis* infection has been reviewed by Lawrence (2004b).

### 15.14.5. *Theileria mutans* infection

*Theileria mutans* occurs in eastern, western and southern Africa and in the Caribbean, where it was introduced in cattle imported from Africa. It typically causes benign or mild bovine theileriosis in cattle, but in eastern Africa pathogenic strains occur that can cause severe clinical illness. *Theileria mutans* also infects African buffalos. It is transmitted transstadially by *Amblyomma* ticks. The principal vector south of central Zimbabwe is *A. hebraeum* and to the north *A. variegatum*. Other tick vectors include *A. astrion*, *A. cohaerens* and *A. gemma*.

Infection with *T. mutans* usually results only in a mild fever and slight swelling of the superficial lymph nodes; however, pathogenic strains, where there is invasion and proliferation of piroplasms in circulating erythrocytes, cause severe anemia, jaundice and sometimes death. The piroplasms of these pathogenic strains can be destroyed by treatment with primaquine, parvaquone and buparvaquone.

*Theileria mutans* infection has been reviewed by Lawrence & Williamson (2004c).

### 15.14.6. *Theileria taurotragi* infection

*Theileria taurotragi*, formerly known as *Cytauxzoon taurotragi*, is a parasite of elands that can be transmitted to cattle, causing benign or mild bovine theileriosis in eastern, central and southern Africa. However, in South Africa it occasionally causes fatal cerebral theileriosis. It is transmitted by rhipicephalid ticks including *R. appendiculatus*, *R. evertsi evertsi* and *R. pulchellus*.

Infection of cattle usually results in a mild fever with slight enlargement of superficial lymph nodes. In South Africa it has caused a fatal nervous condition known as turning sickness, or cerebral theileriosis, discussed in section 15.14.1. *Theileria taurotragi* can cause severe or fatal disease in elands.

*Theileria taurotragi* infection has been reviewed by Lawrence & Williamson (2004b).

## 15.15. Tick Paralysis

Prolonged attachment of some species of ticks can on occasion result in paralysis of the host, caused by neurotoxic substances produced by the salivary glands of the attached engorging tick or ticks, particularly females. More than 40 species of ticks have been reported to cause tick paralysis in various host species.

With regard to the invasive tick species discussed in this book, the following can cause paralysis: *A. cajennense* (in cattle, goats and sheep), *A. ovale* (in humans), *A. rotundatum* (in snakes), *A. variegatum*, *D. auratus*, *D. nuttalli* (in sheep), *Ha. punctata* (in chickens, goats and sheep), *Hy. truncatum* (in humans and sheep), *I. hexagonus*, *I. ricinus*, *R. bursa* (in sheep), *R. evertsi evertsi* (in cattle and sheep), *R. evertsi mimeticus* (in sheep) and *R. simus* (in cattle and sheep).

The first signs of tick paralysis usually occur 4–7 days after tick attachment. In humans, weakness develops in the lower extremities, and an ascending flaccid paralysis progresses to involve the trunk musculature, upper extremities and head within hours or days in the absence of pain. Patients may present with ataxia or respiratory distress. If the tick is not removed, the mortality rate resulting from respiratory paralysis is about 10%. One single tick is sufficient to paralyze and kill an adult human.

Tick paralysis in sheep in southern Africa caused by *R. evertsi* is called spring lamb paralysis. The first signs are a stiff staggering gait developing in the hind legs, even if the ticks are attached to the anterior part of the animal. This leads to complete paralysis of the hindquarters and sternal recumbency. At this stage the animal can only drag itself around using its forelegs. The condition progresses to complete paralysis with lateral recumbency and paralysis of the rectal and urinary sphincters, with passive defecation and urinary incontinence.

Treatment of tick paralysis is simple and spectacular. Detachment or removal of the ticks results in complete disappearance of all clinical signs within 24 hours.

Tick paralysis has been reviewed by Gothe & Neitz (1991) and Mans et al. (2004).

## 15.16. Tick Toxicosis

### 15.16.1. Brown tick toxicosis

Brown tick toxicosis is a condition seen in cattle in southern Africa following heavy infestation with the brown ear tick *R. appendiculatus*. The severity of the condition is dependent on the degree of tick infestation. It is characterized by prolonged fever, edema of the subcutaneous tissues of the ears, eyelids and dewlap, swelling of the regional lymph nodes, anorexia, lacrimation, serous nasal discharge and listlessness. Myiasis is a common complication. Mortality rates are high, and the convalescent period is long.

### 15.16.2. Sweating sickness

Sweating sickness is a tickborne toxicosis of cattle, and also sheep, pigs and dogs, in eastern, central and southern Africa induced by the saliva of the tick *Hy. truncatum* which produces a generalized eczema-like condition of the skin. It is most commonly seen in calves.

In cattle, the affected animal has a high fever, salivation and hyperemia of the visible mucous membranes and skin. These are followed by hyperesthesia, lacrimation and rhinitis. The serous discharges from the eyes and nose may become mucopurulent. A moist eczema appears usually about 72 hours after onset of symptoms. Myiasis is a common complication. Calf mortality rates of 30–100% have been reported. In sheep, the clinical signs are similar to those for cattle except that there is edema of the lips and the eczema is not moist. In pigs, the signs are similar to those for sheep but without edema of the lips. In dogs, clinical signs include fever, hyperemia of mucous membranes and extensive skin necrosis with sloughing of tissue around the site of the tick bites.

## 15.17. Tickborne Encephalitis

Tickborne encephalitis is a life-threatening tickborne disease of humans caused by a flavivirus of the family Flaviviridae. It is endemic in central and eastern Europe, Russia and the Far East, with cases reported from Austria, Belarus, Bulgaria, Croatia, Czech Republic, Denmark, Estonia, Finland, France, Germany, Greece, Hungary, Italy, Latvia, Liechtenstein, Lithuania, Norway, Poland, Romania, Russia, Slovakia, Slovenia, Sweden, Switzerland and Ukraine in Europe, and from China, Japan, Kazakhstan and Russia in Asia. The tickborne encephalitis virus is a single virus species with three genotypes: European, Siberian and Far-Eastern. The disease caused by the European genotype has been called central European encephalitis and that caused by the eastern genotypes Russian spring/summer encephalitis. The natural reservoirs of the virus are small mammals such as rodents (in particular mice of the genus *Apodemus*), hedgehogs, moles and shrews. Humans and other mammals such as cattle, sheep, goats, pigs, deer and dogs can become infected accidentally. The virus is transmitted by the ticks *I. ricinus* (European genotype) and *I. persulcatus* (eastern genotypes) or, less commonly, by drinking unpasteurized milk from goats, cows or sheep or, very rarely, by eating cheese prepared from unpasteurized goat milk. Tick transmission can occur either transovarially or transstadially, and the virus can also be transmitted between co-feeding ticks. Human infections usually occur in the early autumn with the European genotype and in the spring with the eastern genotypes, correlating with the activity of the tick vectors.

About 70% of cases of infection of humans with tickborne encephalitis virus are asymptomatic; however, in the other 30% of clinical cases, the disease can be severe and life-threatening. The initial clinical signs are fever, fatigue, headache and myalgia. This is followed by a weeklong asymptomatic phase only in cases caused by the European genotype. The next phase involves fever and severe headaches and signs of meningitis or meningoencephalitis. These neurological signs include ataxia, cognitive disorders such as impaired concentration and memory, dysphasia, confusion, irritability and paralysis of cranial nerve and respiratory muscles. At this time the patient may be somnolent or unconscious. The fatality rate with the European genotype is 0.5–2% with about 20% of survivors showing neurological sequelae, whereas the fatality rate with the Far-Eastern genotype is 5–35% with higher rates of sequelae. The neurological sequelae seen in survivors include spinal nerve paralysis, neuropsychiatric complaints, dysphasia, ataxia and paresis.

Treatment for patients with tickborne encephalitis is supportive since there is no curative therapy currently available. Immunity following recovery is lifelong; thus inactivated whole-virus vaccines have been developed and used widely in Austria, Germany and Russia.

Tickborne encephalitis has been reviewed by Dumpis et al. (1999), Gritsun et al. (2003), Charrel et al. (2004) and Lindquist & Vapalahti (2008).

## 15.18. Tularemia

Tularemia, also known as rabbit fever, is a tickborne disease of humans and more than 250 species of animals caused by the bacterium *Francisella tularensis*. It is endemic in the temperate regions of the Northern Hemisphere, including North America, much of continental Europe and northern Asia. *Francisella tularensis* is one of the most infectious pathogenic bacteria known, with inoculation or inhalation of as few as 10 organisms able to cause disease. The natural reservoirs of *F. tularensis* are small mammals such as rodents and lagomorphs. A wide variety of animal species can be infected by *F. tularensis*, including primates, domestic livestock (sheep, pigs, horses), carnivores (dogs, cats, minks), rodents, lagomorphs, birds, reptiles, amphibians, fishes and crustaceans. Human cases have been associated with infected rabbits, hares, beavers, muskrats, lemmings, squirrels, cats, sheep, nonhuman primates, pheasants and crayfish. Those at highest risk of infection appear to be people involved in rural activities, such as farming, hunting, trapping, butchering and forest work, that bring them into close contact with infected animals, as well as laboratory staff working with *F. tularensis*.

There are five methods by which *F. tularensis* can be transmitted to humans and other hosts: (1) the bites of infected arthropods (i.e., ticks and flies); (2) direct contact with infected animals or animal products such as while skinning a rabbit; (3) inhalation of infective aerosol droplets or dust such as when working with rodent-contaminated hay, when mowing lawns contaminated with animal carcasses or excreta or when working in a tularemia laboratory; (4) direct contact with contaminated water, food or soil or ingestion of contaminated food or water; and (5) rarely by the bites of infected animals. The primary tick vectors are *D. reticulatus* and *I. ricinus* in Europe and *A. americanum*, *D. andersoni* and *D. variabilis* in North America.

The initial clinical signs of human tularemia include fever, chills and generalized body aches. Seven forms of the disease are recognized in humans. The ulceroglandular form arises typically from the bite of infected ticks or flies or from handling of contaminated carcasses where infection occurs via cuts or abrasions, and it is characterized by a cutaneous ulcer at the inoculation site with regional lymphadenopathy. The glandular form is characterized by regional lymphadenopathy without ulcer formation. The oculoglandular form arises typically following direct contamination of the eye and is characterized by an ulcer on the conjunctiva. The oropharyngeal form arises from drinking contaminated water or ingesting contaminated food or, sometimes, by inhalation of contaminated aerosol droplets; it is characterized by a painful sore throat with pharyngitis, tonsillitis or stomatitis and cervical lymphadenopathy. The gastrointestinal form is characterized by intestinal pain, vomiting and diarrhea, and it can lead to fatal disease with ulceration of the bowel. The pneumonic form is the direct result of inhalation of contaminated aerosols or is secondary to hematogenous spread from a distal site, and it is an acute form that can lead to bronchiolitis and pleuropneumonia. The typhoidal form is a systemic illness in the absence of signs indicating either the site of inoculation or the anatomical localization of infection, and it produces a septicemia without lymphadenopathy or the appearance of an ulcer; patients with this form may be delirious, and shock may develop. Any form may be complicated by hematogenous spread to produce pleuropneumonia and sepsis. The most fatal forms are the pneumonic and typhoidal forms, with fatality rates up to 60% in untreated cases. Currently the fatality rate from human tularemia in the United States is less than 2%.

Tularemia in cats can be an asymptomatic infection, a mild illness with fever and lymphadenopathy or a severe and fatal disease. Clinical signs in cats include fever, anorexia, listlessness, lymphadenopathy, draining abscesses, oral or lingual ulceration, pneumonia and jaundice. Clinical signs in sheep and pigs include fever, stiff gait, diarrhea, weight loss and recumbency, with mortality rates in sheep being 20–50%. In the United States, tularemia produces a well-recognized syndrome of abortion and death in range sheep.

Streptomycin and gentamicin are the drugs of choice for treatment of tularemia. A live attenuated vaccine was used in the old Soviet Union for many years to protect people against tularemia. A similar vaccine was used in the United States to protect laboratory staff routinely working with *F. tularensis*; however, licensing of this vaccine has been difficult since its history is uncertain and the basis of its attenuation is unknown, and thus there is no licensed tularemia vaccine currently available in western countries (Sjostedt, 2007).

Recently *F. tularensis* was listed as a category A bioterrorism agent of public health importance by the U.S. Centers for Disease Control and Prevention (Davis, 2004a). *Francisella tularensis* has been considered a potential biological weapon (Dennis et al., 2006). It was studied at the Japanese germ warfare research units operating in Manchuria between 1932 and 1945. The U.S. military developed weapons to disseminate *F. tularensis* aerosols in the 1950s and 1960s, and by the late 1960s, *F. tularensis* was one of several biological weapons stockpiled by the U.S. military. The United States terminated its biological weapons development program in 1970 by executive order; by 1973, it had destroyed its entire biological arsenal. The Soviet Union also had a biological weapons program involving *F. tularensis* that continued into the early 1990s and resulted in weapons production of *F. tularensis* strains engineered to be resistant to antibiotics and vaccines. A World Health Organization expert committee estimated that an aerosol dispersal of 50 kg of virulent *F. tularensis* over a metropolitan area with 5 million inhabitants would result in 250,000 incapacitating casualties, including 19,000 deaths (Dennis et al., 2006).

Tularemia has been reviewed by Ellis et al. (2002) and Dennis et al. (2006).

# 16

# Risks of Invasive Ticks to the United States

More than 40 years ago, the United States Department of Agriculture (USDA) stated that "without a doubt, the exotic ticks pose a greater potential threat to our livestock industry than any other arthropod" (USDA, 1965a). Later in their review of the threats of foreign arthropod-borne pathogens to livestock in the United States, Bram et al. (2002) outlined the measures adopted to protect U.S. livestock from these foreign pathogens and concluded that "it is clear that US agriculture still faces great risk." They indicated that some of the risks were those resulting from the accelerated movement of people and their goods, the emergence of new diseases, global warming, and the potential for introduction of new pathogens as a consequence of bioterrorism. Others (e.g., Mazzoni et al., 2003) have cited global trade and commerce, in particular the continuing increase in volume of international air transport, as the forces driving dissemination of emerging diseases in human, domestic animal, wildlife and plant populations. There is no doubt that globalization and the increase in speed and volume of international transport of animals have allowed pathogens and their vectors such as ticks to disseminate globally. This global movement of animals has led to the introduction of at least 100 species of invasive ticks into the United States, together at least in some instances with infectious agents that have the potential to threaten agricultural and livestock production, biological diversity and public health.

## 16.1. Establishment of Invasive Ticks in the United States

Many species of invasive ticks have shown the ability to infect new hosts outside their native geographical range once they are transported to the United States or elsewhere through the international trade in live animals. This has been particularly common among reptilian ticks. Ten species of reptilian ticks have been found on new hosts after they have been imported into the United States (Table 16.1), with infestation of new hosts occurring presumably while they were in captivity in the United States. The American tick *A. dissimile* infested three African hosts (the ball python, the flap-necked chameleon and the leopard tortoise) and two Asian hosts (the Burmese python and the king cobra). The African tick *A. exornatum* infested two Asian hosts (Gray's monitor and the water monitor). The African tick *A. flavomaculatum* infested two Asian hosts (the Bengal monitor and the water monitor) and one American host (the giant toad). The African tick *A. latum* infested three American hosts (the boa constrictor, the giant toad and the yellow-footed tortoise) and two Asian hosts (the reticulated python and the water monitor). The African tick *A. marmoreum* infested one host from Indian Ocean islands (the Aldabra giant tortoise), one host from Pacific Ocean islands (the Galapagos giant tortoise), one European host (Hermann's tortoise) and one American host (the yellow-

Table 16.1. Invasive ticks that have spread to exotic host species from different geographical regions

| Invasive tick species | Region of origin of tick | New host species | Region of origin of host | References |
|---|---|---|---|---|
| *Amblyomma cajennense* | Americas | Barbary sheep (*Ammotragus lervia*) | Africa | USDA (1985b, 1987a) |
| | | black rhinoceros (*Diceros bicornis*) | Africa | Wilson & Richard (1984) |
| | | nilgai (*Boselaphus tragocamelus*) | Asia | USDA (1994) |
| *Amblyomma dissimile* | Americas | ball python (*Python regius*) | Africa | Clark & Doten (1995), Burridge (1997) |
| | | Burmese python (*Python molurus*) | Asia | USDA (unpublished data) |
| | | flap-necked chameleon (*Chamaeleo dilepis*) | Africa | Clark & Doten (1995), Burridge (1997) |
| | | king cobra (*Ophiophagus hannah*) | Asia | USDA (1970a, c) |
| | | leopard tortoise (*Stigmochelys pardalis*) | Africa | USDA (unpublished data) |
| | | reticulated python (*Python reticulatus*) | Asia | USDA (unpublished data) |
| *Amblyomma exornatum* | Africa | Gray's monitor (*Varanus olivaceus*) | Asia | Reeves et al. (2006) |
| | | water monitor (*Varanus salvator*) | Asia | Burridge et al. (2000a) |
| *Amblyomma flavomaculatum* | Africa | Bengal monitor (*Varanus bengalensis*) | Asia | UF (unpublished data) |
| | | giant toad (*Bufo marinus*) | Americas | USDA (unpublished data) |
| | | water monitor | Asia | Wilson & Barnard (1985) |
| *Amblyomma latum* | Africa | boa constrictor (*Boa constrictor*) | Americas | USDA (1977b, 1980, 1988), Burridge (2001), Burridge & Simmons (2003) |
| | | giant toad | Americas | USDA (unpublished data) |
| | | reticulated python | Asia | USDA (1994), Clark & Doten (1995), Burridge et al. (2000a), Burridge & Simmons (2003) |
| | | water monitor | Asia | USDA (unpublished data) |
| | | yellow-footed tortoise (*Chelonoidis denticulata*) | Americas | Burridge et al. (2000a), Burridge (2001), Burridge & Simmons (2003) |
| *Amblyomma marmoreum* | Africa | Aldabra giant tortoise (*Aldabrachelys gigantea*) | Indian Ocean | Allan et al. (1998b), Burridge et al. (2000a), Burridge (2001) |
| | | Galapagos giant tortoise (*Chelonoidis elephantopus*) | Pacific Ocean | Burridge et al. (2000a), Burridge (2001) |
| | | Hermann's tortoise (*Testudo hermanni*) | Europe | USDA (unpublished data) |
| | | yellow-footed tortoise | Americas | Allan et al. (1998b), Burridge et al. (2000a), Burridge (2001) |
| *Amblyomma nuttalli* | Africa | boa constrictor | Americas | Burridge et al. (2000a), Burridge (2001) |
| *Amblyomma rotundatum* | Americas | European green toad (*Bufo viridis*) | Eurasia | Doss et al. (1974a) |
| | | Greek tortoise (*Testudo graeca*) | Eurasia | Doss et al. (1974a) |

*continued*

Table 16.1—Continued

| Invasive tick species | Region of origin of tick | New host species | Region of origin of host | References |
|---|---|---|---|---|
| | | leopard tortoise | Africa | USDA (unpublished data) |
| | | yellow-headed temple turtle (*Heosemys annandalii*) | Asia | UF (unpublished data) |
| *Amblyomma sparsum* | Africa | boa constrictor | Americas | USDA (1970a), Wilson & Richard (1984), Wilson & Bram (1998) |
| *Amblyomma transversale* | Africa | Burmese python | Asia | Kaufman (1972) |
| | | reticulated python | Asia | Barnard & Durden (2000) |
| *Amblyomma varanense* | Asia | ball python | Africa | Burridge et al. (2000a) |
| *Amblyomma variegatum* | sub-Saharan Africa | Algerian hedgehog (*Atelerix algirus*) | North Africa | Doss et al. (1974a) |
| | | timber wolf (*Canis lupus*) | North America & Eurasia | Doss et al. (1974a) |
| *Ixodes ricinus* | Eurasia | leopard tortoise | Africa | Wilson & Bram (1998) |
| *Rhipicephalus evertsi* | Africa | nilgai | Asia | Doss et al. (1974b), USDA (1974a), Keirans & Durden (2001) |
| *Rhipicephalus (Boophilus) annulatus* | worldwide | nilgai | Asia | USDA (1980, 1982) |
| *Rhipicephalus (Boophilus) decoloratus* | Africa | common myna (*Acridotheres tristis*) | Asia | Doss et al. (1974a) |

footed tortoise). The African tick *A. nuttalli* infested one American host (the boa constrictor). The American tick *A. rotundatum* infested two Eurasian hosts (the European green toad and the Greek tortoise), one African host (the leopard tortoise) and one Asian host (the yellow-headed temple turtle). The African tick *A. sparsum* infested one American host (the boa constrictor). The African tick *A. transversale* infested two Asian hosts (the Burmese python and the reticulated python). The Asian tick *A. varanense* infested one African host (the ball python). These data demonstrate the ability of invasive ticks to broaden their host ranges when introduced into new geographical areas. Furthermore, three exotic ungulates (Barbary sheep, black rhinoceroses and nilgais) were introduced into game ranches in Texas and Florida where they became infested with invasive ticks (*A. cajennense, R. evertsi* and *R. (B.) annulatus*) present on those ranches (Table 16.1), again demonstrating the ability of invasive ticks to broaden their host ranges when the opportunity arises.

Several species of invasive ticks (*A. auricularium, A. cajennense, A. dissimile, A. exornatum, A. humerale, A. marmoreum, A. ovale, A. rotundatum* and *R. (B.) annulatus*) have been found to have spread to host species native to the United States (Table 16.2). This is particularly apparent in Florida. *Amblyomma dissimile*, which has been established in southern Florida since early last century, has spread to nine species of native snakes as well as native mice, cattle and gopher tortoises. *Amblyomma auricularium* has been found on dogs, feral pigs, armadillos and opossums in Florida since 1989, leading to speculation that this invasive tick species might now be established in the southeastern United States. One worrying feature of these infestations of native species relates to the gopher tortoise, a protected species in Florida. The gopher tortoise (Fig. 16.1) has been found infested with three invasive tick species (*A. dissimile, A. exornatum* and *A. rotundatum*), one of which, *A. exornatum*, is known to be pathogenic for its lizard hosts, causing respiratory distress and even death from suffocation in monitor lizards whose nasal passages are a preferred site of attachment. Very little work has been done experimentally to determine the ability of invasive tick species to infest native American wildlife; however, Oliver et al. (1988) did successfully infest an American turtle and an American snake species with the Asian

Fig. 16.1. A gopher tortoise (*Gopherus polyphemus*), a species native to Florida, found infested with the invasive ticks *Amblyomma dissimile*, *A. exornatum* and *A. rotundatum*. (Photograph courtesy of Ray Ashton.)

tick *A. varanense* (Table 16.2). These data demonstrate the ability of at least some invasive ticks to infest native U.S. host species.

At least three invasive tick species have become established as breeding populations in limited foci in the southern United States (Table 16.3). The reptilian tick *A. dissimile* became established in southern Florida in the early 1900s and has spread to infest native reptiles, particularly snakes. The mammalian tick *A. cajennense*, which infests both animals and humans, became established in southern Texas also in the early 1900s. The amphibian tick *A. rotundatum* became established in southern Florida in the late 1970s and has begun spreading to infest native reptiles.

A further seven species of invasive ticks have become established in the southern United States but have been eradicated through focused efforts (Table 16.3). The two cattle ticks *R. (B.) annulatus* and *R. (B.) microplus* were widely distributed in the southern United States in the early 1900s but were eradicated from the continental United States by 1960 through a costly federal-state eradication program; however, these two ticks continue to pose threats to U.S. cattle industries because of their periodic reintroductions into southern Texas from Mexico. The African ungulate tick *R. evertsi* was found to be established in an exotic wild animal compound in Florida in 1960, but it was successfully eradicated before it could spread to native hosts. The tropical horse tick *D. nitens* was found in established infestations in southern Florida and southern Texas in the 1960s, but it has since been eradicated from both states. Two reptilian ticks, *A. marmoreum* and *A. sparsum*, were found to have established discrete breeding populations in exotic tortoise facilities in Florida in the late 1990s, but immediate actions eradicated these local infestations. The monitor lizard tick *A. komodoense* became established in the Komodo dragon exhibit at a zoo in Florida during the 1990s, but again it was eradicated.

Two other species of invasive ticks have been found recently infesting native species in Florida, suggesting that one or both might now be established in the state (Table 16.3). The armadillo tick *A. auricularium* has been found on a dog, feral pigs, nine-banded armadillos and a Virginian opossum, whereas the reptile tick *A. exornatum* has been found on a gopher tortoise.

Very little work has been published using modeling techniques to estimate the climatic suitability of the United States for invasive tick species. Climate is an important determinant of the distribution of ticks. Sutherst & Maywald (1985) quantified the climate requirements of some important tick species and incorporated them into a climate matching model that calculated an ecoclimatic index to describe the climate suitability of a given location for a given tick species. They showed that the southeastern United States, par-

Table 16.2. Invasive ticks that have spread to host species native to the United States

| Invasive tick species | State | New native host species | References |
|---|---|---|---|
| Amblyomma auricularium | Texas | cattle | USDA (1982) |
| | Florida | dog | USDA (1994) |
| | Florida | feral pig | Allan et al. (2001) |
| | Florida | nine-banded armadillo (*Dasypus novemcinctus*) | Lord & Day (2000) |
| | Florida | Virginian opossum (*Didelphis virginiana*) | USDA (unpublished data) |
| Amblyomma cajennense | Illinois | cattle | USDA (1962a) |
| | Florida | cattle | USDA (1969, 1970a, 1973a, 1981) |
| Amblyomma dissimile | Florida | racer (*Coluber constrictor*) | USDA (1970c) |
| | Florida | cattle | Cooley & Kohls (1944), Bishopp & Trembley (1945), Burridge & Simmons (2003) |
| | Florida | common kingsnake (*Lampropeltis getula*) | Bequaert (1945), Wilson & Kale (1972), Burridge & Simmons (2003) |
| | Florida | corn snake (*Elaphe guttata*) | USDA (unpublished data) |
| | Florida | cotton mouse (*Peromyscus gossypinus*) | Durden et al. (1993), Burridge & Simmons (2003) |
| | Florida | eastern diamondback rattlesnake (*Crotalus adamanteus*) | Bequaert (1945), Burridge & Simmons (2003) |
| | Florida | gopher tortoise (*Gopherus polyphemus*) | Foster et al. (2000) |
| | Texas | "native" snake | USDA (1974a) |
| | Florida | North American rat snake (*Elaphe obsoleta*) | Bequaert (1932), Wilson & Kale (1972), Burridge & Simmons (2003) |
| | Florida | pine snake (*Pituophis melanoleucus*) | Wilson & Kale (1972), Burridge & Simmons (2003) |
| | Florida | pygmy rattlesnake (*Sistrurus miliarus*) | Bequaert (1932), Cooley & Kohls (1944), Bishopp & Trembley (1945), Burridge & Simmons (2003) |
| | Florida | southern indigo snake (*Drymarchon corais*) | Bequaert (1932), Cooley & Kohls (1944), Bishopp & Trembley (1945), Wilson & Kale (1972), Durden et al. (1993), Barnard & Durden (2000), Foster et al. (2000) |
| | Florida | water moccasin (*Agkistrodon piscivorus*) | Foster et al. (2000) |
| Amblyomma exornatum | Florida | gopher tortoise | Berube (2006) |
| Amblyomma humerale | Texas | ornate box turtle (*Terrapene ornata*) | Bilsing & Eads (1947) |
| Amblyomma marmoreum | Florida | dog | Allan et al. (1998b), Burridge et al. (2000a) |
| Amblyomma ovale | Texas | coyote (*Canis latrans*) | USDA (1974b), Keirans & Durden (2001) |

*continued*

Table 16.2—Continued

| | | | |
|---|---|---|---|
| | Iowa | dog | Eddy & Joyce (1942) |
| | Texas | dog | USDA (1972) |
| | Tennessee | hispid cotton rat (*Sigmodon hispidus*) | Durden & Kollars (1992) |
| *Amblyomma rotundatum* | Florida | gopher tortoise | USDA (unpublished data) |
| | Florida | racer (*Coluber constrictor*) | Hanson et al. (2007) |
| *Amblyomma varanense* | - | eastern box turtle[a] (*Terrapene carolina*) | Oliver et al. (1988) |
| | - | North American rat snake[a] | Oliver et al. (1988) |
| *Rhipicephalus (Boophilus) annulatus* | Texas | wapiti (*Cervus elaphus*) | USDA (1994) |

[a]Infested experimentally

Table 16.3. Invasive ticks that have, or may have, become established in limited foci in the United States

| Invasive tick species | Region | References | Comments |
|---|---|---|---|
| *Rhipicephalus (Boophilus) annulatus* | southern United States | Graham & Hourrigan (1977) | Widely distributed in southern U.S. in early 1900s; eradicated from U.S. by 1960; reintroductions to south Texas from Mexico occur periodically |
| *Rhipicephalus (Boophilus) microplus* | southern United States | Graham & Hourrigan (1977) | As for *R. (B.) annulatus* |
| *Amblyomma dissimile* | Florida | Bequaert (1932) | Established in southern Florida since the early 1900s |
| *Amblyomma cajennense* | Texas | Cooley & Kohls (1944) | Established in southern Texas since the early 1900s |
| *Rhipicephalus evertsi* | Florida | Bruce (1962) | Established infestation found in exotic wild animal compound in 1960; eradication of infestation successful |
| *Dermacentor nitens* | Florida & Texas | Strickland & Gerrish (1964) | Established infestations found in southern Florida and southern Texas by 1963; infestations now eradicated |
| *Amblyomma rotundatum* | Florida | Meshaka et al. (2004) | Established in southern Florida since 1979 |
| *Amblyomma marmoreum* | Florida | Allan et al. (1998b), Burridge et al. (2002b) | Established infestations found in exotic tortoise facilities in late 1990s; infestations eradicated |
| *Amblyomma sparsum* | Florida | Burridge et al. (2002b) | Established infestation found in exotic tortoise facility in late 1990s; infestation eradicated |
| *Amblyomma komodoense* | Florida | Burridge et al. (2004b) | Infestation established in Komodo dragon exhibit at zoo in 1990s; infestation eradicated |
| *Amblyomma auricularium* | Florida | USDA (1994), Lord & Day (2000), Allan et al. (2001), USDA (unpublished data) | Found on native animals (dog, feral pig, nine-banded armadillo, Virginian opossum) in Florida, suggesting established in the state |
| *Amblyomma exornatum* | Florida | Berube (2006) | Found on a native wildlife species (gopher tortoise) in central Florida, suggesting possible establishment in the state |

Table 16.4. Non-native vertebrate species established in Florida since 1900

| Non-native species | Year of establishment | Counties with breeding populations | Invasive ticks known to feed on non-native species |
|---|---|---|---|
| **MAMMALS** | | | |
| *Rusa unicolor* (sambar) | 1908 | Franklin | A. javanense, A. testudinarium, D. auratus, Ha. hystricis, Ha. longicornis, Hy. detritum, Hy. marginatum group, R. (B.) microplus |
| *Dasypus novemcinctus* (nine-banded armadillo) | 1920s | All 67 counties | A. auricularium, A. cajennense, A. coelebs, A. humerale, A. oblongoguttatum, A. parvum, A. pictum, A. pseudoconcolor, A. rotundatum |
| *Nasua narica* (northern coati) | 1928 | Dade, Highlands, Palm Beach | A. auricularium, A. ovale |
| *Chlorocebus acthiops* (green monkey) | 1950s | Broward | A. hebraeum, A. pomposum, A. variegatum, I. pilosus, R. appendiculatus, R. evertsi, R. turanicus |
| *Vulpes vulpes* (red fox) | 1950s | All 67 counties | A. variegatum, D. reticulatus, Ha. leachi, Ha. muhsamae, Ha. punctata, Hy. dromedarii, Hy. lusitanicum, I. hexagonus, I. ricinus, R. bursa, R. turanicus, R. (B.) annulatus |
| *Artibeus jamaicensis* (Jamaican fruit-eating bat) | 1983 | ?[a] | A. cajennense |
| **REPTILES** | | | |
| *Leiocephalus carinatus* (northern curly-tailed lizard) | 1935 | Brevard, Broward, Collier, Dade, Martin, Munroe, Palm Beach | A. albopictum, A. dissimile |
| *Ameiva ameiva* (giant ameiva) | 1954 | Broward, Dade, Palm Beach | A. dissimile, A. scutatum |
| *Iguana iguana* (green iguana) | 1966 | Broward, Dade, Lee, Munroe, Palm Beach | A. cajennense, A. dissimile, A. sabanerae, A. scutatum |
| *Ctenosaura pectinata* (Mexican spiny-tailed iguana) | 1972 | Dade | A. dissimile, A. scutatum |
| *Agama agama* (common agama) | 1976 | Broward, Charlotte, Dade, Martin, Seminole | A. nuttalli |
| *Calotes versicolor* (variable agama) | 1978 | St. Lucie | A. varanense |
| *Ctenosaura similis* (black iguana) | 1978 | Broward, Charlotte, Collier, Dade, Lee | A. cajennense, A. dissimile, A. scutatum |
| *Python molurus* (Burmese python) | 1980s | Dade, Munroe | A. crassipes, A. dissimile, A. fuscolineatum, A. javanense, A. latum, A. transversale, A. varanense |
| *Boa constrictor* (boa constrictor) | 1990 | Dade | A. argentinae, A. dissimile, A. latum, A. nuttalli, A. quadricavum, A. rotundatum, A. scutatum, A. sparsum |
| *Mabuya multifasciata* (Indonesian brown mabuya) | 1990 | Dade | A. helvolum |
| *Varanus niloticus* (Nile monitor) | 1990 | Lee | A. exornatum, A. falsomarmoreum, A. flavomaculatum, A. hebraeum, A. latum, A. marmoreum, A. nuttalli, A. sparsum, A. variegatum |
| **AMPHIBIANS** | | | |
| *Bufo marinus* (giant toad) | 1957 | Broward, Dade, Highlands, Hillsborough, Martin, Munroe, Okeechobee, Palm Beach, Pasco, Pinellas | A. argentinae, A. cajennense, A. dissimile, A. flavomaculatum, A. humerale, A. latum, A. rotundatum, A. varium, R. (B.) microplus |

[a]Not reported

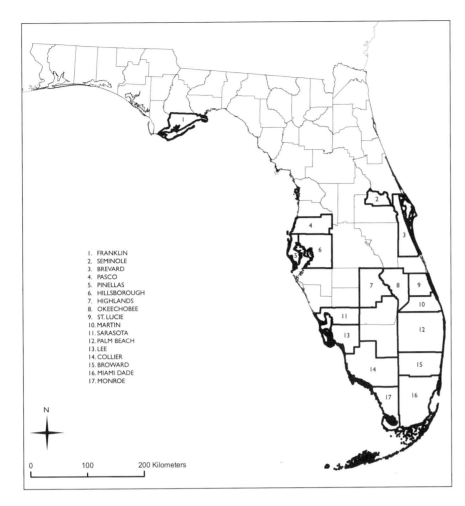

Map 16.1. Florida, showing counties in which non-native vertebrate species have established breeding populations.

ticularly Florida, was suitable for the establishment of *A. variegatum*, *R. appendiculatus* and *R. (B.) microplus*. Later modeling studies by Estrada-Pena (2001) and Estrada-Pena et al. (2005) not only confirmed the habitat suitability of the southeastern United States for *R. (B.) microplus*, but the latter study also showed that the area of habitat climatically suitable for *R. (B.) microplus* in the southeastern United States is expanding. Another modeling study by Estrada-Pena et al. (2007) confirmed the climatic suitability of Florida for *A. variegatum* and found that this tick also has the potential to invade the whole of the Caribbean, wide areas of Colombia and Venezuela, parts of the Mesoamerican corridor and wide areas of Mexico.

A recent report by the Florida Fish & Wildlife Conservation Commission (http://myfwc.com/WILD-LIFEHABITATS/Nonnative_index.htm) showed that a number of non-native vertebrate species have become established in Florida since 1900. Among these are 18 species (6 mammals, 11 reptiles, 1 amphibian) known to act as hosts for some of the invasive ticks introduced into the United States (Table 16.4). The established mammals include carnivores, a deer, an armadillo, a monkey and a bat, the established reptiles lizards and snakes, and the established amphibian a toad, the majority of which have developed breeding populations in central and southern Florida (Map 16.1). Between them, these 18 established vertebrate species are known hosts for no fewer than 55 of the 100 invasive tick species introduced into the United States, demonstrating that established non-native vertebrates form an additional pool of potential hosts for the establishment of invasive ticks, particularly in southern Florida.

The foregoing data demonstrate clearly that some invasive ticks introduced into the United States have the ability to infest new hosts, whether imported species or native species, and to become established in the southern United States, particularly in Florida, as breeding populations; however, information on the

Table 16.5. Invasive ticks reported feeding on humans

| Tick | Stages[a] | | | No. of reports | Vector of human diseases[b] |
|---|---|---|---|---|---|
| I. ricinus | A | N | L | 60 | + |
| A. cajennense | A | N | L | 57 | + |
| Hy. marginatum group | A | N | L | 32 | + |
| A. testudinarium | A | N | L | 25 | - |
| A. ovale | A | | | 22 | - |
| D. auratus | A | N | L | 22 | + |
| I. hexagonus | A | N | L | 22 | + |
| A. variegatum | A | N | L | 21 | + |
| R. (B.) microplus | A | | L | 21 | - |
| A. parvum | A | | | 20 | - |
| D. reticulatus | A | | | 20 | + |
| Ha. punctata | A | | | 19 | + |
| R. bursa | A | | | 19 | - |
| A. oblongoguttatum | A | N | | 17 | - |
| R. pulchellus | A | N | L | 17 | + |
| Hy. anatolicum spp. | A | N | L | 15 | + |
| A. triguttatum | A | N | L | 14 | - |
| Hy. truncatum | A | | | 14 | + |
| Ha. hystricis | A | N | | 13 | + |
| R. turanicus | A | | | 13 | + |
| Hy. detritum | A | | | 12 | + |
| R. appendiculatus | A | N | L | 11 | + |
| A. hebraeum | A | N | L | 10 | + |
| A. incisum | A | N | | 10 | - |
| A. nuttalli | A | N | L | 10 | - |
| Ha. longicornis | A | N | L | 10 | - |
| A. coelebs | A | N | | 9 | - |
| A. longirostre | A | N | | 8 | - |
| I. schillingsi | A | | | 8 | - |
| A. tholloni | A | N | | 7 | - |
| Ha. juxtakochi | A | N | | 7 | - |
| Hy. dromedarii | A | N | L | 7 | - |
| Hy. lusitanicum | A | | | 7 | + |
| R. compositus | A | | | 7 | - |
| R. senegalensis | A | | | 7 | - |
| R. simus | A | | | 7 | + |
| R. (B.) decoloratus | A | | L | 7 | - |
| A. lepidum | A | N | | 6 | - |
| D. nitens | A | | L | 6 | - |
| R. sulcatus | - | | | 6 | - |
| A. marmoreum | A | N | L | 5 | - |
| A. moreliae | - | | | 5 | - |
| D. nuttalli | A | | | 5 | + |
| Ha. leachi | A | | L | 5 | + |
| R. evertsi | A | N | L | 5 | + |
| Ha. elongata | | N | | 4 | - |
| A. dissimile | A | | | 3 | - |
| A. geoemydae | A | N | | 3 | - |
| Hy. aegyptium | | N | | 3 | - |
| Hy. albiparmatum | A | | | 3 | + |
| I. pilosus | A | N | | 3 | - |
| R. (B.) annulatus | A | | | 3 | - |
| A. calcaratum | - | | | 2 | - |
| A. latum | - | | | 2 | - |
| A. rotundatum | - | | | 2 | - |
| B. hydrosauri | - | | | 2 | + |
| D. rhinocerinus | A | | | 2 | - |
| I. luciae | - | | | 2 | - |
| A. gemma | A | | | 1 | - |
| A. pomposum | - | | | 1 | - |
| A. pseudoconcolor | - | | | 1 | - |
| A. sparsum | A | | | 1 | - |
| R. muehlensi | A | | | 1 | - |

continued

[a] A = adult, N = nymph, L = larva, - = not recorded
[b] + = vector role established, - = no vector role established

ability of most other invasive tick species to become established in the United States is totally lacking.

## 16.2. Tickborne Diseases Transmissible to Humans

Fully 63 of the 100 invasive tick species introduced into the United States have been reported on one or more occasions to feed on humans (Table 16.5). Of these, 23 species have been reported to be vectors of human diseases. Two species (*I. ricinus* and *A. cajennense*) are frequent pests of humans, with more than 50 reports of human infestations for each. Another nine species (*Hy. marginatum* group, *A. testudinarium*, *A. ovale*, *D. auratus*, *I. hexagonus*, *A. variegatum*, *R. (B.) microplus*, *A. parvum* and *D. reticulatus*) have been documented feeding on humans in 20–32 reports and a further 15 (*Ha. punctata*, *R. bursa*, *A. oblongoguttatum*, *R. pulchellus*, *Hy. anatolicum* spp., *A. triguttatum*, *Hy. truncatum*, *Ha. hystricis*, *R. turanicus*, *Hy. detritum*, *R. appendiculatus*, *A. hebraeum*, *A. incisum*, *A. nuttalli* and *Ha. longicornis*) in 10–19 reports.

Six invasive tick species have been shown to cause diseases directly in humans. The ticks *A. ovale*, *Hy. truncatum* and *R. simus* can cause tick paralysis in humans (Table 16.6). Another three species have been noted to cause unpleasant local skin reactions; *A. testudinarium* causes skin eruptions, *A. triguttatum* an allergic dermatitis and *A. cajennense* intense itching with uncomfortable indurated lesions.

Invasive tick species are proven vectors of 20 human diseases (Table 16.7), some of which can be life-threatening, such as babesiosis, Brazilian spotted fever, Crimean-Congo hemorrhagic fever, granulocytic anaplasmosis, Omsk hemorrhagic fever, Q fever, tick typhus, tickborne encephalitis and tularemia. Of these 20 diseases, 6 already occur in the United States: babesiosis caused by *B. microti*, Brazilian spotted fever (as Rocky Mountain spotted fever), granulocytic anaplasmosis, Lyme borreliosis, Q fever and tularemia. In addition, the agents causing four of these diseases (Crimean-Congo hemorrhagic fever virus, Omsk

Table 16.6. Diseases caused by invasive ticks

| Diseases | Hosts affected | Ticks |
| --- | --- | --- |
| Respiratory distress | Lizards | A. exornatum |
| Tick paralysis | Humans | A. ovale, Hy. truncatum, R. simus |
| | Cattle | A. cajennense, R. evertsi evertsi, R. simus |
| | Sheep | A. cajennense, A. variegatum, D. auratus, D. nuttalli, Ha. punctata, Hy. truncatum, I. hexagonus, I. ricinus, R. bursa, R. evertsi evertsi, R. evertsi mimeticus, R. simus |
| | Goats | A. cajennense, Ha. punctata |
| | Chickens | Ha. punctata |
| | Snakes | A. rotundatum |
| Tick toxicosis | Cattle, sheep, pigs, dogs | Hy. truncatum |
| | Cattle | R. appendiculatus |

Table 16.7. Diseases transmitted to humans naturally by invasive ticks

| Disease | Causative agent | Tick vectors |
| --- | --- | --- |
| African tick-bite fever | Rickettsia africae | A. hebraeum, A. variegatum |
| Babesiosis | Babesia divergens | I. ricinus |
| Babesiosis | Babesia microti | I. ricinus |
| Babesiosis | Babesia sp. EU1 | I. ricinus |
| Brazilian spotted fever | Rickettsia rickettsii | A. cajennense |
| Crimean-Congo hemorrhagic fever (CCHF) | CCHF virus | A. variegatum, Hy. anatolicum, Hy. marginatum, Hy. rufipes, Hy. truncatum, R. evertsi evertsi, R. evertsi mimeticus, R. pulchellus |
| Eyach fever (EF) | EF virus | I. ricinus |
| Flinders Island spotted fever | Rickettsia honei | B. hydrosauri |
| Granulocytic anaplasmosis | Anaplasma phagocytophilum | I. ricinus |
| Japanese spotted fever | Rickettsia japonica | Ha. hystricis |
| Lyme borreliosis | Borrelia burgdorferi | I. hexagonus, I. ricinus |
| North Asian tick typhus | Rickettsia sibirica sibirica | D. auratus, D. nuttalli, D. reticulatus, Ha. punctata |
| Omsk hemorrhagic fever (OHF) | OHF virus | D. reticulatus |
| Q fever | Coxiella burnetii | A. triguttatum, D. reticulatus, Hy. detritum, Hy. lusitanicum, R. turanicus |
| Rickettsiosis | Rickettsia aeschlimanni | Hy. marginatum |
| Rickettsiosis | Rickettsia helvetica | I. ricinus |
| Tick typhus | Rickettsia conorii | A. variegatum, Ha. leachi, Hy. albiparmatum, Hy. rufipes, Hy. truncatum, I. hexagonus, R. appendiculatus, R. evertsi, R. pulchellus, R. simus, R. turanicus |
| Tickborne encephalitis (TBE) | TBE virus | I. ricinus |
| Tickborne lymphadenopathy | Rickettsia slovaca | D. reticulatus |
| Tularemia | Francisella tularensis | D. reticulatus, I. ricinus |

hemorrhagic fever virus, *Coxiella burnetii* and *Francisella tularensis*) have been listed as potential bioterrorism agents or biological weapons by the U.S. Centers for Disease Control and Prevention.

It can be seen in Table 16.7 that a few invasive tick species are vectors of multiple diseases, making them particular risks to public health. These include *I. ricinus*, a vector of 9 of the 20 human diseases, *D. reticulatus*, a vector of 5, and *A. variegatum*, a vector of 3. Five of the life-threatening human diseases transmitted by the invasive ticks are currently exotic to the United States. They are babesiosis caused by *B. divergens*, Crimean-Congo hemorrhagic fever, Omsk hemorrhagic fever, tick typhus and tickborne encephalitis; *I. ricinus*, *D. reticulatus* and *A. variegatum* transmit one or more of each of these diseases. It is clear, therefore, that from the public health standpoint these three invasive tick species pose definitive risks to human health in the United States.

## 16.3. Tickborne Diseases Transmissible to Cattle

A total of 63 of the 100 invasive tick species introduced into the United States have been reported to feed on cattle, 38 of them commonly. Of those, 24 species have been reported to be vectors of bovine diseases.

Five invasive tick species have been shown to cause diseases directly in cattle. The ticks *A. cajennense*, *R. evertsi evertsi* and *R. simus* can cause tick paralysis in cattle, and *Hy. truncatum* and *R. appendiculatus* can cause tick toxicosis (Table 16.6). Many of the invasive tick species have long mouthparts that can impart considerable damage to the bovine host. Some (*A. hebraeum*, *Hy. rufipes* and *Hy. truncatum*) have been reported to produce tissue damage leading to abscess formation and secondary myiasis, while others cause udder damage resulting in loss of teats or udder quarters (*A. hebraeum*, *A. lepidum* and *A. variegatum*), create damage to hides (*R. (B.) annulatus* and *R. (B.) microplus*), induce lameness (*A. variegatum*) or reduce bovine productivity when infestations are heavy (*A. pomposum*, *A. variegatum*, *R. appendiculatus* and *R. (B.) microplus*).

Invasive tick species are known vectors of many bovine diseases, including heartwater and several forms of anaplasmosis, babesiosis and theileriosis (Table 16.8). The most important form of bovine anaplasmosis, caused by *A. marginale*, is already present in the United States; however, heartwater and the most pathogenic forms of bovine babesiosis and theileriosis are all currently foreign to the United States, although three of the diseases are not far from the U.S. mainland, with heartwater in the eastern Caribbean and babesiosis caused by *B. bigemina* and by *B. bovis* in Mexico. The invasive tick species that transmit these diseases are *Amblyomma* spp. (particularly *A. hebraeum* and *A. variegatum*) for heartwater, *Rhipicephalus* spp. (particularly *R. (B.) annulatus* and *R. (B.) microplus*) for babesiosis caused by *B. bigemina* and by *B. bovis*, *I. ricinus* for babesiosis caused by *B. divergens*, *Rhipicephalus* spp. (especially *R. appendiculatus*) for theilerioses caused by *T. parva*, and *Hyalomma* spp. for theileriosis caused by *T. annulata*. Introduction of any of these diseases into the United States would have profound negative economic impacts for the cattle industries, in terms both of mortality and morbidity on the one hand and of restrictions in international trade on the other.

The USDA has established three categories of foreign arthropod pests and arthropod-borne disease vectors based on potential for introduction, establishment and economic impact (Wilson & Bram, 1998; Bram & George, 2000). Category A is those species with the highest potential for introduction, establishment and economic impact, and it includes six invasive tick species: *A. hebraeum*, *A. variegatum*, *I. ricinus*, *R. appendiculatus*, *R. (B.) annulatus* and *R. (B.) microplus*. Category B is those species that "merit particular concern with respect to introduction, establishment, and economic impact," and it includes unspecified ticks of the genera *Amblyomma*, *Dermacentor*, *Hyalomma*, *Ixodes* and *Rhipicephalus*. Category C is those species "with some potential for introduction, establishment, and economic interest," and it includes unspecified ticks of the family Ixodidae. The risks of introduction of three of these invasive tick species (*A. variegatum*, *R. (B.) annulatus* and *R. B. microplus*) and their associated diseases (heartwater and bovine babesiosis) are so real that the threats that they pose to the U.S. cattle industries and beyond are worthy of more detailed discussion.

### 16.3.1. Risk of introduction of heartwater

Heartwater is an acute tickborne disease of domestic and wild ruminants, including cattle, sheep, goats, deer and some antelopes (see section 15.6). Mortality rates in susceptible species in the United States would be expected to be from 40% to approaching 100%. Its introduction would be devastating for the cattle industries and the deer populations of the United States. Heartwater is a disease of sub-Saharan Africa introduced into the eastern Caribbean when cattle infested with the tropical bont tick *A. variegatum*, the major vector of heartwater, were transported from western Africa to the French West Indies. In 1980, heartwater was diagnosed in Guadeloupe, the first confirmation of the disease outside the African continent (Perreau et al., 1980). Soon thereafter heartwater was found on another French island in the Caribbean, Marie Galante (Uilenberg et al., 1984).

Table 16.8. Diseases of cattle for which invasive ticks are vectors

| Disease | Causative agent | Tick vectors |
| --- | --- | --- |
| Anaplasmosis | *Anaplasma bovis* | *A. variegatum, Hy. anatolicum, Hy. excavatum, R. appendiculatus* |
| Anaplasmosis | *Anaplasma centrale* | *R. simus* |
| Anaplasmosis | *Anaplasma marginale* | *Hy. detritum, Hy. lusitanicum, Hy. rufipes, R. bursa, R. simus, R. (B.) annulatus, R. (B.) microplus* |
| Babesiosis | *Babesia bigemina* | *R. evertsi evertsi, R. (B.) annulatus, R. (B.) decoloratus, R. (B.) microplus* |
| Babesiosis | *Babesia bovis* | *R. (B.) annulatus, R. (B.) microplus* |
| Babesiosis | *Babesia divergens* | *I. ricinus* |
| Babesiosis | *Babesia major* | *Ha. punctata* |
| Babesiosis | *Babesia occultans* | *Hy. rufipes* |
| Borreliosis | *Borrelia theileri* | *R. evertsi evertsi, R. (B.) annulatus, R. (B.) decoloratus, R. (B.) microplus* |
| Corridor disease | *Theileria parva* | *R. appendiculatus* |
| Dermatophilosis (acute) | *Dermatophilus congolensis* | *A. variegatum*[a] |
| East Coast fever | *Theileria parva* | *R. appendiculatus, R. capensis, R. evertsi evertsi, R. pulchellus, R. simus* |
| Heartwater | *Ehrlichia ruminantium* | *A. hebraeum, A. lepidum, A. pomposum, A. variegatum* |
| Pasture fever | *Anaplasma phagocytophilum* | *I. ricinus* |
| Theileriosis | *Theileria buffeli/ T. orientalis* | *Ha. longicornis, Ha. punctata* |
| Theileriosis | *Theileria mutans* | *A. gemma, A. hebraeum, A. variegatum* |
| Theileriosis | *Theileria taurotragi* | *R. appendiculatus, R. evertsi evertsi, R. pulchellus* |
| Theileriosis, benign | *Theileria velifera* | *A. hebraeum, A. lepidum, A. variegatum* |
| Theileriosis, tropical | *Theileria annulata* | *Hy. anatolicum, Hy. excavatum, Hy. detritum, Hy. dromedarii, Hy. lusitanicum, Hy. marginatum* |
| Theileriosis, Zimbabwe | *Theileria parva* | *R. appendiculatus* |
| Trypanosomiasis, benign | *Trypanosoma theileri* | *Hy. anatolicum* |

[a]*A. variegatum* is not known to be a vector of dermatophilosis, but is very closely associated with development of the acute form of the disease.

It was obvious that this situation posed a threat to the animal industries of the United States and, therefore, the author developed a research project to determine the actual distribution of heartwater and its tick vector *A. variegatum* in the Caribbean. The project found that heartwater had spread to a third island, Antigua (Birnie et al., 1985), that *A. variegatum* was widely distributed in the eastern Caribbean from Puerto Rico in the north to St. Lucia and Barbados in the south (Burridge, 1985), and that cattle production on several Caribbean islands was being decimated by acute bovine dermatophilosis, a horrific skin disease experienced only in herds infested with *A. variegatum* ticks (Burridge et al., 1984; Burridge, 1985).

An African bird, the cattle egret (*Bubulcus ibis*), flew across the Atlantic Ocean to reach South America last century, with the first cattle egrets recorded in the New World in Guyana in 1937 (Barnes, 1955). They spread westward across northern South America and into the Caribbean and the United States, with the first report of a cattle egret in Florida in 1948 (Blake, 1961). Cattle egrets have now become a common sight in the Caribbean and the southern United States, especially in the company of cattle, where they feed on insects disturbed by the grazing animals (Fig. 16.2). The risk of introduction of heartwater into the United States from the Caribbean was heightened when it was found that cattle egrets were good hosts for *A. variegatum* nymphs and larvae (Barre et al., 1987, 1988, 1991) and when it was discovered in 1992 that a cattle egret, banded on Guadeloupe, was found on Long Key in Florida (Corn et al., 1993). This latter finding demonstrated the potential for cattle egrets to carry *A. variegatum* ticks from heartwater-endemic islands such as Guadeloupe to the U.S. mainland and provided further evidence that much of the interisland spread of *A. variegatum* in the Caribbean has occurred through the movement of infested migratory birds, in particular, cattle egrets.

Unlike many foreign animal diseases, a tick vector capable of transmitting heartwater already occurs in the United States. Uilenberg (1982) found the Gulf Coast tick *Amblyomma maculatum* to be an experimental vector of heartwater, and Mahan et al. (2000) later showed that *A. maculatum* was as efficient a vector as the two major field vectors, *A. hebraeum* and *A. variegatum*. The Gulf Coast tick is found commonly in the southeastern United States in states such as Alabama, Florida, Georgia, Mississippi, Oklahoma, South

Fig. 16.2. Cattle egrets (*Bubulcus ibis*) with a grazing steer in Florida.

Carolina and Texas (Cooley & Kohls, 1944; Goddard & Norment, 1983). Adult *A. maculatum* feed commonly on cattle, sheep, goats and white-tailed deer (Bishopp & Trembley, 1945), and nymphs have also been found on the domestic species (Hixson, 1940; Bishopp & Trembley, 1945; Ketchum et al., 2005). It seems clear, therefore, that heartwater could spread within the United States, if introduced, through this native tick species.

During the twentieth century, many exotic large mammal species were introduced into private ranches in the United States for hunting, for conservation of endangered species and for esthetic reasons, particularly in Texas and more recently in Florida. Some of these exotic species have even become established as free-ranging unhusbanded herds in Texas. Fifteen species of wild ruminants (African buffalos, black wildebeests, blesboks, blue wildebeests, elands, giraffes, greater kudus, lechwes, sable antelopes, sitatungas, spotted deer, springboks, steenboks, Timor deer and white-tailed deer) have been proven to be susceptible to heartwater (Peter et al., 2002) and, as of 1994, all but two of these species (steenboks and Timor deer) were reported to have been introduced into Texas as free-range exotic hoofed stock (Mungall & Sheffield, 1994). This shows that at least in Texas there is a significant population of exotic wildlife that is free-ranging and susceptible to heartwater.

Many of the wild African ruminants known to be susceptible to heartwater can become subclinical carriers of *E. ruminantium*, capable of infecting ticks that later are able to transmit fatal heartwater to domestic livestock. Carrier species include African buffalos, black wildebeests, blesboks, blue wildebeests, elands, giraffes, greater kudus and sable antelopes (Neitz, 1933, 1935; Andrew & Norval, 1989; Peter et al., 1998, 1999). A Law Enforcement Management Information System (LEMIS) report from the U.S. Fish & Wildlife Service (a wildlife import/export database that catalogs every wildlife shipment declared to the Service) for the period 1992–1997 showed that all eight carrier species were imported into the United States from sub-Saharan Africa on multiple occasions during that five-year period. This demonstrates that African wildlife capable of being subclinically infected with heartwater were recently imported into the United States.

In 1997, the author and his colleagues began investigating the flood of invasive tick species imported into the United States through the international trade in live reptiles. They discovered that tortoises imported from Africa had introduced the tick *A. marmoreum* into central Florida, where it had become established as a breeding population and spread to domestic dogs (Allan et al., 1998b). Previous unpublished work in South Africa had suggested that *A. marmoreum* might transmit heartwater (Bezuidenhout, 1987), and subsequent studies did confirm *A. marmoreum* as a competent experimental vector of the disease (Peter et al., 2000). A follow-up study examined *A. marmoreum* and *A. sparsum* ticks collected from tortoises

Table 16.9. Selected hosts of the invasive tick species which are vectors of heartwater

| Invasive tick species (no. of introductions) | No. of reports for each host | | | | No. of species of reptiles as hosts |
|---|---|---|---|---|---|
| | Cattle | Sheep | Goats | White-tailed deer | |
| **FIELD VECTORS** | | | | | |
| *A. hebraeum* (29) | 59 | 18 | 25 | 0 | 8 |
| *A. lepidum* (3) | 39 | 20 | 18 | 0 | 0 |
| *A. pomposum* (1) | 19 | 7 | 6 | 0 | 1 |
| *A. variegatum* (17) | 188 | 68 | 66 | 1 | 8 |
| **AFRICAN EXPERIMENTAL VECTORS** | | | | | |
| *A. gemma* (16) | 37 | 14 | 14 | 0 | 3 |
| *A. marmoreum* (20) | 8 | 8 | 9 | 0 | 28 |
| *A. sparsum* (18) | 4 | 1 | 2 | 0 | 13 |
| *A. tholloni* (7) | 6 | 4 | 4 | 0 | 3 |
| **AMERICAN EXPERIMENTAL VECTORS** | | | | | |
| *A. cajennense* (62) | 88 | 18 | 18 | 12 | 7 |
| *A. dissimile* (106) | 9 | 7 | 0 | 0 | 68 |

imported into Florida from Africa for evidence of *Ehrlichia ruminantium* infection, the rickettsia causing heartwater, using a PCR assay. It found that, in one shipment imported from Zambia, 15 of 38 *A. sparsum* ticks collected from leopard tortoises were positive for *E. ruminantium* infection (Burridge et al., 2000b). These studies showed that the international trade in live tortoises was allowing potential vectors of heartwater, some infected with the causative agent, to be introduced into the United States.

Analysis of data on those invasive tick species proven to be vectors of heartwater is summarized in Table 16.9. It can be seen that both field and experimental vectors are frequently introduced into the United States. Regarding the common hosts susceptible to heartwater, all 10 invasive tick species have been reported to feed on cattle and sheep, all but one on goats and reptiles, and two on white-tailed deer. Of the predominantly reptilian ticks, *A. marmoreum* will feed on cattle, sheep and goats, and *A. dissimile* on cattle and sheep, whereas *A. sparsum* rarely feeds on domestic livestock but feeds commonly on some wild ungulates susceptible to heartwater such as African buffalos and thus could potentially play a role in transmission of heartwater between wildlife. It is clear that a variety of real and potential vectors of heartwater are being introduced into the United States with some frequency.

Over the years the author has stressed the serious threat that heartwater poses to the livestock and wildlife populations of the United States (Burridge, 1997, 2000; Burridge et al., 2002c). The threat remains because (1) the risk of introduction of infected ticks from the Caribbean is ever present, especially on migratory birds; (2) the risk of introduction of infected ticks on imported reptiles is continuing, particularly through the international trade in live tortoises; and (3) the risk of introduction of infected wild game animals from Africa remains, especially for those that are subclinical carriers of the causative agent. In addition, the United States has a highly susceptible population of domestic livestock (cattle, sheep and goats) and wildlife (native deer and exotic game) and a native tick species (*A. maculatum*) that is a very competent vector of heartwater, indicating that if heartwater were introduced, it could be maintained in the United States through an enzootic cycle involving white-tailed deer (Fig. 16.3) and Gulf Coast ticks (Fig. 16.4). The latter scenario would make heartwater very difficult, if not impossible, to eradicate.

### 16.3.2. Risk of introduction of bovine babesiosis

The ticks *R. (B.) annulatus* and *R. (B.) microplus*, commonly referred to collectively as the cattle fever ticks or fever ticks, were introduced into the New World by explorers and colonists who brought tick-infested cattle with them. By the latter part of the 19th century, bovine babesiosis transmitted by these two parasites

*Above:* Fig. 16.3. A white-tailed deer buck (*Odocoileus virginianus*).

*Left:* Fig. 16.4. An adult Gulf Coast tick (*Amblyomma maculatum*) on a grass stem.

was the major impediment to development of a vibrant cattle industry in the southern United States; consequently, a national campaign to eradicate *R. (B.) annulatus* and *R. (B.) microplus* was launched in 1907. It was successfully completed only in 1960 (Graham & Hourrigan, 1977).

The USDA still has an active cooperative federal-state-industry Cattle Fever Tick Eradication Program in Texas, concentrated in the narrow permanent quarantine or buffer zone along the border with Mexico, from Del Rio southeast for about 800 km to the Gulf of Mexico. The goal of the program is to prevent the reintroduction and establishment of fever ticks and of bovine babesiosis in the continental United States. Historically, the program has worked from a deficit, and inadequate resources for personnel, support, supplies and materials have been ongoing problems (USAHA, 2005). Both *R. (B.) annulatus* and *R. (B.) microplus* remain widespread in Mexico, and incursions of tick-infested cattle, horses and wild ungulates across the Rio Grande River from Mexico into the buffer zone and beyond are a constant challenge to USDA personnel. The United States imports a large number of cattle from Mexico annually; during 1993, for example, 935,000 head of cattle were imported into Texas, with approximately 5% rejected because of tick infestations (Bram & George, 2000). In addition, several thousand stray or smuggled cattle cross the Mexican border annually, with about 90% of those apprehended found to be infested with ticks (Wagner et al., 2002). Since 1967, outbreaks of fever ticks in areas north of the quarantine zone have occurred with increasing frequency. Consequently, it is evident that Texas is at constant and continual risk of introduction of *R. (B.) annulatus* and *R. (B.) microplus* and their associated disease, bovine babesiosis. Recent incidents demonstrate how serious this situation is. By the end of September 2007 there were 67 premises infested with fever ticks in Texas, with 25 located outside the quarantine zone (USAHA, 2007). In July 2008 the preventive cattle fever tick quarantine area in southern Texas was expanded by about 124,000 hectares (307,000 acres) after *Rhipicephalus (Boophilus)* ticks were detected on cattle outside quarantine areas in Starr and Zapata counties. During the previous 12 months, these ticks had been identified on livestock or wildlife on 139 Texas premises. The preventive quarantine area was thus increased to more than 400,000 hectares in southern Texas, in addition to the 200,000 hectares of permanent fever tick quarantine along the Mexican border.

The risk of reestablishment of *R. (B.) annulatus* and *R. (B.) microplus* in Texas and beyond is heightened by a number of factors (George, 1990; Bram et al., 2002). During the eradication program earlier last century, cattle were the only primary hosts for the ticks. Now there are large populations of white-tailed deer throughout the southern United States and, particularly in Texas, large populations of free-ranging exotic ungulates such as nilgais. There is ample evidence that white-tailed deer, elk, red deer and nilgais can all serve as effective hosts for fever ticks, and these ticks were recently also found on spotted deer (USAHA, 2007). Furthermore, it is clear that these wild and exotic animal species are capable of maintaining fever ticks in the absence of cattle. If the ticks spread beyond the quarantine areas in southern Texas, the populations of white-tailed deer will be a significant factor in their dissemination back into their historical range in the southern states. In addition, the free-ranging exotic wildlife populations will make any eradication efforts that much more difficult to accomplish. Another factor favoring the maintenance and spread of *R. (B.) annulatus* and *R. (B.) microplus* infestations is the progressive conversion of the grassland savanna of much of southern Texas and adjacent areas of Mexico to thorn shrubland, a habitat more favorable to the survival of nonparasitic stages of these ticks (Teel, 1991).

Bram et al. (2002) succinctly outlined the potential situation if *R. (B.) annulatus* and *R. (B.) microplus* ticks infected with *Babesia* parasites were to spread back into the southern states. All native yearling and adult cattle would be susceptible to babesiosis, which may cause up to 90% mortality. Restrictions by state and federal regulatory agencies and public attitudes toward use of acaricides would be major obstacles to any effort to eradicate the ticks. Large populations of white-tailed deer would invalidate the pasture vacation option used so successfully in the eradication campaign of the early 1900s, and the abundance of these native wildlife as well as significant populations of free-ranging exotic ungulates would complicate any eradication effort. There seems no doubt that the risk of reintroduction of bovine babesiosis to cattle in the southern United States continues to increase.

## 16.4. Tickborne Diseases Transmissible to Sheep and Goats

Fifty of the 100 invasive tick species introduced into the United States have been reported to feed on both sheep and goats, with 14 species known to be vectors of ovine and/or caprine diseases. In addition, two invasive tick species cause diseases directly in sheep and/or goats. The tick *A. hebraeum* has long mouthparts that can cause sores to develop on the feet of sheep and goats, with subsequent abscess formation. The tick *Hy. truncatum* can also cause foot lameness in sheep, and, as with cattle, it can produce tissue damage leading to secondary myiasis.

Invasive tick species are proven vectors of several ovine and caprine diseases (Table 16.10). Some of these diseases cause mortality in sheep and goat populations and are currently exotic to the United States. They include ovine babesiosis transmitted primarily by *R. bursa*, heartwater transmitted primarily by *A. hebraeum* and *A. variegatum*, louping ill transmitted by *I. ricinus*, Nairobi sheep disease transmitted by *R. appendiculatus*, and malignant theileriosis transmitted by *Hy. anatolicum*. As with cattle, four of the most important invasive tick species for sheep and goats from the vectorial standpoint are *A. hebraeum*, *A. variegatum*, *I. ricinus* and *R. appendiculatus*.

## 16.5. Tickborne Diseases Transmissible to Other Mammals

Only two important diseases are transmitted to other mammals by invasive tick species introduced into the United States. They are canine babesiosis caused by *B. canis*, *B. gibsoni* and *B. rossi* transmitted to dogs by *D. reticulatus*, *Ha. longicornis* and *Ha. leachi*, respectively, and equine piroplasmosis caused by *B. caballi* and *T. equi* and transmitted to horses and other equines by a wide variety of ticks (Table 16.11). Canine babesiosis and equine piroplasmosis are life-threatening diseases of dogs and horses, respectively, and must be prevented from entering the United States. Equine piroplasmosis is a global problem for the equine industry since ex-

Table 16.10. Diseases of sheep and goats for which invasive ticks are vectors

| Disease | Causative agent | Tick vectors |
| --- | --- | --- |
| Anaplasmosis | *Anaplasma ovis* | *D. nuttalli, R. bursa* |
| Babesiosis | *Babesia motasi* | *Ha. punctata, R. bursa* |
| Babesiosis | *Babesia ovis* | *R. bursa* |
| Ehrlichiosis[a] | *Ehrlichia ovina* | *R. bursa, R. evertsi evertsi* |
| Heartwater | *Ehrlichia ruminantium* | *A. hebraeum, A. lepidum, A. pomposum, A. variegatum* |
| Louping ill (LI)[a] | LI virus | *I. ricinus* |
| Nairobi sheep disease (NSD) | NSD virus | *A. variegatum, R. appendiculatus, R. pulchellus, R. simus* |
| Theileriosis, benign | *Theileria ovis* | *Ha. punctata, R. bursa, R. evertsi mimeticus* |
| Theileriosis, benign[a] | *Theileria separata* | *R. bursa, R. evertsi evertsi, R. evertsi mimeticus* |
| Theileriosis, malignant | *Theileria lestoquardi* | *Hy. anatolicum* |
| Tickborne fever[a] | *Anaplasma phagocytophilum* | *I. ricinus* |

[a]Disease of sheep only

Table 16.11. Diseases of other mammals for which invasive ticks are vectors

| Disease | Causative agent | Host species affected | Tick vectors |
| --- | --- | --- | --- |
| Babesiosis | *Babesia canis* | dogs | *D. reticulatus* |
| Babesiosis | *Babesia gibsoni* | dogs | *Ha. longicornis* |
| Babesiosis | *Babesia merionis* | rodents | *Hy. anatolicum, Hy. dromedarii* |
| Babesiosis | *Babesia rossi* | dogs | *Ha. leachi* |
| Babesiosis | *Babesia trautmanni* | pigs | *R. simus* |
| Piroplasmosis | *Babesia caballi* | horses & other equines | *D. nitens, D. nuttalli, D. reticulatus, Hy. anatolicum, Hy. dromedarii, Hy. marginatum, Hy. truncatum, R. bursa, R. evertsi evertsi* |
| Piroplasmosis | *Theileria equi* | horses & other equines | *D. reticulatus, Hy. anatolicum, Hy. excavatum, Hy. detritum, Hy. dromedarii, Hy. marginatum, R. bursa, R. evertsi evertsi, R. evertsi mimeticus, R. turanicus, R. (B.) microplus* |

posed or infected horses are restricted from entering many countries, including the United States. Furthermore, the United States is constantly at risk of introduction of equine piroplasmosis, as can be seen from the incident in Florida in 2008 (see section 15.11).

## 16.6. Ticks and Tickborne Diseases Associated with Reptiles

Yeoman et al. (1967), in their book on the ixodid ticks of Tanzania, stated the following:

> We believe that it is incorrect to regard any species of tick as unimportant. Very little is known about many of them, and since advances remain to be made in our knowledge of the aetiology of many diseases, and since these arthropods have shown themselves to be highly adapted to the transmission of a variety of disease agents, it seems unwise to dismiss any of them as insignificant. Even those that have been recorded only in minute numbers, or from uncommon hosts, may form part of a picture of which even the outline cannot as yet be seen. They may be common on hosts that have still to be examined and that may have an importance, so far unrecognized, as disease reservoirs, or they may ultimately contribute to our knowledge of tick communities and to the definition of ecosystems.

How prophetic that these comments by Yeoman and his colleagues were concerning reptilian ticks that many have dismissed as of little or no relevance to human or animal health. Let us take the example of *Bothriocroton hydrosauri*, a tick of reptiles in Australia. Australian ticks, including those of reptiles, have been well documented since the mid-1900s by Roberts (1953, 1964a, 1964b, 1969, 1970), and extensive long-term ecological studies of *B. hydrosauri* have been conducted by Michael Bull and his colleagues at

Flinders University in South Australia since the 1970s. Despite all those studies, *B. hydrosauri* was considered only a reptilian tick, especially of lizards and snakes, with occasional infestations of echidnas, horses, cattle and bandicoots. It was not until 2003 that the first human infestations were documented (Graves & Stenos, 2003; Whitworth et al., 2003). It has now been determined that *B. hydrosauri* is the arthropod reservoir and putative vector of *Rickettsia honei*, the bacterium that causes Flinders Island spotted fever, a new human disease first discovered in Australia in 1991 (Stewart, 1991) (see section 15.13.3).

Although very little is yet known about the majority of the 31 reptilian invasive ticks introduced into the United States, some are now known to be of significance to human and/or animal health:

- Heavy infestations of *A. exornatum* in the nasal passages can cause suffocation and death in lizards.
- *A. marmoreum* is a competent experimental vector of heartwater.
- *A. sparsum* is an experimental vector of heartwater and has been found to introduce the rickettsial agent causing heartwater into the United States on tortoises imported from Africa.
- *B. hydrosauri* is the arthropod reservoir and putative vector of the rickettsial agent causing Flinders Island spotted fever in humans.
- *A. dissimile* can cause ulcers in snakes and toads and pustular dermatitis in snakes.
- *A. rotundatum* can cause paralysis in snakes.

Clearly, a lot more research needs to be conducted on reptilian ticks to define their significance to human and animal health.

One interesting situation is now evolving in Florida, where an invasive African lizard, the Nile monitor, has become established in southwestern Florida (see section 2.3) and where the invasive African tick, *Amblyomma exornatum*, has been introduced on multiple occasions (see section 3.3). In Africa, one of the preferred hosts of *A. exornatum* is the Nile monitor, on which they can be found in the nasal passages, causing respiratory distress and even death from suffocation. As mentioned in section 16.1, it has been found that *A. exornatum* has spread in Florida to a native and protected species, the gopher tortoise. How will this situation develop? Will the introduced Nile monitors become infested with *A. exornatum* and become a source of infestation for native animals? And will *A. exornatum* become a significant pathogen for native reptiles? This potential problem is worthy of further study.

# 17

# Measures Used to Combat Invasive Ticks

Measures adopted to combat the introduction of invasive ticks into the United States fall into two categories. The first line of defense is precautionary and involves international, federal and state regulations applied typically at the port of entry with the aim of regulating importation of animals. Few of these measures are concerned specifically with introduction of invasive ticks. The second line of defense is reactionary and involves eradication of tick infestations that are not detected at the port of entry and are, in most instances, discovered serendipitously. This begs the question: How many established infestations of invasive ticks have there been in the continental United States not known to regulatory authorities?

## 17.1. Regulation of Global Trade in Animals

The majority of invasive ticks are introduced into the United States on hosts imported through the international trade in animals and animal products, and the U.S. agencies that regulate these importations are guided in part by a number of international agreements, including the Convention on International Trade in Endangered Species of Wild Fauna and Flora (CITES), the Agreement on Sanitary and Phytosanitary Measures of the World Trade Organization (WTO), and the standards of the International Air Transport Association (IATA).

CITES is an international agreement between governments with the objective of ensuring that international trade in wild animals and plants does not threaten their survival. Countries adhere voluntarily to CITES, and it had a membership of 173 countries (including the United States) as of June 2008. CITES regulates wildlife trade by listing species in one of three appendixes according to the degree of protection they need; appendix I includes those species threatened with extinction that can be traded only in exceptional circumstances, appendix II includes those species for which trade must be controlled in order to avoid utilization incompatible with their survival and, in addition, "look-alike" species that are difficult to distinguish from those that are listed otherwise, and appendix III list native species of which individual countries seek to prevent or restrict exploitation. CITES member countries have also adopted a requirement that all live CITES-listed wildlife must be packed and shipped humanely, and CITES implements this requirement by adopting the published standards of IATA.

The Agreement on Sanitary and Phytosanitary Measures of the WTO facilitates international trade while recognizing a nation's right to protect human, animal and plant health. WTO recognizes the World Organization for Animal Health, formerly known as the Office International des Épizooties (OIE), as the international standard-setting body for development of health-related guidelines, standards and recommendations for animal health in member countries, which, as of January 2008, numbered 172, including the United States. OIE is responsible for listing diseases as notifiable so that surveillance and control of their spread through trade can be enhanced. However, few diseases are listed as notifiable in wildlife, partly because OIE has long focused on infectious diseases of livestock, even though wildlife diseases are covered by its authority. The U.S. Department of Agriculture (USDA) is required to participate in the activities of OIE in order for U.S. animals and animal products to compete successfully in the global marketplace.

IATA has regulations concerning transport of live animals with which all member airlines must comply. IATA regulations stipulate that shipments of live animals must be arranged in advance and accompanied by proper documentation. **IATA standards require that animals be packed free of external parasitic infestations, including mites, ticks and leeches, that can readily be seen under normal lighting conditions.** This requirement is intended to provide for humane and healthy transport of animals rather than to prevent the introduction of ticks and other external parasites into the importing country.

The U.S. agencies that regulate animal importations are also guided by two federal laws, the Endangered Species Act and the Lacey Act. The Endangered Species Act makes it illegal for any person subject to U.S. jurisdiction to import, export, deliver, receive, carry, transport, ship, sell or offer for sale in interstate commerce any species of animal or plant listed as threatened or endangered pursuant to the Act. The Lacey Act, as amended in 1981, prohibits the import, export, transport, sale, receipt, acquisition or purchase of wildlife or plants taken, possessed or sold in violation of any wildlife law, treaty or regulation of the United States or in violation of any Indian tribal law. The Act also prohibits the import, export, transport, acquisition, receipt, sale or purchase in interstate or foreign commerce of any wild animal taken, possessed, transported or sold in violation of any foreign law. Furthermore, the original Lacey Act is still in effect and prohibits any person from knowingly causing or permitting any wild animal to be transported to the United States under inhumane or unhealthful conditions.

Four U.S. agencies from four different federal departments regulate animal importations. They are the Animal and Plant Health Inspection Service (APHIS) of USDA, the Fish and Wildlife Service (FWS) of the Department of the Interior, the Centers for Disease Control and Prevention (CDC) of the Department of Health and Human Services, and the National Marine Fisheries Service of the Department of Commerce.

### 17.1.1. The U.S. Department of Agriculture

APHIS is the agency within USDA charged with the responsibility of enforcing laws and developing and enforcing regulations pertaining to the importation of animals and animal products that threaten U.S. agriculture, thereby preventing the entry of potentially devastating livestock diseases and helping to maintain overseas markets for U.S. agriculture, forestry and fisheries exports worth over $62 billion in fiscal year 2005 (http://www.aphis.usda.gov/import_export). To accomplish this responsibility, APHIS enforces laws and regulations regarding the importation of farm animals, poultry and certain other birds. In addition, APHIS inspectors screen travelers entering the United States from abroad to intercept and destroy items that could be the source of pathogenic organisms or vectors. Regulatory officials also inspect cargo ships, cargo planes, truck shipments, trains, and livestock and wildlife carriers (Bram & George, 2000).

Cattle are not permitted to be imported into the United States from countries affected with bovine spongiform encephalopathy ("mad cow disease"), foot-and-mouth disease or rinderpest. Cattle to be imported from other countries (except from Canada and Mexico) **must be maintained in a tick-free environment for a minimum of 60 days prior to export from the country of origin with treatment for ticks with an approved acaricide within 10 days of export,** must have an official health certificate and a USDA import permit, and upon importation must be placed in quarantine for 30 days at a USDA animal import center.

Horses to be imported must be accompanied by a veterinary health certificate issued by a veterinary officer of the national government of the exporting country that states, among other things, that **the horses have been inspected and found to be free of ticks** and other ectoparasites. Upon importation, the horses must be placed in quarantine at a USDA-operated quarantine facility for a period of time depending on the country of origin. Horses to be imported from Afghanistan, Albania, Australia, Bermuda, British Virgin Islands, European Union countries, Hong Kong, Iceland, Israel, Japan, Jordan, Lebanon, Mexico, New Zealand, Norway, Pakistan, Russia, South Korea, Switzerland and Turkey require a 3-day quarantine while they are tested for dourine (*Trypanosoma equiperdum* infection), glanders (*Pseudomonas mallei* infection), equine piroplasmosis and equine infectious anemia. Horses to be imported from all countries and territories in the Western Hemisphere except Bermuda, British Virgin Islands, Canada and Mexico require a 7-day quarantine since they are not recognized by the USDA as being free from Venezuelan equine encephalitis or screwworm. Horses to be imported from countries affected with African horse sickness (Oman, Saudi Arabia, Yemen and all countries in Africa except Morocco) require a 60-day quarantine.

Poultry are defined by USDA to include chickens, doves, ducks, geese, grouse, guineafowl, partridges, peafowl, pheasants, pigeons, quail, swans and turkeys. When imported, they must be accompanied by a veterinary health certificate endorsed by a veterinary officer of the agency responsible for animal health of the national government of the exporting country and by a USDA animal import permit. Upon importation, poultry must be placed in quarantine for 30 days at a USDA animal import center.

Ratites are flightless birds, including ostriches, emus, cassowaries and rheas. Before importation, **they must be treated with a pesticide to kill any ticks and other ectoparasites 8–14 days prior to shipment.** When imported, they must be accompanied by a veterinary health certificate issued by a veterinary officer of the agency responsible for animal health of the national government of the exporting country and by a USDA ratite import permit. Upon importation, ratites must be placed in quarantine for 30 days at a USDA animal import center **where they will be treated for ticks and other ectoparasites.**

Commercial birds are defined by USDA as birds that are imported for resale, breeding, public display or any other purpose, but they do not include pet birds, zoological birds, research birds or performing or theatrical birds. Pet birds are defined by the USDA as birds owned for the personal pleasure of their owners and not intended for resale. Commercial and pet birds to be imported are subject to similar regulations as poultry, including requirements for a veterinary health certificate, a USDA animal import permit and 30-day quarantine in a USDA animal import center.

USDA has guidelines for the importation of animal trophies. Fully taxidermy finished trophies have unrestricted entry provided they have been professionally cleaned, processed and preserved. Similarly, hides and skins of ruminants, swine and birds are eligible to be imported unrestricted provided they meet certain criteria. For example, ruminant hides and skins have unrestricted entry if one or more of the following conditions is met: the hide/skin has been processed into a finished product such as a rug or jacket, the hide/skin was derived from a ruminant that originated from a country free of foot-and-mouth disease and rinderpest, the hide/skin was flint-dried, the hide/skin has been tanned, the hide/skin has been pickled, or the hide/skin has been limed and dehaired.

APHIS, through a memorandum of understanding dating back to 1978, has an agreement with FWS in control of vectors and diseases that pose threats to domestic poultry and livestock. While wild birds and mammals are covered by this agreement, the situation with reptiles remains unclear. As a result of this confusing situation, tick-infested reptiles imported into the United States are seldom subjected to veterinary inspection (Clark & Doten, 1995).

From time to time, USDA promulgates new temporary rules and regulations when a new disease threat to agriculture is discovered. Such an interim rule was published in 2000, prohibiting the importation of leopard tortoises, African spurred tortoises and Bell's hinged tortoises (USDA, 2000). This action was taken because these three chelonian species were commonly imported into the United States and **often infested with tick vectors of heartwater.** Another interim rule was published in 1989 prohibiting the importation of ratites (USDA, 1989). This action was deemed necessary **to prevent the introduction of ticks capable of transmitting several exotic diseases following the discovery of 10 species of non-native ticks on ostriches imported from Africa and Europe.**

### 17.1.2. The U.S. Fish and Wildlife Service

FWS is the agency within the Department of the Interior charged with the responsibility of enforcing laws and regulations pertaining to the importation of wildlife. Wildlife is defined by FWS as any living or dead wild animal, its parts and products made from it, and includes not only mammals, birds, reptiles, amphibians and fishes, but also invertebrates such as arthropods, mollusks and coelenterates. The Office of Law Enforcement of FWS regulates the importation of wildlife by enforcing the terms of CITES and several federal laws that protect wildlife such as the Lacey Act, the Endangered Species Act and the Wild Bird Conservation Act. These regulations are part of an international conservation effort to protect wildlife subjected to trade, with special emphasis on endangered species.

Anyone importing wildlife must have an import license, which is valid for one year, and file a Declaration for Importation or Exportation of Fish or Wildlife form with FWS at an authorized port of entry before receiving clearance for U.S. Customs to release the wildlife shipment. Reptiles and amphibians are imported into the United States through this system, and they are exempt from the requirement of a USDA import permit. Although FWS regulates the importation of birds protected by CITES and the Wild Bird Conservation Act, including parrots, parakeets, macaws, lories and cockatoos, certain members of the parrot family are exempt from these regulations, namely budgerigars, cockatiels and rosy-faced lovebirds.

### 17.1.3. The Centers for Disease Control and Prevention

CDC is the agency within the Department of Health and Human Services charged with the responsibility of enforcing regulations pertaining to the importation of dogs, cats, turtles, primates and other animals, including wildlife, capable of causing human disease. CDC does not require general certificates of health for importation of pets into the United States.

Pet dogs to be imported must have a certificate showing that they have been vaccinated against rabies at least 30 days prior to importation if they are coming from a country where rabies is endemic. Pet dogs are subject to inspection at the port of entry and may be denied entry if they have evidence of an infectious disease transmissible to humans. If the dog appears ill upon entry, further examination by a licensed veterinarian might be required. Pet cats are subject to the same regulations as pet dogs except that there is no requirement for proof of rabies vaccination.

CDC regulates the importation of small turtles, with those with a carapace length less than 10.16 cm (4 in) prohibited from importation for any commercial purpose. This regulation was implemented in 1975 after it was discovered that small turtles frequently transmitted salmonellosis to humans, particularly to young children.

CDC has placed importation restrictions on other animals because they are known to harbor agents capable of causing diseases in humans. Some examples are monkeys, bats, rodents, civets and birds. Monkeys and other nonhuman primates cannot be imported as pets under any circumstance. Bats, known to carry rabies and histoplasmosis, can be imported only with permits from CDC and FWS, but permits will not be granted for importing bats as pets. In 2003 CDC and the U.S. Food and Drug Administration issued a joint order, later replaced by an interim final rule, prohibiting the importation of all African rodents as part of the public health response to the first reported outbreak of human monkeypox in the United States (see section 2.3). In 2004 CDC issued an order for an immediate ban on the import of all civets because they can potentially infect humans with severe acute respiratory syndrome (SARS). Between 2004 and 2007 CDC and USDA placed embargoes on the importation of birds and unprocessed bird products from specific African, Asian and European countries where the highly pathogenic avian influenza H5N1 virus occurs in domestic poultry because of the potential of the virus to cause illness in humans.

### 17.1.4. The National Marine Fisheries Service

The National Marine Fisheries Service (NMFS) of the National Oceanic and Atmospheric Administration is the agency within the Department of Commerce charged with the responsibility of enforcing regulations pertaining to the importation of marine mammals, sea turtles and other threatened and endangered marine species. It conducts this charge by implementing the Marine Mammal Protection Act and the Endangered Species Act.

### 17.1.5. Overview of U.S. regulatory agencies

Examination of data presented above demonstrates the complexity of the regulatory process of importing animals into the United States. The process involves four different departments of the federal government: the Departments of Agriculture, Interior, Health and Human Services, and Commerce. This situation is further complicated by the emphases of the agencies involved in the regulatory process within each department that influence the thoroughness with which imported animals will be examined for ticks. APHIS is focused primarily on the health of farm animals and poultry and CDC on human health, whereas FWS and NMFS are concerned primarily with protection of wildlife and enforcement of CITES regulations and federal laws. Consequently, FWS and NMFS are less concerned about health issues than APHIS or CDC.

Regulation of importation of domestic animals is fairly straightforward. Importation of farm animals, poultry, and pet and commercial birds is regulated by APHIS and importation of pet dogs, pet cats and primates by CDC; however, regulation of importation of wildlife is much more complex. Importation of most terrestrial mammalian wildlife is regulated by FWS, but if human health issues are implicated, such as with bats, civets and African rodents, CDC becomes involved. Importation of avian wildlife is regulated also by FWS, but again if human health issues are implicated, as with avian influenza, CDC becomes involved, and if agricultural issues are implicated, as with the tick-infested ostriches imported in 1989, APHIS becomes involved. The system reaches its height of complexity with reptiles. Importation of terrestrial reptiles is regulated by FWS and of marine reptiles like sea turtles by NMFS; however, importation of small turtles is regulated by CDC because of potential problems with human salmonellosis and, if agricultural issues are implicated, as with the heartwater vector ticks imported on African tortoises, APHIS becomes involved. As a consequence of this convoluted process, imported reptiles are not examined adequately, if at all, for tick infestations.

### 17.1.6. National Reptile Improvement Plan

It became apparent by 2000 that invasive ticks capable of introducing fatal diseases of livestock such as heartwater were slipping through the federal regulatory process and becoming established in reptile facilities in the United States. In consequence a meeting was convened in Florida, involving the author as one of the veterinary specialists, to develop methods to minimize the risk of introduction of foreign tick-borne diseases through the international trade in reptiles (Reaser et al., 2008). Following that meeting, the Pet Industry Joint Advisory Council (PIJAC) held meetings to evaluate the feasibility of developing best management practices to minimize the risk of introduction of unwanted parasites on imported reptiles. These meetings culminated in 2003 in the adoption of the National Reptile Improvement Plan: Best Management Practices for Reptile Trade and Hobby. According to PIJAC's website (www.pijac.org), the National Reptile Improvement Plan is established as a function of a subcommittee of PIJAC, obtains its infrastructure and staff support from PIJAC, mandates that every participant is required to obtain all necessary permits, licenses and other authorizations as required by federal and/or state law, and is voluntary and open to any person, business or other entity desiring to participate. According to Reaser et al. (2008), about 75% of reptiles and amphibians imported into the United States are being distributed through facilities accredited by the National Reptile Improvement Plan. PIJAC is the organization that focuses on education, information and government issues that involve or impact the pet industry, with the goal of ensuring the availability of companion animals to sustain the industry. It represents all aspects of the pet industry, and membership in its companion animal segment includes breeders, importers/exporters, wholesalers and hobbyists.

Whereas PIJAC is an advocate for the pet reptile industry, other prominent American organizations are strongly opposed to the industry. The Humane Society of the United States recommends that reptiles not be kept as pets by the general public (Franke & Telecky, 2001), and the Smithsonian Institution has a no herp-as-pet policy, discouraging anyone from keeping a reptile as a pet (http://vertebrates.si.edu/herps/herps_reptilesaspets.htm).

### 17.1.7. Reptile smuggling

The measures taken to regulate the flow of animals imported legally into the United States that might be carrying invasive ticks have been outlined above. Unfortunately, illegal sale of animals and plants is considered second only to drugs as the most profitable form of contraband trafficking activity in the United States (Franke & Telecky, 2001), and reptiles form a thriving and lucrative part of this illegal trade. The illegal trade in reptiles has been reviewed by Franke & Telecky (2001) and a book, *The Lizard King* (Christy, 2008), recently explored the vast criminal enterprise that supplies reptiles to pet shops, collectors and zoos worldwide. These two publications highlight the enormous extent of the smuggling of reptiles around the world and indicate the weaknesses in international agreements and federal laws that allow smugglers to continue their illegal operations.

Some 40-plus years ago, in the 1960s, smugglers supplied several well-known U.S. zoos with rare but illegal reptiles in direct violation of the Lacey Act and foreign laws. This and other smuggling operations had become so widespread by 1979 that President Jimmy Carter directed the Justice Department to establish a new wildlife section after a massive illegal trade in wildlife was uncovered. He also directed the Departments of Agriculture, Commerce, Interior, Treasury and Justice to investigate illegal wildlife trade aggressively and to prosecute violators as white-collar criminals. As a result, the Wildlife and Marine Resources Section was set up within the Justice Department to take on responsibility for this new interagency wildlife enforcement program.

This action by President Carter was followed in 1981 by amendment of the Lacey Act, the oldest national wildlife statute of the United States, to make it a federal crime to import or export wildlife in violation of any state, federal, Indian or **foreign** law, with penalties for violating foreign wildlife laws becoming penalties under U.S. law; however, most prosecutors lack the will and the resources to prosecute foreign-law cases under the Lacey Act because they require an understanding of foreign laws, locating of foreign witnesses and shipping records, and sometimes extradition of smugglers.

CITES, the international trade agreement for wildlife, seeks to stop the commercial flow of species threatened with extinction and to modify the flow of species considered potentially vulnerable but still abundant enough that their regulated commercial trade can be allowed; however, one important exception allows unlimited commercial trade in potentially vulnerable species that are born in captivity, and this exception has been exploited by smugglers. All a smuggler needed to turn an illegally poached reptile into a "legal" one under CITES was a government official somewhere in the world who would say that the animal had been captive-borne in his or her country and stamp the official paperwork. Such government officials were not hard to find; in 1994, the CITES Secretariat in Geneva sent out a notice alerting parties that four countries (Myanmar, Seychelles, Tanzania and United Arab Emirates) were issuing false captive-bred permits for Indian star tortoises. The response of the U.S. government to this notice was that, if an exporting country said an animal was captive-bred, then the U.S. government was not going to question it.

## 17.2. Eradication of Infestations of Invasive Ticks

From time to time invasive ticks are not detected at the port of entry and become established as breeding populations at facilities within the continental United States. In these situations it is necessary to launch a program to eradicate the ticks and thus eliminate the invasive threat. The following are examples of such eradication programs.

### 17.2.1. Eradication of a *Rhipicephalus evertsi* infestation from a wild animal compound

It was during routine surveillance activities for *Rhipicephalus (Boophilus) microplus* ticks in Florida in 1960 that an inspector found eight *R. evertsi* ticks on an eland that had just died in a wild animal compound in Boca Raton, Florida (Bruce, 1962). The wild animal compound was immediately placed under federal and state quarantine. Before eradication procedures commenced, a sample of wild animals on the compound was immobilized and examined for ticks. A total of 151 *R. evertsi* adults was collected from 23 animals, with 21 of the ticks being females. Treatment of individual wild animals with acaricide was considered impractical, so treatment of the entire 53 hectare compound with an acaricide was chosen as the preferred method of tick eradication. The compound received six applications of DDT suspension spray at three-week intervals, using a mist blower to spray open ground and a conventional power sprayer for wooded areas. The DDT treatment was supplemented with two applications of granular dieldrin around the perimeter of the compound in a double-fenced band six meters wide.

After treatment of the compound, a sample of ani-

Fig. 17.1. *Amblyomma marmoreum* ticks feeding on an Aldabra giant tortoise (*Aldabrachelys gigantea*).

mals was examined for ticks, and 72 *R. evertsi* were found, including one female. It was decided, therefore, to continue the eradication program by making four more applications of DDT to the compound. In addition, the 251 mammals were sprayed with dioxathion, with their ears treated with pine oil containing lindane, and the 72 ostriches were dusted with DDT. After completion of this phase of the eradication program, two subsequent examinations of immobilized animals were made and no ticks were found. These negative results were considered evidence that *R. evertsi* had been successfully eradicated from the wild animal compound in Florida.

The source of the *R. evertsi* infestation was thought to be zebras or elands imported from Africa (Bram & George, 2000); however, based on the number of ticks and the number of species infested, it was believed that the infestation had been established in the wild animal compound for at least three years. It was indeed fortunate that there was no evidence of spread of *R. evertsi* outside the compound for, if there was and if native wildlife had become infested, eradication would in all likelihood have been impossible.

### 17.2.2. Eradication of an *Amblyomma sparsum* infestation from a tortoise facility

During investigations of invasive tick infestations in Florida by the author and his colleagues, it became evident that numerous shipments of infested reptiles had been imported and that at least two invasive tick species, *A. marmoreum* and *A. sparsum*, both known experimental vectors of heartwater, had already become established as breeding populations (Burridge et al., 2002b). *Amblyomma marmoreum* had become established on three premises containing exotic tortoise species. At one premises, African spurred tortoises had introduced *A. marmoreum*; on another, leopard tortoises were the initial host from which the ticks spread to Aldabra giant tortoises (Fig. 17.1), yellow-footed tortoises, a Galapagos giant tortoise and domestic dogs. *Amblyomma sparsum* had become established at one tortoise-breeding premises where the hosts were leopard tortoises. Consequently, protocols were developed to eradicate these tick infestations to minimize the risk that such invasive tick vectors of heartwater would spread to native fauna and thus become established as indigenous tick species, as had happened in past years in Florida with the iguana tick *A. dissimile* (Bequaert, 1932) and the rotund toad tick *A. rotundatum* (Oliver et al., 1993).

The eradication protocol finally adopted has been described in detail by the author (Burridge, 2005). All the tortoises on the infested premises were treated with Provent-A-Mite (Pro Products, Mahopac, New York), a permethrin product specifically formulated for use on reptiles (Fig. 17.2), and then removed to a

Fig. 17.2. Treating a tortoise with an acaricide spray.

Fig. 17.3. Treating tortoise housing with an acaricide spray.

tick-free area. Provent-A-Mite is the only formulation of permethrin that has received a Section 18 exempt registration from the U.S. Environmental Protection Agency and is available commercially to treat reptiles and their environments. Next, the premises were sprayed on two occasions two weeks apart by a licensed pest control company with a cyfluthrin product specifically formulated for premises treatment, ensuring that all surface areas, including housing and burrows, were treated (Fig. 17.3). Finally, one week after the second premises treatment, sentinel tortoises, such as Hermann's tortoises, which roam freely and whose skin area is easy to inspect, were placed on the treated premises for 10 days to ensure that all ticks had been killed. Using this protocol, the *A. sparsum* infestation was eradicated successfully from the tortoise-breeding facility in Florida, with sentinel tortoises remaining tick-free when introduced again four months after premises treatment.

### 17.2.3. Eradication of an *Amblyomma komodoense* infestation from a zoo exhibit

The author was asked in 2001 to assist with control of an *A. komodoense* tick infestation in the Komodo dragon exhibit at the Miami Metro Zoo in Florida. The zoo had acquired a pair of Komodo dragons from Indonesia in 1994. They were heavily infested with ticks upon arrival and, over the years, the tick infestation became established in the Komodo dragon exhibit, with an estimated average of 50 ticks collected from the exhibit every month during the summer season. The zoo did not want acaricides applied directly to the Komodo dragons because of their protected status and their strength, and thus an eradication protocol different from that used with tortoises had to be developed.

The eradication protocol adopted for use in the Komodo dragon exhibit at the zoo has been described in detail by the author (Burridge et al., 2004b; Burridge, 2005). The Komodo dragons were all secured in their indoor enclosures. The outdoor enclosure was sprayed with Provent-A-Mite, paying special attention to suspected tick-molting areas such as the base of walls, between boards, under ledges and rocks, around the base of basking spots, and in cracks and holes. The acaricide was allowed to dry. The Komodo dragons were then released into their treated outdoor enclosure. The indoor enclosures, including hide boxes, were then sprayed with Provent-A-Mite. Finally, two small test areas, one in the indoor enclosure and one in the outdoor enclosure, were selected and sprayed with Provent-A-Mite. Every two weeks, 15 sentinel house crickets (*Acheta domestica*) were placed in an overturned plastic container in each test area and examined after 15 minutes to determine whether all had been killed by the acaricide. When the test area failed to kill the sentinel insects within 15 minutes, that enclosure was retreated with Provent-A-Mite. Using this technique, it was found necessary to retreat the indoor enclosures only every 8–10 weeks, whereas the outdoor enclosure had to be retreated every two weeks, presumably in part because of the intense ultraviolet light and frequent rainfalls occurring in the outdoor area. The Komodo dragons could not be inspected thoroughly for ticks without immobilization, so tick presence on the lizards was monitored visually from a distance. During the week after the initial treatment, visual inspections revealed that 300 ticks had crawled from their concealed indoor and outdoor off-host sites and died. The number of ticks seen in the exhibit declined until week 26, when no ticks were seen either on any of the Komodo dragons or in their environment for a period of two weeks. Over the next 21-month period, no ticks were seen in the Komodo dragon exhibit, demonstrating that the acaricidal protocol had eradicated the tick infestation.

This protocol is the one recommended for eradication of tick infestations of all lizards and snakes because direct application of acaricide should be avoided, especially in the case of endangered and dangerous species (Fig. 17.4).

Fig. 17.4. Treating a snake container with an acaricide spray.

# 18

# Actions Needed to Minimize Introduction of Invasive Ticks

The preceding chapters have provided a detailed account of the invasive ticks introduced into the continental United States and the diseases they cause or transmit to both humans and animals, with a discussion of the measures currently employed to combat the risks associated with these introductions. The discussion has highlighted some of the weaknesses in the measures adopted to keep invasive ticks and their associated diseases out of the United States. The most significant weaknesses are outlined below, together with suggestions for actions needed to minimize the real and potential impacts of further introductions of invasive ticks.

## 18.1. Regulatory Actions

### 18.1.1. CITES regulations

CITES permits one important exception to its regulations, allowing unlimited commercial trade in potentially vulnerable species if they are certified as born in captivity. This loophole has been exploited by smugglers and others who, in collaboration with government officials of the country of origin, have obtained falsely issued captive-bred permits and imported many wild-caught reptiles illegally into the United States (see section 17.1.7). Since wild-caught reptiles are more likely to harbor ticks, measures need to be taken to prevent this exception to CITES regulations from being used for illegal trade.
**Action needed: Modify CITES regulations to ensure that wild-caught reptiles cannot be certified as born in captivity.**

### 18.1.2. U.S. regulations

The regulatory process of importing animals into the United States is complex, with various agencies within four different government departments (Agriculture, Commerce, Health and Human Services, and Interior) involved, only one of which (Agriculture) has any regulations specifically addressing ticks. Moreover, only two of the federal departments (Agriculture and Health and Human Services) are involved primarily in regulations relating to health issues (see section 17.1.5). As a result of this complex and inefficient regulatory process, wildlife, and especially reptiles, are not adequately examined for tick infestations upon introduction into the United States. With no one agency with the clear leadership authority to protect the United States from invasive ticks, it will be only a matter of time before invasive ticks introduce diseases that have tragic effects on the nation's economy or its public or animal health. The regulatory process for importation of animals into the United States needs to be streamlined. The optimal solution would appear to be giving leadership authority to USDA, since it already has the experience and an administrative structure in place to control introduction of ticks.
**Action needed: Streamline the U.S. regulatory process for animal importations so that one federal agency has the sole leadership authority.**

## 18.2. Working Groups

Several working groups have been created both nationally and regionally in the United States to combat the problems associated with invasive species. The most prominent of these is the National Invasive Species Council (see chapter 2). Most if not all working groups include invertebrates in the lists of invasive species, often citing insects as examples; however, few if any list ticks as invasive species. As a consequence, invasive ticks have received no consideration as invasive species that can cause economic or environmental harm or harm to human or animal health. The data presented in this book show this to be a glaring omission in need of immediate correction.

**Action needed: Include ticks as important invasive species for consideration by national and regional working groups such as the National Invasive Species Council.**

## 18.3. Reptile Importations

The enormity of the numbers of invasive ticks introduced in the United States on imported reptiles was first documented by two veterinary medical officers of USDA (Clark & Doten, 1995). They examined 349 shipments containing 117,690 reptiles from 22 countries imported through Miami International Airport in Florida during a three-month period in 1994–95. They recovered ticks from one or more animals in 97 (27.8%) of the shipments. In discussion of their results, they concluded that "no class of birds or wild mammals or domestic animals are permitted entry into the United States with such an absence of veterinary inspection and control as is the case with reptiles. . . . The gap in regulatory oversight of reptile importation should be closed promptly and effectively." Clark & Doten make the following recommendations: "Veterinary inspection should be required of all live reptiles imported into the United States. . . . Imported reptiles should be subjected to USDA, APHIS quarantine at the port of entry into the U.S. . . . Reptile importers and dealers should be subject to the same APHIS licensing requirements as are mandatory for dealers and importers of exotic birds and domestic animals. . . . All shipping and packing materials associated with importation of exotic reptiles should be required to be disposed of in such a manner as to preclude dissemination of ticks and tick eggs that may contaminate such materials." These recommendations were never implemented by USDA.

The scale of the threats posed by invasive ticks introduced into the United States on imported reptiles was soon reinforced by the studies of the author and his colleagues (Allan et al., 1998b; Burridge et al., 2000a, 2002a, 2006; Simmons & Burridge, 2000, 2002; Simmons et al., 2002). The initial discoveries that experimentally proven tick vectors of heartwater (*A. marmoreum* and *A. sparsum*) were being introduced into Florida on a regular basis on tortoises imported from Africa led USDA to promulgate an interim rule in 2000 prohibiting the importation of three species of tortoises (see section 17.1.1). It is interesting to note that a few years earlier in 1995 two USDA veterinarians had reported that all four shipments of leopard tortoises imported into Florida during a three-month period were infested with *A. marmoreum* ticks and had warned of the danger of introducing heartwater through the international trade in live reptiles (Clark & Doten, 1995), yet on that occasion USDA took no action. It is also worthy of note that both the interim rule concerning importation of tortoises (USDA, 2000) and that concerning importation of ostriches (USDA, 1989) were reactionary rather than precautionary.

While the interim rule concerning importation of tortoises is likely to have reduced the numbers of a few species of invasive ticks that gain entry to the United States, it has had no impact on the vast majority of species of invasive ticks introduced on imported reptiles. There is a clear need for APHIS and FWS to work more closely together in regulation of reptile importations, with FWS enforcing CITES regulations and federal laws protecting wildlife, followed by inspections of shipments by USDA inspectors. As Clark & Doten (1995) have pointed out, and as the author and his colleagues have witnessed when working at Miami International Airport, tick-infested reptiles imported through that airport are seldom subjected to veterinary inspection. Some have suggested that reptiles should be treated for tick infestations in the exporting country before shipment and that a tick-free certificate be issued by the exporting country, allowing federal officials in the United States to refuse entry to those reptiles found tick infested (USAHA, 2007). If shipments of reptiles are found infested with ticks, they should be treated at the shipper's expense before the shipment is released, using treatment protocols described in sections 17.2.2/3.

USDA has been urged to accept responsibility for addressing the problem of introduction of invasive ticks on imported reptiles and to work with affected industries and affected states to implement control measures (USAHA, 2005). Reptiles infested with invasive ticks should not be allowed to continue to be imported without restriction. It is time for the USDA and cooperating federal and state agencies to act to reduce the risk of entry of foreign diseases through the importation of tick-infested reptiles (USAHA, 2007).

Much of the ineffective regulation of reptile importations is due to lack of knowledge of the international trade in reptiles in general and of reptilian ticks and their disease associations in particular. For example, USDA recently conducted a pathway analysis to determine the pathways of introduction of heartwater into the United States. The analysis demonstrated that legal and illegal reptile importations into the United States are feasible pathways for release of tick vectors of heartwater, but it concluded that the importance of such introductions could not be determined because of lack of data (USAHA, 2007). There is an urgent need to develop collaborative research projects between scientists at universities and those at USDA, FWS and CDC to define which invasive tick species are of significance to human and animal health in the United States, with initial priority given to those tick species frequently imported on reptiles and capable of establishment in U.S. habitats.

**Actions needed: Modify procedures for regulation of reptile importations to allow FWS and USDA to work closely together so that shipments are examined adequately for tick infestations. Develop collaborative research projects between universities and federal agencies to determine the health significance of invasive tick species introduced into the United States on imported reptiles.**

## 18.4. USDA Reports

From 1962 to 1994, USDA published annual reports on cooperative tick eradication programs and on tick surveillance for the years 1962–1989 that provided useful information on the invasive ticks introduced into the United States. Then, following reports of invasive ticks on ostriches (Mertins & Schlater, 1991) and on reptiles imported through Miami International Airport (Clark & Doten, 1995), no further reports giving detailed data on invasive ticks imported into the United States have been published by USDA. USDA should resurrect publication of annual reports of ticks imported into the United States, a task that should be simple given that only one laboratory, the USDA National Veterinary Services Laboratory in Ames, Iowa, confirms all such tick identifications. With such information, state and local government agencies and scientists would be informed of any important trends in tick importations. Furthermore, USDA annual reports could act as a forum for dissemination of information on invasive ticks, including commentary on new data on disease associations. It is unlikely that the problems associated with reptilian invasive ticks outlined in this book would have received adequate attention if the author and his colleagues had not published the results of their work, given the lack of a medium for dissemination of information on invasive ticks by USDA and the lack of response by USDA to the results of its own reptilian tick survey presented by Clark and Doten (1995). However, if USDA annual tick reports had been available, the trends in introduced reptile ticks would have been immediately obvious.

**Action needed: Revive publication of USDA annual tick surveillance reports, including commentary on trends in ticks introduced into the United States and on new data concerning diseases associated with invasive ticks.**

## 18.5. Eradication of the Tropical Bont Tick from the Caribbean

The tropical bont tick *A. variegatum* is an invasive tick of great concern to the United States since it is well established on several islands in the eastern Caribbean, has been introduced into the United States on 17 occasions on domestic animals, African wildlife, animal trophies, ostriches, humans and a monitor lizard, and is associated with a variety of important diseases of both humans and domestic livestock (see section 4.1). For many years, it has been known that the presence of heartwater and its tick vector *A. variegatum* in the eastern Caribbean pose significant threats to the livestock industries and deer populations of the United States (see section 16.3.1). Additionally, it would be expected that, if *A. variegatum* were to become established in

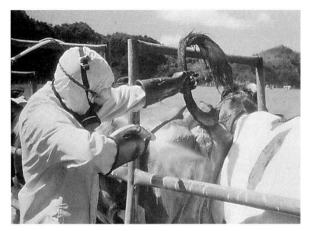

Fig. 18.1. Spraying cattle with acaricides during the Puerto Rican tropical bont tick eradication program manually (*top*) and using a spray race (*bottom*) in the early 1980s. (Photographs courtesy of the U.S. Department of Agriculture.)

the United States, the horrific skin disease acute bovine dermatophilosis would soon become a problem for U.S. cattle owners much as it did in the Caribbean (see section 15.5). Furthermore, any importation of *A. variegatum* has the potential to introduce the new human disease African tick-bite fever into the United States (see section 15.13.1). It is no wonder, therefore, that several attempts have been made to eradicate *A. variegatum* infestations in the Caribbean.

In 1967, *A. variegatum* was discovered on cattle on St. Croix in the U.S. Virgin Islands (Hourrigan et al., 1969). An *A. variegatum* eradication program was immediately launched. Cattle, sheep and goats were dipped in an acaricide (coumaphos) at weekly intervals, infested premises were sprayed with another acaricide (carbaryl) at three-week intervals and, where the premises were inaccessible to motor vehicles, aerial spraying of carbaryl was used. With this two-pronged approach of animal and ground treatments with acaricides, *A. variegatum* was eradicated from the infested area on St. Croix by the end of April 1968. Infestations of *A. variegatum* reappeared on St. Croix in 1970 and again in 1985 (Barre et al., 1995) but, because of lack of funds, the affected farm in the latter outbreak was only placed in quarantine with infested animals subjected to acaricide treatment (USAHA, 2005). By 2003 the number of infested farms had increased to eight. Then in 2005 a cooperative agreement was initiated between USDA and the U.S. Virgin Islands to eradicate *A. variegatum* from St. Croix, but as of October 2006, there were still 10 infested premises on St. Croix (USAHA, 2007).

In 1974, *A. variegatum* was discovered on cattle on Puerto Rico (Garris, 1987), but it was not until 1981 that an *A. variegatum* eradication program began (Garris et al., 1989). The neighboring islands of Vieques and Culebra were found to be infested with *A. variegatum* in 1981 and 1986, respectively, and these two islands were added to the eradication program. The eradication protocol comprised spraying of cattle, goats, sheep, horses and dogs with acaricides at 14-day intervals (Fig. 18.1). With this approach of systematic acaricidal treatments, *A. variegatum* was eradicated from Puerto Rico and Vieques by 1985 and 1986, respectively; however, additional *A. variegatum* males were found on Culebra in 1987 (Garris et al., 1989).

Despite these eradication programs, *A. variegatum* reappeared both on St. Croix in 1970 and 1985 and on Puerto Rico in 1992 (Bokma & Shaw, 1993). These findings demonstrated the need for a more regional approach to eradication of a tick that occurred on multiple islands in the eastern Caribbean. Consequently, the Caribbean Amblyomma Program was established in 1994 to eradicate *A. variegatum* from the anglophone Caribbean islands through a plan based on biweekly treatment of animals with flumethrin pour-on (Pegram et al., 1996), with a complementary program established in 1995 to eradicate the tick from the francophone Caribbean islands (Barre et al., 1996). By 2002, six of the islands (Anguilla, Barbados, Dominica, Montserrat, St. Kitts and St. Lucia) had attained the status of provisional freedom from *A. variegatum* (Pegram et al., 2004); however, Barbados, Dominica, St. Kitts and St. Lucia experienced new foci of infestation, and in 2006, the Caribbean Amblyomma Program was terminated for financial reasons without attaining its goal of eradication of *A. variegatum* from the Caribbean region (Pegram, 2006).

The demise of the Caribbean Amblyomma Program will assuredly result in the continued spread of *A. variegatum* within the Caribbean, increasing the probability that this tick and its associated diseases will reach the U.S. mainland. It is vital, therefore, that the international effort to eradicate *A. variegatum* from the Caribbean be resurrected as soon as possible; however, any future follow-up must learn from the shortcomings of the past program, which was, in effect, three separate programs: (1) the Caribbean Amblyomma Program, an international effort involving the independent Caribbean nations and the British and Dutch Caribbean territories; (2) the French program involving the French Caribbean territories; and (3) the U.S. program involving the American Caribbean territories. Furthermore, the American program was in reality a series of programs reacting to new foci of infestations, particularly on St. Croix. These past programs were not coordinated and were doomed to failure when dealing with a tick such as *A. variegatum*, which has a very wide host range (having been found on 157 species of animals), has been established on some of the islands for more than a century and can move freely between islands on dogs, livestock and birds. The ease with which *A. variegatum* can reinfest islands can be seen from the data summarized above, where new foci of infestation were found on St. Croix, Puerto Rico, Barbados, Dominica, St. Kitts and St. Lucia following apparently successful eradication efforts. Dominica, which lies between Guadeloupe and Martinique, continues to be reinfested from the French territories (USAHA, 2006). It is clear that, for any future *A. variegatum* eradication program in the Caribbean, all infested islands must be included in the program under one authority if there is to be any hope of total tick eradication.

Another issue to be considered in any future regional eradication program is the use of novel technologies. All eradication efforts thus far in the Caribbean have relied on regular treatments of livestock with acaricides applied either as dips, sprays or pour-ons. While these methods are effective in killing ticks on treated animals, they have little to no impact on ticks in the environment. Furthermore, the disappointing results of past efforts indicate a need to explore alter-

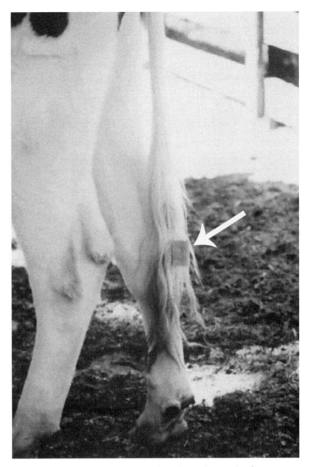

Fig. 18.2. A tick decoy attached to the tail switch of a cow.

native technologies. Such a technology is the tick decoy, which kills ticks on the animal as well as those it attracts from the environment and remains effective for up to three months. Tick decoys are plastic tags impregnated with a pheromone to attract ticks from the environment and an acaricide to kill them; both the pheromone and the acaricide are released slowly to provide a three-month active life for the decoy. The decoy can be attached to the animal as a tail tag (Fig. 18.2), as an ear tag or on a neck collar; tail and ear tags are used primarily with cattle and collar tags with sheep and goats. Tick decoys were first developed for control of the South African bont tick *A. hebraeum* with successful trials conducted in Zimbabwe (Norval et al., 1996; Allan et al., 1996) and were later modified for control of the tropical bont tick *A. variegatum* with trials conducted in Guadeloupe (Allan et al., 1998a; Sonenshine et al., 2000). Trials should be conducted in the Caribbean to compare acaricides applied as pour-ons (or sprays or dips) with acaricides applied as tick

Fig. 18.3. AppliGator self-medicating applicator attached to the lip of a feeding trough in South Africa; an AppliGator being primed with acaricide for tick control (*top*) and cattle feeding from trough with their necks rubbing on the AppliGator, allowing passive transfer of acaricide to the animals as they feed (*bottom*).

decoys to determine the optimal method for the treatment of livestock in eradication of *A. variegatum*. Such trials should be done before any new regional tropical bont tick eradication program is undertaken.

A problem encountered by eradication programs in the Caribbean has been treatment of feral and other livestock that were difficult or impossible to catch or restrain. The Caribbean Amblyomma Program used the Duncan applicator (Duncan, 1989, 1992) to treat such hosts with limited success. Other self-medicating applicators have been developed to control ticks on wild and feral animals; these include the 4-poster deer treatment bait station developed by USDA scientists (Pound et al., 2000) and the AppliGator developed by scientists at the University of Florida (Burridge et al., 2004a). The AppliGator (Fig. 18.3) has several advantages over the Duncan applicator and the 4-poster device, including easy portability because of to its light weight, lack of movable parts or valves (which can malfunction), lack of protruding parts (which can become entangled with horns or antlers), convenience of use with groups of animals, and simplicity of design and low cost of production. All hosts of *A. variegatum* adults must be treated if an eradication program is to be successful; therefore, trials should be conducted in the Caribbean to compare the various self-medicating applicators in treatment of feral animals for *A. variegatum* infestation. Again, such trials should be done before any new regional tropical bont tick eradication program is undertaken.

**Actions needed: Restore a well-funded international program to eradicate the tropical bont tick from the Caribbean as a matter of urgency, involving all infested islands under one authority and utilizing novel tick control technologies where appropriate.**

## 18.6. Prevention of Reestablishment of Cattle Fever Ticks in the United States

The cattle fever ticks *R. (B.) annulatus* and *R. (B.) microplus* remain constant threats to the U.S. cattle industries as a result of their continued presence in Mexico. Tick-infested cattle, horses and wild ungulates often cross the Rio Grande River from Mexico into Texas and, consequently, it is not surprising that each year cattle fever ticks are found on native cattle and sometimes wildlife in Texas counties bordering Mexico (see section 16.3.2). These problems are a constant challenge to USDA personnel.

Bram et al. (2002) described with disturbing clarity the potential situation if *R. (B.) annulatus* and/or *R. (B.) microplus* ticks infected with *Babesia* parasites were to spread back into the southern states. All native yearling and adult cattle would be susceptible to babesiosis, which could cause up to 90% mortality (see sections 15.2.1/2). Restrictions by state and federal regulatory agencies and public attitudes to use of acaricides would be major obstacles to any efforts to eradicate the ticks, and the abundance of native wildlife such as white-tailed deer as well as significant populations of free-ranging exotic ungulates such as nilgais, both of which have become frequent hosts for cattle fever ticks, would complicate any eradication effort. Furthermore, populations of *R. (B.) microplus* in Mexico have developed resistance to many of the available acaricides. These complications have led George et al. (2002) to believe that a successful campaign to re-eradicate widespread populations of cattle fever ticks would be difficult, if not impossible.

In 2006 a National Fever Tick Eradication Strategic Plan was developed and approved by USDA. The plan called for the prevention of entry of cattle fever ticks into the United States, the enhancement and maintenance of effective surveillance to rapidly detect cattle fever tick incursions, the eradication of infestations resulting from cattle fever tick incursions, the identification and procurement of the tools and knowledge necessary to keep the United States free of cattle fever ticks, and the fostering of collaboration and cooperation with Mexico to eliminate cattle fever ticks in areas of Mexico that impact the United States (USAHA, 2007). The estimated cost of full implementation of all elements of the Strategic Plan is approximately $8 million a year for five years, but the plan had not been funded as of October 2007.

**Action needed: Implement fully and without further delay the National Fever Tick Eradication Strategic Plan to prevent reestablishment of cattle fever ticks and bovine babesiosis in the southern United States.**

# Appendix 1. Invasive Ticks Introduced into the Continental United States

| Tick | Year | Host | State[a] | Origin of host | Original reference |
|---|---|---|---|---|---|
| Amblyomma albopictum | ?[b] | iguana | FL | Hispaniola | Keirans & Durden (2001) |
| Amblyomma argentinae | ? | reptile | ? | ? | Becklund (1968) |
| Amblyomma cajennense | 1913 | capybara (Hydrochoerus hydrochoeris) | PA | ? | Snetsinger (1968) |
| | 1915 | horses | GA/IL | Guatemala | Becklund (1968) |
| | 1965 | equines | FL/TX | ? | USDA (1966a) |
| | 1966 | ? | AZ | ? | USDA (1967a) |
| | 1967 | equine | FL | ? | USDA (1968a) |
| | 1967 | frozen beef | NV | Nicaragua | USDA (1967b) |
| | 1967 | horse | FL | Costa Rica | USDA (1967b) |
| | 1967 | inanimate object | NV | ? | USDA (1968a) |
| | ? | cattle | TX | Mexico | Becklund (1968) |
| | ? | horse | LA | ? | Becklund (1968) |
| | ? | horse | TX | ? | Becklund (1968) |
| | 1968 | equine | TX | ? | USDA (1969) |
| | 1968 | horse | FL | Honduras | USDA (1969) |
| | 1969 | equine | TX | ? | USDA (1970a) |
| | 1969 | palm fronds | TX | Mexico | USDA (1970c) |
| | 1969 | plants | TX | Mexico | USDA (1970c) |
| | 1970 | cattle | TX | ? | USDA (1971) |
| | 1970 | ferns | MD | Guatemala | USDA (1970c) |
| | 1970 | human | TX | ? | USDA (1971) |
| | 1970 | inanimate object | MD | ? | USDA (1971) |
| | 1970 | inanimate object | TX | ? | USDA (1971) |
| | 1971 | cattle | TX | ? | USDA (1972) |
| | 1971 | inanimate objects | FL/TX | ? | USDA (1972) |
| | 1972 | equine | FL | ? | USDA (1973a) |
| | 1972 | inanimate objects | LA/TX | ? | USDA (1973a) |
| | 1973 | cattle | TX | Mexico | USDA (1974a) |
| | 1973 | horse | FL | Honduras | USDA (1974b) |
| | 1973 | inanimate object | TX | ? | USDA (1974a) |
| | 1973 | palm frond | TX | ? | USDA (1974a) |

[a] AK = Alaska, AL = Alabama, AZ = Arizona, CA = California, CO = Colorado, CT = Connecticut, DC = District of Columbia, DE = Delaware, FL = Florida, GA = Georgia, ID = Idaho, IL = Illinois, IN = Indiana, KS = Kansas, KY = Kentucky, LA = Louisiana, MA = Massachusetts, MD = Maryland, ME = Maine, MI = Michigan, MN = Minnesota, MO = Missouri, MS = Mississippi, MT = Montana, NC = North Carolina, ND = North Dakota, NE = Nebraska, NH = New Hampshire, NJ = New Jersey, NM = New Mexico, NV = Nevada, NY = New York, OH = Ohio, OK = Oklahoma, OR = Oregon, PA = Pennsylvania, SC = South Carolina, TN = Tennessee, TX = Texas, UT = Utah, VA = Virginia, WA = Washington
[b] ? = data not reported

| Tick | Year | Host | State[a] | Origin of host | Original reference |
|---|---|---|---|---|---|
| | 1973 | three-toed sloth | IL | Panama | USDA (1974a) |
| | 1974 | cattle | TX | ? | USDA (1975) |
| | 1975 | inanimate object | TX | ? | USDA (1977a) |
| | 1975 | orchid | TX | ? | USDA (1977a) |
| | 1976 | fern | FL | ? | USDA (1977b) |
| | 1976 | house plant | FL | ? | USDA (1977b) |
| | 1976 | orchid | FL | ? | USDA (1977b) |
| | 1977 | limes | TX | ? | USDA (1978) |
| | 1977 | ? | OK | ? | USDA (1978) |
| | 1978 | annasas (fruit) | TX | ? | USDA (1980) |
| | 1978 | baggage | LA | ? | USDA (1980) |
| | 1978 | cattle | TX | ? | USDA (1980) |
| | 1978 | palm frond | TX | ? | USDA (1980) |
| | 1979 | cattle | TX | ? | USDA (1981) |
| | 1983 | cattle | TX | ? | USDA (1985a) |
| | 1984 | cattle | TX | ? | USDA (1985b) |
| | 1985 | human | PA | ? | USDA (1987a) |
| | 1985 | orchid | FL | ? | USDA (1987a) |
| | 1986 | cattle | TX | ? | USDA (1987b) |
| | 1986 | plant | FL | ? | USDA (1987b) |
| | 1987 | cattle | TX | ? | USDA (1988) |
| | 1988 | cattle | TX | ? | USDA (1994) |
| | 1988 | human | CO | ? | USDA (1994) |
| | 1988 | plant material | DE | ? | USDA (1994) |
| | 1989 | cattle | TX | ? | USDA (1994) |
| | 1989 | horse | TX | ? | USDA (1994) |
| | 1989 | plantain | DE | ? | USDA (1994) |
| | ? | human | CT | Jamaica | Keirans & Durden (2001) |
| | ? | ? | FL | ? | Berube (2006) |
| Amblyomma calcaratum | 1985 | ground (flannel drag) | KY | ? | Bloemer et al. (1987) |
| Amblyomma chabaudi | 2001 | spider tortoise (Pyxis arachnoides) | FL | Madagascar | Simmons & Burridge (2002) |
| Amblyomma clypeolatum | 1987 | star tortoise (Geochelone elegans) | TX | ? | USDA (1988) |
| | ? | tortoise | NY | ? | Keirans & Durden (2001) |
| Amblyomma coelebs | ? | seeds | LA | Costa Rica | Keirans & Durden (2001) |
| Amblyomma compressum | ? | pangolin | OH | Togo | Keirans & Durden (2001) |
| Amblyomma crassipes | 1988 | Bengal monitor (Varanus bengalensis) | FL | Thailand | Burridge & Simmons (2003) |
| | ? | lizard | FL | ? | Keirans & Durden (2001) |
| Amblyomma dissimile | 1954 | boa constrictor (Boa constrictor) | TX | South America | Eads et al. (1956) |
| | 1962 | snake | TX | ? | USDA (1962a) |
| | 1964 | iguana | FL | ? | USDA (1965a) |
| | 1965 | alligator | UT | ? | USDA (1965b) |
| | 1965 | boa constrictors | KS/NJ | ? | USDA (1966a) |
| | 1965 | iguana | IN | ? | USDA (1965b) |

| Tick | Year | Host | State[a] | Origin of host | Original reference |
|---|---|---|---|---|---|
| | 1967 | anaconda | IL | ? | USDA (1967b) |
| | 1967 | boa constrictor | CA | ? | USDA (1968a) |
| | 1967 | snake | FL | Colombia | USDA (1968a) |
| | 1967 | ? | FL | ? | USDA (1968a) |
| | ? | boa constrictor | PA | ? | Snetsinger (1968) |
| | ? | iguana | PA | ? | Snetsinger (1968) |
| | ? | Neuwied's false boa (*Pseudoboa neuwiedii*) | PA | ? | Snetsinger (1968) |
| | 1968 | boa constrictor | WA | ? | USDA (1970b) |
| | 1968 | iguana | AL | ? | USDA (1969) |
| | 1968 | reptile | OK | ? | USDA (1970b) |
| | 1968 | snake | OK | ? | USDA (1969) |
| | 1969 | anaconda | AL | ? | USDA (1970b) |
| | 1969 | boa constrictor | AL | ? | USDA (1970b) |
| | 1969 | boa constrictors | CA/TX | ? | USDA (1970a) |
| | 1969 | boa constrictor | WA | Colombia | USDA (1970a) |
| | 1969 | iguanas | CA/TN | ? | USDA (1970a) |
| | 1969 | iguana | IL | Central America | USDA (1970a) |
| | 1969 | iguana | WA | Colombia | USDA (1970a) |
| | 1969 | snakes | AL/CA | ? | USDA (1970a) |
| | 1969 | snake | AZ | Colombia | USDA (1970a) |
| | 1970 | black racer (*Alsophis ater*) | MD | ? | USDA (1970c) |
| | 1970 | boa constrictor | CA | ? | USDA (1970c) |
| | 1970 | boa constrictor | WA | Colombia | USDA (1970c) |
| | 1970 | iguana | FL | Colombia | USDA (1970c) |
| | 1971 | boa constrictor | IL | Colombia | USDA (1972) |
| | 1971 | boa constrictor | MD | ? | USDA (1972) |
| | 1972 | boa constrictors | GA/NM | ? | USDA (1973a) |
| | 1972 | boa constrictor | ID | Colombia | USDA (1973a) |
| | 1972 | boa constrictor | OR | ? | Burridge & Simmons (2003) |
| | 1972 | iguanas | OR/VA | ? | USDA (1973a) |
| | 1973 | boa constrictor | MI | ? | USDA (1974a) |
| | 1973 | iguanas | MD/MN | ? | USDA (1974a) |
| | 1973 | snake | TX | ? | USDA (1974a) |
| | 1974 | boa constrictor | NY | ? | USDA (1975) |
| | 1974 | iguana | CA | ? | USDA (1975) |
| | 1975 | snakes | KS/MI | ? | USDA (1977a) |
| | 1976 | iguana | CO | ? | USDA (1977b) |
| | 1979 | iguana | OH | ? | USDA (1981) |
| | 1980 | Mexican spiny-tailed iguana (*Ctenosaura pectinata*) | FL | ? | UF (unpublished data) |
| | 1981 | rainbow boa (*Epicrates cenchria*) | CT | Colombia | Anderson et al. (1984) |
| | 1985 | iguana | SC | Venezuela | USDA (1987a) |
| | 1987 | iguana | AZ | ? | USDA (1988) |
| | 1995 | black iguana (*Ctenosaura similis*) | FL | Nicaragua | Clark & Doten (1995) |
| | 1995 | boa constrictors | FL | Colombia/ Nicaragua/ Suriname | Clark & Doten (1995) |

| Tick | Year | Host | State[a] | Origin of host | Original reference |
|---|---|---|---|---|---|
| | 1995 | Central American wood turtle (*Rhinoclemmys pulcherrima*) | FL | Nicaragua | Clark & Doten (1995) |
| | 1995 | Ecuadorean desert tegu (*Dicrodon heterolepis*) | FL | Peru | Clark & Doten (1995) |
| | 1995 | giant toad (*Bufo marinus*) | FL | Nicaragua | Clark & Doten (1995) |
| | 1995 | green basilisk (*Basiliscus plumifrons*) | FL | Nicaragua | Clark & Doten (1995) |
| | 1995 | green iguanas (*Iguana iguana*) | FL | Colombia/ El Salvador/ Guatemala/ Peru/ Suriname | Clark & Doten (1995) |
| | 1995 | Oaxacan spiny-tailed iguana (*Ctenosaura quinquecarinatus*) | FL | Nicaragua | Clark & Doten (1995) |
| | 1995 | narrow-bridged mud turtle (*Kinosternon angustipons*) | FL | Nicaragua | Clark & Doten (1995) |
| | 1995 | smooth-helmeted iguana (*Corytophanes cristatus*) | FL | Nicaragua | Clark & Doten (1995) |
| | 1997 | anaconda | FL | ? | Burridge (1997) |
| | 1997 | boa constrictor | FL | ? | Burridge (1997) |
| | 1997 | chicken snake | FL | ? | Burridge (1997) |
| | 1997 | green iguana | FL | Costa Rica | Burridge (1997) |
| | 1997 | rainbow boa | FL | ? | Burridge (1997) |
| | 1997 | tegu lizard | FL | ? | Burridge (1997) |
| | 1997 | tortoise | FL | ? | Burridge (1997) |
| | 1999 | red-tailed boa (*Boa constrictor*) | FL | South America | USDA (unpublished data) |
| | 1999 | iguana | FL | ? | USDA (unpublished data) |
| | 2000 | green pampas rat snake (*Philodryas baroni*) | FL | Argentina | USDA (unpublished data) |
| | ? | alligator | UT | South America | Keirans & Durden (2001) |
| | ? | anaconda | MT | ? | Keirans & Durden (2001) |
| | ? | boa constrictors | AK/CA/NY/OH/PA | ? | Keirans & Durden (2001) |
| | ? | boa constrictor | CT | Colombia | Keirans & Durden (2001) |
| | ? | boa constrictor | FL | Central America | Keirans & Durden (2001) |
| | ? | iguana | CA | Colombia | Keirans & Durden (2001) |
| | ? | iguanas | DC/LA/ND/OH | ? | Keirans & Durden (2001) |
| | ? | iguana | NY | Honduras | Keirans & Durden (2001) |
| | ? | iguana | TX | Mexico | Keirans & Durden (2001) |
| | ? | rat snake | NY | ? | Keirans & Durden (2001) |
| | ? | reptile | PA | Paraguay | Keirans & Durden (2001) |

| Tick | Year | Host | State[a] | Origin of host | Original reference |
|---|---|---|---|---|---|
| | ? | snakes | CA/OH | South America | Keirans & Durden (2001) |
| | 2001 | Mexican spiny-tailed iguana | FL | Mexico | Burridge & Simmons (2003) |
| | 2001 | scorpion mud turtle (Kinosternon scorpioides) | FL | Guatemala | Burridge & Simmons (2003) |
| | 2001 | yellow anaconda (Eunectes notaeus) | FL | Guyana | Burridge & Simmons (2003) |
| | 2002 | green spiny lizard (Sceloporus malachitinus) | FL | ? | USDA (unpublished data) |
| Amblyomma exornatum | 1969 | mamba | MD | ? | USDA (1970a) |
| | 1974 | lizard | NJ | ? | USDA (1975) |
| | 1982 | Nile monitor (Varanus niloticus) | FL | ? | UF (unpublished data) |
| | 1984 | Nile monitor | FL | ? | UF (unpublished data) |
| | 1988 | monitor lizard | NE | ? | USDA (1994) |
| | 1994 | white-throated monitor (Varanus albigularis) | CA | Tanzania | Bayless & Simmons (2000) |
| | 1995 | African savanna monitor (Varanus exanthematicus) | FL | Tanzania | Clark & Doten (1995) |
| | 1995 | Nile monitor | FL | Tanzania | Clark & Doten (1995) |
| | ? | African savanna monitor | ? | ? | Williams & Bayless (1998) |
| | ? | Nile monitor | ? | ? | Williams & Bayless (1998) |
| | ? | white-throated monitor | FL | Tanzania | Williams & Bayless (1998) |
| | 1998 | African savanna monitor | FL | ? | Burridge et al. (2000a) |
| | 1998 | ball python | FL | ? | Burridge et al. (2000a) |
| | 1998 | white-throated monitor | FL | Tanzania | Burridge et al. (2000a) |
| | 1999 | white-throated monitor | FL | Tanzania | Burridge et al. (2000a) |
| | 2001 | African savanna monitor | FL | Togo | UF (unpublished data) |
| | 2001 | great plated lizard (Gerrhosaurus major) | FL | Tanzania | Burridge & Simmons (2003) |
| | 2001 | white-throated monitor | FL | ? | USDA (unpublished data) |
| | ? | Nile monitor | FL | ? | UF (unpublished data) |
| Amblyomma falsomarmoreum | 1974 | lizard | AZ | ? | USDA (1975) |
| | 1974 | leopard tortoise (Stigmochelys pardalis) | AZ | Kenya | USDA (1975) |
| Amblyomma fimbriatum | 1988 | monitor | NM | ? | Burridge & Simmons (2003) |
| | 2000 | harlequin monitor (Varanus rudicollis) | FL | ? | USDA (unpublished data) |
| | 2002 | amethystine python (Morelia amethistina) | FL | Indonesia | USDA (unpublished data) |
| Amblyomma flavomaculatum | 1978 | ball python | NY | Ghana | Kim et al. (1978) |
| | 1983 | Nile monitor | GA | Ghana | Wilson & Barnard (1985) |
| | 1989 | lizard | FL | ? | USDA (1994) |

| Tick | Year | Host | State[a] | Origin of host | Original reference |
|---|---|---|---|---|---|
| | 1989 | monitor | FL | ? | USDA (1994) |
| | 1994 | white-throated monitor | CA | Tanzania | Bayless & Simmons (2000) |
| | 1995 | African savanna monitors | FL | Benin/Ghana/Togo | Clark & Doten (1995) |
| | 1995 | ? | OH | ? | Anon (1996) |
| | 1997 | African rock python (*Python sebae*) | FL | ? | Burridge et al. (2000a) |
| | ? | African savanna monitor | ? | ? | Williams & Bayless (1998) |
| | ? | Nile monitor | ? | ? | Williams & Bayless (1998) |
| | ? | white-throated monitor | FL | Tanzania | Williams & Bayless (1998) |
| | 1998 | African savanna monitor | FL | Togo | Burridge et al. (2000a) |
| | 1998 | ball python | FL | ? | Burridge et al. (2000a) |
| | 1998 | white-throated monitor | FL | ? | Burridge et al. (2000a) |
| | 1999 | African savanna monitors | FL | Tanzania/Togo | Burridge et al. (2000a) |
| | 1999 | ball python | FL | ? | Burridge et al. (2000a) |
| | 2000 | African savanna monitor | FL | ? | USDA (unpublished data) |
| | ? | ball python | NY | Togo | Keirans & Durden (2001) |
| | ? | lizard | NY | Togo | Keirans & Durden (2001) |
| | ? | lizard | OH | ? | Keirans & Durden (2001) |
| | 2001 | African savanna monitor | FL | Togo | USDA (unpublished data) |
| | 2004 | African savanna monitor | FL | ? | USDA (unpublished data) |
| | 2005 | African savanna monitor | FL | ? | Burridge et al. (2006) |
| | 2005 | lizard | FL | ? | Burridge et al. (2006) |
| *Amblyomma fuscolineatum* | ? | Burmese python (*Python molurus*) | NY | ? | Kaufman (1972) |
| *Amblyomma geayi* | 1969 | sloth | PA | ? | USDA (1970a) |
| | 1973 | three-toed sloth | IL | Panama | USDA (1974a) |
| *Amblyomma gemma* | 1956 | rhinoceros | TX | ? | USDA (1963) |
| | 1964 | animal hide | MD | ? | USDA (1965a) |
| | 1965 | black rhinoceros | MI | East Africa | USDA (1966b) |
| | 1967 | zebra hide | FL | ? | USDA (1968a) |
| | ? | giraffe (*Giraffa camelopardalis*) | NY | Kenya | Becklund (1968) |
| | ? | zebra | NY | Kenya | Becklund (1968) |
| | 1969 | giraffe hide | TX | Kenya | USDA (1970c) |
| | 1969 | zebras | NJ/NY | Kenya | USDA (1970c) |
| | 1970 | zebra | NY | ? | USDA (1971) |
| | 1974 | animal trophy | LA | ? | USDA (1975) |
| | 1974 | rhinoceros | NC | ? | USDA (1975) |
| | 1989 | ostrich (*Struthio camelus*) | IL | Tanzania | Mertins & Schlater (1991), USDA (1994) |

| Tick | Year | Host | State[a] | Origin of host | Original reference |
|---|---|---|---|---|---|
| | 1989 | ostriches | NY/TX | Tanzania | Mertins & Schalter (1991), USDA (1994) |
| | ? | rhinoceros | TX | ? | Keirans & Durden (2001) |
| *Amblyomma geoemydae* | 2000 | black-breasted leaf turtle (*Geoemyda splengleri*) | NY | Asia | Simmons & Burridge (2000) |
| | 2000 | jagged-shelled turtle (*Cuora mouhotii*) | NY | Asia | Simmons & Burridge (2000) |
| | ? | turtle | CA | China | Keirans & Durden (2001) |
| | 2002 | giant Asian pond turtle (*Heosemys grandis*) | NY | Asia | Burridge & Simmons (2003) |
| | 2002 | spiny terrapin (*Heosemys spinosa*) | FL | ? | UF (unpublished data) |
| | 2003 | jagged-shelled turtle | MA | southeast Asia | UF (unpublished data) |
| *Amblyomma hebraeum* | 1962 | white rhinoceros (*Ceratotherium simum*) | NY | South Africa | USDA (1964b), Diamant (1965) |
| | 1963 | white rhinoceros | OK | South Africa | Diamant (1965) |
| | 1963 | white rhinoceros | OK | Namibia | USDA (1964a) |
| | 1966 | African elephant | FL | ? | Wilson & Bram (1998) |
| | 1966 | rhinoceroses | CA/TX | ? | USDA (1966b) |
| | 1968 | antelope hide | TX | ? | USDA (1968b) |
| | 1970 | rhinoceros | TX | ? | USDA (1971) |
| | 1973 | white rhinoceros | VA | ? | USDA (1974a) |
| | 1977 | animal hide | CA | ? | USDA (1978) |
| | 1979 | African buffalo (*Syncerus caffer*) | CO | ? | USDA (1981) |
| | 1980 | eland (*Tarotragus oryx*) | CO | ? | USDA (1982) |
| | 1981 | zebra hide | NC | ? | USDA (1982) |
| | 1984 | black rhinoceros (*Diceros bicornis*) | TX | ? | USDA (1985b) |
| | 1984 | rhinoceros hide | ME | ? | USDA (1985b) |
| | 1985 | animal hide | CO | Africa | USDA (1987a) |
| | 1986 | animal hide | MS | ? | USDA (1987b) |
| | 1987 | animal hides | CO/GA | ? | USDA (1988) |
| | 1988 | animal hide | MS | ? | USDA (1994) |
| | 1989 | animal hides | AZ/GA/MS | ? | USDA (1994) |
| | 1994 | giraffe | ? | ? | Wilson & Bram (1998) |
| | ? | black rhinoceros | MO | South Africa | Keirans & Durden (2001) |
| | ? | African buffalo | DC | Africa | Keirans & Durden (2001) |
| | ? | human | CT | South Africa | Keirans & Durden (2001) |
| | ? | zebra | MD | Africa | Keirans & Durden (2001) |
| | 2001 | human | FL | southern Africa | Burridge et al. (2002a) |
| *Amblyomma helvolum* | 1970 | snake | MD | Thailand | USDA (1970c) |
| | 1976 | water monitor | FL | ? | UF (unpublished data) |
| | 1984 | water monitor | FL | ? | UF (unpublished data) |
| | 1995 | reticulated python (*Python reticulatus*) | FL | Indonesia | Clark & Doten (1995) |

| Tick | Year | Host | State[a] | Origin of host | Original reference |
|---|---|---|---|---|---|
| | 1995 | water monitor | FL | Indonesia | Clark & Doten (1995) |
| | 2001 | king cobra (*Ophiophagus hannah*) | MI | ? | Simmons et al. (2002) |
| *Amblyomma humerale* | 1999 | yellow-footed tortoise (*Chelonoidis denticulata*) | FL | Guyana | Simmons & Burridge (2000) |
| | 2001 | yellow-footed tortoise | FL | ? | USDA (unpublished data) |
| *Amblyomma incisum* | 1964 | South American tapir (*Tapirus terrestris*) | NY | ? | USDA (1964b) |
| *Amblyomma javanense* | ? | ? | ? | ? | USDA (1965a) |
| | 1968 | anteater | OK | Thailand | USDA (1970b) |
| | 1968 | pangolin | OK | India | USDA (1969) |
| | 1971 | pangolins | IL/NY | Thailand | USDA (1972, 1973b) |
| | 1973 | Indian pangolin (*Manis crassicaudatus*) | OK | India | USDA (1974a) |
| *Amblyomma komodoense* | 1994 | Komodo dragon (*Varanus komodoensis*) | FL | Indonesia | Burridge & Simmons (2003) |
| | ? | Komodo dragon | DC | Indonesia | Keirans & Durden (2001) |
| *Amblyomma kraneveldi* | 2005 | reticulated python | FL | ? | Burridge et al. (2006) |
| *Amblyomma latum* | 1969 | black mamba (*Dendroaspis polylepis*) | MD | ? | USDA (1970a) |
| | 1973 | snake | AZ | ? | USDA (1974a) |
| | 1976 | cobra | FL | ? | USDA (1977b) |
| | 1978 | ball python | NY | Ghana | Kim et al. (1978) |
| | 1978 | snake | IL | ? | USDA (1980) |
| | 1980 | snake | FL | ? | USDA (1982) |
| | 1983 | ball pythons | CT/GA | West Africa | Anderson et al. (1984), Wilson & Barnard (1985) |
| | 1984 | ball python | TX | ? | Wilson & Barnard (1985) |
| | 1986 | ball python | MS | ? | USDA (1987b) |
| | ? | ball python | CA | Ghana | Hammond & Dorsett (1988) |
| | 1989 | ball python | CO | ? | USDA (1994) |
| | 1989 | snakes | NC/TX | ? | USDA (1994) |
| | ? | ball python | MD | ? | Robbins (1990) |
| | 1992 | ball python | TN | ? | Durdens & Kollars (1992) |
| | 1994 | white-throated monitor | CA | Tanzania | Bayless & Simmons (2000) |
| | 1995 | African rock python | FL | Togo | Clark & Doten (1995) |
| | 1995 | African savanna monitor | FL | Benin | Clark & Doten (1995) |
| | 1995 | ball pythons | FL | Benin/Ghana/Togo | Clark & Doten (1995) |
| | 1995 | great plated lizard | FL | Tanzania | Clark & Doten (1995) |
| | 1995 | human | FL | ? | Clark & Doten (1995) |
| | 1997 | African rock python | FL | ? | Burridge et al. (2000a) |
| | 1997 | ball python | FL | Togo | Burridge et al. (2000a) |
| | 1998 | African rock python | FL | ? | UF (unpublished data) |
| | 1998 | ball python | FL | Benin | USDA (unpublished data) |
| | 1998 | Gabon viper (*Bitis gabonica*) | FL | Tanzania | Burridge et al. (2000a) |

| Tick | Year | Host | State[a] | Origin of host | Original reference |
|---|---|---|---|---|---|
| | 1998 | white-throated monitor | FL | ? | Burridge et al. (2000a) |
| | 1999 | ball python | FL | Africa | USDA (unpublished data) |
| | 1999 | puff adder (*Bitis arietans*) | FL | ? | Burridge et al. (2000a) |
| | 2000 | ball python | FL | ? | USDA (unpublished data) |
| | 2000 | ball python | LA | western Africa | UF (unpublished data) |
| | 2000 | great plated lizard | FL | ? | USDA (unpublished data) |
| | ? | ball python | FL | Africa | Keirans & Durden (2001) |
| | ? | ball pythons | GA/KS/MD | western Africa | Keirans & Durden (2001) |
| | ? | ball pythons | IL/IN/KY/NH/OH/TN | ? | Keirans & Durden (2001) |
| | ? | ball python | NY | Togo | Keirans & Durden (2001) |
| | ? | python | SC | ? | Keirans & Durden (2001) |
| | ? | snake | CA | Tanzania | Keirans & Durden (2001) |
| | ? | snake | PA | ? | Keirans & Durden (2001) |
| | 2001 | ball python | FL | Togo | USDA (unpublished data) |
| | 2001 | forest cobra (*Naja melanoleuca*) | FL | ? | USDA (unpublished data) |
| | 2002 | black-necked spitting cobra (*Naja nigricollis*) | FL | western Africa | USDA (unpublished data) |
| | 2002 | puff adder | MI | ? | UF (unpublished data) |
| | 2004 | African burrowing python (*Calabaria reinhardtii*) | FL | ? | USDA (unpublished data) |
| | 2004 | African savanna monitor | FL | ? | USDA (unpublished data) |
| | 2004 | ball python | FL | ? | USDA (unpublished data) |
| | 2005 | ball python | FL | ? | Burridge et al. (2006) |
| | 2005 | Nile monitor | FL | ? | Burridge et al. (2006) |
| | 2005 | ball python | FL | ? | Burridge et al. (2006) |
| *Amblyomma lepidum* | 1965 | zebra hide | MD | ? | USDA (1965b) |
| | 1989 | ostriches | NY/TX | Tanzania | Mertins & Schlater (1991), USDA (1994) |
| *Amblyomma longirostre* | 1929 | white-eyed vireo (*Vireo griseus*) | TX | ? | Eads et al. (1956) |
| | 1947 | northern cardinal (*Cardinalis cardinalis*) | TN | ? | Durden & Kollars (1992) |
| | 1953 | summer tanager (*Piranga rubra*) | PA | ? | Snetsinger (1968) |
| | 1966 | porcupine | NJ | ? | USDA (1967a) |
| | 1967 | porcupine | NJ | ? | USDA (1967b) |
| | ? | Brazilian tree porcupine (*Coendou prehensilis*) | MI | ? | Becklund (1968) |
| *Amblyomma marmoreum* | 1965 | leopard tortoise | FL | ? | USDA (1965b) |

| Tick | Year | Host | State[a] | Origin of host | Original reference |
|---|---|---|---|---|---|
| | 1987 | leopard tortoise | TX | ? | USDA (1988) |
| | 1991 | leopard tortoise | FL | ? | Wilson & Bram (1998) |
| | 1994 | leopard tortoise | ? | ? | Wilson & Bram (1998) |
| | 1995 | leopard tortoise | FL | South Africa | Clark & Doten (1995) |
| | 1995 | ? | ? | ? | Wilson & Bram (1998) |
| | 1996 | leopard tortoise | ? | ? | Wilson & Bram (1998) |
| | 1997 | Bell's hinged tortoise (*Kinixys belliana*) | ? | ? | Burridge (1997) |
| | 1997 | Karoo Cape tortoise (*Homopus femoralis*) | ? | ? | Wilson & Bram (1998) |
| | 1997 | leopard tortoise | FL | Mozambique | Allan et al. (1998b) |
| | 1998 | African spurred tortoise (*Geochelone sulcata*) | FL | central Africa | Burridge et al. (2000a) |
| | 1998 | leopard tortoise | FL | Mozambique | Burridge et al. (2000a) |
| | 1999 | African spurred tortoise | FL | ? | Burridge et al. (2000a) |
| | 1999 | leopard tortoise | FL | ? | Burridge et al. (2000a) |
| | 2000 | African spurred tortoise | FL | ? | USDA (unpublished data) |
| | 2000 | leopard tortoise | FL | Mozambique | USDA (unpublished data) |
| | ? | leopard tortoise | FL | Zambia | UF (unpublished data) |
| | ? | tortoises | FL/NY | ? | Keirans & Durden (2001) |
| | ? | tortoise | MD | South Africa | Keirans & Durden (2001) |
| *Amblyomma moreliae* | 1973 | python | TX | ? | USDA (1974a) |
| | 1988 | skink | NM | ? | Burridge & Simmons (2003) |
| *Amblyomma multipunctum* | 1968 | tapir | CA | Ecuador | USDA (1968b) |
| *Amblyomma nodosum* | 1969 | anteater | NC | ? | USDA (1970a) |
| | 1973 | three-toed sloth | IL | Panama | USDA (1974a) |
| | 1976 | anteater | IL | ? | USDA (1977b) |
| | 1986 | pineapples | FL | ? | USDA (1987b) |
| | 1987 | anteater | TX | ? | USDA (1988) |
| | 2002 | anteater | FL | ? | USDA (unpublished data) |
| *Amblyomma nuttalli* | 1967 | hedgehog | NJ | France | USDA (1968b) |
| | 1968 | hedgehog | NY | Africa | USDA (1969) |
| | 1977 | hedgehog | NY | ? | USDA (1978) |
| | 1978 | ball python | NY | Ghana | Kim et al. (1978) |
| | 1986 | hedgehog | NY | ? | USDA (1987b) |
| | 1989 | hedgehogs | FL/MS | ? | USDA (1994) |
| | 1995 | African savanna monitors | FL | Benin/Togo | Clark & Doten (1995) |
| | 1995 | common agama (*Agama agama*) | FL | Ghana | Clark & Doten (1995) |
| | 1995 | Home's hingeback tortoise (*Kinixys homeana*) | FL | Togo | Clark & Doten (1995) |
| | 1995 | Nile monitor | FL | Togo | Clark & Doten (1995) |
| | 1997 | African savanna monitor | FL | Togo | Burridge et al. (2000a) |
| | 1997 | ball python | FL | Togo | Burridge et al. (2000a) |
| | 1998 | ball python | FL | Benin | Burridge et al. (2000a) |
| | 1998 | Bell's hinged tortoise | FL | ? | Burridge et al. (2000a) |

| Tick | Year | Host | State[a] | Origin of host | Original reference |
|---|---|---|---|---|---|
| | 1998 | Gabon viper | FL | ? | Burridge et al. (2000a) |
| | 1999 | African savanna monitor | FL | Togo | UF (unpublished data) |
| | 1999 | Bell's hinged tortoise | FL | Benin | Burridge et al. (2000a) |
| | 2000 | ball python | FL | ? | USDA (unpublished data) |
| | ? | ball python | NY | Togo | Keirans & Durden (2001) |
| | ? | lizard | NY | Togo | Keirans & Durden (2001) |
| | ? | lizard | OK | western Africa | Keirans & Durden (2001) |
| | ? | python | NV | Africa | Keirans & Durden (2001) |
| | ? | tortoise | CA | ? | Keirans & Durden (2001) |
| | 2001 | African savanna monitor | FL | Togo | USDA (unpublished data) |
| | 2001 | Bell's hinged tortoise | FL | Ghana | USDA (unpublished data) |
| | 2002 | Speke's hinged tortoise (*Kinixys spekii*) | MA | ? | UF (unpublished data) |
| | 2004 | African savanna monitor | FL | ? | USDA (unpublished data) |
| | 2005 | African savanna monitor | FL | ? | Burridge et al. (2006) |
| *Amblyomma oblongoguttatum* | ? | human | MI | Venezuela | Walker et al. (1998) |
| *Amblyomma ovale* | ? | ocelot (*Leopardus pardalis*) | CA | Ecuador | Becklund (1968) |
| | 1968 | cougar (*Puma concolor*) | WA | Ecuador | USDA (1970a) |
| | 1988 | dog | NY | ? | USDA (1994) |
| *Amblyomma parvum* | ? | armadillo | OH | Paraguay | Keirans & Durden (2001) |
| | 2005 | human | FL | ? | Berube (2006) |
| *Amblyomma pictum* | 1972 | anteater | NY | ? | USDA (1973a) |
| *Amblyomma pomposum* | ? | topi (*Damaliscus lunatus*) | NY | Germany | Becklund (1968) |
| *Amblyomma pseudoconcolor* | 1973 | six-banded armadillo (*Euphractus sexcinctus*) | IL | Bolivia | USDA (1974a) |
| | ? | armadillo | OH | Paraguay | Keirans & Durden (2001) |
| *Amblyomma quadricavum* | 1979 | Haitian boa (*Epicrates striatus*) | CT | ? | Anderson et al. (1981b) |
| *Amblyomma rhinocerotis* | 1967 | rhinoceros | AL | Namibia | USDA (1967b) |
| | 1989 | rhinoceros | TX | ? | USDA (1994) |
| *Amblyomma rotundatum* | 1964 | iguana | IL | ? | USDA (1965a) |
| | 1965 | snake | MD | ? | USDA (1966a) |
| | 1967 | red-tailed boa | IL | Colombia | USDA (1968b) |
| | 1968 | anaconda | MD | ? | USDA (1969) |
| | 1968 | boa constrictor | IL | ? | USDA (1969) |
| | 1997 | yellow-footed tortoise | FL | ? | USDA (unpublished data) |
| | 1998 | boa constrictor | FL | ? | UF (unpublished data) |
| | 1998 | giant toad | FL | ? | USDA (unpublished data) |
| | 1998 | yellow-footed tortoise | FL | ? | UF (unpublished data) |

| Tick | Year | Host | State[a] | Origin of host | Original reference |
|---|---|---|---|---|---|
| | 1999 | box turtle | FL | ? | USDA (unpublished data) |
| | 2000 | emerald tree boa (*Corallus caninus*) | FL | Guyana | USDA (unpublished data) |
| | 2001 | boa constrictor | FL | Guyana | USDA (unpublished data) |
| | 2001 | Suriname toad (*Pipa pipa*) | FL | Suriname | USDA (unpublished data) |
| | 2005 | rainbow boa | FL | ? | USDA (unpublished data) |
| *Amblyomma sabanerae* | 1995 | Central American wood turtle | FL | Nicaragua | Clark & Doten (1995) |
| | 1995 | narrow-bridged mud turtle | FL | Nicaragua | Clark & Doten (1995) |
| | 1997 | yellow-footed tortoise | FL | ? | Burridge et al. (2000a) |
| | 1998 | emerald tree boa | FL | ? | Burridge et al. (2000a) |
| | 1998 | red-footed tortoise (*Chelonoidis carbonaria*) | FL | ? | Burridge et al. (2000a) |
| | ? | reptile | IN | Mexico | Keirans & Durden (2001) |
| *Amblyomma scutatum* | 1964 | iguana | IL | ? | Burridge & Simmons (2003) |
| | 1975 | iguana | NY | ? | USDA (1977a) |
| | 1995 | smooth-helmeted iguana | FL | Nicaragua | Clark & Doten (1995) |
| | ? | iguana | NY | Honduras | Keirans & Durden (2001) |
| | 2005 | black iguana | FL | Honduras | Burridge et al. (2006) |
| *Amblyomma sparsum* | 1969 | tortoise | CA | Kenya | Burridge & Simmons (2003) |
| | 1970 | reptile | AZ | Kenya | USDA (1970c) |
| | 1971 | tortoise | OR | South Africa | USDA (1972) |
| | 1974 | tortoises | AZ/MN | ? | Burridge & Simmons (2003) |
| | 1989 | leopard tortoise | CA | ? | USDA (1994) |
| | 1989 | rhinoceros | TX | ? | USDA (1994) |
| | 1991 | tortoise | SC | ? | Wilson & Bram (1998) |
| | 1992 | ball python | TN | ? | Durden & Kollars (1992) |
| | 1992 | tortoise | ? | Tanzania | Clark & Doten (1995) |
| | 1994 | monitor | FL | ? | Clark & Doten (1995), Wilson & Bram (1998) |
| | 1999 | leopard tortoise | FL | Zambia | Burridge et al. (2000b) |
| | 2000 | African spurred tortoise | FL | ? | UF (unpublished data) |
| | 2000 | leopard tortoise | FL | Zambia | USDA (unpublished data) |
| | 2000 | leopard tortoise | PA | ? | USDA (unpublished data) |
| | ? | ball python | TN | western Africa | Keirans & Durden (2001) |
| | ? | tortoise | NY | Tanzania | Keirans & Durden (2001) |
| | 2001 | leopard tortoise | FL | ? | USDA (unpublished data) |
| *Amblyomma splendidum* | 1972 | elephant trophy | AZ | Africa | USDA (1973a) |

| Tick | Year | Host | State[a] | Origin of host | Original reference |
|---|---|---|---|---|---|
| *Amblyomma sylvaticum* | 1990 | tortoise | ? | South Africa | USDA (1990) |
| *Amblyomma testudinarium* | 1987 | rhinoceros | CA | ? | USDA (1988) |
| | ? | rhinoceros | DC | ? | Keirans & Durden (2001) |
| *Amblyomma tholloni* | 1965 | African elephant | FL | ? | USDA (1966a) |
| | 1965 | African elephant | TX | Uganda | USDA (1966a) |
| | 1973 | elephant trophy | CO | Africa | USDA (1974b) |
| | 1973 | game hide | LA | Africa | USDA (1974a) |
| | 1974 | elephant hide | CA | ? | USDA (1975) |
| | 1974 | African elephant | TN | ? | USDA (1975) |
| | ? | ruminant hide | AK | Zimbabwe | Keirans & Durden (2001) |
| *Amblyomma transversale* | 1973 | royal python (*Python regius*) | NY | Africa | USDA (1974b) |
| | ? | ball python | OH | ? | Keirans & Durden (2001) |
| | ? | python | NY | Togo | Keirans & Durden (2001) |
| | ? | python | TX | ? | Keirans & Durden (2001) |
| *Amblyomma triguttatum* | ? | human | GA | Australia | Keirans & Durden (2001) |
| *Amblyomma trimaculatum* | 1976 | water monitor | FL | ? | UF (unpublished data) |
| | 1978 | Gray's monitor (*Varanus olivaceus*) | FL | Philippines | Burridge & Simmons (2003) |
| | 1988 | Bengal monitor (*Varanus bengalensis*) | FL | Thailand | Burridge & Simmons (2003) |
| | 1995 | peachthroat monitor (*Varanus jobiensis*) | FL | Indonesia | Clark & Doten (1995) |
| | ? | blue tree monitor (*Varanus macraei*) | SC | ? | Reeves et al. (2002) |
| | 2005 | blue-tailed monitor (*Varanus doreanus*) | FL | New Guinea | Burridge et al. (2006) |
| *Amblyomma varanense* | 1969 | Indian cobra (*Naja naja*) | MD | ? | Burridge & Simmons (2003) |
| | 1969 | monitor | CA | India | USDA (1970a) |
| | 1974 | Bengal monitor | FL | ? | UF (unpublished data) |
| | 1978 | water monitor | FL | ? | UF (unpublished data) |
| | 1979 | python | GA | ? | USDA (1981) |
| | 1982 | Sumatran short-tailed python (*Python curtus*) | GA | ? | Wilson & Barnard (1985) |
| | 1984 | water monitor | FL | ? | UF (unpublished data) |
| | 1988 | Bengal monitor | FL | ? | UF (unpublished data) |
| | 1991 | Bengal monitor | FL | ? | UF (unpublished data) |
| | 1995 | water monitor | FL | Indonesia | Burridge & Simmons (2003) |
| | 1995 | white-lipped python (*Leiopython albertisii*) | FL | Indonesia | Burridge & Simmons (2003) |
| | 1998 | harlequin monitor (*Varanus rudicollis*) | FL | ? | Burridge et al. (2000a) |
| | ? | cobra | CA | Thailand | Keirans & Durden (2001) |
| | ? | lizard | FL | Indonesia | Keirans & Durden (2001) |

| Tick | Year | Host | State[a] | Origin of host | Original reference |
|---|---|---|---|---|---|
| | ? | lizard | FL | New Guinea | Keirans & Durden (2001) |
| | ? | python | FL | Indonesia | Keirans & Durden (2001) |
| | 2001 | harlequin monitor | FL | ? | USDA (unpublished data) |
| | 2001 | water monitor | FL | Indonesia | USDA (unpublished data) |
| | 2002 | water monitor | FL | ? | USDA (unpublished data) |
| | 2005 | green tree python (*Morelia viridis*) | FL | ? | Burridge et al. (2006) |
| Amblyomma variegatum | 1965 | black rhinoceros | MI | eastern Africa | USDA (1966b) |
| | 1971 | African buffalo hide | CA | ? | USDA (1972) |
| | 1973 | wildlife trophy | CO | Africa | USDA (1974b) |
| | 1974 | rhinoceros | NC | ? | USDA (1975) |
| | 1979 | kudu | CO | ? | USDA (1981) |
| | 1980 | eland | CO | ? | Wilson & Bram (1998) |
| | 1989 | ostrich | NY | Tanzania | Mertins & Schlater (1991), USDA (1994) |
| | 1989 | ostrich | TX | Tanzania | Mertins & Schlater (1991), Wilson & Bram (1998) |
| | 1993 | cattle | ? | ? | Wilson & Bram (1998) |
| | 1994 | horse | ? | ? | Wilson & Bram (1998) |
| | 1994 | sheep | ? | ? | Wilson & Bram (1998) |
| | 1997 | African savanna monitor | FL | Ghana | Wilson & Bram (1998) |
| | ? | giraffe | NY | Africa | Keirans & Durden (2001) |
| | ? | human | AZ | Tanzania | Keirans & Durden (2001) |
| | ? | human | NY | Kenya | Keirans & Durden (2001) |
| | ? | human | PA | Africa | Keirans & Durden (2001) |
| | ? | zebra | NY | Kenya | Keirans & Durden (2001) |
| Amblyomma varium | 1966 | sloth | CA | ? | USDA (1967a) |
| | 1971 | southern two-toed sloth (*Choloepus didactylus*) | IL | Peru | USDA (1972) |
| | 1972 | southern two-toed sloth | IL | Panama | USDA (1973a) |
| Bothriocroton concolor | 1973 | echidna | IL | Australia | USDA (1974a) |
| Bothriocroton hydrosauri | ? | echidna | IL | ? | Keirans & Durden (2001) |
| Dermacentor auratus | 1986 | plant | IL | ? | USDA (1987b) |
| | ? | humans | AZ/ME | Nepal | Keirans & Durden (2001) |
| Dermacentor nitens | 1941 | donkey | MD | Haiti | Becklund (1968) |
| | 1946 | deer | LA | Honduras | Becklund (1968) |
| | 1958 | horses | FL | Mexico/ Puerto Rico | Becklund (1968) |
| | 1960 | horses | FL | Cuba/ Venezuela | Becklund (1968) |
| | 1960 | horse | FL | ? | Becklund (1968) |

| Tick | Year | Host | State[a] | Origin of host | Original reference |
|---|---|---|---|---|---|
| | 1962 | cattle | IL | ? | USDA (1962a) |
| | 1963 | equine | LA | ? | USDA (1964a) |
| | 1963 | ocelot | CA | Ecuador | Becklund (1968) |
| | 1964 | horse | TX | Cuba/ Guatemala/ Puerto Rico/Venezuela | USDA (1965a) |
| | 1965 | horse | FL | Colombia | USDA (1966a) |
| | 1965 | horse | TX | Peru | USDA (1966a) |
| | 1966 | cattle | TX | ? | USDA (1967a) |
| | 1966 | equine | FL | ? | USDA (1967a) |
| | 1967 | horses | FL | Colombia/ Costa Rica/ Jamaica | USDA (1967b, 1968b) |
| | 1968 | equines | MS/TX | ? | USDA (1969) |
| | 1968 | horse | FL | Dominican Republic | USDA (1968b) |
| | 1968 | horse | TN | Puerto Rico | USDA (1969) |
| | 1969 | horses | FL | Dominican Republic/ Honduras/ South America | USDA (1970a, b) |
| | 1969 | horse | MS | Puerto Rico | USDA (1970b) |
| | 1969 | horse | NJ | ? | USDA (1970a) |
| | 1970 | equine | AL | ? | USDA (1971) |
| | 1970 | horse | FL | Nicaragua | USDA (1970c) |
| | 1971 | equine | FL | ? | USDA (1972) |
| | 1972 | horse | FL | ? | USDA (1973a) |
| | 1973 | horses | FL | Belize/ Venezuela | USDA (1974a, b) |
| | 1975 | horse | TX | ? | USDA (1977a) |
| | 1987 | cattle | TX | ? | USDA (1988) |
| | 1987 | horse | FL | ? | USDA (1988) |
| | 1989 | horse | TX | ? | USDA (1994) |
| | 1999 | horse | FL | Colombia | USAHA (2005) |
| | ? | horses | FL | Cuba/Guatemala/ Jamaica | Keirans & Durden (2001) |
| | ? | horses | TX | Puerto Rico/ Venezuela | Keirans & Durden (2001) |
| Dermacentor nuttalli | 1982 | animal hide | IL | ? | USDA (1983) |
| Dermacentor reticulatus | ? | hedgehog | NY | France | USDA (1962b) |
| | 1961 | oryx | ? | ? | Wilson & Bram (1998) |
| | 1985 | horse | NY | France | USDA (1987a) |
| | ? | ? | ? | ? | Keirans & Durden (2001) |
| Dermacentor rhinocerinus | 1969 | rhinoceros | FL | ? | USDA (1970a) |
| | 1973 | white rhinoceros | VA | ? | USDA (1974a) |
| | 1989 | rhinoceros | TX | ? | USDA (1994) |
| Haemaphysalis elongata | ? | tenrec | OH | Madagascar | Keirans & Durden (2001) |
| Haemaphysalis hoodi | 1984 | grey parrot (Psittacus erithacus) | NY | ? | USDA (1985b) |
| Haemaphysalis hystricis | 1971 | tapir | MA | ? | USDA (1972) |
| Haemaphysalis juxtakochi | 1978 | palm frond | LA | ? | USDA (1980) |
| Haemaphysalis leachi | ? | hedgehog | ? | ? | USDA (1962b) |
| | 1965 | lion (Panthera leo) | TX | ? | USDA (1965b) |
| | 1975 | hedgehog | NY | ? | USDA (1977a) |
| | 1983 | dog | CT | South Africa | Anderson et al. (1984) |

| Tick | Year | Host | State[a] | Origin of host | Original reference |
|---|---|---|---|---|---|
| | ? | jackal | MO | South Africa | Keirans & Durden (2001) |
| Haemaphysalis longicornis | 1969 | horse | NJ | ? | USDA (1970a) |
| | 1982 | horse | CA | ? | USDA (1983) |
| | 1983 | horses | CA | ? | USDA (1985a, 1994) |
| | 1992 | horse | ? | ? | Wilson & Bram (1998) |
| | 1995 | horse | ? | ? | Wilson & Bram (1998) |
| | 1997 | horse | ? | ? | Wilson & Bram (1998) |
| Haemaphysalis muhsamae | 1966 | bat-eared fox (Otocyon megalotis) | NY | ? | USDA (1966b) |
| | ? | hedgehog | NY | Kenya | Becklund (1968) |
| Haemaphysalis punctata | 1989 | ostriches | CA/NY | Portugal | Mertins & Schlater (1991), USDA (1994), Wilson & Bram (1998) |
| | ? | pig | FL | ? | Wilson & Bram (1998) |
| Hyalomma aegyptium | 1970 | tortoise | FL | Italy | USDA (1970c) |
| | 1975 | tortoise | NY | ? | USDA (1977a) |
| | 1995 | Egyptian tortoise (Testudo kleinmanni) | FL | Egypt | Clark & Doten (1995) |
| | 1995 | Greek tortoise (Testudo graeca) | FL | Egypt | Clark & Doten (1995) |
| | 2002 | Greek tortoise | FL | ? | USDA (unpublished data) |
| | ? | tortoise | NC | northern Africa | Keirans & Durden (2001) |
| | ? | ? | ? | ? | Keirans & Durden (2001) |
| Hyalomma albiparmatum | ? | giraffe | NY | ? | Becklund (1968) |
| | ? | rhinoceros | TX | ? | Becklund (1968) |
| | 1969 | giraffe hide | TX | Kenya | USDA (1970a) |
| | 1976 | zebra | NJ | ? | USDA (1977b) |
| | 1989 | ostrich | TX | Tanzania | Mertins & Schlater (1991), USDA (1994) |
| Hyalomma anatolicum spp.[c] | ? | horse | NY | Jordan | Becklund (1968) |
| | ? | onager (Equus hemionus) | NY | Iran | Becklund (1968) |
| | 1968 | horse | NJ | Spain | USDA (1968b) |
| | 1984 | baggage | DE | ? | USDA (1985b) |
| Hyalomma detritum | 1969 | horse | FL | Portugal | USDA (1970a) |
| Hyalomma dromedarii | ? | camel | NY | ? | Becklund (1968) |
| Hyalomma impressum | 1984 | rhinoceros hide | ME | ? | USDA (1985b) |
| Hyalomma lusitanicum | 1989 | animal hide | MS | ? | USDA (1994) |
| | 1989 | ostrich | NY | Portugal | Mertins & Schlater (1991), USDA (1994) |
| Hyalomma marginatum group[d] | 1966 | horse | NJ | Spain | USDA (1966b) |
| | 1966 | horse | NY | Spain | USDA (1967a) |
| | ? | zebra | NY | Germany | Becklund (1968) |
| | 1972 | zebra | NY | ? | USDA (1973a) |
| | 1973 | zebra | NJ | Africa | USDA (1974b) |
| | 1979 | animal hide | CA | ? | USDA (1981) |
| | 1981 | virgin cork | MD | ? | USDA (1982) |

[c] Hyalomma anatolicum or Hy. excavatum (see section 7.3)
[d] Hyalomma marginatum or Hy. rufipes (see section 7.5)

| Tick | Year | Host | State[a] | Origin of host | Original reference |
|---|---|---|---|---|---|
| | 1989 | ostriches | CA/IN/NY/TX | ? | Mertins & Schlater (1991), USDA (1994), Wilson & Bram (1998) |
| | 1989 | rhinoceros | TX | ? | USDA (1994) |
| | ? | human | ? | Greece | Keirans & Durden (2001) |
| | ? | ? | ? | ? | Keirans & Durden (2001) |
| *Hyalomma truncatum* | ? | oryx | NY | Namibia | Becklund (1968) |
| | 1973 | wildlife trophy | CO | Africa | USDA (1974b) |
| | 1974 | rhinoceros | NC | ? | USDA (1975) |
| | 1979 | animal hide | CA | ? | USDA (1981) |
| | 1985 | animal hide | MS | Africa | USDA (1987a) |
| | 1987 | animal hide | AZ | ? | USDA (1988) |
| | 1987 | ostrich | NY | ? | USDA (1988) |
| | 1988 | animal hide | MS | ? | USDA (1994) |
| | 1989 | ostrich | NY | Angola | Mertins & Schlater (1991), USDA (1994) |
| | 1989 | rhinoceros | TX | ? | USDA (1994) |
| | ? | zebra | NY | Germany | Keirans & Durden (2001) |
| *Ixodes hexagonus* | ? | hedgehogs | NY | France/ United Kingdom | USDA (1962b), Becklund (1968) |
| | 1976 | hedgehog | NY | ? | USDA (1977b) |
| | 1985 | virgin cork | MD | Portugal | USDA (1987a) |
| *Ixodes luciae* | 1965 | greyish mouse-opossum (*Marmosa canescens*) | TX | Mexico | Eads et al. (1966) |
| *Ixodes pilosus* | 1983 | plant | NY | South Africa | USDA (1985a) |
| | 1984 | cut flowers | NY | ? | USDA (1985b) |
| | 1985 | cut flowers | NY | South Africa | USDA (1987a) |
| *Ixodes ricinus* | ? | horse | NY | Germany | Becklund (1968) |
| | ? | lizard | PA | ? | Snetsinger (1968) |
| | 1973 | donkey | NY | ? | Wilson & Bram (1998) |
| | 1988 | horse | CA | ? | USDA (1994) |
| | 1989 | horse | CA | ? | USDA (1994) |
| | ? | horse | GA | Switzerland | Keirans & Durden (2001) |
| | ? | human | CA | United Kingdom | Keirans & Durden (2001) |
| *Ixodes schillingsi* | 1967 | hyrax | IL | ? | USDA (1968a) |
| *Rhipicephalus appendiculatus* | 1966 | zebra | NJ | ? | USDA (1966b) |
| | 1967 | wool | SC | South Africa | USDA (1968a) |
| | 1969 | lion hide | TX | Kenya | USDA (1970a) |
| | 1972 | animal trophy | CO | Africa | USDA (1973a) |
| | 1973 | animal trophy | CO | ? | USDA (1974a) |
| | 1973 | zebra | NJ | Africa | USDA (1974b) |
| | 1974 | animal trophies | CO | ? | USDA (1975) |
| | 1986 | animal hide | MS | ? | USDA (1987b) |
| | 1988 | animal hide | MS | ? | USDA (1994) |
| | ? | ? | SC | ? | Keirans & Durden (2001) |

212 / Appendix 1

| Tick | Year | Host | State[a] | Origin of host | Original reference |
|---|---|---|---|---|---|
| *Rhicephalus bursa* | 1966 | horse | NJ | Spain | USDA (1966b) |
| | ? | Cretan goat (*Capra aegagrus*) | NY | Italy | Becklund (1968) |
| *Rhipicephalus capensis* | 1985 | cut flowers | NY | South Africa | USDA (1987a) |
| *Rhipicephalus compositus* | 1983 | plant | NY | ? | USDA (1985a) |
| *Rhicephalus evertsi* | 1960 | Abyssinian ass (*Equus asinus*) | FL | ? | Becklund (1968) |
| | 1960 | camel | FL | ? | Becklund (1968) |
| | 1960 | eland | FL | ? | USDA (1974a) |
| | 1960 | zebra | FL | Kenya | USDA (1962b) |
| | 1960 | zebra | NY | ? | USDA (1962b) |
| | 1961 | hartebeest (*Alcelaphus buselaphus*) | ? | ? | Wilson & Bram (1998) |
| | 1961 | oryx | ? | ? | Wilson & Bram (1998) |
| | 1961 | zebra | NY | ? | Wilson & Bram (1998) |
| | 1965 | eland | NJ | ? | USDA (1966a) |
| | 1966 | eland | NJ | ? | USDA (1967a) |
| | 1966 | giraffe | NJ | ? | USDA (1967a) |
| | 1966 | sable antelope (*Hippotragus niger*) | NJ | Germany | USDA (1967a) |
| | 1966 | zebras | NJ/NY | ? | USDA (1966b, 1967a) |
| | 1967 | zebra | NJ | Kenya | USDA (1967b) |
| | 1967 | zebra | NJ | South Africa | USDA (1968b) |
| | 1967 | zebra | NY | ? | USDA (1968a) |
| | ? | gemsbok (*Oryx gazella*) | NY | ? | Becklund (1968) |
| | ? | hartebeest | NY | ? | Becklund (1968) |
| | ? | oryx | NY | ? | Becklund (1968) |
| | 1968 | zebra | NJ | South Africa | USDA (1968b) |
| | 1968 | zebra | NY | ? | USDA (1969) |
| | 1969 | impala (*Aepyceros melampus*) | NJ | ? | USDA (1970a) |
| | 1969 | zebra | NJ | Kenya | USDA (1970c) |
| | 1969 | zebras | NY | Kenya/ Tanzania | USDA (1970a, c) |
| | 1970 | zebra | NJ | Kenya | USDA (1970c) |
| | 1971 | zebra | NJ | Africa | USDA (1973b) |
| | 1971 | zoo ruminants | NJ | Namibia | USDA (1973b) |
| | 1972 | animal trophy | CO | ? | USDA (1973a) |
| | 1972 | zebra | NJ | ? | USDA (1973b) |
| | 1972 | zebra | NY | Namibia | USDA (1973a) |
| | 1973 | animal trophy | CO | Africa | USDA (1974b) |
| | 1973 | zebra | NJ | Africa | USDA (1974b) |
| | 1974 | black wildebeest (*Connochaetes gnou*) | NJ | ? | USDA (1975) |
| | 1974 | zebra | NJ | ? | USDA (1975) |
| | 1975 | zebra | NJ | ? | USDA (1977a) |
| | 1976 | zebra | NJ | ? | USDA (1977b) |
| | 1979 | kudu | CO | ? | USDA (1981) |
| | 1981 | zebra hide | NC | ? | USDA (1982) |
| | 1983 | zebra | NY | ? | USDA (1985a) |
| | 1984 | rhinoceros hide | ME | ? | USDA (1985b) |
| | 1987 | animal hide | AZ | ? | USDA (1988) |
| | 1988 | animal hide | MS | ? | USDA (1994) |
| | 1989 | animal hides | AZ/GA | ? | USDA (1994) |

| Tick | Year | Host | State[a] | Origin of host | Original reference |
|---|---|---|---|---|---|
| | 1990 | horse | ? | ? | Wilson & Bram (1998) |
| | ? | Abyssinian ass | FL | Africa | Keirans & Durden (2001) |
| | ? | antelope | NY | Namibia | Keirans & Durden (2001) |
| | ? | camel | FL | Africa | Keirans & Durden (2001) |
| | ? | eland | FL | Africa | Keirans & Durden (2001) |
| | ? | gemsbok | NY | Namibia | Keirans & Durden (2001) |
| | ? | hartebeest | NY | Germany | Keirans & Durden (2001) |
| | ? | oryx | OH | Germany | Keirans & Durden (2001) |
| | ? | zebras | FL/NJ/NY | Kenya | Keirans & Durden (2001) |
| | ? | zebra | NY | Germany | Keirans & Durden (2001) |
| | ? | zebras | NY | Namibia/ Tanzania | Keirans & Durden (2001) |
| | ? | zebra × horse | FL | Africa | Keirans & Durden (2001) |
| *Rhipicephalus kochi* | 1984 | cut flowers | NY | ? | USDA (1985b) |
| *Rhipicephalus muehlensi* | 1970 | animal trophy | CA | ? | USDA (1970c) |
| | 1970 | nyala (*Tragelaphus angasii*) | CA | Africa | USDA (1970c) |
| *Rhipicephalus pulchellus* | 1958 | giraffe | ? | ? | Wilson & Bram (1998) |
| | 1960 | zebra | FL | ? | USDA (1962b) |
| | 1961 | zebra | ? | ? | Wilson & Bram (1998) |
| | 1965 | black rhinoceros | MI | eastern Africa | USDA (1966b) |
| | 1966 | zebras | NJ/NY | ? | USDA (1966b, 1967a) |
| | 1967 | ruminant hide | FL | Kenya | USDA (1968a) |
| | 1967 | zebra | NJ | Kenya | USDA (1967b) |
| | 1967 | zebra | NY | ? | USDA (1968a) |
| | 1967 | zebra hide | FL | ? | USDA (1968a) |
| | ? | giraffe | NY | ? | Becklund (1968) |
| | ? | zebra | FL | Kenya | Becklund (1968) |
| | 1969 | lion hide | TX | Kenya | USDA (1970a) |
| | 1969 | zebra | NJ | Kenya | USDA (1970c) |
| | 1969 | zebras | NY | Kenya/ Tanzania | USDA (1970a, c) |
| | 1970 | zebra | NJ | Kenya | USDA (1970c) |
| | 1970 | zebra | NY | ? | USDA (1971) |
| | 1971 | zebra | NJ | ? | USDA (1973b) |
| | 1972 | zebras | NJ/NY | ? | USDA (1973a, b) |
| | 1972 | zebra hide | CO | ? | USDA (1973a) |
| | 1973 | animal trophy | CO | Africa | USDA (1974b) |
| | 1973 | zebra | NJ | Africa | USDA (1974b) |
| | 1974 | animal trophy | CO | ? | USDA (1975) |
| | 1974 | rhinoceros | NC | ? | USDA (1975) |
| | 1974 | zebra | NJ | ? | USDA (1975) |
| | 1976 | zebra | NJ | ? | USDA (1977b) |

| Tick | Year | Host | State[a] | Origin of host | Original reference |
|---|---|---|---|---|---|
| | ? | eland | PA | eastern Africa | Keirans & Durden (2001) |
| | ? | giraffes | NJ/NY | ? | Keirans & Durden (2001) |
| | ? | human | MI | Kenya | Keirans & Durden (2001) |
| | ? | rhinoceros | MI | eastern Africa | Keirans & Durden (2001) |
| | ? | zebra | MD | Africa | Keirans & Durden (2001) |
| | ? | zebra | NY | Tanzania | Keirans & Durden (2001) |
| | ? | zebra | ? | Germany | Keirans & Durden (2001) |
| *Rhipicephalus senegalensis* | 1988 | animal hide | MS | ? | USDA (1994) |
| *Rhipicephalus simus* | 1965[e] | black rhinoceros | MI | eastern Africa | USDA (1966b) |
| | 1969[e] | giraffe hide | TX | ? | USDA (1970a) |
| | 1969[e] | lion hide | TX | Kenya | USDA (1970a) |
| | 1979[e] | human | CT | Kenya | Anderson et al. (1981a) |
| | 1983 | warthog hide | MS | South Africa | USDA (1985a) |
| | 1989[e] | animal hide | MS | ? | USDA (1994) |
| | ? | jackal | MO | South Africa | Keirans & Durden (2001) |
| *Rhipicephalus sulcatus* | 1973 | hedgehog | NY | ? | USDA (1974a) |
| *Rhipicephalus turanicus* | 1989 | ostriches | NY | Botswana/ Portugal/ Tanzania | Mertins & Schlater (1991), USDA (1994) |
| *Rhipicephalus (Boophilus) annulatus* | 1962 | cattle | ? | Mexico | USDA (1962b) |
| | 1963 | cattle | TX | ? | USDA (1964a) |
| | 1963 | equine | TX | ? | USDA (1964a) |
| | 1964 | cattle | TX | Mexico | USDA (1965a) |
| | 1964 | horse | TX | Mexico | USDA (1965a) |
| | 1965 | cattle | TX | ? | USDA (1966a) |
| | 1965 | horse | TX | ? | USDA (1966a) |
| | 1966 | cattle | TX | ? | USDA (1967a) |
| | 1966 | equine | TX | ? | USDA (1967a) |
| | 1967 | cattle | TX | ? | USDA (1968a) |
| | 1968 | cattle | TX | Mexico | USDA (1969) |
| | 1969 | cattle | TX | Mexico | USDA (1970a) |
| | 1970 | cattle | TX | Mexico | USDA (1970c) |
| | 1970 | equine | TX | ? | USDA (1971) |
| | 1971 | cattle | TX | ? | USDA (1972) |
| | 1971 | equine | CA | ? | USDA (1972) |
| | 1972 | cattle | TX | ? | USDA (1973a) |
| | 1972 | deer | TX | ? | USDA (1973a) |
| | 1973 | cattle | TX | Mexico | USDA (1974b) |
| | 1973 | cattle | TX | ? | USDA (1974a) |
| | 1973 | deer | TX | ? | USDA (1974a) |
| | 1974 | cattle | TX | ? | USDA (1975) |
| | 1975 | cattle | TX | ? | USDA (1977a) |
| | 1977 | cattle | AZ/TX | ? | USDA (1978) |

[e] probably tick misidentifications (see section 3.37)

| Tick | Year | Host | State[a] | Origin of host | Original reference |
|---|---|---|---|---|---|
| | 1977 | deer | TX | ? | USDA (1978) |
| | 1978 | cattle | TX | ? | USDA (1980) |
| | 1979 | cattle | TX | ? | USDA (1981) |
| | 1979 | deer hide | TX | ? | USDA (1981) |
| | 1980 | cattle | TX | ? | USDA (1982) |
| | 1981 | cattle | TX | ? | USDA (1982) |
| | 1981 | horse | NY | ? | USDA (1982) |
| | 1982 | cattle | TX | ? | USDA (1983) |
| | 1983 | cattle | TX | ? | USDA (1985a) |
| | 1983 | deer | TX | ? | USDA (1985a) |
| | 1983 | horse | TX | ? | USDA (1985a) |
| | 1984 | cattle | TX | ? | USDA (1985b) |
| | 1984 | deer hide | TX | ? | USDA (1985b) |
| | 1985 | cattle | TX | ? | USDA (1987a) |
| | 1985 | deer hide | TX | ? | USDA (1987a) |
| | 1986 | cattle | TX | ? | USDA (1987b) |
| | 1986 | deer hide | TX | ? | USDA (1987b) |
| | 1987 | cattle | CA/TX | ? | USDA (1988) |
| | 1987 | deer hide | TX | ? | USDA (1988) |
| | 1988 | cattle | TX | ? | USDA (1994) |
| | 1988 | deer hide | TX | ? | USDA (1994) |
| | 1989 | cattle | TX | ? | USDA (1994) |
| | 1989 | deer hide | TX | ? | USDA (1994) |
| | 1989 | horse | TX | ? | USDA (1994) |
| *Rhipicephalus (Boophilus) decoloratus* | ? | giraffe | ? | ? | USDA (1962b) |
| | 1965 | giraffe | NJ | ? | USDA (1966a) |
| | 1965 | hartebeest | NJ | ? | USDA (1966a) |
| | ? | giraffe | NY | Africa | Becklund (1968) |
| | ? | impala | NY | Kenya | Becklund (1968) |
| | 1972 | animal trophy | CO | Africa | USDA (1973a) |
| | 1973 | zebra | NJ | ? | USDA (1974b) |
| | 1974 | zebra | ? | ? | Wilson & Bram (1998) |
| | 1976 | zebra | NJ | ? | USDA (1977b) |
| | 1985 | animal hide | MS | ? | USDA (1987a) |
| | 1989 | animal hide | AZ | ? | USDA (1994) |
| *Rhipicephalus (Boophilus) microplus* | 1962 | beef hindquarters | ? | ? | USDA (1962b) |
| | 1962 | horse | ? | ? | USDA (1962b) |
| | 1963 | cattle | TX | ? | USDA (1964a) |
| | 1964 | beef hindquarters | ? | ? | USDA (1965a) |
| | 1964 | cattle | ? | ? | USDA (1965a) |
| | 1964 | horse | ? | ? | USDA (1965a) |
| | 1966 | cattle | TX | ? | USDA (1967a) |
| | 1967 | cattle | TX | ? | USDA (1968a) |
| | ? | cattle hides | FL | Cuba | Becklund (1968) |
| | ? | hair | NY | Brazil | Becklund (1968) |
| | ? | horses | FL | Cuba/ Guatemala | Becklund (1968) |
| | ? | refrigerated beef | FL | Costa Rica/ Honduras | Becklund (1968) |
| | ? | refrigerated beef | TX | Mexico | Becklund (1968) |
| | 1968 | cattle | TX | ? | USDA (1969) |

| Tick | Year | Host | State[a] | Origin of host | Original reference |
|---|---|---|---|---|---|
| | 1969 | cattle | TX | ? | USDA (1970c) |
| | 1970 | cattle | TX | ? | USDA (1970c) |
| | 1970 | equine | TX | ? | USDA (1971) |
| | 1971 | cattle | TX | ? | USDA (1972) |
| | 1972 | cattle | TX | ? | USDA (1973a) |
| | 1972 | equine | FL | ? | USDA (1973a) |
| | 1973 | cattle | TX | Mexico | USDA (1974a) |
| | 1973 | horse | FL | Belize | USDA (1974b) |
| | 1974 | cattle | TX | ? | USDA (1975) |
| | 1975 | cattle | TX | ? | USDA (1977a) |
| | 1976 | cattle | TX | ? | USDA (1977b) |
| | 1977 | cattle | TX | ? | USDA (1978) |
| | 1977 | two-toed sloth | NY | ? | USDA (1978) |
| | 1978 | cattle | TX | ? | USDA (1980) |
| | 1979 | cattle | TX | ? | USDA (1981) |
| | 1980 | cattle | TX | ? | USDA (1982) |
| | 1982 | cattle | TX | ? | USDA (1983) |
| | 1983 | cattle | TX | ? | USDA (1985a) |
| | 1984 | cattle | TX | ? | USDA (1985b) |
| | 1985 | cattle | TX | ? | USDA (1987a) |
| | 1986 | cattle | TX | ? | USDA (1987b) |
| | 1986 | equine | AZ | ? | USDA (1987b) |
| | 1987 | cattle | TX | ? | USDA (1988) |
| | 1987 | plant | IL | ? | USDA (1988) |
| | 1988 | cattle | TX | ? | USDA (1994) |
| | 1988 | deer hide | TX | ? | USDA (1994) |
| | 1989 | cattle | TX | ? | USDA (1994) |
| | 1989 | deer hide | TX | ? | USDA (1994) |

# Appendix 2. Invasive Ticks by Host

Tick genera are abbreviated as follows: *A*. = *Amblyomma*, *B*. = *Bothriocroton*, *D*. = *Dermacentor*, *Ha*. = *Haemaphysalis*, *Hy*. = *Hyalomma*, *I*. = *Ixodes*, *R*. = *Rhipicephalus*, *R*. *(B.)* = *Rhipicephalus (Boophilus)*. Stages of ticks found on hosts are given in parentheses after name of tick, where recorded, with A = adult, N = nymph and L = larva. Tick species are underlined to indicate their more common hosts.

## Class Amphibia (amphibians)

ORDER ANURA

Family Bufonidae (toads)

*Bufo arenarum* (**common toad**): *A. rotundatum*
*Bufo calamita* (**natterjack toad**): *A. rotundatum*
*Bufo granulosus* (**common lesser toad**): *A. rotundatum*
*Bufo marinus* (**giant toad**): *A. argentinae* (AN), *A. cajennense*, *A. dissimile* (AL), *A. flavomaculatum* (A), *A. humerale*, *A. latum* (A), *A. rotundatum* (ANL), *A. varium*, *R. (B.) microplus* (A)
*Bufo peltocephalus* (**Tschudi's toad**): *A. dissimile*, *A. rotundatum*, *D. nitens* (A)
*Bufo schneideri* (**Schneider's toad**): *A. dissimile*, *A. rotundatum* (A)
*Bufo viridis* (**European green toad**): *A. rotundatum*

Family Pipidae (pipas and clawed frogs)

*Pipa pipa* (**Suriname toad**): *A. rotundatum* (A)

## Class Aves (birds)

ORDER ANSERIFORMES

Family Anatidae (swans, geese and ducks)

*Anas platyrhynchos* (**common mallard**): *Ha. longicornis*, *Hy. marginatum* group,[a] *I. ricinus*
*Anser anser* (**greylag goose**): *I. ricinus*
*Aythya ferina* (**common pochard**): *I. ricinus*
*Cairina moschata* (**muscovy duck**): *A. variegatum*
*Dendrocygna viduata* (**white-faced tree duck**): *A. variegatum* (A), *R. senegalensis* (A)
*Netta rufina* (**red-crested pochard**): *I. ricinus*
*Plectropterus gambiensis* (**spur-winged goose**): *A. variegatum* (N)
*Somateria mollissima* (**common eider**): *I. ricinus* (N)

ORDER APODIFORMES

Family Apodidae (swifts)

*Apus apus* (**common swift**): *I. ricinus*

ORDER APTERYGIFORMES

Family Apterygidae (kiwis)

*Apteryx australis* (**brown kiwi**): *Ha. longicornis* (NL)

ORDER CAPRIMULGIFORMES

Family Caprimulgidae (nightjars)

*Caprimulgus europaeus* (**Eurasian nightjar**): *Ha. punctata*, *Hy. marginatum* group, *I. ricinus*, *R. turanicus*
*Caprimulgus fossii* (**square-tailed nightjar**): *Hy. marginatum* group (N)
*Caprimulgus pectoralis* (**fiery-necked nightjar**): *A. hebraeum*
*Caprimulgus rufigena* (**rufous-cheeked nightjar**): *A. hebraeum*, *A. marmoreum* (N), *Hy. marginatum* group (NL)

*Caprimulgus tristigma* (**freckled nightjar**): *Hy. marginatum* group (N)

ORDER CHARADRIIFORMES

Family Burhinidae (stone curlews)

*Burhinus capensis* (**spotted thick-knee**): *A. hebraeum, Hy. marginatum* group (A), *Hy. truncatum* (A), *R. turanicus* (A)

*Burhinus oedicnemus* (**stone curlew**): *A. lepidum* (A), *Ha. punctata, Hy. anatolicum* spp.,[b] *Hy. marginatum* group (NL)

Family Charadriidae (plovers and lapwings)

*Charadrius alexandrinus* (**Kentish plover**): *Hy. marginatum* group

*Charadrius asiaticus* (**Caspian plover**): *A. variegatum* (N), *Hy. marginatum* group (NL)

*Charadrius dubius* (**little ringed plover**): *Hy. marginatum* group

*Charadrius pecuarius* (**Kittlitz's plover**): *Hy. marginatum* group (N)

*Pluvianus aegyptius* (**Egyptian plover**): *Hy. dromedarii*

*Pluvialis apricaria* (**Eurasian golden plover**): *Hy. marginatum* group, *I. hexagonus, I. ricinus* (NL)

*Vanellus coronatus* (**crowned lapwing**): *Hy. marginatum* group (NL)

*Vanellus gregarius* (**social lapwing**): *Ha. punctata*

*Vanellus vanellus* (**northern lapwing**): *Ha. punctata, Hy. marginatum* group, *I. ricinus* (N)

Family Glareolidae (pratincoles and coursers)

*Rhinoptilus africanus* (**double-banded courser**): *Hy. marginatum* group (N)

Family Haematopodidae (oystercatchers)

*Haematopus ostralegus* (**Eurasian oystercatcher**): *Ha. punctata* (NL), *I. ricinus* (NL)

Family Laridae (gulls)

*Larus argentatus* (**herring gull**): *R. turanicus* (A)

*Larus canus* (**common gull**): *I. ricinus* (NL)

*Larus marinus* (**great black-backed gull**): *Ha. punctata*

*Larus ridibundus* (**black-headed gull**): *Ha. punctata* (A), *I. ricinus* (NL)

*Sterna hirundo* (**common tern**): *Hy. marginatum* group

Family Scolopacidae (sandpipers and snipes)

*Gallinago gallinago* (**common snipe**): *I. ricinus* (L)

*Numenius arquata* (**curlew**): *Ha. punctata* (AN), *I. ricinus* (NL)

*Philomachus pugnax* (**ruff**): *Ha. punctata* (L), *I. ricinus* (N)

*Scolopax rusticola* (**Eurasian woodcock**): *Ha. punctata, I. ricinus* (NL)

*Tringa hypoleucos* (**common sandpiper**): *Ha. punctata*

ORDER CICONIIFORMES

Family Ardeidae (herons and bitterns)

*Ardea melanocephala* (**black-headed heron**): *A. lepidum* (N)

*Ardeola ralloides* (**squacco heron**): *Ha. punctata*

*Bubulcus ibis* (**cattle egret**): *A. variegatum* (NL), *Hy. marginatum* group (A), *Hy. aegyptium, R. (B.) decoloratus* (A), *R. (B.) microplus* (L)

*Cochlearius cochlearia* (**boat-billed heron**): *A. dissimile*

*Dupetor flavicollis* (**black bittern**): *A. geoemydae* (L)

*Egretta alba* (**great white egret**): *A. variegatum* (L)

Family Ciconiidae (storks)

*Ciconia abdimii* (**Abdim's stork**): *A. variegatum* (NL), *Ha. hoodi* (A), *R. sulcatus* (A)

*Ciconia ciconia* (**white stork**): *A. lepidum, Hy. truncatum* (L)

*Ciconia nigra* (**black stork**): *A. lepidum* (AN)

*Ephippiorhynchus senegalensis* (**saddlebill stork**): *R. turanicus* (A)

*Leptoptilos crumeniferus* (**marabou stork**): *A. lepidum* (A), *Hy. marginatum* group, *R. turanicus* (A)

Family Scopidae (hammerheads)

*Scopus umbretta* (**hamerkop**): *A. hebraeum*

Family Threskiornithidae (ibises and spoonbills)

*Bostrychia hagedash* (**hadada ibis**): *A. marmoreum* (N)

*Theristicus caudatus* (**buff-necked ibis**): *A. cajennense* (N)

ORDER COLIIFORMES

Family Coliidae (mousebirds)

*Colius colius* (white-backed mousebird): *R. appendiculatus*, *R. evertsi*

ORDER COLUMBIFORMES

Family Columbidae (pigeons)

*Columba eversmanni* (pale-backed pigeon): *Hy. marginatum* group
*Columba livia* (common pigeon): *Ha. punctata*, *Hy. marginatum* group, *I. ricinus*
*Columba oenas* (stock dove): *I. ricinus*
*Columba palumbus* (wood pigeon): *I. ricinus* (NL)
*Columba rupestris* (hill pigeon): *Ha. punctata*
*Columbina passerina* (common ground dove): *A. variegatum* (L)
*Leptotila verreauxi* (white-tipped dove): *A. coelebs* (L)
*Oena capensis* (long-tailed dove): *Hy. marginatum* group (L)
*Streptopelia capicola* (ring-necked dove): *A. marmoreum* (L)
*Streptopelia decaocto* (collared dove): *I. ricinus*
*Streptopelia orientalis* (oriental turtle dove): *Ha. punctata*
*Streptopelia senegalensis* (laughing dove): *Hy. marginatum* group, *Hy. truncatum*
*Streptopelia turtur* (European turtle dove): *A. lepidum* (L), *Ha. punctata*, *Hy. aegyptium* (N), *Hy. marginatum* group (NL), *I. ricinus*
*Treron calvus* (African green pigeon): *Ha. hoodi*
*Turtur abyssinicus* (black-billed wood dove): *A. lepidum* (L)
*Turtur chalcospilos* (emerald-spotted wood dove): *Ha. hoodi*

Family Pteroclididae (sandgrouses)

*Pterocles burchelli* (Burchell's sandgrouse): *A. hebraeum* (L)
*Pterocles orientalis* (black-bellied sandgrouse): *Ha. punctata*
*Pterocles personatus* (Madagascar sandgrouse): *A. variegatum* (N)
*Pterocles quadricinctus* (four-banded sandgrouse): *Hy. truncatum* (L)
*Syrrhaptes paradoxus* (Pallas' sandgrouse): *Ha. punctata* (A)

ORDER CORACIIFORMES

Family Alcedinidae (kingfishers)

*Alcedo atthis* (common kingfisher): *Hy. marginatum* group (N)
*Ceyx erithacus* (black-backed kingfisher): *A. geoemydae* (N)
*Halcyon albiventris* (brown-hooded kingfisher): *R. appendiculatus*
*Lacedo pulchella* (banded kingfisher): *A. geoemydae* (N)

Family Bucerotidae (hornbills)

*Anthracoceros malayanus* (Asian black hornbill): *A. geoemydae* (N)
*Bucorvus abyssinicus* (Abyssinian ground hornbill): *A. variegatum* (AN), *Ha. hoodi* (A)
*Ceratogymna albotibialis* (white-thighed hornbill): *R. sulcatus* (A)
*Tockus erythrorhynchus* (red-billed hornbill): *A. hebraeum*, *A. variegatum* (L), *Hy. marginatum* group (NL)
*Tockus flavirostris* (eastern yellow-billed hornbill): *A. hebraeum*, *Hy. truncatum* (L)
*Tockus nasutus* (African grey hornbill): *A. hebraeum*, *A. variegatum* (AN), *Ha. hoodi* (A), *Hy. marginatum* group (L), *Hy. truncatum*

Family Coraciidae (rollers)

*Coracias caudata* (lilac-breasted roller): *Hy. marginatum* group (N), *Hy. truncatum* (L)
*Coracias garrulus* (European roller): *A. hebraeum* (L), *A. nuttalli*, *Ha. punctata*, *Hy. aegyptium*, *Hy. marginatum* group (N), *Hy. truncatum* (L), *I. ricinus*
*Coracias spatulata* (racket-tailed roller): *A. hebraeum*, *Hy. marginatum* group
*Eurystomus glaucurus* (broad-billed roller): *Hy. marginatum* group

Family Meropidae (bee-eaters)

*Merops apiaster* (bee-eater): *Ha. punctata*, *Hy. marginatum* group

*Merops pusillus* (**little bee-eater**): *Hy. marginatum* group (L)

*Merops superciliosus* (**blue-cheeked bee-eater**): *Hy. marginatum* group (NL)

Family Momotidae (motmots)

*Baryphthengus martii* (**rufous motmot**): *A. longirostre* (NL)

*Baryphthengus ruficapillus* (**rufous-capped motmot**): *A. cajennense* (N), *A. calcaratum* (N), *A. coelebs* (N), *A. nodosum* (N)

*Momotus momota* (**blue-crowned motmot**): *A. cajennense* (N), *A. coelebs* (N)

Family Phoeniculidae (wood-hoopoes)

*Phoeniculus purpureus* (**green woodhoopoe**): *A. variegatum* (L), *Hy. marginatum* group (NL)

*Rhinopomastus cyanomelas* (**common scimitar bill**): *Hy. marginatum* group (L)

Family Upupidae (hoopoes)

*Upupa africana* (**African hoopoe**): *A. sparsum* (N)

*Upupa epops* (**Eurasian hoopoe**): *A. sparsum*, *Ha. punctata*, *Hy. aegyptium*, *Hy. detritum* (N), *Hy. marginatum* group (NL), *I. ricinus*, *R. bursa*

ORDER CUCULIFORMES

Family Cuculidae (cuckoos)

*Centropus burchelli* (**Burchell's coucal**): *Ha. hoodi* (A)

*Centropus grillii* (**black coucal**): *Ha. hoodi* (A)

*Centropus monachus* (**blue-headed coucal**): *Ha. hoodi* (A)

*Centropus senegalensis* (**Senegal coucal**): *A. hebraeum* (NL), *A. nuttalli* (NL), *A. variegatum* (NL), *Ha. hoodi* (AN), *Hy. marginatum* group (NL), *Hy. truncatum* (L), *R. sulcatus* (A)

*Centropus sinensis* (**greater coucal**): *A. geoemydae* (N), *Hy. marginatum* group (L)

*Centropus superciliosus* (**white-browed coucal**): *A. hebraeum*, *A. variegatum* (N), *Ha. hoodi* (ANL)

*Centropus toulou* (**Madagascar coucal**): *A. hebraeum*, *A. variegatum* (N), *Ha. hoodi* (A)

*Ceuthmochares aereus* (**yellowbill**): *A. variegatum*

*Cuculus canorus* (**common cuckoo**): *Ha. punctata*, *Hy. marginatum* group (NL), *I. ricinus*

*Oxylophus levaillantii* (**Levaillant's cuckoo**): *Hy. marginatum* group

*Saurothera merlini* (**great lizard-cuckoo**): *A. cajennense*

Family Musophagidae (turacos)

*Corythaixoides concolor* (**grey go-away-bird**): *A. hebraeum*

*Corythaixoides personatus* (**bare-faced go-away-bird**): *Ha. hoodi* (A)

*Tauraco persa* (**Guinea turaco**): *A. compressum* (A), *A. splendidum*, *A. tholloni* (N)

ORDER FALCONIFORMES

Family Accipitridae (hawks and eagles)

*Accipiter badius* (**little banded sparrowhawk**): *A. hebraeum*

*Accipiter gentilis* (**northern goshawk**): *I. ricinus*

*Accipiter melanoleucus* (**black goshawk**): *A. hebraeum*

*Accipiter nisus* (**Eurasian sparrowhawk**): *I. ricinus* (NL)

*Aquila chrysaetos* (**golden eagle**): *Ha. punctata*, *Hy. detritum* (N), *I. ricinus* (A)

*Aquila nipalensis* (**steppe eagle**): *R. turanicus* (A)

*Aquila pomarina* (**lesser spotted eagle**): *I. ricinus*

*Aquila rapax* (**tawny eagle**): *A. hebraeum*, *Hy. marginatum* group

*Buteo buteo* (**Eurasian buzzard**): *Ha. punctata*, *Hy. marginatum* group (NL), *I. ricinus* (N), *R. turanicus* (A)

*Buteo magnirostris* (**roadside hawk**): *A. cajennense* (A)

*Buteo platypterus* (**broad-winged hawk**): *A. longirostre* (L)

*Circaetus gallicus* (**short-toed eagle**): *Hy. marginatum* group (L), *R. turanicus* (A)

*Circaetus pectoralis* (**black-breasted harrier-eagle**): *A. exornatum*

*Circus aeruginosus* (**Eurasian marsh harrier**): *Hy. marginatum* group (L)

*Circus cyaneus* (**hen harrier**): *Ha. punctata*, *Hy. marginatum* group

*Circus macrourus* (**pale harrier**): *Hy. marginatum* group (NL)

*Circus pygargus* (**Montagu's harrier**): *Ha. punctata*

*Gyps africanus* (**African white-backed vulture**): *A. variegatum, Hy. truncatum* (A)

*Hieraaetus wahlbergi* (**Wahlberg's eagle**): *R. turanicus* (A)

*Melierax canorus* (**pale chanting goshawk**): *R. evertsi* (A)

*Melierax metabates* (**dark chanting goshawk**): *Hy. marginatum* group (N)

*Milvus migrans* (**black kite**): *A. variegatum, Hy. marginatum* group (NL), *I. ricinus, R. turanicus* (A)

*Milvus milvus* (**red kite**): *Hy. marginatum* group (L), *I. ricinus* (N)

*Necrosyrtes monachus* (**hooded vulture**): *A. tholloni* (N), *A. variegatum* (N)

*Neophron percnopterus* (**Egyptian vulture**): *Ha. hoodi* (A), *Hy. marginatum* group

*Pernis apivorus* (**European honey buzzard**): *Hy. marginatum* group, *I. ricinus*

*Polemaetus bellicosus* (**martial eagle**): *Hy. marginatum* group (L)

*Torgos tracheliotus* (**lappet-faced vulture**): *A. gemma* (N)

*Trigonoceps occipitalis* (**white-headed vulture**): *R. turanicus* (A)

Family Cathartidae (New World vultures)

*Coragyps atratus* (**black vulture**): *A. oblongoguttatum* (N)

Family Falconidae (falcons)

*Falco biarmicus* (**lanner falcon**): *Ha. hoodi* (A)

*Falco cherrug* (**saker falcon**): *Hy. marginatum* group, *I. ricinus*

*Falco columbarius* (**merlin**): *I. ricinus* (L)

*Falco naumanni* (**lesser kestrel**): *Ha. punctata, Hy. marginatum* group (NL)

*Falco subbuteo* (**Eurasian hobby**): *Ha. punctata, I. ricinus*

*Falco tinnunculus* (**common kestrel**): *A. variegatum* (N), *Ha. punctata, Hy. aegyptium, Hy. dromedarii, Hy. marginatum* group (NL), *I. hexagonus, I. ricinus* (NL)

*Falco vespertinus* (**red-footed falcon**): *Hy. marginatum* group

Family Sagittariidae (secretarybirds)

*Sagittarius serpentarius* (**secretarybird**): *A. variegatum* (N), *Hy. truncatum* (N), *R. turanicus* (A)

ORDER GALLIFORMES

Family Cracidae (guans and curassows)

*Crax rubra* (**globose curassow**): *A. oblongoguttatum* (AN)

*Penelope obscura* (**dusky-legged guan**): *A. longirostre* (A)

*Penelope purpurascens* (**crested guan**): *A. cajennense*

*Penelope superciliaris* (**rusty-margined guan**): *A. cajennense* (N), *A. ovale*

Family Phasianidae (pheasants and quails)

*Agelastes meleagrides* (**white-breasted guineafowl**): *Ha. hoodi* (N)

*Agelastes niger* (**black guineafowl**): *A. nuttalli* (L), *R. (B.) decoloratus*

*Alectoris barbara* (**Barbary partridge**): *Hy. marginatum* group (N), *I. ricinus* (A)

*Alectoris chukar* (**chukar**): *Hy. marginatum* group (A)

*Alectoris graeca* (**rock partridge**): *Ha. punctata* (NL), *Hy. aegyptium, Hy. marginatum* group, *R. bursa*

*Alectoris rufa* (**red-legged partridge**): *Ha. punctata, Hy. lusitanicum, Hy. marginatum* group (NL)

*Ammoperdix griseogularis* (**see-see partridge**): *Ha. punctata, Hy. marginatum* group

*Bonasa bonasia* (**hazel grouse**): *I. ricinus*

*Colinus virginianus* (**bobwhite quail**): *A. cajennense*

*Coturnix adansonii* (**blue quail**): *Ha. hoodi* (A)

*Coturnix coturnix* (**common quail**): *Ha. hoodi* (A), *Ha. punctata* (NL), *Hy. aegyptium, Hy. anatolicum* spp., *Hy. detritum, Hy. dromedarii, Hy. marginatum* group (NL), *I. ricinus, R. appendiculatus*

*Coturnix delegorguei* (**harlequin quail**): *A. hebraeum*

*Francolinus adspersus* (**red-billed francolin**): *Hy. marginatum* group (L)

*Francolinus africanus* (**grey-winged francolin**): *A. marmoreum* (L)

*Francolinus ahantensis* (**ahanta francolin**): *A. nuttalli* (N), *Ha. hoodi* (AN)

*Francolinus albogularis* (**white-throated francolin**): *Ha. hoodi* (A)

***Francolinus bicalcaratus*** (**double-spurred francolin**): *A. nuttalli* (NL), *A. variegatum* (N), <u>*Ha. hoodi*</u> (ANL), *Hy. marginatum* group (L)

***Francolinus capensis*** (**Cape francolin**): *A. marmoreum* (NL)

***Francolinus clappertoni*** (**Clapperton's francolin**): *A. nuttalli* (N), *A. variegatum* (N), *Ha. hoodi* (A), *Hy. marginatum* group (NL), *Hy. truncatum*

***Francolinus francolinus*** (**black francolin**): *Ha. punctata, I. ricinus*

***Francolinus icterorhynchus*** (**Heuglin's francolin**): *A. nuttalli* (N), *Ha. hoodi* (A)

***Francolinus leucoscepus*** (**yellow-necked spurfowl**): *A. variegatum, Ha. hoodi* (A), *Hy. marginatum* group (L), *R. kochi* (A), *R. pulchellus* (A)

***Francolinus levaillantoides*** (**Orange River francolin**): *Hy. marginatum* group (L)

***Francolinus natalensis*** (**Natal francolin**): *A. hebraeum, A. nuttalli* (L), *Ha. hoodi* (A), *Hy. truncatum*

***Francolinus sephaena*** (**crested francolin**): *A. hebraeum* (NL), *A. marmoreum* (NL), *Ha. hoodi* (AN), *Hy. marginatum* group (NL), *Hy. truncatum* (L), *R. appendiculatus* (NL), *R. evertsi* (L), *R. turanicus* (L), *R. (B.) decoloratus* (L)

***Francolinus squamatus*** (**scaly francolin**): *Ha. hoodi* (A)

***Francolinus swainsonii*** (**Swainson's spurfowl**): *A. hebraeum* (NL), *A. variegatum, Ha. hoodi* (A), *Hy. marginatum* group (NL), *Hy. truncatum* (L)

***Gallus gallus*** (**chicken/red junglefowl**): *A. cajennense, A. fimbriatum* (A), *A. hebraeum, A. testudinarium,* <u>*A. variegatum*</u> (ANL), *D. auratus* (N), <u>*Ha. hoodi*</u> (ANL), *Ha. longicornis* (AL), *Ha. punctata* (NL), *Hy. detritum* (N), *Hy. marginatum* group (NL), *I. ricinus* (AN), *R. appendiculatus* (A), *R. (B.) microplus*

***Guttera plumifera*** (**plumed guineafowl**): *A. nuttalli* (NL)

***Guttera pucherani*** (**crested guineafowl**): *A. hebraeum, A. nuttalli, Ha. hoodi* (AN)

***Lagopus lagopus*** (**red grouse**): <u>*I. ricinus*</u> (ANL)

***Lophura bulweri*** (**Bulwer's pheasant**): *A. fimbriatum* (N)

***Meleagris gallopavo*** (**turkey**): *A. cajennense* (N), *A. hebraeum, A. marmoreum* (N), *A. nuttalli, A. oblongoguttatum, A. splendidum, A. variegatum* (N), *Ha. hoodi* (A), *Ha. longicornis* (A), *Ha. punctata* (A), *Hy. marginatum* group, *Hy. truncatum, I. ricinus, R. appendiculatus* (A), *R. (B.) annulatus*

***Numida meleagris*** (**helmeted guineafowl**): *A. gemma, A. hebraeum* (NL), <u>*A. marmoreum*</u> (NL), *A. splendidum* (NL), <u>*A. variegatum*</u> (NL), *Ha. hoodi* (A), <u>*Hy. marginatum* group</u> (NL), *Hy. truncatum* (NL), *R. appendiculatus* (ANL), *R. evertsi* (L), *R. simus* (L), *R. (B.) decoloratus* (AL)

***Pavo cristatus*** (**Indian peafowl**): *A. longirostre* (NL)

***Perdix dauurica*** (**daurian partridge**): *Ha. punctata*

***Perdix perdix*** (**grey partridge**): *Ha. hoodi* (A), *Ha. punctata, Hy. aegyptium* (ANL), *Hy. marginatum* group, *I. ricinus* (NL)

***Phasianus colchicus*** (**common pheasant**): *Ha. longicornis* (A), *Ha. punctata* (ANL), *Hy. marginatum* group, <u>*I. ricinus*</u> (ANL), *R. (B.) microplus*

***Rhizothera longirostris*** (**long-billed partridge**): *A. geoemydae* (N)

***Rollulus rouloul*** (**crested partridge**): *A. testudinarium*

***Tetrao tetrix*** (**black grouse**): *Ha. punctata, I. ricinus* (NL)

***Tetrao urogallus*** (**capercaillie**): *I. ricinus* (NL)

ORDER GRUIFORMES

Family Cariamidae (seriemas)

***Cariama cristata*** (**red-legged seriema**): *A. cajennense* (N), *A. longirostre* (NL)

Family Gruidae (cranes)

***Anthropoides virgo*** (**demoiselle crane**): *Hy. detritum, Hy. marginatum* group

***Balearica pavonina*** (**black-crowned crane**): *Hy. marginatum* group (L)

***Grus grus*** (**common crane**): *Ha. punctata, Hy. marginatum* group

Family Otididae (bustards)

***Afrotis afra*** (**black bustard**): *Hy. marginatum* group (L), *R. sulcatus* (A)

***Afrotis afraoides*** (**white-quilled bustard**): *Hy. marginatum* group (NL)

***Ardeotis kori*** (**kori bustard**): *A. hebraeum, A. lepidum* (A), *Hy. marginatum* group (NL), *Hy. truncatum*

*Eupodotis senegalensis* (**Senegal bustard**): *A. variegatum* (NL), *Hy. marginatum* group (L)

*Lissotis melanogaster* (**black-bellied bustard**): *A. variegatum* (NL), *R. kochi* (A)

*Neotis cafra* (**Stanley's bustard**): *R. turanicus* (A)

*Neotis denhami* (**Denham's bustard**): *A. lepidum* (A), *A. variegatum* (N), *Hy. truncatum* (A), *R. turanicus* (A)

*Otis tarda* (**great bustard**): *D. reticulatus*, *Ha. punctata*, *Hy. marginatum* group (A)

*Tetrax tetrax* (**little bustard**): *Ha. punctata*, *Hy. marginatum* group

Family Rallidae (rails)

*Crecopsis egregia* (**African crake**): *Ha. hoodi* (A)

*Crex crex* (**corn crake**): *Ha. punctata* (N), *Hy. detritum* (NL), *I. hexagonus*, *I. ricinus* (NL)

*Fulica atra* (**common coot**): *I. ricinus* (NL)

*Gallinula chloropus* (**common moorhen**): *Ha. punctata*, *Hy. marginatum* group (L), *Hy. truncatum* (L)

*Gallirallus philippensis* (**buff-banded rail**): *Ha. longicornis* (L)

*Laterallus albigularis* (**white-throated crake**): *A. longirostre* (L)

*Porzana porzana* (**spotted crake**): *Hy. marginatum* group (N), *I. ricinus*

ORDER PASSERIFORMES

Family Aegithalidae (long-tailed tits)

*Aegithalos caudatus* (**long-tailed tit**): *I. ricinus* (L)

Family Alaudidae (larks)

*Alauda arvensis* (**Eurasian skylark**): *Ha. longicornis* (L), *Ha. punctata* (NL), *Hy. marginatum* group, *I. ricinus* (NL)

*Alauda gulgula* (**oriental skylark**): *Hy. marginatum* group

*Calandrella acutirostris* (**Hume's lark**): *Hy. marginatum* group

*Calandrella brachydactyla* (**greater short-toed lark**): *Hy. marginatum* group (N)

*Calandrella cinerea* (**red-capped lark**): *Ha. punctata*, *Hy. marginatum* group (NL), *R. evertsi*

*Calandrella rufescens* (**lesser short-toed lark**): *Hy. marginatum* group

*Certhilauda curvirostris* (**long-billed lark**): *Hy. marginatum* group (NL)

*Chersomanes albofasciata* (**spike-heeled lark**): *A. marmorcum* (NL), *Hy. marginatum* group (NL)

*Eremophila alpestris* (**horned lark**): *Ha. punctata*

*Galerida cristata* (**crested lark**): *Ha. punctata* (NL), *Hy. marginatum* group (NL), *Hy. truncatum* (N), *I. ricinus* (A), *R. turanicus*

*Galerida magnirostris* (**large-billed lark**): *A. marmoreum* (L), *Hy. marginatum* group (NL)

*Lullula arborea* (**wood lark**): *Ha. punctata*

*Melanocorypha bimaculata* (**bimaculated lark**): *Ha. punctata*, *Hy. marginatum* group

*Melanocorypha calandra* (**calandra lark**): *Ha. punctata*, *Hy. marginatum* group

*Mirafa africana* (**rufous-naped lark**): *A. marmoreum* (L), *Ha. hoodi* (A), *Hy. marginatum* group (N)

*Mirafa cheniana* (**latakoo lark**): *A. marmoreum* (NL), *Hy. marginatum* group (NL)

*Mirafra passerina* (**monotonous lark**): *Hy. marginatum* group (L)

*Mirafra rufocinnamomea* (**flappet lark**): *Ha. hoodi* (A)

*Mirafra sabota* (**sabota lark**): *Hy. marginatum* group (L)

*Spizocorys conirostris* (**pink-billed lark**): *Hy. marginatum* group (NL)

Family Bombycillidae (waxwings)

*Bombycilla garrulus* (**Bohemian waxwing**): *I. ricinus*

Family Certhiidae (holarctic treecreepers)

*Certhia brachydactyla* (**short-toed treecreeper**): *I. ricinus* (L)

*Certhia familiaris* (**Eurasian treecreeper**): *I. ricinus* (N)

Family Conopophagidae (gnateaters)

*Conopophaga lineata* (**rufous gnateater**): *A. cajennense* (L), *A. calcaratum* (N), *A. coelebs* (N), *A. longirostre* (N), *A. nodosum* (NL)

Family Corvidae (crows)

*Corvus albus* (**pied crow**): *Hy. truncatum*

*Corvus corax* (**raven**): *I. ricinus*

*Corvus corone* (**carrion crow**): *Ha. punctata, Hy. marginatum* group (N), *I. ricinus* (NL)

*Corvus frugilegus* (**rook**): *Ha. punctata* (ANL), *Hy. marginatum* group (NL), *I. ricinus* (NL), *R. bursa, R. turanicus*

*Corvus monedula* (**jackdaw**): *Ha. punctata* (N), *Hy. marginatum* group, *I. ricinus* (NL)

*Cyanocorax chrysops* (**plush-crested jay**): *Ha. hystricis* (NL)

*Garrulus glandarius* (**Eurasian jay**): *D. reticulatus* (NL), *Ha. longicornis* (N), *Ha. punctata* (L), *Hy. detritum* (NL), *Hy. marginatum* group (NL), *I. ricinus* (NL)

*Nucifraga caryocatactes* (**nutcracker**): *I. ricinus* (NL)

*Pica pica* (**black-billed magpie**): *Ha. punctata* (NL), *Hy. aegyptium, Hy. marginatum* group (NL), *I. hexagonus* (N), *I. ricinus* (ANL)

*Platylophus galericulatus* (**crested jay**): *A. geoemydae* (N)

*Ptilostomus afer* (**piapiac**): *A. variegatum* (N), *Ha. hoodi* (ANL)

Family Cotingidae (cotingas)

*Querula purpurata* (**purple-throated fruitcrow**): *A. longirostre* (N)

Family Cracticidae (butcherbirds)

*Gymnorhina tibicen* (**Australian magpie**): *Ha. longicornis* (L)

Family Dendrocolaptidae (woodcreepers)

*Campylorhamphus facularius* (**black-billed scythebill**): *A. longirostre* (L)

*Dendrocincla turdina* (**thrush-like woodcreeper**): *A. longirostre* (L)

*Dendrocolaptes platyrostris* (**planalto woodcreeper**): *A. nodosum* (N)

*Lepidocolaptes fuscus* (**lesser woodcreeper**): *A. longirostre* (L)

*Lepidocolaptes souleyetii* (**streak-headed woodcreeper**): *A. longirostre* (NL)

*Lepidocolaptes squamatus* (**scaled woodcreeper**): *A. longirostre* (L)

*Sittasomus griseicapillus* (**olivaceous woodcreeper**): *A. cajennense, A. longirostre* (NL)

*Xiphorhynchus guttatus* (**buff-throated woodcreeper**): *A. longirostre* (NL)

*Xiphorhynchus lachrymosus* (**black-striped woodcreeper**): *A. longirostre* (N)

Family Dicruridae (drongos)

*Dicrurus adsimilis* (**fork-tailed drongo**): *Hy. marginatum* group (L)

Family Emberizidae (buntings and New World sparrows)

*Arremon flavirostris* (**saffron-billed sparrow**): *A. cajennense* (N), *A. nodosum* (N), *A. ovale* (NL), *Ha. juxtakochi* (L)

*Arremonops conirostris* (**black-striped sparrow**): *A. ovale*

*Calcarius lapponicus* (**Lapland bunting**): *I. ricinus*

*Cardinalis cardinalis* (**northern cardinal**): *A. longirostre*

*Cyanocompsa brissonii* (**ultramarine grosbeak**): *A. longirostre*

*Cyanocompsa cyanoides* (**blue-black grosbeak**): *A. longirostre* (L)

*Dacnis cayana* (**blue dacnis**): *A. longirostre* (N)

*Emberiza bruniceps* (**red-headed bunting**): *Ha. punctata*

*Emberiza buchanani* (**grey-necked bunting**): *Hy. marginatum* group

*Emberiza caesia* (**Cretzschmar's bunting**): *Hy. aegyptium, Hy. marginatum* group (L)

*Emberiza cia* (**rock bunting**): *Ha. punctata, Hy. marginatum* group, *I. ricinus* (L)

*Emberiza cioides* (**meadow bunting**): *I. ricinus*

*Emberiza cirlus* (**cirl bunting**): *Ha. punctata* (NL)

*Emberiza citrinella* (**yellowhammer**): *Ha. punctata* (L), *Hy. marginatum* group, *I. ricinus* (N)

*Emberiza flaviventris* (**African golden-breasted bunting**): *A. marmoreum* (L)

*Emberiza hortulana* (**ortolan bunting**): *Ha. punctata, Hy. marginatum* group (N), *I. ricinus*

*Emberiza leucocephalos* (**pine bunting**): *Ha. punctata*

*Emberiza melanocephala* (**black-headed bunting**): *Ha. punctata, Hy. marginatum* group (NL)

*Emberiza rustica* (**rustic bunting**): *I. ricinus*

*Emberiza schoeniclus* (**reed bunting**): *I. ricinus* (NL)

*Emberiza spodocephala* (**black-faced bunting**): *Ha. longicornis* (L)

*Emberiza stewarti* (**chestnut-breasted bunting**): *Hy. marginatum* group

*Embernagra longicauda* (**pale-throated pampa finch**): *A. cajennense*

*Euphonia pectoralis* (**chestnut-bellied euphonia**): *A. longirostre* (NL)

*Euphonia violacea* (**violaceous euphonia**): *A. longirostre* (N)

*Habia rubica* (**red-crowned ant-tanager**): *A. longirostre* (NL), *A. nodosum* (N)

*Haplospiza unicolor* (**uniform finch**): *A. cajennense*

*Miliaria calandra* (**corn bunting**): *Hy. marginatum* group, *I. ricinus*

*Pipraeidea melanonota* (**fawn-breasted tanager**): *A. longirostre* (N)

*Piranga rubra* (**summer tanager**): *A. longirostre*

*Pyrrhocoma ruficeps* (**chestnut-headed tanager**): *A. cajennense* (L), *Ha. juxtakochi* (A)

*Ramphocelus bresilius* (**Brazilian tanager**): *A. longirostre* (NL)

*Ramphocelus sanguinolentus* (**crimson-collared tanager**): *A. longirostre* (N)

*Saltator albicollis* (**Lesser Antillean saltator**): *A. longirostre* (N)

*Saltator maximus* (**buff-throated saltator**): *A. longirostre* (NL)

*Saltator similis* (**green-winged saltator**): *A. cajennense*, *A. calcaratum* (N), *A. longirostre* (NL)

*Saltatricula multicolor* (**many-colored chaco-finch**): *A. parvum* (NL)

*Schistochlamys ruficapillus* (**cinnamon tanager**): *A. cajennense*

*Sicalis flaveola* (**saffron finch**): *A. parvum* (L)

*Tachyphonus coronatus* (**ruby-crowned tanager**): *A. cajennense* (L), *A. longirostre* (NL)

*Tachyphonus cristatus* (**flame-crested tanager**): *A. longirostre* (N)

*Tachyphonus delatrii* (**tawny-crested tanager**): *A. longirostre*

*Tangara seledon* (**green-headed tanager**): *A. longirostre* (N)

*Tersina viridis* (**swallow tanager**): *A. longirostre* (N)

*Thraupis sayaca* (**sayaca tanager**): *A. longirostre* (L)

*Tiaris bicolor* (**black-faced grassquit**): *A. variegatum* (L)

*Trichothraupis melanops* (**black-goggled tanager**): *A. cajennense*, *A. longirostre* (NL), *A. nodosum* (N)

Family Formicariidae (antbirds)

*Cercomacra tyrannina* (**dusky antbird**): *A. longirostre* (L)

*Cymbilaimus lineatus* (**fasciated antskrike**): *A. longirostre*

*Dysithamnus mentalis* (**plain antvireo**): *A. cajennense* (N), *A. longirostre* (NL), *A. nodosum* (N)

*Formicivora rufa* (**rusty-backed antwren**): *A. nodosum* (N)

*Mackenziaena severa* (**tufted antshrike**): *A. longirostre* (N)

*Pyriglena leucoptera* (**white-shouldered fire-eye**): *A. cajennense*, *A. calcaratum* (N)

*Taraba major* (**great antshrike**): *A. longirostre* (L)

*Thamnophilus bridgesi* (**black-hooded antshrike**): *A. longirostre* (N)

*Thamnophilus caerulescens* (**variable antshrike**): *A. cajennense*, *A. longirostre* (N), *A. nodosum* (N)

*Thamnophilus punctatus* (**eastern slaty-antshrike**): *A. cajennense* (N), *A. coelebs* (NL), *A. nodosum* (NL), *A. ovale* (N)

Family Fringillidae (chaffinches)

*Carduelis cannabina* (**Eurasian linnet**): *Ha. punctata*, *Hy. marginatum* group, *I. ricinus* (N)

*Carduelis carduelis* (**European goldfinch**): *Hy. marginatum* group (N), *I. ricinus*

*Carduelis chloris* (**European greenfinch**): *Hy. marginatum* group (NL), *I. ricinus* (NL)

*Carduelis flammea* (**redpoll**): *I. ricinus* (NL)

*Carduelis spinus* (**Eurasian siskin**): <u>*I. ricinus*</u> (NL)

*Carpodacus erythrinus* (**common rosefinch**): *I. ricinus* (N)

*Coccothraustes coccothraustes* (**hawfinch**): *Ha. punctata*, *Hy. marginatum* group (NL), <u>*I. ricinus*</u> (NL)

*Fringilla coelebs* (**chaffinch**): *Ha. punctata* (NL), *Hy. marginatum* group (NL), <u>*I. ricinus*</u> (NL)

*Fringilla montifringilla* (**brambling**): <u>*I. ricinus*</u> (NL)

*Loxia curvirostra* (**common crossbill**): *I. ricinus* (N)

*Mycerobas carnipes* (**white-winged grosbeak**): *Ha. punctata*

*Pyrrhula pyrrhula* (**bullfinch**): <u>*I. ricinus*</u> (NL)

*Serinus atrogularis* (**black-throated canary**): *Hy. marginatum* group (L)

*Serinus canaria* (**canary**): *I. ricinus*

*Serinus mozambicus* (**green singing finch**): *Hy. marginatum* group (N)

*Serinus serinus* (**European serin**): *I. ricinus* (N)

Family Furnariidae (ovenbirds)

*Anabazenops fuscus* (**white-collared foliage-gleaner**): *A. longirostre* (L)

*Automolus leucophthalmus* (**white-eyed foliage-gleaner**): *A. cajennense*, *A. longirostre* (NL)

*Automolus ochrolaemus* (**buff-throated foliage-gleaner**): *A. longirostre* (L)

*Cichlocolaptes leucophrus* (**pale-browed treehunter**): *A. cajennense* (L)

*Philydor atricapillus* (**black-capped foliage-gleaner**): *A. longirostre* (N)

*Philydor rufus* (**buff-fronted foliage-gleaner**): *A. longirostre* (N)

*Synallaxis ruficapilla* (**rufous-capped spinetail**): *A. longirostre* (N)

*Synallaxis spixi* (**chicli spinetail**): *A. longirostre* (N)

*Syndactyla rufosuperciliata* (**buff-browed foliage-gleaner**): *A. longirostre* (N)

Family Hirundinidae (swallows)

*Delichon urbica* (**house martin**): *I. ricinus* (A)

*Hirundo culcullata* (**greater striped swallow**): *R. evertsi* (A)

*Hirundo rustica* (**barn swallow**): *Ha. punctata*, <u>*Hy. marginatum* group</u> (ANL), *I. ricinus* (N)

*Riparia riparia* (**bank swallow**): *I. hexagonus*, *I. ricinus*

Family Icteridae (icterids)

*Cacicus cela* (**yellow-rumped cacique**): *A. longirostre*

*Cacicus uropygialis* (**scarlet-rumped cacique**): *A. longirostre* (N)

*Icterus chrysater* (**yellow-backed oriole**): *A. longirostre* (N)

*Icterus jamacaii* (**campo oriole**): *A. longirostre*

*Quiscalus lugubris* (**Carib grackle**): *A. variegatum* (NL)

Family Laniidae (shrikes)

*Dryoscopus cubla* (**black-backed puffback**): *Hy. marginatum* group

*Eurocephalus anguitimens* (**white-crowned shrike**): *Hy. marginatum* group (N)

*Laniarius atrococcineus* (**crimson-breasted gonolek**): *Hy. marginatum* group (NL)

*Laniarius erythrogaster* (**black-headed gonolek**): *Ha. hoodi* (AN)

*Laniarius ferrugineus* (**southern boubou**): *Ha. hoodi* (A), *Hy. marginatum* group (NL)

*Lanius collaris* (**common fiscal**): *Hy. marginatum* group (NL)

*Lanius collurio* (**red-backed shrike**): *Ha. punctata*, <u>*Hy. marginatum* group</u> (NL), <u>*I. ricinus*</u> (NL)

*Lanius cristatus* (**brown shrike**): *Ha. punctata*, *Hy. marginatum* group, *I. ricinus*

*Lanius excubitor* (**great grey shrike**): *I. ricinus*

*Lanius meridionalis* (**southern grey shrike**): *Hy. marginatum* group (N)

*Lanius minor* (**lesser grey shrike**): *Ha. punctata*, *Hy. marginatum* group

*Lanius nubicus* (**masked shrike**): *Hy. marginatum* group (NL)

*Lanius senator* (**woodchat shrike**): *Hy. marginatum* group (NL)

*Prionops plumatus* (**straight-crested helmetshrike**): *Hy. marginatum* group (NL), *R. appendiculatus* (A)

*Tchagra australis* (**brown-crowned tchagra**): *Hy. marginatum* group (ANL), *R. evertsi* (N)

*Tchagra senegala* (**black-headed shrike**): *A. hebraeum*, *A. variegatum* (N), *Ha. hoodi* (A)

*Telophorus zeylonus* (**bokmakierie bushshrike**): *Hy. marginatum* group (NL)

Family Motacillidae (wagtails and pipits)

*Anthus campestris* (**tawny pipit**): *Ha. punctata*, *Hy. marginatum* group (NL), *I. ricinus*

*Anthus cervinus* (**red-throated pipit**): *A. variegatum* (N), *Hy. marginatum* group (NL)

*Anthus cinnamomeus* (**African pipit**): *Hy. marginatum* group (L)

*Anthus gustavi* (**pechora pipit**): *I. ricinus*

*Anthus leucophrys* (**plain-backed pipit**): *A. variegatum* (N), *Ha. hoodi* (A)

*Anthus pratensis* (**meadow pipit**): *D. reticulatus* (N), *Ha. punctata*, *Hy. marginatum* group (N), <u>*I. ricinus*</u> (ANL)

*Anthus similis* (**long-billed pipit**): *Hy. marginatum* group (NL)

*Anthus spinoletta* (**water pipit**): *Ha. punctata, Hy. marginatum* group (L), *I. ricinus*

*Anthus trivialis* (**tree pipit**): *A. nuttalli* (N), *A. variegatum* (N), *Ha. punctata* (NL), *Hy. aegyptium, Hy. anatolicum* spp., *Hy. marginatum* group (NL), *I. ricinus* (ANL), *R. bursa*

*Anthus vaalensis* (**buffy pipit**): *Ha. hoodi* (NL), *Hy. marginatum* group (L)

*Macronyx croceus* (**yellow-throated longclaw**): *A. variegatum* (L), *Ha. hoodi* (A), *Hy. marginatum* group (NL)

*Motacilla alba* (**white wagtail**): *Ha. punctata* (A), *Hy. anatolicum* spp. (N), *Hy. marginatum* group (NL), *I. ricinus* (N)

*Motacilla capensis* (**Cape wagtail**): *Hy. marginatum* group (N)

*Motacilla cinerea* (**grey wagtail**): *Ha. punctata, Hy. marginatum* group, *I. ricinus* (NL)

*Motacilla flava* (**yellow wagtail**): *Ha. punctata, Hy. marginatum* group (NL), *I. ricinus* (NL)

Family Muscicapidae (Old World flycatchers)

*Acrocephalus arundinaceus* (**great reed warbler**): *Ha. punctata, Hy. marginatum* group (L), *I. ricinus*

*Acrocephalus beaticatus* (**African reed warbler**): *Ha. hoodi* (A)

*Acrocephalus dumetorum* (**Blyth's reed warbler**): *Ha. punctata*

*Acrocephalus palustris* (**marsh warbler**): *Ha. punctata, Hy. marginatum* group (NL), *I. ricinus* (NL)

*Acrocephalus schoenobaenus* (**sedge warbler**): *Ha. punctata, Hy. marginatum* group (AN), *I. ricinus* (N)

*Acrocephalus scirpaceus* (**Eurasian reed warbler**): *Ha. punctata* (N), *Hy. marginatum* group (ANL), *I. ricinus* (ANL)

*Batis pririt* (**pririt batis**): *A. marmoreum* (L)

*Bradornis mariquensis* (**mariqua flycatcher**): *Hy. marginatum* group (N)

*Calamonastes stierlingi* (**Stierling's wren-warbler**): *Hy. marginatum* group (N)

*Catharus fuscescens* (**veery**): *A. sabanerae* (N)

*Catharus minimus* (**grey-cheeked thrush**): *A. humerale* (N)

*Cercomela familiaris* (**familiar chat**): *Hy. marginatum* group (NL)

*Cercotrichas barbata* (**miombo scrub robin**): *Ha. muhsamae* (A)

*Cercotrichas coryphaeus* (**Karoo scrub robin**): *A. marmoreum* (L), *Hy. marginatum* group (NL)

*Cercotrichas galactotes* (**rufous bush robin**): *Hy. marginatum* group (NL)

*Cercotrichas leucophrys* (**red-backed scrub robin**): *Ha. hoodi*

*Cercotrichas paena* (**Kalahari scrub robin**): *A. hebraeum* (L), *A. marmoreum* (N)

*Cercotrichas quadrivirgata* (**bearded scrub robin**): *Ha. muhsamae* (A)

*Cisticola aridulus* (**desert cisticola**): *Hy. marginatum* group (L)

*Cisticola brachypterus* (**siffling cisticola**): *Ha. hoodi*

*Cisticola chinianus* (**rattling cisticola**): *Ha. hoodi*

*Cisticola natalensis* (**croaking cisticola**): *Ha. hoodi* (N)

*Cisticola textrix* (**tink-tink cisticola**): *A. marmoreum* (N), *Hy. marginatum* group (NL)

*Cisticola tinniens* (**tinkling cisticola**): *Ha. hoodi*

*Cossypha albicapilla* (**white-crowned robin-chat**): *Hy. marginatum* group (N)

*Cossypha caffra* (**Cape robin-chat**): *A. marmoreum* (N), *Hy. marginatum* group (NL)

*Erithacus rubecula* (**European robin**): *Ha. punctata* (NL), *Hy. marginatum* group (N), *I. ricinus* (NL)

*Ficedula albicollis* (**collared flycatcher**): *Hy. marginatum* group (NL), *I. ricinus* (N)

*Ficedula hypoleuca* (**European pied flycatcher**): *Hy. marginatum* group (NL), *I. ricinus* (ANL)

*Ficedula parva* (**red-breasted flycatcher**): *I. ricinus* (L)

*Hippolais icterina* (**icterine warbler**): *I. ricinus* (NL)

*Hippolais pallida* (**eastern olivaceous warbler**): *Hy. marginatum* group (L)

*Hippolais polyglotta* (**melodious warbler**): *Ha. punctata, Hy. marginatum* group, *I. ricinus* (N)

*Locustella fluviatilis* (**Eurasian river warbler**): *I. ricinus* (N)

*Locustella luscinioides* (**Savi's warbler**): *Hy. marginatum* group

*Locustella naevia* (**grasshopper warbler**): *I. ricinus*

*Luscinia luscinia* (**thrush nightingale**): *A. nuttalli*

(N), *Ha. punctata* (NL), *Hy. aegyptium* (NL), *Hy. marginatum* group (NL), <u>*I. ricinus*</u> (NL)

***Luscinia megarhynchos*** (**common nightingale**): *Ha. punctata*, *Hy. marginatum* group (NL), <u>*I. ricinus*</u> (NL)

***Luscinia svecica*** (**bluethroat**): *Ha. punctata*, *Hy. marginatum* group (N), *I. ricinus* (NL)

***Melocichla mentalis*** (**moustached grass warbler**): *A. nuttalli* (N), *Ha. hoodi* (N)

***Monticola gularis*** (**white-throated rock thrush**): *Ha. hoodi* (A)

***Monticola saxatilis*** (**rufous-tailed rock thrush**): *Ha. punctata*, <u>*Hy. marginatum* group</u> (NL)

***Monticola solitarius*** (**blue rock thrush**): *Ha. punctata* (L), *Hy. marginatum* group (NL)

***Muscicapa striata*** (**spotted flycatcher**): *Ha. punctata* (NL), *Hy. aegyptium* (L), <u>*Hy. marginatum* group</u> (L), <u>*I. ricinus*</u> (NL)

***Myrmecocichla aethiops*** (**northern anteater-chat**): *Hy. marginatum* group (L)

***Myrmecocichla arnotti*** (**white-headed black chat**): *Ha. hoodi* (L)

***Myrmecocichla formicivora*** (**southern anteater-chat**): *Hy. marginatum* group (NL)

***Myrmecocichla nigra*** (**sooty chat**): *A. variegatum* (N)

***Napothera macrodactyla*** (**large wren-babbler**): *A. geoemydae* (N)

***Oenanthe finschii*** (**Finsch's wheatear**): *Hy. marginatum* group (NL)

***Oenanthe hispanica*** (**black-eared wheatear**): <u>*Hy. marginatum* group</u> (NL)

***Oenanthe isabellina*** (**isabelline wheatear**): *Ha. punctata*, *Hy. aegyptium*, *Hy. marginatum* group (NL), *R. bursa*

***Oenanthe oenanthe*** (**northern wheatear**): *Ha. hoodi* (N), *Ha. punctata*, *Hy aegyptium*, <u>*Hy. marginatum* group</u> (NL), <u>*I. ricinus*</u> (NL)

***Oenanthe pictata*** (**variable wheatear**): *Ha. punctata*

***Oenanthe pileata*** (**capped wheatear**): *Ha. hoodi*

***Oenanthe pleschanka*** (**pied wheatear**): *Ha. punctata* (N), *Hy. marginatum* group (NL)

***Phoenicurus auroreus*** (**daurian redstart**): *Ha. punctata*

***Phoenicurus ochruros*** (**black redstart**): *Hy. marginatum* group (N), *I. ricinus* (L)

***Phoenicurus phoenicurus*** (**common redstart**): <u>*Ha. punctata*</u> (ANL), *Hy. aegyptium* (N), *Hy. anatolicum* spp. (A), <u>*Hy. marginatum* group</u> (NL), <u>*I. ricinus*</u> (NL)

***Phylloscopus collybita*** (**chiffchaff**): *Ha. punctata*, *Hy. marginatum* group (L), <u>*I. ricinus*</u> (NL)

***Phylloscopus nitidus*** (**bright green warbler**): *Hy. marginatum* group

***Phylloscopus sibilatrix*** (**wood warbler**): *Hy. marginatum* group, *I. ricinus* (NL)

***Phylloscopus trochiloides*** (**greenish warbler**): *I. ricinus* (NL)

***Phylloscopus trochilus*** (**willow warbler**): *Ha. punctata* (NL), *Hy. aegyptium* (L), *Hy. marginatum* group (NL), <u>*I. ricinus*</u> (NL)

***Prinia flavicans*** (**black-chested prinia**): *A. marmoreum* (L), *Hy. marginatum* group (N)

***Regulus ignicapillus*** (**firecrest**): *I. ricinus*

***Regulus regulus*** (**goldcrest**): *I. ricinus* (NL)

***Saxicola rubetra*** (**whinchat**): *Ha. punctata* (L), *Hy. aegyptium* (N), *Hy. marginatum* group (NL), <u>*I. ricinus*</u> (L)

***Saxicola torquata*** (**stonechat**): *Ha. punctata*, *Hy. marginatum* group (NL), *I. ricinus*

***Scotocerca inquieta*** (**scrub warbler**): *Hy. marginatum* group

***Sigelus silens*** (**fiscal flycatcher**): *Hy. marginatum* group (NL)

***Stachyris poliocephala*** (**grey-headed babbler**): *A. geoemydae* (N)

***Sylvia althaea*** (**Hume's whitethroat**): *Hy. marginatum* group

***Sylvia atricapilla*** (**blackcap**): *Hy. marginatum* group (NL), <u>*I. ricinus*</u> (NL)

***Sylvia borin*** (**garden warbler**): *Hy. marginatum* group, <u>*I. ricinus*</u> (NL)

***Sylvia cantillans*** (**subalpine warbler**): *Hy. marginatum* group (NL)

***Sylvia communis*** (**common whitethroat**): *Ha. punctata* (N), *Hy. aegyptium* (N), *Hy. marginatum* group (NL), <u>*I. ricinus*</u> (ANL)

***Sylvia curruca*** (**lesser whitethroat**): *Ha. punctata*, *Hy. aegyptium* (N), *Hy. marginatum* group (NL), <u>*I. ricinus*</u> (NL)

***Sylvia hortensis*** (**orphean warbler**): *Hy. marginatum* group (L)

***Sylvia melanothorax*** (**Cyprus warbler**): *Hy. marginatum* group (N)

*Sylvia nisoria* (**barred warbler**): *Ha. punctata, Hy. marginatum* group (N), *I. ricinus* (NL)

*Sylvia subcaeruleum* (**rufous-vented warbler**): *Hy. marginatum* group (NL)

*Terpsiphone viridis* (**African paradise flycatcher**): *Hy. marginatum* group (N)

*Thamnolaea cinnamomeiventris* (**mocking cliff-chat**): *Hy. marginatum* group (ANL)

*Turdoides affinis* (**yellow-billed babbler**): *Hy. marginatum* group (L)

*Turdoides jardineii* (**arrow-marked babbler**): *Ha. hoodi, Ha. muhsamae, R. evertsi*

*Turdoides leucopygius* (**white-rumped babbler**): *A. hebraeum* (L)

*Turdus albicollis* (**white-necked thrush**): *A. cajennense* (N), *A. calcaratum* (N), *A. longirostre* (NL), *Ha. juxtakochi* (NL)

*Turdus amaurochalinus* (**creamy-bellied thrush**): *A. coelebs* (N), *A. longirostre* (N), *A. nodosum* (N)

*Turdus cardis* (**Japanese thrush**): *Ha. longicornis* (N)

*Turdus falcklandii* (**austral thrush**): *A. argentinae* (A)

*Turdus fumigatus* (**cocoa thrush**): *A. longirostre* (N)

*Turdus grayi* (**clay-colored thrush**): *A. longirostre* (N)

*Turdus iliacus* (**redwing**): *I. ricinus* (NL)

*Turdus leucomelas* (**pale-breasted thrush**): *A. cajennense*

*Turdus libonyanus* (**kurrichane thrush**): *Hy. marginatum* group (NL)

*Turdus merula* (**blackbird**): *A. lepidum* (N), *Ha. punctata* (NL), *Hy. detritum* (N), *Hy. marginatum* group (NL), *I. ricinus* (ANL), *R. turanicus*

*Turdus nudigenis* (**yellow-eyed thrush**): *A. longirostre*

*Turdus olivaceus* (**Cape thrush**): *A. nuttalli* (N), *Ha. hoodi* (A), *Hy. marginatum* group (L)

*Turdus philomelos* (**song thrush**): *Ha. longicornis, Ha. punctata* (N), *Hy. detritum* (NL), *Hy. marginatum* group (NL), *I. ricinus* (NL)

*Turdus pilaris* (**fieldfare**): *Ha. punctata, I. ricinus* (NL)

*Turdus ruficollis* (**dark-throated thrush**): *Ha. punctata*

*Turdus rufiventris* (**rufous-bellied thrush**): *A. cajennense, A. longirostre* (N), *Ha. juxtakochi* (NL)

*Turdus subalaris* (**eastern slaty thrush**): *A. longirostre* (N), *A. ovale* (N)

*Turdus torquatus* (**ring ouzel**): *I. ricinus* (N)

*Turdus viscivorus* (**mistle thrush**): *Ha. punctata* (N), *Hy. marginatum* group (N), *I. ricinus* (NL)

*Zoothera marginata* (**dark-sided thrush**): *A. geoemydae* (N)

Family Oriolidae (Old World orioles)

*Oriolus larvatus* (**African black-headed oriole**): *Hy. marginatum* group (L)

*Oriolus oriolus* (**Eurasian golden oriole**): *Hy. marginatum* group (N), *I. ricinus*

Family Paridae (true tits)

*Parus afer* (**grey tit**): *Hy. marginatum* group (ANL)

*Parus ater* (**coal tit**): *I. ricinus* (L)

*Parus caeruleus* (**blue tit**): *Hy. marginatum* group (N), *I. hexagonus* (N), *I. ricinus* (NL)

*Parus cinerascens* (**ashy tit**): *Hy. marginatum* group (N)

*Parus cristatus* (**crested tit**): *I. ricinus*

*Parus cyanus* (**azure tit**): *Ha. punctata*

*Parus griseiventris* (**miombo tit**): *Hy. marginatum* group

*Parus major* (**great tit**): *Ha. punctata* (NL), *Hy. marginatum* group (N), *I. ricinus* (NL)

*Parus montanus* (**willow tit**): *I. ricinus* (NL)

*Parus niger* (**black tit**): *Hy. marginatum* group (N)

*Parus palustris* (**marsh tit**): *I. ricinus* (NL)

Family Parulidae (New World warblers)

*Basileuterus culicivorus* (**golden-crowned warbler**): *A. longirostre* (NL), *A. nodosum* (N)

*Basileuterus flaveolus* (**flavescent warbler**): *A. cajennense, A. nodosum* (N)

*Basileuterus hypoleucus* (**white-bellied warbler**): *A. cajennense*

*Wilsonia canadensis* (**Canada warbler**): *A. longirostre* (L)

Family Pipridae (manakins)

*Chiroxiphia caudata* (**blue manakin**): *A. cajennense, A. calcaratum* (N), *A. longirostre* (NL)

*Chiroxiphia linearis* (**long-tailed manakin**): *A. longirostre* (L)

***Ilicura militaris* (pin-tailed manakin):** *A. longirostre* (NL)

***Manacus manacus* (white-bearded manakin):** *A. longirostre* (NL), *A. nodosum* (N)

***Pipra fasciicauda* (band-tailed manakin):** *A. longirostre* (N)

***Schiffornis virescens* (greenish schiffornis):** *A. cajennense*

Family Pittidae (pittas)

***Pitta brachyura* (Indian pitta):** *A. geoemydae* (NL)

***Pitta caerulea* (giant pitta):** *A. geoemydae*

***Pitta moluccensis* (blue-winged pitta):** *A. geoemydae* (L)

***Pitta reichenowi* (green-breasted pitta):** *A. tholloni* (ANL)

***Pitta sordida* (hooded pitta):** *A. geoemydae* (NL)

Family Ploceidae (weavers)

***Euplectes macroura* (yellow-backed whydah):** *Ha. hoodi* (N)

***Euplectes orix* (red bishop):** *Hy. marginatum* group (NL)

***Passer domesticus* (house sparrow):** *Ha. longicornis* (NL), *Ha. punctata* (ANL), *Hy. marginatum* group (N), *I. hexagonus*, *I. ricinus* (NL), *R. (B.) microplus*

***Passer griseus* (grey-headed sparrow):** *Hy. marginatum* group (NL), *Hy. truncatum* (L)

***Passer hispaniolensis* (Spanish sparrow):** *Ha. punctata*, *Hy. anatolicum* spp., *Hy. marginatum* group (AL), *I. ricinus*

***Passer montanus* (Eurasian tree sparrow):** *Ha. punctata* (NL), *Hy. detritum* (L), *Hy. marginatum* group (N), *I. hexagonus*, *I. ricinus* (NL)

***Petronia petronia* (rock sparrow):** *Ha. punctata*, *Hy. marginatum* group

***Petronia superciliaris* (yellow-throated petronia):** *Hy. marginatum* group (N), *R. appendiculatus*

***Plocepasser mahali* (white-browed sparrow-weaver):** *A. hebraeum* (L), *Hy. marginatum* group (NL)

***Ploceus cucullatus* (Layard's black-headed weaver):** *Hy. marginatum* group (NL)

***Ploceus melanocephalus* (black-headed weaver):** *Hy. marginatum* group (NL)

***Ploceus velatus* (southern masked weaver):** *Hy. marginatum* group (NL)

***Quelea quelea* (red-billed quelea):** *Hy. marginatum* group (L), *R. evertsi*

***Vidua macroura* (pin-tailed whydah):** *A. sparsum* (L)

Family Prunellidae (accentors)

***Prunella modularis* (dunnock):** *Ha. punctata* (NL), <u>*I. ricinus*</u> (ANL)

Family Pycnonotidae (bulbuls)

***Nicator chloris* (yellow-spotted nicator):** *A. hebraeum*

***Pycnonotus barbatus* (common bulbul):** *Hy. marginatum* group (L)

***Pycnonotus nigricans* (black-fronted bulbul):** *Hy. marginatum* group (NL)

***Pycnonotus xanthopygos* (white-spectacled bulbul):** *Hy. marginatum* group (L)

Family Sittidae (nuthatches)

***Sitta europaea* (Eurasian nuthatch):** <u>*I. ricinus*</u> (NL)

***Sitta neumayer* (rock nuthatch):** *Ha. punctata*, *Hy. marginatum* group

***Sitta tephronota* (eastern rock nuthatch):** *Hy. marginatum* group

Family Sturnidae (starlings and mynas)

***Acridotheres tristis* (common myna):** *Ha. punctata*, *Hy. marginatum* group (L), *R. (B.) decoloratus*

***Cinnyricinclus leucogaster* (violet-backed starling):** *Hy. marginatum* group (N)

***Creatophora cinerea* (wattled starling):** *Hy. marginatum* group (NL), *Hy. truncatum* (L)

***Lamprotornis caudatus* (long-tailed glossy starling):** *A. variegatum* (L), *Hy. marginatum* group

***Lamprotornis chalybaeus* (greater blue-eared glossy starling):** *A. hebraeum*

***Lamprotornis chloropterus* (lesser blue-eared glossy starling):** *Ha. hoodi* (A)

***Lamprotornis corruscus* (black-bellied glossy starling):** *Hy. marginatum* group (L)

***Lamprotornis mevesii* (Meves' glossy starling):** *Ha. hoodi* (A), *Hy. marginatum* group (NL), *R. appendiculatus*

***Onychognathus morio* (red-winged starling):** *Ha. hoodi* (A), *Hy. marginatum* group

***Onychognathus tristramii*** (**Tristram's grackle**): *Hy. marginatum* group

***Spreo bicolor*** (**African pied starling**): *Hy. marginatum* group (NL)

***Sturnus pagodarum*** (**Brahminy starling**): *Hy. marginatum* group (N)

***Sturnus roseus*** (**rose-colored starling**): *Ha. punctata*, *Hy. marginatum* group (N), *R. turanicus*

***Sturnus vulgaris*** (**common starling**): *Ha. punctata* (NL), *Hy. marginatum* group, <u>*I. ricinus*</u> (ANL), *R. bursa*

Family Troglodytidae (wrens)

***Thryothorus atrogularis*** (**black-throated wren**): *A. longirostre* (L)

***Thryothorus longirostris*** (**long-billed wren**): *A. longirostre*

***Thryothorus modestus*** (**plain wren**): *A. longirostre* (L)

***Thryothorus nigricapillus*** (**bay wren**): *A. longirostre* (NL)

***Troglodytes troglodytes*** (**winter wren**): *I. hexagonus*, <u>*I. ricinus*</u> (NL)

Family Tyrannidae (tyrant flycatchers)

***Attila rufus*** (**grey-hooded attila**): *A. longirostre* (N)

***Casiornis rufa*** (**rufous casiornis**): *A. nodosum* (N)

***Cnemotriccus fuscatus*** (**fuscous flycatcher**): *A. longirostre* (N), *A. nodosum* (N)

***Elaenia parvirostris*** (**small-billed elaenia**): *A. longirostre* (N)

***Empidonax flaviventris*** (**yellow-billed flycatcher**): *A. longirostre* (N)

***Empidonax traillii*** (**willow flycatcher**): *A. longirostre* (NL)

***Hemitriccus margaritaceiventer*** (**pearly-vented tody-tyrant**): *A. nodosum* (NL)

***Hemitriccus nidipendulus*** (**hangnest tody-tyrant**): *A. longirostre* (N)

***Lathrotriccus euleri*** (**Euler's flycatcher**): *A. cajennense*, *A. longirostre* (N)

***Leptopogon amaurocephalus*** (**sepia-capped flycatcher**): *A. longirostre* (NL)

***Mionectes oleagineus*** (**ochre-bellied flycatcher**): *A. longirostre* (L)

***Mionectes rufiventris*** (**grey-hooded flycatcher**): *A. cajennense*, *A. longirostre* (L)

***Myiarchus ferox*** (**short-crested flycatcher**): *A. longirostre* (N)

***Myiodynastes maculatus*** (**streaked flycatcher**): *A. longirostre* (N)

***Myiopagis viridicata*** (**greenish elaenia**): *A. longirostre* (L)

***Phylloscartes ventralis*** (**mottle-checked tyrannulet**): *A. cajennense*, *A. longirostre* (N)

***Platyrinchus mystaceus*** (**white-throated spadebill**): *A. cajennense*, *A. longirostre* (NL), *A. nodosum* (N)

***Rhynchocyclus brevirostris*** (**eye-ringed flatbill**): *A. longirostre* (N)

***Tolmomyias sulphurescens*** (**yellow-olive flycatcher**): *A. cajennense*, *A. longirostre* (AN)

Family Vireonidae (vireos)

***Cyclarhis gujanensis*** (**rufous-browed peppershrike**): *A. cajennense*, *A. longirostre* (N)

***Vireo griseus*** (**white-eyed vireo**): *A. longirostre* (N)

***Vireo olivaceus*** (**red-eyed vireo**): *A. longirostre* (N)

ORDER PELECANIFORMES

Family Pelecanidae (pelicans)

***Pelecanus occidentalis*** (**brown pelican**): *A. variegatum*

Family Phalacrocoracidae (cormorants)

***Phalacrocorax africanus*** (**long-tailed cormorant**): *R. evertsi* (L), *R. simus* (A), *R. (B.) decoloratus* (L)

***Phalacrocorax neglectus*** (**bank cormorant**): *Hy. marginatum* group (L)

ORDER PICIFORMES

Family Bucconidae (puffbirds)

***Malacoptila panamensis*** (**white-whiskered puffbird**): *A. longirostre* (N)

***Malacoptila striata*** (**crescent-chested puffbird**): *A. coelebs* (N), *A. longirostre* (NL)

Family Capitonidae (barbets)

***Megalaima mystacophanos*** (**flame-fronted barbet**): *A. geoemydae* (N)

***Trachyphonus vaillantii*** (**crested barbet**): *R. appendiculatus*

Family Picidae (woodpeckers)

***Campethera abingoni* (golden-tailed woodpecker):** *R. appendiculatus*

***Celeus flavescens* (blond-crested woodpecker):** *A. longirostre* (N)

***Dendrocopos major* (great spotted woodpecker):** *Ha. punctata, Hy. marginatum* group, *I. ricinus* (NL)

***Dendrocopos medius* (middle spotted woodpecker):** *I. ricinus*

***Dendrocopos minor* (lesser spotted woodpecker):** *I. ricinus*

***Dendropicos namaquus* (bearded woodpecker):** *A. hebraeum* (N)

***Jynx torquilla* (Eurasian wryneck):** *Hy. marginatum* group (L), *I. ricinus*

***Picoides tridactylus* (Eurasian three-toed woodpecker):** *I. ricinus*

***Picus viridis* (Eurasian green woodpecker):** *I. ricinus*

Family Ramphastidae (toucans)

***Ramphastos dicolorus* (red-breasted toucan):** *A. longirostre* (N)

***Ramphastos toco* (toco toucan):** *A. cajennense*

ORDER PSITTACIFORMES

Family Psittacidae (parrots)

***Melopsittacus undulatus* (budgerigar):** *Ha. longicornis* (N)

***Psittacus erithacus* (grey parrot):** *Ha. hoodi*

ORDER RHEIFORMES

Family Rheidae (rheas)

***Rhea americana* (greater rhea):** *A. parvum* (A)

ORDER STRIGIFORMES

Family Strigidae (owls)

***Aegolius funereus* (boreal owl):** *Hy. marginatum* group (A)

***Asio capensis* (African marsh owl):** *A. variegatum* (L), *Hy. marginatum* group (NL)

***Asio flammeus* (short-eared owl):** *A. lepidum* (N), *Hy. marginatum* group, *I. ricinus*

***Asio otus* (long-eared owl):** *Ha. punctata, I. ricinus* (NL), *R. bursa*

***Athene noctua* (little owl):** *Hy. aegyptium, Hy. marginatum* group (NL), *I. ricinus*

***Bubo africanus* (African eagle owl):** *Hy. marginatum* group (ANL), *Hy. truncatum* (L), *R. (B.) decoloratus* (A)

***Bubo bubo* (Eurasian eagle owl):** *Hy. lusitanicum* (A), *Hy. marginatum* group (N), *I. ricinus*

***Bubo lacteus* (giant eagle owl):** *A. variegatum* (N), *Ha. muhsamae* (A), *Hy. marginatum* group (L)

***Glaucidium sjostedti* (Sjostedt's owlet):** *A. nuttalli* (N)

***Otus scops* (common scops owl):** *Hy. marginatum* group (NL)

***Ptilopsis leucotis* (white-faced scops owl):** *Hy. marginatum* group (AN)

***Pulsatrix koeniswaldiana* (tawny-browed owl):** *A. longirostre* (N)

***Strix aluco* (tawny owl):** *Hy. aegyptium, I. ricinus*

***Strix butleri* (desert owl):** *I. ricinus*

***Strix woodfordii* (African wood owl):** *A. variegatum* (N)

Family Tytonidae (barn and bay owls)

***Tyto alba* (barn owl):** *Hy. marginatum* group (NL), *Hy. truncatum, I. ricinus* (N)

ORDER STRUTHIONIFORMES

Family Struthionidae (ostriches)

***Struthio camelus* (ostrich):** *A. gemma* (A), *A. hebraeum* (ANL), *A. lepidum* (A), *A. variegatum* (A), *Ha. punctata* (A), *Hy. albiparmatum* (A), *Hy. dromedarii* (A), *Hy. impressum, Hy. lusitanicum* (A), *Hy. marginatum* group (A), *Hy. truncatum* (A), *R. appendiculatus, R. pulchellus* (A), *R. turanicus* (A)

ORDER TINAMIFORMES

Family Tinamidae (tinamous)

***Nothura boraquira* (white-bellied nothura):** *A. pseudoconcolor* (NL)

***Nothura maculosa* (spotted nothura):** *A. pseudoconcolor* (ANL)

***Rhynchotus rufescens* (red-winged tinamou):** *A. pseudoconcolor* (NL)

# Class Mammalia (mammals)

ORDER AFROSORICIDA[c]

Family Tenrecidae (Madagascar hedgehogs)

***Echinops telfairi* (lesser hedgehog tenrec):** *Ha. elongata* (A)

***Hemicentetes semispinosus* (lowland streaked tenrec):** *Ha. elongata* (ANL)

***Microgale talazaci* (Talazac's shrew-tenrec):** *Ha. elongata* (NL)

***Setifer setosus* (greater hedgehog tenrec):** *Ha. elongata* (AN)

***Tenrec ecaudatus* (common tenrec):** *A. variegatum* (N), *Ha. elongata* (AL)

ORDER ARTIODACTYLA[d]

Family Bovidae (antelopes, cattle, bisons, buffalos, goats and sheep)

***Aepyceros melampus* (impala):** *A. gemma* (A), *A. hebraeum* (ANL), *A. lepidum* (A), *A. marmoreum* (ANL), *A. sparsum* (A), *A. tholloni* (N), *A. variegatum* (ANL), *D. rhinocerinus* (A), *Hy. albiparmatum* (A), *Hy. marginatum* group, *Hy. truncatum* (AL), *I. pilosus* (A), *R. appendiculatus* (ANL), *R. evertsi* (ANL), *R. kochi* (ANL), *R. muehlensi* (ANL), *R. pulchellus* (AN), *R. simus* (AL), *R. sulcatus* (A), *R. (B.) decoloratus* (ANL)

***Alcelaphus buselaphus* (hartebeest):** *A. gemma* (A), *A. lepidum* (A), *A. pomposum* (AN), *A. variegatum* (ANL), *Hy. albiparmatum*, *Hy. marginatum* group (A), *Hy. truncatum* (AN), *R. appendiculatus* (AN), *R. compositus* (A), *R. evertsi* (ANL), *R. kochi* (A), *R. muehlensi* (A), *R. pulchellus* (ANL), *R. senegalensis* (A), *R. simus* (A), *R. sulcatus* (A), *R. (B.) decoloratus* (ANL)

***Ammodorcas clarkei* (Clarke's gazelle):** *R. pulchellus*

***Ammotragus lervia* (Barbary sheep):** *A. cajennense*, *Hy. dromedarii* (A), *Hy. marginatum* group

***Antidorcas marsupialis* (springbok):** *A. hebraeum* (A), *A. marmoreum* (L), *Hy. marginatum* group (A), *Hy. truncatum* (A), *I. pilosus* (A), *R. appendiculatus* (AN), *R. evertsi* (ANL), *R. (B.) decoloratus* (A)

***Bison bison* (American bison):** *R. (B.) annulatus*

***Bison bonasus* (European bison):** *D. reticulatus*, *I. hexagonus*, *I. ricinus*

***Bos gaurus* (gaur):** *A. testudinarium* (AN), *A. varanense*, *R. (B.) microplus* (AN)

***Bos javanicus* (banteng):** *Hy. marginatum* group, *R. (B.) annulatus*, *R. (B.) microplus*

***Bos mutus* (yak):** *Ha. longicornis*

***Bos taurus* (cattle):** *A. auricularium* (A), *A. cajennense* (ANL), *A. coelebs*, *A. dissimile* (ANL), *A. exornatum*, *A. falsomarmoreum*, *A. fimbriatum*, *A. gemma* (ANL), *A. hebraeum* (ANL), *A. lepidum* (ANL), *A. marmoreum* (ANL), *A. moreliae*, *A. nodosum*, *A. nuttalli* (AN), *A. oblongoguttatum* (A), *A. ovale* (A), *A. parvum* (ANL), *A. pomposum* (AN), *A. pseudoconcolor* (A), *A. rhinocerotis* (A), *A. sparsum* (AN), *A. splendidum* (AN), *A. testudinarium* (AN), *A. tholloni* (NL), *A. trimaculatum*, *A. triguttatum* (ANL), *A. variegatum* (ANL), *B. hydrosauri*, *D. auratus* (ANL), *D. nitens* (AN), *D. nuttalli* (A), *D. reticulatus* (A), *D. rhinocerinus* (A), *Ha. juxtakochi* (ANL), *Ha. longicornis* (ANL), *Ha. punctata* (AN), *Hy. aegyptium*, *Hy. albiparmatum* (A), *Hy. anatolicum* spp. (ANL), *Hy. detritum* (ANL), *Hy. dromedarii* (AN), *Hy. impressum* (A), *Hy. lusitanicum* (A), *Hy. marginatum* group (ANL), *Hy. truncatum* (ANL), *I. hexagonus* (A), *I. pilosus* (A), *I. ricinus* (ANL), *R. appendiculatus* (ANL), *R. bursa* (ANL), *R. capensis* (A), *R. compositus* (A), *R. evertsi* (ANL), *R. kochi* (A), *R. muehlensi* (A), *R. pulchellus* (ANL), *R. senegalensis* (A), *R. simus* (A), *R. sulcatus* (A), *R. turanicus* (AN), *R. (B.) annulatus* (ANL), *R. (B.) decoloratus* (ANL), *R. (B.) microplus* (ANL)

***Boselaphus tragocamelus* (nilgai):** *A. cajennense*, *D. auratus* (N), *Hy. marginatum* group, *R. evertsi*, *R. (B.) annulatus* (A), *R. (B.) microplus*

***Bubalus arnee* (wild Asiatic buffalo):** *Hy. anatolicum* spp., *R. (B.) annulatus*

***Bubalus bubalis* (water buffalo):** *A. cajennense*, *A. clypeolatum* (A), *A. helvolum* (A), *A. oblongoguttatum*, *A. parvum* (A), *A. pseudoconcolor* (A), *A. testudinarium* (AN), *A. varanense*, *A. variegatum*, *D. auratus* (NL), *D. nitens*, *D. reticulatus*, *Ha. hystricis* (A), *Ha. punctata*, *Hy. anatolicum* spp. (AN), *Hy. detritum* (A), *Hy. dromedarii* (A), *Hy. marginatum* group (A), *I. ricinus*, *R. appendiculatus* (A), *R. bursa*, *R. evertsi* (A), *R. turanicus* (A), *R. (B.) annulatus*, *R. (B.) decoloratus* (AN), *R. (B.) microplus* (ANL)

***Capra aegagrus*** (**Cretan goat**): *R. bursa*

***Capra caucasia*** (**East Caucasian tur**): *I. ricinus*

***Capra hircus*** (**goat**): <u>*A. cajennense*</u> (AN), *A. exornatum*, <u>*A. gemma*</u> (A), <u>*A. hebraeum*</u> (ANL), <u>*A. lepidum*</u> (ANL), *A. marmoreum* (NL), *A. nuttalli* (AN), *A. oblonguguttatum*, *A. parvum* (ANL), *A. pomposum* (A), *A. sparsum* (AN), *A. splendidum*, *A. testudinarium*, *A. tholloni* (NL), *A. triguttatum* (N), <u>*A. variegatum*</u> (ANL), *D. nitens*, *D. nuttalli*, *D. reticulatus* (AN), *D. rhinocerinus* (A), *Ha. hoodi* (A), *Ha. juxtakochi* (A), <u>*Ha. longicornis*</u> (ANL), <u>*Ha. punctata*</u> (AN), *Hy. aegyptium*, *Hy. albiparmatum* (A), <u>*Hy. anatolicum* spp.</u> (ANL), <u>*Hy. detritum*</u> (AN), *Hy. dromedarii* (A), *Hy. impressum* (A), *Hy. lusitanicum* (A), <u>*Hy. marginatum* group</u> (A), <u>*Hy. truncatum*</u> (ANL), *I. hexagonus* (A), *I. pilosus* (ANL), <u>*I. ricinus*</u> (ANL), <u>*R. appendiculatus*</u> (ANL), <u>*R. bursa*</u> (ANL), *R. compositus* (A), <u>*R. evertsi*</u> (ANL), *R. kochi* (A), *R. muehlensi* (A), <u>*R. pulchellus*</u> (ANL), <u>*R. senegalensis*</u> (A), <u>*R. simus*</u> (A), *R. sulcatus* (A), *R. turanicus* (AN), <u>*R. (B.) annulatus*</u> (ANL), <u>*R. (B.) decoloratus*</u> (ANL), <u>*R. (B.) microplus*</u> (ANL)

***Capra ibex*** (**ibex**): *Hy. anatolicum* spp. (A), *Hy. marginatum* group (A), *I. ricinus* (AN), *R. (B.) annulatus* (ANL)

***Capra pyrenaica*** (**Spanish ibex**): *Ha. punctata* (A)

***Capricornis crispus*** (**Japanese serow**): *A. testudinarium*

***Capricornis sumatraensis*** (**mainland serow**): *A. crassipes*, *A. varanense*

***Capricornis swinhoei*** (**Taiwan serow**): *A. testudinarium*

***Cephalophus dorsalis*** (**bay duiker**): *A. spendidum* (ANL), *R. appendiculatus* (A), *R. senegalensis* (A), *R. sulcatus*

***Cephalophus natalensis*** (**Natal duiker**): *A. hebraeum* (L), *A. marmoreum* (NL), *I. pilosus* (A), *R. appendiculatus* (AN), *R. evertsi* (ANL), *R. kochi* (A), *R. muehlensi* (AN), *R. pulchellus* (AN)

***Cephalophus niger*** (**black duiker**): *A. compressum* (A), *A. nuttalli* (N), *R. senegalensis* (A)

***Cephalophus rufilatus*** (**red-flanked duiker**): *A. variegatum* (AN), *R. senegalensis* (A)

***Cephalophus silvicultor*** (**yellow-backed duiker**): *R. kochi* (A)

***Connochaetes gnou*** (**black wildebeest**): *A. marmoreum* (NL), *Hy. marginatum* group (A), *Hy. truncatum* (A), *R. evertsi* (ANL), *R. pulchellus* (A), <u>*R. (B.) decoloratus*</u> (ANL)

***Connochaetes taurinus*** (**blue wildebeest**): *A. gemma* (A), *A. hebraeum* (ANL), *A. lepidum*, *A. variegatum* (A), *Ha. leachi* (A), *Hy. albiparmatum*, *Hy. marginatum* group (A), <u>*Hy. truncatum*</u> (A), <u>*R. appendiculatus*</u> (ANL), <u>*R. evertsi*</u> (ANL), *R. kochi* (A), *R. pulchellus* (AN), *R. senegalensis* (A), *R. simus* (A), <u>*R. (B.) decoloratus*</u> (ANL)

***Damaliscus lunatus*** (**tsessebe**): *A. gemma*, *A. hebraeum* (AN), *A. lepidum* (AN), <u>*A. pomposum*</u> (AN), *A. variegatum* (ANL), *Ha. muhsamae* (A), *Hy. truncatum* (A), *R. appendiculatus* (ANL), *R. evertsi* (ANL), *R. kochi* (A), *R. pulchellus* (A), *R. (B.) decoloratus* (ANL)

***Damaliscus pygargus*** (**blesbok/bontebok**): *A. hebraeum*, *A. marmoreum* (ANL), *Hy. marginatum* group (AL), *Hy. truncatum* (ANL), *I. pilosus* (ANL), *R. appendiculatus* (ANL), *R. capensis* (A), <u>*R. evertsi*</u> (ANL), *R. (B.) decoloratus* (AL)

***Eudorcas rufifrons*** (**red-fronted gazelle**): *A. variegatum* (AN), *R. sulcatus* (A), *R. (B.) decoloratus*

***Eudorcas thomsonii*** (**Thomson's gazelle**): *A. lepidum* (A), *A. variegatum* (AN), *R. appendiculatus* (A), *R. compositus*, *R. evertsi* (ANL), *R. pulchellus* (ANL), *R. turanicus* (A)

***Gazella arabica*** (**Arabian gazelle**): *R. turanicus*

***Gazella dorcas*** (**dorcas gazelle**): *A. tholloni*

***Gazella gazella*** (**mountain gazelle**): *A. variegatum* (A), *Hy. anatolicum* spp. (A), *Hy. marginatum* group (A), *R. bursa* (ANL), *R. turanicus* (A), *R. (B.) annulatus* (ANL), *R. (B.) decoloratus*

***Gazella spekei*** (**Speke's gazelle**): *R. pulchellus* (A)

***Gazella subgutturosa*** (**black-tailed gazelle**): *Hy. detritum* (A), *R. turanicus*

***Hippotragus equinus*** (**roan antelope**): *A. hebraeum*, *A. lepidum* (A), <u>*A. pomposum*</u> (A), *A. splendidum* (A), <u>*A. variegatum*</u> (AN), *D. rhinocerinus* (A), *Hy. marginatum* group (A), <u>*Hy. truncatum*</u> (A), *I. pilosus* (A), *R. appendiculatus* (A), *R. compositus* (A), *R. evertsi* (A), *R. kochi* (A), *R. muehlensi* (A), *R. senegalensis* (A), *R. simus* (A), *R. turanicus* (A), *R. (B.) annulatus*, <u>*R. (B.) decoloratus*</u> (AN)

***Hippotragus leucophaeus*** (**bluebuck**): *I. pilosus* (A)

***Hippotragus niger*** (**sable antelope**): *A. hebraeum*, <u>*A. pomposum*</u> (A), *A. sparsum* (A), *A. transversale* (N), <u>*A. variegatum*</u> (AN), *Hy. marginatum* group

(A), *Hy. truncatum* (A), *R. appendiculatus* (ANL), *R. capensis*, *R. evertsi* (ANL), *R. kochi* (A), *R. muehlensi* (A), *R, pulchellus* (A), *R. simus* (A), *R. (B.) decoloratus* (ANL)

***Kobus ellipsiprymnus* (waterbuck):** *A. gemma* (A), *A. hebraeum* (AN), *A. pomposum*, *A. sparsum* (A), *A. splendidum* (N), *A. variegatum* (ANL), *Hy. truncatum* (A), *I. pilosus*, *R. appendiculatus* (ANL), *R. compositus* (A), *R. evertsi* (ANL), *R. kochi* (A), *R. muehlensi* (A), *R. pulchellus* (A), *R. simus* (A), *R. sulcatus* (A), *R. (B.) decoloratus* (ANL)

***Kobus kob* (kob):** *A. variegatum* (A), *Hy. truncatum*, *R. appendiculatus* (AN), *R. evertsi* (A), *R. senegalensis* (A), *R. sulcatus*

***Kobus leche* (lechwe):** *A. pomposum* (A), *A. variegatum* (AN), *Hy. truncatum* (A), *R. appendiculatus* (ANL), *R. evertsi* (ANL), *R. simus* (A), *R. turanicus* (A), *R. (B.) decoloratus* (ANL)

***Kobus vardonii* (puku):** *A. variegatum* (AN), *R. appendiculatus* (ANL), *R. compositus* (A), *R. evertsi* (ANL), *R. (B.) decoloratus* (AL)

***Litocranius walleri* (gerenuk):** *A. lepidum* (A), *A. variegatum* (N), *R. evertsi* (ANL), *R. muehlensi* (A), *R. pulchellus* (AN), *R. (B.) decoloratus* (A)

***Madoqua guentheri* (Guenther's dikdik):** *A. variegatum* (AN), *Ha. muhsamae*, *R. appendiculatus*, *R. pulchellus* (N)

***Madoqua kirkii* (Kirk's dikdik):** *A. variegatum* (AN), *R. appendiculatus* (AN), *R. evertsi* (AN), *R. kochi* (A), *R. pulchellus* (N)

***Madoqua piacentinii* (silver dikdik):** *R. pulchellus* (A)

***Madoqua saltiana* (Salt's dikdik):** *R. pulchellus* (A)

***Nanger dama* (addra gazelle):** *Hy. anatolicum* spp. (A), *R. (B.) decoloratus* (A)

***Nanger granti* (Grant's gazelle):** *A. gemma* (A), *A. lepidum* (AN), *A. variegatum* (AN), *Hy. albiparmatum*, *Hy. marginatum* group (A), *Hy. truncatum* (A), *R. appendiculatus* (A), *R. compositus*, *R. evertsi* (ANL), *R. muehlensi* (A), *R. pulchellus* (AN), *R. turanicus* (A), *R. (B.) decoloratus* (A)

***Nanger soemmerringii* (Soemmerring's gazelle):** *R. pulchellus* (AN)

***Neotragus moschatus* (suni):** *A. exornatum*, *A. hebraeum* (ANL), *A. variegatum* (A), *R. appendiculatus* (A), *A. kochi* (A), *A. muehlensi* (ANL)

***Neotragus pygmaeus* (royal antelope):** *A. nuttalli* (N), *A. splendidum* (L), *R. senegalensis* (A)

***Oreotragus oreotragus* (klipspringer):** *A. hebraeum*, *A. variegatum* (N), *R. appendiculatus* (AN), *R. evertsi*, *R. kochi* (A), *R. simus* (A)

***Oryx beisa* (beisa oryx):** *A. gemma* (A), *A. lepidum* (A), *A. sparsum* (A), *Hy. truncatum* (A), *R. pulchellus*

***Oryx dammah* (Sahara oryx):** *I. hexagonus*

***Oryx gazella* (gemsbok):** *A. hebraeum* (A), *A. marmoreum* (NL), *Hy. albiparmatum* (A), *Hy. marginatum* group (A), *Hy. truncatum* (A), *I. pilosus* (ANL), *R. appendiculatus* (A), *R. capensis* (A), *R. evertsi* (ANL), *R. muehlensi* (A), *R. simus* (A), *R. turanicus* (A), *R. (B.) decoloratus* (A)

***Oryx leucoryx* (Arabian oryx):** *Hy. dromedarii* (A)

***Ourebia ourebi* (oribi):** *A. nuttalli* (N), *A. pomposum*, *A. variegatum* (ANL), *R. appendiculatus* (AN), *R. evertsi* (ANL), *R. kochi* (A), *R. senegalensis* (A), *R. simus*, *R. sulcatus* (A)

***Ovis ammon* (Asian wild sheep):** *D. reticulatus*, *Hy. anatolicum* spp., *Hy. detritum*, *Hy. marginatum* group, *I. ricinus* (ANL), *R. bursa* (A), *R. turanicus* (A)

***Ovis aries* (domestic sheep):** *A. cajennense* (A), *A. dissimile* (A), *A. falsomarmoreum*, *A. gemma* (A), *A. hebraeum* (ANL), *A. lepidum* (AN), *A. marmoreum* (ANL), *A. parvum* (AN), *A. pomposum* (AN), *A. rhinocerotis*, *A. sparsum* (N), *A. splendidum* (A), *A. testudinarium* (A), *A. tholloni* (NL), *A. triguttatum* (ANL), *A. variegatum* (ANL), *D. auratus*, *D. nitens*, *D. nuttalli*, *D. reticulatus* (A), *D. rhinocerinus* (A), *Ha. longicornis* (AN), *Ha. punctata* (ANL), *Hy. aegyptium* (A), *Hy. albiparmatum* (A), *Hy. anatolicum* spp. (ANL), *Hy. detritum* (AN), *Hy. dromedarii* (AN), *Hy. impressum* (A), *Hy. lusitanicum* (A), *Hy. marginatum* group (ANL), *Hy. truncatum* (ANL), *I. hexagonus* (AN), *I. pilosus* (ANL), *I. ricinus* (ANL), *R. appendiculatus* (ANL), *R. bursa* (ANL), *R. capensis*, *R. compositus* (A), *R. evertsi* (ANL), *R. kochi* (A), *R. muehlensi* (A), *R. pulchellus* (A), *R. senegalensis* (A), *R. simus* (A), *R. sulcatus* (A), *R. turanicus* (AN), *R. (B.) annulatus* (ANL), *R. (B.) decoloratus* (ANL), *R. (B.) microplus* (AN)

***Ovis orientalis* (mouflon):** *Ha. punctata* (A), *Hy. anatolicum* spp. (A), *Hy. marginatum* group (A), *I. ricinus*, *R. bursa* (A), *R. turanicus* (A)

*Pelea capreolus* (**grey rhebok**): *A. hebraeum* (A), *A. marmoreum* (NL), *I. pilosus* (ANL), *R. evertsi* (ANL)

*Philantomba maxwellii* (**Maxwell's duiker**): *A. splendidum* (L), *R. senegalensis* (A)

*Philantomba monticola* (**blue duiker**): *A. variegatum*, *I. pilosus* (A), *R. appendiculatus*, *R. evertsi*

*Pseudois nayaur* (**blue sheep**): *A. varanense* (A)

*Raphicerus campestris* (**steenbok**): *A. hebraeum* (ANL), *A. marmoreum* (NL), *A. pomposum*, *A. variegatum* (AN), *Hy. truncatum* (A), *R. appendiculatus* (AN), *R. evertsi* (ANL), *R. kochi* (A), *R. muehlensi* (A), *R. pulchellus* (N), *R. simus* (A), *R. sulcatus* (A), *R. (B.) decoloratus* (ANL)

*Raphicerus melanotis* (**Cape grybok**): *A. hebraeum* (NL), *I. pilosus*, *R. evertsi* (AN), *R. kochi* (A), *R. sulcatus* (A)

*Raphicerus sharpei* (**Sharpe's grysbok**): *A. variegatum* (L), *I. pilosus* (A), *R. appendiculatus* (NL), *R. evertsi* (A), *R. kochi* (A)

*Redunca arundinum* (**southern reedbuck**): *A. hebraeum* (ANL), *A. marmoreum* (L), *A. pomposum*, *A. variegatum* (AN), *I. pilosus* (A), *R. appendiculatus* (ANL), *R. evertsi* (ANL), *R. kochi* (A), *R. muehlensi* (AN), *R. simus* (A), *R. (B.) decoloratus* (ANL)

*Redunca fulvorufula* (**mountain reedbuck**): *A. marmoreum* (NL), *Hy. truncatum* (A), *R. appendiculatus* (A), *R. evertsi* (ANL)

*Redunca redunca* (**bohor reedbuck**): *A. variegatum* (ANL), *R. appendiculatus* (ANL), *R. kochi* (A), *R. senegalensis* (A), *R. sulcatus*

*Rupicapra rupicapra* (**alpine chamois**): *Ha. punctata*, *I. ricinus* (AN)

*Saiga tatarica* (**saiga antelope**): *R. turanicus*

*Sylvicapra grimmia* (**grey duiker**): *A. hebraeum* (ANL), *A. lepidum* (A), *A. marmoreum* (L), *A. nuttalli*, *A. pomposum*, *A. variegatum* (AN), *Hy. marginatum* group (A), *Hy. truncatum* (A), *I. pilosus* (A), *R. appendiculatus* (ANL), *R. compositus*, *R. evertsi* (ANL), *R. kochi* (A), *R. muehlensi* (AN), *R. pulchellus*, *R. senegalensis* (A), *R. simus* (A), *R. sulcatus* (A), *R. turanicus* (A), *R. (B.) decoloratus* (ANL), *R. (B.) microplus* (A)

*Syncerus caffer* (**African buffalo**): *A. gemma* (A), *A. hebraeum* (ANL), *A. lepidum* (A), *A. marmoreum* (NL), *A. pomposum* (A), *A. rhinocerotis* (A), *A. sparsum* (AN), *A. splendidum* (AN), *A. tholloni* (AN), *A. variegatum* (AN), *D. rhinocerinus* (A), *Hy. albiparmatum* (A), *Hy. impressum*, *Hy. marginatum* group (A), *Hy. truncatum* (A), *I. pilosus* (A), *R. appendiculatus* (ANL), *R. compositus* (A), *R. evertsi* (ANL), *R. kochi* (A), *R. muehlensi* (ANL), *R. pulchellus* (AN), *R. senegalensis* (AN), *R. simus* (A), *R. sulcatus* (A), *R. (B.) decoloratus* (ANL)

*Taurotragus derbianus* (**giant eland**): *A. variegatum* (AN), *R. (B.) annulatus* (A), *R. (B.) decoloratus* (A)

*Taurotragus oryx* (**eland**): *A. gemma* (A), *A. hebraeum* (ANL), *A. lepidum* (A), *A. marmoreum* (L), *A. nuttalli* (A), *A. pomposum* (A), *A. rhinocerotis* (A), *A. sparsum* (A), *A. splendidum* (A), *A. variegatum* (AN), *D. rhinocerinus* (A), *Hy. albiparmatum* (A), *Hy. impressum*, *Hy. marginatum* group (A), *Hy. truncatum* (A), *I. pilosus* (ANL), *R. appendiculatus* (ANL), *R. capensis* (A), *R. compositus* (A), *R. evertsi* (ANL), *R. kochi* (A), *R. muehlensi* (A), *R. pulchellus* (AN), *R. senegalensis* (A), *R. simus* (A), *R. turanicus* (A), *R. (B.) decoloratus* (ANL)

*Tragelaphus angasii* (**nyala**): *A. hebraeum* (ANL), *A. variegatum* (AN), *Hy. albiparmatum* (A), *Hy. truncatum* (A), *I. pilosus* (AL), *R. appendiculatus* (ANL), *R. evertsi* (ANL), *R. kochi* (AN), *R. muehlensi* (ANL), *R. simus* (A), *R. (B.) decoloratus* (ANL)

*Tragelaphus eurycerus* (**bongo**): *R. senegalensis* (A)

*Tragelaphus imberbis* (**lesser kudu**): *A. gemma* (A), *A. lepidum* (A), *A. variegatum* (A), *R. evertsi* (A), *R. muehlensi* (A), *R. pulchellus* (AN)

*Tragelaphus scriptus* (**bushbuck**): *A. compressum* (A), *A. gemma*, *A. hebraeum* (ANL), *A. marmoreum* (NL), *A. pomposum* (A), *A. variegatum* (ANL), *Ha. hoodi*, *Hy. marginatum* group (A), *Hy. truncatum* (A), *I. pilosus* (A), *R. appendiculatus* (ANL), *R. bursa*, *R. compositus* (A), *R. evertsi* (ANL), *R. kochi* (AN), *R. muehlensi* (AN), *R. pulchellus* (A), *R. senegalensis* (A), *R. simus* (A), *R. sulcatus*, *R. (B.) decoloratus* (ANL)

*Tragelaphus spekii* (**sitatunga**): *A. variegatum* (A), *R. appendiculatus* (A), *R. capensis*, *R. compositus* (A), *R. evertsi*, *R. (B.) decoloratus* (A)

*Tragelaphus strepsiceros* (**greater kudu**): *A. gemma* (A), *A. hebraeum* (ANL), *A. marmoreum* (NL), *A. pomposum* (A), *A. sparsum* (A), *A. tholloni* (A), *A. variegatum* (ANL), *Hy. albiparmatum* (A), *Hy. marginatum* group (A), *Hy. truncatum* (A), *I. pilosus* (AN), *R. appendiculatus* (ANL), *R. evertsi* (ANL),

*R. kochi* (AN), *R. muehlensi* (ANL), *R. pulchellus* (A), R. simus (A), *R. sulcatus* (A), *R. turanicus* (A), R. (B.) decoloratus (ANL)

Family Camelidae (camelids)

**Camelus bactrianus (bactrian camel):** *D. reticulatus, Hy. dromedarii*

**Camelus dromedarius (camel):** *A. falsomarmoreum,* A. gemma (A), A. lepidum (A), *A. sparsum* (A), A. variegatum (A), *D. reticulatus, Ha. punctata* (A), *Hy. aegyptium, Hy. albiparmatum,* Hy. anatolicum spp. (AN), Hy. detritum (A), Hy. dromedarii (AN), *Hy. impressum* (A), *Hy. lusitanicum* (A), Hy. marginatum group (A), Hy. truncatum (A), *R. appendiculatus* (A), *R. bursa* (A), R. evertsi (A), *R. kochi* (A), R. pulchellus (A), *R. turanicus* (A), *R. (B.) annulatus, R. (B.) decoloratus* (A), *R. (B.) microplus* (AN)

**Lama glama (llama):** *I. ricinus*

**Vicugna pacos (alpaca):** *Ha. juxtakochi*

Family Cervidae (deer)

**Alces alces (moose):** *D. reticulatus, I. ricinus* (ANL)

**Axis axis (spotted deer):** *D. auratus* (NL), *Hy. aegyptium, Hy. marginatum* group, *R. (B.) annulatus, R. (B.) microplus* (A)

**Axis porcinus (hog deer):** *D. auratus* (N)

**Blastocerus dichotomus (marsh deer):** *A. cajennense, A. longirostre* (A), *A. ovale* (A), *D. nitens* (A), *R. (B.) microplus* (A)

**Capreolus capreolus (roe deer):** *D. reticulatus* (AN), *Ha. punctata* (A), *Hy. detritum* (A), *Hy. marginatum* group (A), *I. hexagonus* (N), I. ricinus (ANL), *R. bursa, R. turanicus* (AN)

**Cervus elaphus (red deer):** D. reticulatus (AN), *Ha. longicornis* (A), *Ha. punctata* (A), *Hy. anatolicum* spp., *Hy. detritum, Hy. lusitanicum, Hy. marginatum* group (A), I. ricinus (ANL), *R. bursa, R. turanicus, R. (B.) annulatus* (ANL), *R. (B.) microplus*

**Cervus nippon (sika deer):** *A. testudinarium* (AN), *D. reticulatus, Ha. longicornis* (AN), *I. ricinus* (N), *R. bursa*

**Dama dama (fallow deer):** *D. reticulatus, Ha. longicornis, Ha. punctata, Hy. lusitanicum* (A), I. ricinus (ANL), *R. bursa* (A)

**Hydropotes inermis (Chinese water deer):** *I. ricinus*

**Mazama americana (red brocket deer):** *A. cajennense, A. calcaratum* (A), *A. coelebs, A. oblongoguttatum, A. ovale* (A), *A. parvum, Ha. juxtakochi* (ANL)

**Mazama bororo (Sao Paulo bororo):** *A. incisum* (N), *Ha. juxtakochi*

**Mazama gouazoubira (brown brocket deer):** *A. cajennense* (AN), *A. incisum* (N), *A. ovale* (A), *A. parvum* (AN), *D. nitens, Ha. juxtakochi* (ANL), *I. luciae, R. (B.) microplus* (A)

**Muntiacus muntjak (barking deer):** *A. testudinarium* (A), *D. auratus* (NL), *Ha. hystricis, R. (B.) microplus* (L)

**Muntiacus reevesi (Chinese muntjac):** *Ha. hystricis, I. ricinus*

**Odocoileus virginianus (white-tailed deer):** A. cajennense (A), *A. oblongoguttatum* (A), *A. ovale, A. parvum* (A), *A. variegatum,* D. nitens (AN), *Ha. juxtakochi* (A), *R. (B.) annulatus* (A), R. (B.) microplus

**Ozotoceros bezoarticus (pampas deer):** *Ha. juxtakochi, R. (B.) microplus* (A)

**Rusa timorensis (Timor deer):** *A. trimaculatum* (A), *A. variegatum* (AN), *Ha. longicornis, R. (B.) microplus* (ANL)

**Rusa unicolor (sambar):** *A. javanense,* A. testudinarium (AN), *D. auratus* (AN), *Ha. hystricis* (A), *Ha. longicornis, Hy. detritum, Hy. marginatum* group, *R. (B.) microplus* (AN)

Family Giraffidae (giraffes and okapis)

**Giraffa camelopardalis (giraffe):** A. gemma (A), A. hebraeum (ANL), *A. lepidum, A. marmoreum* (N), *A. nuttalli* (A), *A. sparsum* (A), *A. variegatum* (A), *Hy. albiparmatum* (A), Hy marginatum group (A), Hy. truncatum (A), R. appendiculatus (ANL), *R. compositus* (A), R. evertsi (ANL), *R. kochi* (A), *R. muehlensi* (A), *R. pulchellus* (A), *R. senegalensis* (A), *R. simus* (A), R. (B.) decoloratus (ANL)

Family Hippopotamidae (hippopotamuses)

**Hippopotamus amphibius (hippopotamus):** *A. hebraeum, A. rhinocerotis* (A), *A. sparsum* (A), A. tholloni (ANL), *A. variegatum* (ANL), *D. rhinocerinus, Hy. truncatum* (A), *R. appendiculatus* (N), *R. simus*

Family Moschidae

**Moschus moschiferus (Siberian musk deer):** *I. ricinus*

Family Suidae (pigs)

***Hylochoerus meinertzhageni* (giant forest hog)**: *A. compressum* (A), *R. appendiculatus* (N), *R. compositus* (A), *R. sulcatus*

***Phacochoerus africanus* (warthog)**: *A. exornatum*, *A. flavomaculatum*, *A. gemma* (AN), <u>*A. hebraeum*</u> (ANL), *A. marmoreum* (L), *A. nuttalli* (A), <u>*A. pomposum*</u> (AN), *A. rhinocerotis* (A), *A. sparsum* (A), *A. tholloni* (N), <u>*A. variegatum*</u> (ANL), *Hy. albiparmatum* (A), *Hy. dromedarii*, *Hy. impressum* (A), <u>*Hy. marginatum*</u> group (A), <u>*Hy. truncatum*</u> (A), *I. pilosus* (A), <u>*R. appendiculatus*</u> (ANL), *R. bursa*, <u>*R. compositus*</u> (A), *R. evertsi* (ANL), *R. kochi* (A), *R. muehlensi* (A), *R. pulchellus* (AN), <u>*R. senegalensis*</u> (A), <u>*R. simus*</u> (A), *R. turanicus* (A), *R. (B.) decoloratus* (ANL)

***Potamochoerus larvatus* (bush pig)**: *A. hebraeum* (ANL), *A. pomposum* (A), *A. sparsum*, *A. splendidum* (AN), *A. tholloni* (A), *A. variegatum* (ANL), *Hy. truncatum* (A), *I. pilosus* (A), <u>*R. appendiculatus*</u> (ANL), <u>*R. compositus*</u> (A), *R. evertsi* (A), *R. kochi* (A), *R. muehlensi* (AN), *R. pulchellus* (A), <u>*R. simus*</u> (A), *R. turanicus* (A), *R. (B.) decoloratus*

***Potamochoerus porcus* (red river hog)**: *Ha. leachi*, *Hy. truncatum*, *R. senegalensis*

***Sus barbatus* (bearded pig)**: *A. testudinarium* (A)

***Sus celebensis* (Sulawesi wild pig)**: *A. testudinarium*

***Sus philippensis* (Philippine warty pig)**: *A. testudinarium* (A)

***Sus scrofa* (domestic pig/wild boar)**: *A. auricularium*, <u>*A. cajennense*</u> (AN), *A. coelebs*, *A. geoemydae* (AN), *A. hebraeum* (A), *A. helvolum* (A), *A. javanense* (A), *A. lepidum* (A), *A. oblongoguttatum* (AN), *A. ovale* (A), *A. parvum* (A), *A. splendidum* (AN), <u>*A. testudinarium*</u> (ANL), *A. triguttatum* (ANL), *A. varanense* (AN), <u>*A. variegatum*</u> (ANL), <u>*D. auratus*</u> (AN), *D. nitens*, *D. nuttalli*, <u>*D. reticulatus*</u> (A), <u>*Ha. hystricis*</u> (ANL), *Ha. longicornis* (A), *Ha. punctata* (A), *Hy. aegyptium*, <u>*Hy. anatolicum*</u> spp. (A), *Hy. detritum* (A), *Hy. dromedarii* (A), *Hy. impressum* (A), *Hy. lusitanicum* (A), <u>*Hy. marginatum*</u> group (AN), *Hy. truncatum* (A), *I. hexagonus* (A), *I. pilosus*, <u>*I. ricinus*</u> (A), *R. appendiculatus* (AN), *R. bursa* (A), *R. compositus* (A), *R. evertsi* (AN), *R. pulchellus* (A), <u>*R. senegalensis*</u> (A), *R. simus* (A), *R. sulcatus*, *R. turanicus* (A), *R. (B.) annulatus* (A), *R. (B.) decoloratus* (A), <u>*R. (B.) microplus*</u> (A)

***Sus verrucosus* (Javan pig)**: *A. testudinarium* (A)

Family Tayassuidae (peccaries)

***Catagonus wagneri* (chacoan peccary)**: *A. cajennense* (A), *A. parvum* (A)

***Pecari tajacu* (collared peccary)**: <u>*A. cajennense*</u> (ANL), *A. coelebs* (AN), *A. oblongoguttatum* (A), *A. ovale* (A), *A. parvum* (N), *Ha. juxtakochi* (NL)

***Tayassu pecari* (white-lipped peccary)**: *A. cajennense* (ANL), *A. oblongoguttatum* (A), *A. ovale* (A), *R. (B.) microplus* (A)

Family Tragulidae (mouse deer)

***Hyemoschus aquaticus* (water chevrotain)**: *A. nuttalli*

***Moschiola meminna* (Indian chevrotain)**: *D. auratus* (NL)

***Tragulus javanicus* (Javan mouse-deer)**: *A. testudinarium*

ORDER CARNIVORA (CARNIVORES)

Family Canidae (dogs, wolves, coyotes, jackals and foxes)

***Canis adustus* (side-striped jackal)**: *A. hebraeum* (L), *A. variegatum* (ANL), <u>*Ha. leachi*</u> (A), *Ha. muhsamae* (A), *R. appendiculatus* (AN), *R. evertsi* (N), *R. kochi* (A), *R. pulchellus* (AN), *R. senegalensis*, *R. simus* (AN), *R. sulcatus* (A), *R. turanicus* (A)

***Canis aureus* (golden jackal)**: *A. variegatum* (ANL), *Ha. leachi* (AN), *Ha. muhsamae* (A), *Hy. anatolicum* spp., *Hy. detritum*, *R. senegalensis* (A), *R. turanicus* (A), *R. (B.) decoloratus* (A)

***Canis familiaris* (dog)**: *A. auricularium* (AN), <u>*A. cajennense*</u> (ANL), *A. calcaratum* (A), *A. crassipes*, *A. dissimile*, *A. exornatum*, *A. gemma* (A), <u>*A. hebraeum*</u> (ANL), *A. humerale*, *A. javanense*, <u>*A. lepidum*</u> (AN), *A. marmoreum* (ANL), *A. nuttalli* (N), <u>*A. oblongoguttatum*</u> (A), <u>*A. ovale*</u> (ANL), <u>*A. parvum*</u> (A), *A. pictum*, *A. pomposum* (A), *A. pseudoconcolor* (A), *A. sparsum* (A), *A. splendidum* (NL), *A. testudinarium* (AN), *A. tholloni* (A), <u>*A. triguttatum*</u> (AN), *A. varanense*, <u>*A. variegatum*</u> (ANL), *A. varium*, *D. auratus* (AN), *D. nitens* (AN), *D. nuttalli* (AN), <u>*D. reticulatus*</u> (AN), *Ha. hoodi* (A), <u>*Ha. hystricis*</u> (AN), *Ha. juxtakochi* (AN), <u>*Ha. leachi*</u> (ANL), <u>*Ha. longicornis*</u> (AN), *Ha. muhsamae* (A),

*Ha. punctata* (ANL), *Hy. aegyptium*, *Hy. albiparmatum* (A), *Hy. anatolicum spp.* (A), *Hy. detritum* (AN), *Hy. dromedarii* (ANL), *Hy. impressum*, *Hy. lusitanicum*, *Hy. marginatum* group (A), *Hy. truncatum* (AN), *I. hexagonus* (ANL), *I. luciae*, *I. pilosus* (ANL), *I. ricinus* (ANL), *R. appendiculatus* (ANL), *R. bursa* (A), *R. compositus* (A), *R. evertsi* (ANL), *R. kochi* (A), *R. muehlensi* (A), *R. pulchellus* (AN), *R. senegalensis* (A), *R. simus* (AN), *R. sulcatus* (AN), *R. turanicus* (AN), *R. (B.) annulatus* (A), *R. (B.) decoloratus* (AN), *R. (B.) microplus* (ANL)

***Canis latrans*** **(coyote):** *A. cajennense* (ANL), *A. ovale*

***Canis lupus*** **(timber wolf):** *A. variegatum*, *D. nuttalli*, *D. reticulatus* (A), *Ha. punctata*, *I. hexagonus*, *I. ricinus* (A), *R. turanicus*

***Canis mesomelas*** **(black-backed jackal):** *A. hebreaum* (ANL), *A. marmoreum* (NL), *A. variegatum* (AN), *Ha. leachi* (A), *Ha. muhsamae* (A), *Hy. marginatum* group (A), *I. pilosus* (ANL), *R. appendiculatus* (AN), *R. evertsi* (L), *R. muehlensi* (A), *R. pulchellus* (AN), *R. simus* (AL), *R. turanicus* (A), *R. (B.) decoloratus* (A)

***Cerdocyon thous*** **(forest fox):** *A. auricularium* (A), *A. cajennense* (ANL), *A. longirostre*, *A. ovale* (A), *A. parvum* (A), *D. nitens* (AN), *R. (B.) microplus* (AN)

***Chrysocyon brachyurus*** **(maned wolf):** *A. auricularium*, *A. cajennense* (AN), *A. ovale* (A), *A. parvum* (A), *R. (B.) microplus* (N)

***Cuon alpinus*** **(Asiatic wild dog):** *A. testudinarium* (AN)

***Lycalopex gymnocerus*** **(pampa fox):** *A. cajennense*, *A. ovale* (A), *A. parvum* (AN)

***Lycalopex sechurae*** **(sechura fox):** *D. nitens*

***Lycalopex vetulus*** **(field fox):** *A. cajennense* (N)

***Lycaon pictus*** **(African hunting dog):** *A. hebraeum* (ANL), *A. marmoreum* (N), *A. variegatum*, *Ha. muhsamae* (A), *R. appendiculatus* (A), *R. evertsi* (N), *R. pulchellus* (A), *R. senegalensis* (A), *R. simus* (AL), *R. turanicus* (A), *R. (B.) decoloratus* (N)

***Nyctereutes procyonoides*** **(raccoon dog):** *Ha. longicornis*, *I. ricinus* (NL)

***Otocyon megalotis*** **(bat-eared fox):** *A. marmoreum* (L), *A. variegatum* (AN), *Ha. muhsamae* (A), *Hy. truncatum* (A), *I. pilosus* (L), *R. appendiculatus* (AN), *R. pulchellus* (AN), *R. turanicus* (A)

***Speothos venaticus*** **(bush dog):** *A. cajennense* (AN), *A. ovale* (A)

***Urocyon cinereoargenteus*** **(grey fox):** *A. auricularium* (N), *A. cajennense* (N), *A. oblongoguttatum* (A), *A. ovale* (A), *A. parvum* (A), *R. (B.) microplus*

***Vulpes chama*** **(Cape fox):** *A. hebraeum* (ANL), *R. capensis* (A), *R. simus* (A)

***Vulpes corsak*** **(corsac fox):** *D. nuttalli*, *R. turanicus*

***Vulpes lagopus*** **(Arctic fox):** *I. ricinus*

***Vulpes pallida*** **(pale fox):** *R. sulcatus* (A)

***Vulpes rueppellii*** **(sand fox):** *Hy. dromedarii* (A)

***Vulpes vulpes*** **(red fox):** *A. variegatum* (N), *D. reticulatus* (A), *Ha. leachi* (AL), *Ha. muhsamae*, *Ha. punctata* (AN), *Hy. dromedarii* (N), *Hy. lusitanicum*, *I. hexagonus* (N), *I. ricinus* (ANL), *R. bursa*, *R. turanicus* (A), *R. (B.) annulatus*

Family Felidae (cats)

***Acinonyx jubatus*** **(cheetah):** *A. hebraeum* (ANL), *A. lepidum* (A), *A. marmoreum* (NL), *A. variegatum* (AN), *Hy. truncatum* (A), *R. appendiculatus* (ANL), *R. compositus* (A), *R. evertsi* (AL), *R. pulchellus* (A), *R. simus* (AL), *R. turanicus* (A), *R. (B.) decoloratus* (L)

***Caracal caracal*** **(African caracal):** *A. hebraeum* (NL), *A. marmoreum* (NL), *A. variegatum* (N), *Ha. muhsamae* (A), *Hy. marginatum* group (L), *I. pilosus* (ANL), *R. evertsi* (NL), *R. simus* (AL), *R. turanicus* (A), *R. (B.) decoloratus* (ANL)

***Catopuma temminckii*** **(Asian golden cat):** *A. testudinarium*

***Felis catus*** **(cat):** *A. auricularium*, *A. cajennense*, *A. marmoreum* (NL), *A. nuttalli* (NL), *A. oblongoguttatum*, *A. ovale* (A), *A. parvum* (A), *A. testudinarium*, *A. triguttatum* (L), *A. variegatum* (ANL), *D. nitens* (A), *D. nuttalli* (N), *D. reticulatus* (A), *Ha. leachi* (AN), *Ha. longicornis* (ANL), *Ha. muhsamae* (A), *Ha. punctata* (AN), *Hy. anatolicum* spp., *Hy. dromedarii* (NL), *Hy. marginatum* group (A), *Hy. truncatum* (A), *I. hexagonus* (ANL), *I. pilosus* (AN), *I. ricinus* (AN), *R. appendiculatus* (ANL), *R. bursa* (A), *R. evertsi* (N), *R. simus* (A), *R. sulcatus* (A), *R. turanicus* (A), *R. (B.) annulatus*, *R. (B.) decoloratus* (A), *R. (B.) microplus*

***Felis chaus*** **(jungle cat):** *D. auratus* (N), *Ha. leachi* (A), *R. turanicus* (A)

***Felis nigripes*** **(black-footed cat):** *R. turanicus* (A)

***Felis silvestris*** (**wild cat**): *A. lepidum* (A), *A. marmoreum* (L), *Ha. leachi* (A), *Ha. muhsamae* (A), *Ha. punctata, Hy. truncatum* (A), *I. hexagonus, I. pilosus* (A), *I. ricinus* (AN), *R. appendiculatus* (N), *R. evertsi* (N), *R. pulchellus* (AN), *R. simus* (A), *R. turanicus* (A)

***Leopardus colocolo*** (**Chilean pampa cat**): *R. (B.) microplus* (AN)

***Leopardus geoffroyi*** (**Geoffroy's cat**): *A. parvum* (A)

***Leopardus pardalis*** (**ocelot**): *A. cajennense* (NL), *A. ovale* (AN), *A. parvum* (A), *D. nitens, R. (B.) microplus*

***Leopardus tigrinus*** (**little spotted cat**): *A. cajennense* (A), *A. longirostre* (AN), *D. nitens* (A), *R. (B.) microplus*

***Leopardus wiedii*** (**margay**): *A. ovale* (A)

***Leptailurus serval*** (**serval**): *A. hebraeum* (L), *A. variegatum* (ANL), *Ha. leachi* (A), *Ha. muhsamae* (A), *I. pilosus, R. appendiculatus* (AN), *R. evertsi* (N), *R simus* (A), *R. sulcatus* (A), *R. turanicus* (A)

***Lynx pardinus*** (**Iberian lynx**): *I. ricinus, R. turanicus*

***Neofelis nebulosa*** (**clouded leopard**): *Ha. hystricis* (A)

***Panthera leo*** (**lion**): *A. gemma* (A), *A. hebraeum* (ANL), *A. marmoreum* (NL), *A. pomposum* (N), *A. rhinocerotis* (A), *A. sparsum* (A), *A. variegatum* (AN), <u>*Ha. leachi*</u> (A), *Hy. albiparmatum* (A), *Hy. impressum, Hy. marginatum* group (A), <u>*Hy. truncatum*</u> (ANL), *R. appendiculatus* (ANL), <u>*R. compositus*</u> (A), *R. evertsi* (AL), *R. kochi* (A), <u>*R. pulchellus*</u> (ANL), *R. senegalensis* (A), <u>*R. simus*</u> (ANL), <u>*R. sulcatus*</u> (A), *R. turanicus* (A), *R. (B.) decoloratus* (AL), *R. (B.) microplus*

***Panthera onca*** (**jaguar**): *A. cajennense* (ANL), *A. coelebs* (N), *A. oblongoguttatum* (A), <u>*A. ovale*</u> (A), *A. parvum* (A), *D. nitens* (AN), *R. (B.) microplus* (AN)

***Panthera pardus*** (**leopard**): *A. hebraeum* (ANL), *A. marmoreum* (NL), *A. nuttalli* (NL), *A. pomposum, A. tholloni* (A), *A. variegatum* (AN), *D. auratus* (AN), <u>*Ha. leachi*</u> (AL), *Ha. muhsamae* (A), *Hy. albiparmatum* (A), *Hy. marginatum* group (AN), *Hy. truncatum* (A), *I. pilosus, I. ricinus, R. appendiculatus* (AL), *R. compositus* (A), *R. evertsi* (A), *R. kochi* (A), *R. muehlensi* (A), *R. pulchellus* (AN), *R. senegalensis* (A), <u>*R. simus*</u> (A), <u>*R. sulcatus*</u> (A), *R. turanicus* (A), *R. (B.) decoloratus* (ANL), *R. (B.) microplus*

***Panthera tigris*** (**tiger**): <u>*A. testudinarium*</u> (NL), *D. auratus* (AN), <u>*Ha. hystricis*</u> (A)

***Pardofelis marmorata*** (**marbled cat**): *A. geoemydae* (N), *Ha. hystricis* (N)

***Prionailurus bengalensis*** (**leopard cat**): *A. testudinarium* (NL)

***Prionailurus viverrinus*** (**fishing cat**): *D. auratus*

***Puma concolor*** (**cougar/Florida panther**): *A. cajennense* (ANL), *A. coelebs* (N), *A. ovale* (AN), *A. parvum* (A), *D. nitens* (A), *I. ricinus, R. (B.) microplus* (AN)

***Puma yagouaroundi*** (**jaguarundi**): *A. ovale* (AN), *A. parvum* (A)

Family Herpestidae (mongooses)

***Atilax paludinosus*** (**marsh mongoose**): *Ha. muhsamae* (A), *R. appendiculatus* (N), *R. evertsi, R. simus* (A)

***Cynictis penicillata*** (**yellow mongoose**): *A. exornatum, A. hebraeum* (L), *A. marmoreum* (L), *Ha. muhsamae* (A), *Hy. marginatum* group (L), *R. appendiculatus, R. evertsi* (NL), *R. turanicus* (L)

***Galerella pulverulenta*** (**Cape grey mongoose**): *Ha. muhsamae. I. pilosus*

***Galerella sanguinea*** (**slender mongoose**): *A. variegatum* (NL), *Ha. muhsamae* (A), *Hy. truncatum* (L), *R. appendiculatus* (NL), *R. evertsi* (L), *R. muehlensi* (N), *R. pulchellus* (A), *R. (B.) decoloratus* (L)

***Helogale parvula*** (**dwarf mongoose**): *A. hebraeum, Ha. muhsamae* (A)

***Herpestes brachyurus*** (**short-tailed mongoose**): *A. varanense* (A)

***Herpestes edwardsi*** (**Indian grey mongoose**): *D. auratus* (N)

***Herpestes ichneumon*** (**Egyptian mongoose**): *A. hebraeum* (L), *A. nuttalli* (N), *A. variegatum* (N), *Ha. muhsamae* (A), *I. hexagonus* (N), *R. appendiculatus* (NL), *R. sulcatus, R. turanicus* (A)

***Herpestes javanicus*** (**Javan mongoose**): *A. crassipes, A. testudinarium* (N), *A. variegatum* (NL), *D. auratus* (N), *Ha. hystricis, R. (B.) microplus*

***Ichneumia albicauda*** (**white-tailed mongoose**): *A. gemma, A. hebraeum* (L), *A. marmoreum* (L), *A. variegatum* (AN), *Ha. leachi* (AN), <u>*Ha. muhsamae*</u> (AN), *Hy. truncatum* (L), *I. pilosus, R. appendiculatus* (ANL), *R. muehlensi* (A), *R. pulchellus* (A)

***Mungos mungo*** (**banded mongoose**): *A. hebraeum*

(L), *A. marmoreum* (L), *A. nuttalli* (NL), *A. pomposum* (A), *A. tholloni* (A), *Ha. leachi* (A), *Ha. muhsamae* (AN), *I. pilosus, R. appendiculatus* (NL), *R. evertsi, R. senegalensis* (N), *R. simus* (ANL)

***Suricata suricatta* (meerkat):** *A. hebraeum* (A), *Ha. muhsamae* (ANL), *I. pilosus* (A), *R. appendiculatus* (AL), *R. evertsi*

Family Hyaenidae (hyaenas)

***Crocuta crocuta* (spotted hyena):** *A. hebraeum* (NL), *A. lepidum* (AN), *A. marmoreum* (N), *A. variegatum* (N), *Ha. leachi* (A), *Ha. muhsamae* (AN), *R. appendiculatus* (AN), *R. pulchellus, R. simus* (A), *R. sulcatus, R. (B.) decoloratus* (L)

***Hyaena brunnea* (brown hyena):** *Ha. muhsamae* (A), *Hy. dromedarii* (A), *R. appendiculatus* (A), *R. simus* (A), *R. turanicus* (A)

***Hyaena hyaena* (striped hyena):** *A. javanense* (AN), *Ha. leachi* (A), *Ha. muhsamae* (A), *R. appendiculatus* (A), *R. pulchellus* (AN)

***Proteles cristata* (aardwolf):** *A. hebraeum* (A), *A. marmoreum* (L), *A. variegatum, Ha. muhsamae* (A), *I. pilosus* (L), *R. evertsi* (L), *R. pulchellus* (AN), *R. simus* (A), *R. turanicus* (A)

Family Mephitidae (skunks)

***Conepatus semistriatus* (striped hog-nosed skunk):** *A. auricularium* (A), *A. ovale* (A)

***Mydaus marchei* (Palawan stink badger):** *A. testudinarium* (N)

Family Mustelidae (weasels, badgers and otters)

***Arctonyx collaris* (hog badger):** *A. crassipes, D. auratus* (AN), *Ha. hystricis* (A)

***Eira barbara* (tayra):** *A. longirostre* (A), *A. oblongoguttatum, A. ovale* (AN)

***Galictis cuja* (lesser grison):** *A. ovale* (A), *A. parvum* (A), *A. pseudoconcolor*

***Galictis vittata* (greater grison):** *A. auricularium* (A), *A. ovale* (A)

***Gulo gulo* (wolverine):** *I. hexagonus* (ANL)

***Hydrictis maculicollis* (speckle-throated otter):** *Ha. muhsamae* (A)

***Ictonyx striatus* (striped polecat):** *Ha. muhsamae* (A), *R. appendiculatus* (NL)

***Lontra longicaudis* (long-tailed otter):** *A. ovale* (A)

***Lutra lutra* (Eurasian otter):** *I. hexagonus* (ANL), *I. ricinus* (AN)

***Lutrogale perspicillata* (Indian smooth-coated otter):** *A. varanense* (NL)

***Martes flavigula* (yellow-throated marten):** *A. testudinarium* (AN), *Ha. hystricis* (ANL)

***Martes foina* (beech marten):** *Hy. marginatum* group, *I. hexagonus, I. ricinus* (A), *R. turanicus*

***Martes martes* (European pine marten):** *I. hexagonus* (N), *I. ricinus* (AL)

***Martes zibellina* (sable):** *I. hexagonus*

***Meles meles* (Eurasian badger):** *D. nuttalli, Ha. longicornis* (A), *Ha. punctata* (N), <u>*I. hexagonus*</u> (ANL), <u>*I. ricinus*</u> (ANL), *R. turanicus* (A)

***Mellivora capensis* (honey badger):** *A. hebraeum* (ANL), *Ha. muhsamae* (A), *Hy. truncatum* (A), *R. appendiculatus* (ANL), *R. evertsi* (ANL), *R. simus* (A), *R. (B.) decoloratus* (ANL)

***Melogale personata* (Burmese ferret badger):** *A. testudinarium* (N), *D. auratus* (NL), *Ha. hystricis* (N)

***Mustela erminea* (stoat):** *Ha. longicornis,* <u>*I. hexagonus*</u> (ANL), <u>*I. ricinus*</u> (ANL)

***Mustela eversmanii* (steppe polecat):** *D. reticulatus* (NL). *Ha. punctata, I. ricinus* (AL), *R. turanicus*

***Mustela furo* (ferret):** *Ha. punctata* (AL), *I. hexagonus* (ANL), *I. ricinus* (N)

***Mustela lutreola* (European mink):** *I. hexagonus, I. ricinus*

***Mustela nivalis* (weasel):** *D. reticulatus* (NL), *Ha. longicornis, Ha. punctata* (N), *Hy. anatolicum* spp., *Hy. lusitanicum,* <u>*I. hexagonus*</u> (ANL), <u>*I. ricinus*</u> (NL), *R. turanicus* (N)

***Mustela nudipes* (Malayan weasel):** *A. testudinarium*

***Mustela putorius* (European polecat):** *D. nuttalli, D. reticulatus* (N), *Ha. longicornis, Hy. lusitanicum,* <u>*I. hexagonus*</u> (ANL), *I. ricinus* (ANL), *R. bursa, R. turanicus*

***Neovison vison* (American mink):** *I. hexagonus* (ANL), *I. ricinus* (A)

***Poecilogale albinucha* (African striped weasel):** *Ha. muhsamae* (A)

***Taxidea taxus* (American badger):** *A. cajennense*

***Vormela peregusna* (marbled polecat):** *Hy. anatolicum* spp., *Hy. detritum, R. bursa, R. turanicus*

Family Nandiniidae (African palm civets)

***Nandinia binotata*** **(African palm civet):** *Ha. muhsamae* (A)

Family Procyonidae (raccoons)

***Nasua narica*** **(northern coati):** *A. auricularium* (A), *A. ovale* (A)

***Nasua nasua*** **(South American coati):** *A. auricularium* (A), <u>*A. cajennense*</u> (A), *A. oblongoguttatum*, <u>*A. ovale*</u> (AN), *A. parvum* (A), *A. rotundatum* (A), *Ha. juxtakochi* (N)

***Procyon cancrivorus*** **(crab-eating raccoon):** *A. cajennense* (AN), *A. calcaratum* (A), *A. oblongoguttatum*, <u>*A. ovale*</u> (A), *A. parvum* (A)

***Procyon lotor*** **(northern raccoon):** *A. cajennense* (N), *A. ovale* (A)

Family Ursidae (bears)

***Helarctos malayanus*** **(Malayan sun bear):** *A. testudinarium* (A), *D. auratus* (N), *Ha. hystricis* (A)

***Melursus ursinus*** **(sloth bear):** *D. auratus* (A)

***Ursus arctos*** **(brown bear):** *D. reticulatus* (A), *I. ricinus*, *R. turanicus*

***Ursus thibetanus*** **(Asian black bear):** *A. testudinarium* (A), *D. auratus* (A), *Ha. hystricis* (A)

Family Viverridae (civets and genets)

***Arctogalidia trivirgata*** **(small-toothed palm civet):** *A. geoemydae* (A), *A. testudinarium*, *A. varanense*, *D. auratus* (N)

***Civettictis civetta*** **(African civet):** *A. hebraeum* (ANL), *A. marmoreum* (NL), *A. nuttalli* (N), *A. variegatum* (AN), <u>*Ha. leachi*</u> (ANL), <u>*Ha. muhsamae*</u> (A), *I. pilosus*, *R. appendiculatus* (ANL), *R. compositus* (A), *R. evertsi* (L), *R. kochi* (N), *R. muehlensi* (N), *R. senegalensis* (A), *R. simus* (ANL), *R. sulcatus* (A), *R. turanicus* (A), *R. (B.) decoloratus* (NL)

***Genetta genetta*** **(small-spotted genet):** *A. hebraeum* (L), *A. nuttalli* (L), *A. variegatum* (L), *Ha. muhsamae* (A), *I. pilosus* (NL), *I. ricinus*, *R. appendiculatus* (N), *R. pulchellus* (A), *R. senegalensis* (A)

***Genetta maculata*** **(panther genet):** *A. hebraeum*, *Ha. leachi* (A), *Ha. muhsamae* (A), *I. pilosus* (A), *R. appendiculatus*, *R. sulcatus* (A)

***Genetta thierryi*** **(false genet):** *Ha. muhsamae*

***Genetta tigrina*** **(large-spotted genet):** *A. hebraeum* (ANL), *A. marmoreum* (NL), *A. variegatum* (NL), *Ha. muhsamae* (A), *R. appendiculatus* (ANL), *R. pulchellus* (A), *R. simus* (L), *R. (B.) decoloratus* (L)

***Hemigalus derbyanus*** **(banded palm civet):** *A. geoemydae* (N), *A. testudinarium*

***Paguma larvata*** **(masked palm civet):** *A. crassipes*, *D. auratus* (N)

***Paradoxurus hermaphroditus*** **(Asian palm civet):** *A. testudinarium* (NL), *D. auratus* (N), *Ha. hystricis* (N)

***Viverra tangalunga*** **(Malayan civet):** *A. testudinarium*

***Viverra zibetha*** **(large Indian civet):** *A. crassipes*, *A. testudinarium* (NL)

***Viverricula indica*** **(small Indian civet):** *D. auratus* (N), *R. (B.) microplus*

ORDER CHIROPTERA (BATS)

Family Noctilionidae (bulldog bats)

***Noctilio albiventris*** **(lesser bulldog bat):** *A. scutatum* (N)

Family Nycteridae (slit-faced bats)

***Nycteris hispida*** **(hairy slit-faced bat):** *A. nuttalli* (N)

Family Phyllostomidae (American leaf-nosed bats)

***Artibeus jamaicensis*** **(Jamaican fruit-eating bat):** *A. cajennense* (NL)

***Artibeus lituratus*** **(great fruit-eating bat):** *A. longirostre* (N)

***Carollia brevicauda*** **(silky short-tailed bat):** *A. parvum* (A)

***Carollia perspicillata*** **(Seba's short-tailed bat):** *A. oblongoguttatum* (A), *Ha. juxtakochi* (A)

***Carollia subrufa*** **(grey short-tailed bat):** *A. parvum* (A)

***Choeroniscus minor*** **(lesser long-tailed bat):** *A. cajennense* (A)

***Macrophyllum macrophyllum*** **(long-legged bat):** *A. calcaratum* (A)

***Platyrrhinus helleri*** **(Heller's broad-nosed bat):** *A. cajennense* (A)

Family Rhinolophidae (horseshoe bats)

***Rhinolophus clivosus*** **(Arabian horseshoe bat):** *D. reticulatus*

*Rhinolophus euryale* (**Mediterranean horseshoe bat**): *I. ricinus* (A)

*Rhinolophus ferrumequinum* (**greater horseshoe bat**): *I. hexagonus, I. ricinus*

Family Rhinopomatidae (long-tailed bats)

*Rhinopoma hardwickii* (**lesser mouse-tailed bat**): *R. pulchellus* (A)

Family Vespertilionidae (vespertilionid bats)

*Miniopterus schreibersii* (**Schreiber's long-fingered bat**): *D. reticulatus*

*Myotis myotis* (**greater mouse-eared bat**): *I. hexagonus*

*Neoromicia capensis* (**Cape serotine**): *R. pulchellus* (A)

ORDER CINGULATA[e]

Family Dasypodidae (armadillos)

*Cabassous centralis* (**northern naked-tailed armadillo**): *A. auricularium* (AN)

*Cabassous tatouay* (**greater naked-tailed armadillo**): *A. pseudoconcolor*

*Cabassous unicinctus* (**southern naked-tailed armadillo**): *A. auricularium, A. pseudoconcolor*

*Catyptophractus retusus* (**Burmeister's armadillo**): *A. pseudoconcolor* (AN)

*Chaetophractus vellerosus* (**lesser hairy armadillo**): *A. auricularium, A. parvum* (A), *A. pseudoconcolor*

*Chaetophractus villosus* (**larger hairy armadillo**): *A. auricularium* (A), *A. pseudoconcolor*

*Dasypus hybridus* (**southern long-nosed armadillo**): *A. auricularium, A. pseudoconcolor*

*Dasypus kappleri* (**greater long-nosed armadillo**): *A. auricularium* (A), *A. cajennense, A. parvum* (A)

*Dasypus novemcinctus* (**nine-banded armadillo**): <u>*A. auricularium*</u> (ANL), *A. cajennense* (AN), *A. coelebs* (N), *A. humerale, A. oblongoguttatum, A. parvum* (A), *A. pseudoconcolor* (A), *A. rotundatum*

*Dasypus sabanicola* (**northern long-nosed armadillo**): *A. auricularium* (A)

*Dasypus septemcinctus* (**seven-banded armadillo**): *A. auricularium*

*Euphractus sexcinctus* (**six-banded armadillo**): <u>*A. auricularium*</u> (AN), *A. cajennense, A. nodosum* (A), <u>*A. pseudoconcolor*</u> (A)

*Priodontes maximus* (**giant armadillo**): *A. cajennense* (A), *A. pseudoconcolor*

*Tolypeutes matacus* (**southern three-banded armadillo**): *A. auricularium* (AN), *A. parvum* (AN), *A. pseudoconcolor* (A), *Ha. juxtakochi*

*Tolypeutes tricinctus* (**Brazilian three-banded armadillo**): *A. auricularium*

*Zaedyus pickiy* (**pichi**): *A. auricularium* (AN), *A. pseudoconcolor* (ANL)

ORDER DASYUROMORPHIA[f]

Family Myrmecobiidae (banded anteater)

*Myrmecobius fasciatus* (**banded anteater**): *A. triguttatum*

ORDER DIDELPHIMORPHIA[f]

Family Didelphidae (American opossums)

*Caluromys derbianus* (**Central American woolly opossum**): *A. geayi* (N)

*Caluromys lanatus* (**western woolly opossum**): *I. luciae* (A)

*Caluromys philander* (**bare-tailed woolly opossum**): *I. luciae* (L)

*Chironectes minimus* (**water opossum**): *A. oblongoguttatum*

*Didelphis albiventris* (**white-eared opossum**): *A. cajennense, A. coelebs* (N), *A. ovale* (N), *A. parvum* (NL), *A. pseudoconcolor, I. luciae*

*Didelphis aurita* (**big-eared opossum**): *A. cajennense, I. luciae*

*Didelphis marsupialis* (**southern opossum**): *A. auricularium* (A), *A. cajennense* (AN), *A. humerale* (N), <u>*I. luciae*</u> (AN)

*Didelphis virginiana* (**Virginian opossum**): *A. auricularium* (N)

*Lutreolina crassicaudata* (**little water opossum**): *A. auricularium, A. cajennense*

*Marmosa canescens* (**greyish mouse-opossum**): *I. luciae*

*Marmosa murina* (**murine mouse-opossum**): *I. luciae* (NL)

*Marmosa robinsoni* (**Robinson's mouse opossum**): *A. ovale* (N), *A. sabanerae* (A), *I. luciae* (NL)

*Metachirus nudicaudatus* (**brown four-eyed opossum**): *A. cajennense, I. luciae* (A)

*Micoureus demerarae* (**ashy opossum**): *I. luciae* (N)

*Monodelphis brevicaudata* (**red-legged short-tailed opossum**): *I. luciae* (NL)

*Philander andersoni* (**black four-eyed opossum**): *I. luciae* (A)

*Philander opossum* (**grey four-eyed opossum**): *A. auricularium* (ANL), *A. geayi*, *I. luciae* (A)

*Thylamys pusillus* (**small fat-tailed opossum**): *A. scutatum*

ORDER DIPROTODONTIA[f]

Family Macropodidae (wallabies and kangaroos)

*Macropus agilis* (**agile wallaby**): *A. triguttatum*

*Macropus dorsalis* (**black-striped wallaby**): *A. triguttatum* (ANL), *Ha. longicornis*

*Macropus eugenii* (**dama wallaby**): *A. triguttatum*

*Macropus fuliginosus* (**western grey kangaroo**): *A. triguttatum* (N), *B. concolor* (AN)

*Macropus giganteus* (**eastern grey kangaroo**): *A. triguttatum* (ANL)

*Macropus parryi* (**pretty-face wallaby**): *A. triguttatum*

*Macropus robustus* (**common wallaroo**): *A. triguttatum* (A), *Ha. longicornis*

*Macropus rufus* (**red kangaroo**): *A. triguttatum* (ANL)

*Onychogalea fraenata* (**bridled nailtail wallaby**): *A. triguttatum*

*Wallabia bicolor* (**swamp wallaby**): *A. triguttatum*

Family Phalangeridae (possums and cuscuses)

*Trichosurus vulpecula* (**common brushtail possum**): *Ha. longicornis* (NL)

Family Phascolarctidae (koalas)

*Phascolarctos cinereus* (**koala**): *A. triguttatum* (A)

Family Pseudocheiridae

*Pseudocheirus peregrinus* (**Queensland ringtail**): *Ha. longicornis*

ORDER ERINACEOMORPHA[c]

Family Erinaceidae (hedgehogs and gymnures)

*Atelerix albiventris* (**four-toed hedgehog**): *A. nuttalli* (ANL), *A. variegatum* (ANL), *Ha. muhsamae* (AN), *Hy. dromedarii* (A), *Hy. impressum* (L), *Hy. marginatum* group (NL), *Hy. truncatum* (NL)

*Atelerix algirus* (**Algerian hedgehog**): *A. variegatum*, *Hy. anatolicum* spp. (A), *I. hexagonus* (A), *R. turanicus* (A)

*Atelerix frontalis* (**southern African hedgehog**): *A. hebraeum* (A), *A. marmoreum*, *A. nuttalli* (A), *A. variegatum*, *Ha. muhsamae* (A), *Hy. marginatum* group (L), *I. pilosus*, *R. appendiculatus*, *R. simus* (A)

*Atelerix sclateri* (**Somali hedgehog**): *R. pulchellus* (A)

*Erinaceus concolor* (**eastern European hedgehog**): *R. turanicus* (A)

*Erinaceus europaeus* (**northern hedgehog**): *D. reticulatus* (NL), *Ha. longicornis* (A), *Ha. punctata* (ANL), *Hy. aegyptium* (ANL), *Hy. anatolicum* spp. (A), *Hy. lusitanicum*, *Hy. marginatum* group, <u>*I. hexagonus*</u> (ANL), <u>*I. ricinus*</u> (ANL), *R. bursa*, *R. turanicus* (ANL)

*Hemiechinus auritus* (**long-eared hedgehog**): *Hy. anatolicum* spp. (N), *Hy. dromedarii* (N), *Hy. marginatum* group (N), *I. ricinus*, *R. turanicus* (A), *R. (B.) annulatus*

*Paraechinus aethiopicus* (**desert hedgehog**): *Hy. anatolicum* spp. (N), *Hy. dromedarii* (AN), *Hy. marginatum* group (N), *R. turanicus* (A)

*Paraechinus hypomelas* (**Brandt's hedgehog**): *R. turanicus* (A)

*Paraechinus micropus* (**Indian hedgehog**): *R. turanicus* (A)

ORDER HYRACOIDEA (HYRAXES)

Family Procaviidae

*Dendrohyrax arboreus* (**southern tree hyrax**): *Ha. muhsamae* (A)

*Dendrohyrax dorsalis* (**western tree hyrax**): *A. compressum* (N)

*Heterohyrax brucei* (**yellow-spotted hyrax**): *Ha. muhsamae* (A), *R. appendiculatus* (NL), *R. evertsi* (AN)

*Procavia capensis* (**Cape rock hyrax**): *A. hebraeum*, *A. marmoreum* (NL), *Hy. marginatum* group (NL), *Hy. truncatum* (AL), *R. appendiculatus* (ANL), *R. evertsi* (AL)

***Procavia habessinicus*** (**Ethiopian rock hyrax**): *Ha. muhsamae* (A)

***Procavia ruficeps*** (**red-headed rock hyrax**): *R. sulcatus*

ORDER LAGOMORPHA (LAGOMORPHS)

Family Leporidae (hares and rabbits)

***Lepus brachyurus*** (**Japanese hare**): *Ha. longicornis*

***Lepus capensis*** (**Cape hare**): *A. gemma* (N), *A. hebraeum*, *A. marmoreum* (NL), *A. variegatum* (ANL), *Ha. muhsamae* (A), *Ha. punctata*, *Hy. anatolicum* spp. (NL), *Hy. dromedarii* (N), *Hy. lusitanicum*, <u>*Hy. marginatum* group</u> (NL), <u>*Hy. truncatum*</u> (ANL), *I. ricinus* (A), *R. appendiculatus* (ANL), *R. bursa*, <u>*R. evertsi*</u> (ANL), *R. pulchellus* (ANL), *R. senegalensis* (A), *R. simus* (N), *R. sulcatus* (A), *R. turanicus* (A), *R. (B.) annulatus*, *R. (B.) decoloratus* (A)

***Lepus europaeus*** (**brown hare**): *D. reticulatus* (ANL), *Ha. longicornis* (ANL), <u>*Ha. punctata*</u> (ANL), *Hy. aegyptium* (ANL), *Hy. anatolicum* spp., *Hy. detritum*, *Hy. lusitanicum* *Hy. marginatum* group (NL), *I. hexagonus*, <u>*I. ricinus*</u> (ANL), *R. bursa* (N), *R. turanicus*, *R. (B.) microplus*

***Lepus fagani*** (**Ethiopian hare**): *R. pulchellus* (A)

***Lepus mandschuricus*** (**Manchurian hare**): *Hy. marginatum* group

***Lepus microtis*** (**African savanna hare**): *A. variegatum* (N), *Hy. impressum* (N), *Hy. marginatum* group, *Hy. truncatum* (A), *R. appendiculatus* (ANL), *R. evertsi* (NL), *R. kochi* (AN), *R. pulchellus* (NL), *R. simus* (AN), *R. sulcatus* (A), *R. turanicus* (A), *R. (B.) decoloratus* (A)

***Lepus nigricollis*** (**Indian hare**): *D. auratus* (N), *Hy. marginatum* group (ANL), *R. turanicus* (A)

***Lepus saxatilis*** (**scrub hare**): <u>*A. hebraeum*</u> (NL), <u>*A. marmoreum*</u> (NL), *A. sparsum*, *Ha. hoodi* (A), *Ha. muhsamae* (A), <u>*Hy. marginatum* group</u> (ANL), <u>*Hy. truncatum*</u> (NL), *I. pilosus* (ANL), <u>*R. appendiculatus*</u> (ANL), <u>*R. evertsi*</u> (ANL), *R. kochi* (ANL), *R. muehlensi* (NL), *R. simus* (NL), *R. sulcatus* (A), *R. turanicus* (ANL), *R. (B.) decoloratus* (AL)

***Lepus sinensis*** (**Chinese hare**): *Ha. hystricis*

***Lepus tibetanus*** (**desert hare**): *R. turanicus*

***Lepus timidus*** (**mountain hare**): *Ha. punctata* (AN), *Hy. detritum* (A), *I. hexagonus*, <u>*I. ricinus*</u> (ANL)

***Lepus tolai*** (**tolai hare**): *D. nuttalli*, *Ha. punctata*, *Hy. anatolicum* spp., *Hy. detritum*, *Hy. dromedarii* (A), *R. bursa*, *R. turanicus*

***Oryctolagus cuniculus*** (**rabbit**): *A. cajennense*, *A. triguttatum* (NL), *D. reticulatus* (ANL), *Ha. longicornis* (AN), *Ha. punctata* (ANL), *Hy. anatolicum* spp., *Hy. dromedarii*, *Hy. lusitanicum* (ANL), *Hy. marginatum* group, *I. hexagonus* (A), <u>*I. ricinus*</u> (ANL), *R. appendiculatus*, *R. bursa* (A), *R. evertsi* (A), *R. turanicus* (A), *R. (B.) annulatus*, *R. (B.) microplus*

***Pentalagus furnessi*** (**amami rabbit**): *A. testudinarium* (AL), *Ha. hystricis* (AN)

***Pronolagus crassicaudatus*** (**Natal red hare**): *A. sparsum*, *Ha. muhsamae* (A), *I. pilosus*, *R. appendiculatus*, *R. evertsi*

***Pronolagus randensis*** (**rand red hare**): *Hy. marginatum* group (NL), *R. sulcatus*

***Pronolagus rupestris*** (**common red rock rabbit**): *A. marmoreum* (NL), *Hy. marginatum* group (NL), *Hy. truncatum* (NL), *R. evertsi* (ANL)

***Sylvilagus brasiliensis*** (**forest rabbit**): *Ha. juxtakochi*

***Sylvilagus floridanus*** (**eastern cottontail**): *A. parvum* (ANL), *Ha. juxtakochi* (A)

Family Ochotonidae (mouse hares)

***Ochotona dauurica*** (**daurian pika**): *D. nuttalli*

***Ochotona macrotis*** (**large-eared pika**): *I. ricinus*

ORDER MACROSCELIDEA (ELEPHANT SHREWS)

Family Macroscelididae

***Elephantulus brachyrhynchus*** (**short-snouted elephant-shrew**): *Hy. truncatum* (L), *R. evertsi*, *R. simus*

***Elephantulus intufi*** (**bushveld elephant-shrew**): *Hy. marginatum* group

***Elephantus myurus*** (**eastern rock elephant-shrew**): *A. marmoreum* (NL), *Hy. marginatum* group (NL), *Hy. truncatum* (NL), *R. appendiculatus* (L), *R. evertsi* (NL)

***Elephantus rufescens*** (**rufous elephant-shrew**): *A. sparsum* (N), *R. evertsi* (NL)

***Elephantus rupestris*** (**rock elephant-shrew**): *A. marmoreum*, *Ha. muhsamae*, *I. pilosus*, *R. appendiculatus*

***Macroscelides proboscideus*** (**short-eared elephant-shrew**): *R. evertsi*

***Petrodromus tetradactylus* (four-toed elephant-shrew):** *R. appendiculatus* (NL), *R. kochi* (ANL), *R. pulchellus*, *R. simus* (L)

***Rhynchocyon chrysopygus* (golden-rumped elephant-shrew):** *R. appendiculatus*

***Rhynchocyon petersi* (black-and-rufous elephant-shrew):** *R. simus*

ORDER MONOTREMATA (MONOTREMES)

Family Ornithorhynchidae (duck-billed platypuses)

***Ornithorhynchus anatinus* (duck-billed platypus):** *A. triguttatum*

Family Tachyglossidae (spiny anteaters)

***Tachyglossus aculeatus* (short-beaked echidna):** *A. triguttatum*, <u>*B. concolor*</u> (A), *B. hydrosauri* (AN)

***Zaglossus bruijni* (long-beaked echidna):** *B. concolor* (A)

ORDER PERAMELEMORPHIA[f]

Family Peramelidae (bandicoots)

***Isoodon macrourus* (northern brown bandicoot):** *Ha. longicornis* (N)

ORDER PERISSODACTYLA[d]

Family Equidae (horses, zebras and asses)

***Equus asinus* (donkey):** *A. cajennense* (N), *A. gemma* (A), *A. hebraeum* (AN), *A. humerale* (A), *A. lepidum* (A), *A. parvum* (A), *A. pomposum* (A), <u>*A. variegatum*</u> (ANL), <u>*D. nitens*</u> (ANL), *D. reticulatus* (A), *D. rhinocerinus* (A), *Ha. longicornis*, <u>*Ha. punctata*</u> (A), *Hy. aegyptium*, <u>*Hy. anatolicum* spp.</u> (AN), *Hy. detritum* (A), <u>*Hy. dromedarii*</u> (A), <u>*Hy. marginatum* group</u> (AN), *Hy. truncatum* (A), *I. hexagonus*, *I. pilosus*, *I. ricinus* (A), *R. appendiculatus* (ANL), <u>*R. bursa*</u> (ANL), *R. compositus* (A), <u>*R. evertsi*</u> (AN), *R. kochi* (A), *R. pulchellus* (A), *R. simus* (A), *R. sulcatus*, *R. turanicus* (A), *R. (B.) annulatus* (AN), *R. (B.) decoloratus* (AN), *R. (B.) microplus*

***Equus asinus* × *Equus caballus* (mule):** *A. cajennense* (AN), *A. gemma* (A), *A. hebraeum*, *A. lepidum* (A), *A. ovale* (A), *A. parvum* (A), *A. pomposum* (A), *A. variegatum* (AN), <u>*D. nitens*</u> (ANL), *Ha. juxtakochi*, *Hy. anatolicum* spp. (A), *Hy. detritum* (A), *Hy. dromedarii*, *Hy. marginatum* group (A), *Hy. truncatum* (A), *I. pilosus*, *I. ricinus*, *R. appendiculatus* (ANL), *R. bursa*, *R. evertsi* (A), *R. pulchellus* (A), *R. simus* (A), *R. turanicus*, *R. (B.) annulatus*, *R. (B.) decoloratus*, *R. (B.) microplus*

***Equus burchelli* (Burchell's zebra):** *A. gemma* (AN), *A. hebraeum* (ANL), *A. lepidum* (A), *A. pomposum* (A), <u>*A. variegatum*</u> (AN), *Hy. albiparmatum* (A), <u>*Hy. marginatum* group</u> (A), <u>*Hy. truncatum*</u> (A), <u>*R. appendiculatus*</u> (ANL), <u>*R. evertsi*</u> (ANL), *R. kochi* (A), *R. muehlensi* (A), *R. pulchellus* (ANL), <u>*R. simus*</u> (A), *R. turanicus* (A), <u>*R. (B.) decoloratus*</u> (ANL)

***Equus caballus* (horse):** *A. auricularium* (A), <u>*A. cajennense*</u> (ANL), *A. coelebs* (A), *A. fimbriatum*, *A. gemma* (A), *A. hebraeum* (AN), *A. lepidum* (A), *A. marmoreum* (L), *A. moreliae*, *A. oblongoguttatum* (A), *A. ovale* (A), <u>*A. parvum*</u> (AN), *A. pomposum* (A), *A. sparsum* (A), *A. testudinarium* (A), *A. tholloni*, *A. trimaculatum*, <u>*A. triguttatum*</u> (ANL), <u>*A. variegatum*</u> (ANL), *B. hydrosauri*, *D. auratus* (N), <u>*D. nitens*</u> (ANL), <u>*D. nuttalli*</u> (A), <u>*D. reticulatus*</u> (AN), *Ha. juxtakochi* (AN), <u>*Ha. longicornis*</u> (ANL), <u>*Ha. punctata*</u> (ANL), <u>*Hy. anatolicum* spp.</u> (AN), <u>*Hy. detritum*</u> (ANL), <u>*Hy. dromedarii*</u>, *Hy. impressum* (A), *Hy. lusitanicum*, <u>*Hy. marginatum* group</u> (AN), <u>*Hy. truncatum*</u> (AN), *I. hexagonus* (ANL), *I. pilosus*, <u>*I. ricinus*</u> (ANL), <u>*R. appendiculatus*</u> (ANL), <u>*R. bursa*</u> (AN), *R. capensis* (A), <u>*R. evertsi*</u> (ANL), *R. kochi* (A), <u>*R. pulchellus*</u> (A), *R. senegalensis* (A), *R. simus* (A), *R. sulcatus*, *R. turanicus* (AN), <u>*R. (B.) annulatus*</u> (ANL), <u>*R. (B.) decoloratus*</u> (ANL), <u>*R. (B.) microplus*</u> (AN)

***Equus grevyi* (Grevy's zebra):** *A. gemma* (A), *A. lepidum* (A), *Hy. marginatum* group, *Hy. truncatum* (A), *R. evertsi* (AN), *R. pulchellus* (ANL)

***Equus hemionus* (Asian wild ass):** *D. nuttalli*, *Hy. anatolicum* spp., *Hy. detritum*, *R. turanicus*

***Equus przewalskii* (Mongolian wild horse):** *D. nuttalli*

***Equus zebra* (mountain zebra):** *A. marmoreum* (NL), *A. variegatum*, <u>*Hy. marginatum* group</u> (A), *Hy. truncatum* (A), *R. capensis* (A), <u>*R. evertsi*</u> (ANL)

Family Rhinocerotidae (rhinoceroses)

***Ceratotherium simum* (white rhinoceros):** *A. gemma*, <u>*A. hebraeum*</u> (ANL), *A. rhinocerotis* (A), *A. tholloni* (A), *A. variegatum* (A), <u>*D. rhinocerinus*</u> (A), *Hy.*

*marginatum* group (A), *Hy. truncatum* (A), *R. appendiculatus*, *R. evertsi* (A), *R. muehlensi*, *R. simus* (A)

***Dicerorhinus sumatrensis*** **(Sumatran rhinoceros):** *A. testudinarium* (A)

***Diceros bicornis*** **(black rhinoceros):** *A. cajennense* (A), *A. gemma* (A), *A. hebraeum* (ANL), *A. lepidum* (A), *A. nuttalli* (A), *A. pomposum* (A), *A. rhinocerotis* (A), *A. sparsum* (A), *A. tholloni* (A), *A. variegatum* (A), *D. rhinocerinus* (A), *Hy. albiparmatum* (A), *Hy. impressum*, *Hy. marginatum* group (A), *Hy. truncatum* (A), *R. appendiculatus* (A), *R. capensis* (A), *R. compositus* (A), *R. evertsi* (A), *R. kochi* (A), *R. muehlensi* (A), *R. pulchellus* (AN), *R. senegalensis* (A), *R. simus* (A)

***Rhinoceros sondaicus*** **(Javan rhinoceros):** *A. testudinarium* (A)

***Rhinoceros unicornis*** **(great Indian rhinoceros):** *D. auratus* (AN)

Family Tapiridae (tapirs)

***Tapirus bairdii*** **(Central American tapir):** *A. cajennense* (A), *A. coelebs* (A), *A. humerale* (A), *A. oblongoguttatum* (A), *A. ovale* (A), *D. nitens*, *Ha. juxtakochi* (A), *R. (B.) microplus*

***Tapirus indicus*** **(Asian tapir):** *A. testudinarium* (A)

***Tapirus pinchaque*** **(Andean tapir):** *A. multipunctum* (A)

***Tapirus terrestris*** **(South American tapir):** *A. cajennense* (A), *A. calcaratum* (ANL), *A. coelebs* (A), *A. incisum* (AN), *A. multipunctum* (A), *A. nodosum*, *A. oblongoguttatum* (A), *A. ovale* (A), *A. parvum* (A), *A. pseudoconcolor* (A), *Ha. juxtakochi* (A), *R. (B.) microplus* (A)

ORDER PHOLIDOTA (PANGOLINS)

Family Manidae

***Manis crassicaudata*** **(Indian pangolin):** *A. geoemydae* (A), *A. javanense* (ANL)

***Manis gigantea*** **(giant ground pangolin):** *A. compressum* (ANL), *R. senegalensis* (A)

***Manis javanica*** **(Malayan pangolin):** *A. javanense* (ANL), *A. varanense*

***Manis pentadactyla*** **(Chinese pangolin):** *A. javanense* (A), *A. testudinarium*

***Manis temminckii*** **(Cape pangolin):** *A. compressum*, *A. hebraeum* (A), *A. variegatum*, *R. appendiculatus* (AN), *R. simus* (A)

***Manis tetradactyla*** **(black-bellied pangolin):** *A. compressum* (AL)

***Manis tricuspis*** **(tree pangolin):** *A. compressum* (ANL), *A. flavomaculatum*, *A. latum*, *A. pomposum* (A)

ORDER PILOSA[e]

Family Bradypodidae (three-toed tree sloths)

***Bradypus torquatus*** **(maned sloth):** *A. varium* (AN)

***Bradypus tridactylus*** **(pale-throated sloth):** *A. geayi* (A), *A. longirostre* (N), *A. varium* (AN), *R. (B.) microplus*

***Bradypus variegatus*** **(brown-throated sloth):** *A. geayi* (AN), *A. varium* (A)

Family Cyclopedidae

***Cyclopes didactylus*** **(pygmy anteater):** *A. calcaratum* (A), *A. humerale* (N)

Family Megalonychidae (two-toed tree sloths)

***Choloepus didactylus*** **(southern two-toed sloth):** *A. varium* (A)

***Choloepus hoffmanni*** **(Hoffmann's two-toed sloth):** *A. calcaratum* (A), *A. geayi* (A), *A. varium* (A)

Family Myrmecophagidae (anteaters)

***Myrmecophaga tridactyla*** **(giant anteater):** *A. auricularium* (A), *A. cajennense* (AN), *A. calcaratum* (AN), *A. coelebs*, *A. longirostre*, *A. nodosum* (A), *A. oblongoguttatum*, *A. parvum* (A), *A. pictum* (A), *A. pseudoconcolor*, *A. scutatum*

***Tamandua mexicana*** **(northern tamandua):** *A. auricularium*, *A. calcaratum* (A)

***Tamandua tetradactyla*** **(collared anteater):** *A. auricularium* (A), *A. cajennense* (AN), *A. calcaratum* (A), *A. nodosum* (A), *A. oblongoguttatum*, *A. parvum* (A), *A. pictum* (A), *A. rotundatum*

ORDER PRIMATES

Family Atelidae (howler, spider and woolly monkeys)

***Alouatta caraya*** **(black howler):** *A. cajennense* (N)

***Alouatta seniculus*** **(red howler):** *A. cajennense* (A)

*Ateles paniscus* (red-faced black spider monkey): *A. cajennense*

Family Cebidae (New World monkeys)

*Cebus apella* (black-capped capuchin): *A. cajennense* (N), *A. ovale* (A), *A. parvum* (A)

*Cebus capucinus* (white-faced capuchin): *A. oblongoguttatum*

*Cebus olivaceus* (weeper capuchin): *Ha. juxtakochi* (A)

Family Cercopithecidae (Old World monkeys)

*Cercopithecus albogularis* (Syke's monkey): *I. schillingsi* (A)

*Cercopithecus cephus* (moustached monkey): *I. schillingsi* (A)

*Cercopithecus mitis* (blue monkey): *I. schillingsi*

*Cercopithecus mona* (mona monkey): *A. variegatum* (N)

*Cercopithecus petaurista* (lesser spot-nosed guenon): *R. senegalensis* (A)

*Chlorocebus aethiops* (green monkey): *A. hebraeum* (L), *A. pomposum*, *A. variegatum* (NL), *I. pilosus*, *R. appendiculatus* (ANL), *R. evertsi* (L), *R. turanicus* (A)

*Colobus guereza* (eastern black-and-white colobus): *I. schillingsi* (A)

*Colobus polykomos* (king colobus): *Ha. leachi* (A), *I. schillingsi* (ANL)

*Erythrocebus patas* (patas monkey): *A. variegatum* (NL), *Hy. marginatum* group (L)

*Macaca cyclopis* (Formosan rock macaque): *R. (B.) microplus*

*Macaca radiata* (bonnet macaque): *D. auratus* (NL)

*Papio cynocephalus* (yellow baboon): *R. evertsi* (A), *R. pulchellus* (A)

*Papio ursinus* (chacma baboon): *A. hebraeum* (L), *Hy. truncatum* (L), *R. appendiculatus* (ANL), *R. evertsi* (AL)

*Piliocolobus badius* (red colobus): *A. nuttalli* (N), *I. schillingsi* (NL)

*Semnopithecus entellus* (common langur): *D. auratus* (NL)

Family Galagidae (bushbabies)

*Galago senegalensis* (lesser bushbaby): *A. variegatum* (ANL)

*Galagoides zanzibaricus* (Zanzibar bushbaby): *I. schillingsi*

*Otolemur crassicaudatus* (greater bushbaby): *A. hebraeum* (NL), *A. variegatum* (N), *I. schillingsi* (NL), *R. appendiculatus* (AL)

Family Hominidae (human beings and great apes)

*Gorilla gorilla* (gorilla): *R. appendiculatus* (A)

*Homo sapiens* (human): <u>*A. cajennense*</u> (ANL), *A. calcaratum*, *A. coelebs* (AN), *A. dissimile* (A), *A. gemma* (A), *A. geoemydae* (AN), <u>*A. hebraeum*</u> (ANL), *A. incisum* (AN), *A. latum*, *A. lepidum* (AN), *A. longirostre* (AN), *A. marmoreum* (ANL), *A. moreliae*, *A. nuttalli* (ANL), <u>*A. oblongoguttatum*</u> (AN), <u>*A. ovale*</u> (A), <u>*A. parvum*</u> (A), *A. pomposum*, *A. rotundatum*, *A. sparsum* (A), <u>*A. testudinarium*</u> (ANL), *A. tholloni* (AN), <u>*A. triguttatum*</u> (ANL), <u>*A. variegatum*</u> (ANL), *B. hydrosauri*, <u>*D. auratus*</u> (ANL), *D. nitens* (AL), *D. nuttalli* (A), <u>*D. reticulatus*</u> (A), *D. rhinocerinus* (A), *Ha. elongata* (N), <u>*Ha. hystricis*</u> (AN), <u>*Ha. juxtakochi*</u> (AN), *Ha. leachi* (AL), <u>*Ha. longicornis*</u> (ANL), <u>*Ha. punctata*</u> (A), *Hy. aegyptium* (AN), *Hy. albiparmatum* (A), <u>*Hy. anatolicum* spp.</u> (ANL), <u>*Hy. detritum*</u> (A), *Hy. dromedarii* (ANL), *Hy. lusitanicum* (A), <u>*Hy. marginantum*</u> group (ANL), <u>*Hy. truncatum*</u> (A), <u>*I. hexagonus*</u> (ANL), *I. luciae*, *I. pilosus* (AN), <u>*I. ricinus*</u> (ANL), <u>*I. schillingsi*</u> (A), <u>*R. appendiculatus*</u> (ANL), <u>*R. bursa*</u> (A), *R. compositus* (A), *R. evertsi* (ANL), *R. muehlensi* (A), <u>*R. pulchellus*</u> (ANL), *R. senegalensis* (A), *R. simus* (A), *R. sulcatus*, <u>*R. turanicus*</u> (A), *R. (B.) annulatus* (A), *R. (B.) decoloratus* (AL), <u>*R. (B.) microplus*</u> (AL)

Family Indriidae (sifakas, avahis and indris)

*Propithecus verreauxi* (Verreaux's sifaka): *Ha. hoodi*

ORDER PROBOSCIDEA (ELEPHANTS)

Family Elephantidae

*Elephas maximus* (Asian elephant): *A. testudinarium* (A)

*Loxodonta africana* (African elephant): *A. gemma* (A), *A. hebraeum*, *A. pomposum*, *A. sparsum* (A), *A. splendidum* (A), <u>*A. tholloni*</u> (ANL), *A. variegatum* (AN), *D. rhinocerinus* (A), *Hy. marginatum* group, *Hy. truncatum* (A), *R. appendiculatus* (A), *R. evertsi*

(A), *R. kochi* (A), *R. muehlensi* (A), *R. pulchellus* (A), *R. senegalensis* (A), *R. simus* (A)

ORDER RODENTIA (RODENTS)

Family Anomaluridae (scaly-tailed squirrels)

***Anomalurus derbianus*** (**Lord Derby's flying squirrel**): *A. exornatum*

***Zenkerella insignis*** (**flightless scaly-tailed squirrel**): *A. variegatum*

Family Bathyergidae (African mole-rats)

***Bathyergus suillus*** (**Cape dune mole-rat**): *Ha. muhsamae* (A)

Family Castoridae (beavers)

***Castor fiber*** (**Eurasian beaver**): *I. hexagonus* (ANL), *I. ricinus*

Family Caviidae (cavies and Patagonian hares)

***Cavia aperea*** (**Brazilian guinea-pig**): *A. cajennense*

***Dolichotis salinicola*** (**salt-desert cavy**): *A. auricularium* (AN), *A. parvum* (N)

***Galea musteloides*** (**common yellow-toothed cavy**): *A. parvum* (NL)

***Hydrochoerus hydrochaeris*** (**capybara**): *A. auricularium* (A), <u>*A. cajennense*</u> (ANL), *A. coelebs* (A), <u>*A. dissimile*</u>, *A. incisum* (A), *A. oblongoguttatum* (A), *A. parvum* (A), *A. rotundatum*, *D. nitens*, *R. (B.) annulatus* (A), *R. (B.) microplus* (A)

***Kerodon rupestris*** (**rock cavy**): *A. parvum* (A)

Family Chinchillidae (viscachas and chinchillas)

***Lagostomus maximus*** (**plains viscacha**): *A. auricularium* (AN), *A. cajennense* (N), *A. parvum* (A)

Family Cricetidae (voles, hamsters and lemmings)

***Allocricetulus curtatus*** (**Mongolian hamster**): *D. nuttalli*

***Allocricetulus eversmanni*** (**Eversmann's hamster**): *I. ricinus*

***Altocola argentatus*** (**silver mountain vole**): *D. nuttalli*

***Arvicola amphibius*** (**European water vole**): *D. reticulatus* (NL), *Ha. punctata*, *I. ricinus* (L), *R. turanicus*

***Blanfordimys afghanus*** (**Afghan vole**): *R. turanicus*

***Chionomys nivalis*** (**European snow vole**): *I. ricinus* (NL), *R. turanicus*

***Cricetulus barabensis*** (**striped dwarf hamster**): *D. nuttalli*

***Cricetulus longicaudatus*** (**lesser long-tailed hamster**): *D. nuttalli*

***Cricetulus migratorius*** (**grey dwarf hamster**): *Ha. punctata*, *I. ricinus*, *R. turanicus*

***Cricetus cricetus*** (**black-bellied hamster**): *D. nuttalli*, *Hy. aegyptium*, *I. hexagonus*, *I. ricinus* (ANL), *R. bursa*, *R. turanicus*

***Ellobius talpinus*** (**northern mole-vole**): *Hy. anatolicum* spp.

***Holochilus brasiliensis*** (**web-footed marsh rat**): *A. ovale* (N)

***Holochilus sciureus*** (**marsh rat**): *A. ovale*

***Lagurus lagurus*** (**steppe lemming**): *I. ricinus*

***Lasiopodomys brandtii*** (**Brandt's vole**): *D. nuttalli*

***Lemmus lemmus*** (**Norway lemming**): *I. ricinus*

***Lophiomys imhausi*** (**crested rat**): *I. schillingsi* (A)

***Mesocricetus auratus*** (**golden hamster**): *I. ricinus*

***Microtus agrestis*** (**field vole**): *D. reticulatus* (NL), <u>*I. ricinus*</u> (NL)

***Microtus arvalis*** (**common vole**): *D. nuttalli*, *D. reticulatus* (NL), *Ha. punctata* (N), <u>*I. ricinus*</u> (NL), *R. turanicus*

***Microtus duodecimcostatus*** (**Mediterranean pine vole**): *R. turanicus*

***Microtus fortis*** (**reed vole**): *D. nuttalli*

***Microtus gregalis*** (**narrow-headed vole**): *D. nuttalli* (L), *D. reticulatus* (NL)

***Microtus guentheri*** (**Guenther's vole**): *R. turanicus*

***Microtus majori*** (**Major's pine vole**): *I. ricinus*

***Microtus multiplex*** (**alpine pine vole**): *I. ricinus*

***Microtus oeconomus*** (**root vole**): *D. nuttalli*, *D. reticulatus* (NL), *I. ricinus* (L)

***Microtus socialis*** (**social vole**): *Hy. dromedarii*, *I. ricinus*, *R. bursa*, *R. turanicus*

***Microtus subterraneus*** (**European pine vole**): *D. reticulatus* (NL), *I. ricinus* (L)

***Myodes glareolus*** (**bank vole**): *D. reticulatus* (NL), *Ha. punctata* (NL), <u>*I. ricinus*</u> (NL)

***Myodes rufocanus*** (**grey red-backed vole**): *D. nuttalli*, *I. ricinus*

***Myodes rutilus*** (**northern red-backed vole**): *D. nuttalli*

*Necromys lasiurus* (**hairy-tailed bolo mouse**): *A. cajennense*

*Neodon sikimensis* (**Sikkim vole**): *R. (B.) microplus*

*Oecomys bicolor* (**bicolored arboreal rice rat**): *I. luciae* (L)

*Oecomys concolor* (**unicolored arboreal rice rat**): *I. luciae* (NL)

*Oligoryzomys eliurus* (**Brazilian pygmy rice rat**): *A. cajennense*

*Oligoryzomys fulvescens* (**fulvous pygmy rice rat**): *I. luciae* (L)

*Oligoryzomys microtis* (**small-eared pygmy rice rat**): *I. luciae* (N)

*Ondatra zibethicus* (**muskrat**): *I. ricinus*

*Oryzomys albigularis* (**Tome's rice rat**): *I. luciae* (L)

*Oryzomys megacephalus* (**large-headed rice rat**): *A. humerale*, *A. longirostre* (N), *A. ovale* (N), *I. luciae* (NL)

*Oryzomys perenensis* (**western Amazonian oryzomys**): *I. luciae* (N)

*Oryzomys talamancae* (**Talamancan rice rat**): *A. ovale* (A)

*Oryzomys xantheolus* (**yellowish rice rat**): *I. luciae* (A)

*Oryzomys yunganus* (**Yungas rice rat**): *I. luciae* (N)

*Oxymycterus roberti* (**Robert's hocicudo**): *A. cajennense*

*Peromyscus gossypinus* (**cotton mouse**): *A. dissimile* (L)

*Phodopus roborovskii* (**desert hamster**): *D. nuttalli*

*Phodopus sungorus* (**striped hairy-footed hamster**): *D. nuttalli*

*Sigmodon alstoni* (**Alston's cotton rat**): *A. auricularium*

*Sigmodon hispidus* (**hispid cotton rat**): *A. auricularium* (AN), *A. cajennense* (A), *A. ovale* (A), *A. parvum* (ANL), *I. luciae* (NL)

*Zygodontomys brevicauda* (**short-tailed cane mouse**): *A. ovale* (AN), *I. luciae* (N)

Family Cuniculidae (pacas)

*Cuniculus paca* (**paca**): *A. cajennense*, *A. coelebs* (A), *A. nodosum*, *A. oblongoguttatum* (A), *A. ovale* (A), *D. nitens* (AN), *I. luciae*

Family Dasyproctidae (agoutis)

*Dasyprocta azarae* (**Azara's agouti**): *A. cajennense*

*Dasyprocta fuliginosa* (**black agouti**): *Ha. juxtakochi* (NL)

*Dasyprocta leporina* (**Brazilian agouti**): *A. cajennense* (A), *A. oblongoguttatum*, *A. scutatum*, *Ha. juxtakochi* (NL)

*Dasyprocta punctata* (**Central American agouti**): *A. oblongoguttatum* (A)

*Dasyprocta variegata* (**brown agouti**): *A. cajennense*

Family Dipodidae (jerboas)

*Allactaga elater* (**small five-toed jerboa**): *Hy. anatolicum* spp.

*Allactaga severtzovi* (**Severtzov's jerboa**): *Hy. dromedarii* (A)

*Allactaga sibirica* (**Mongolian five-toed jerboa**): *D. nuttalli*

*Allactaga williamsi* (**Williams' jerboa**): *I. ricinus*

*Dipus sagitta* (**northern three-toed jerboa**): *D. nuttalli*

*Jaculus jaculus* (**lesser Egyptian jerboa**): *Hy. anatolicum* spp. (NL), *Hy. dromedarii* (N), *I. ricinus*

*Sicista betulina* (**northern birch mouse**): *D. reticulatus* (NL), *I. ricinus*

*Sicista napaea* (**Altai birch mouse**): *I. ricinus*

Family Echimyidae (spiny rats)

*Proechimys canicollis* (**Colombian spiny rat**): *A. ovale* (NL)

*Proechimys guyannensis* (**Cayenne spiny rat**): *A. humerale*, *A. longirostre* (N), *A. ovale* (NL), *I. luciae*

*Proechimys semispinosus* (**Tome's spiny rat**): *A. ovale* (AN)

Family Erethizontidae (New World porcupines)

*Chaetomys subspinosus* (**bristle-spined rat**): *A. longirostre* (A)

*Coendou bicolor* (**bicolor-spined porcupine**): *A. longirostre* (A)

*Coendou prehensilis* (**Brazilian tree porcupine**): *A. geayi* (A), *A. longirostre* (A)

*Coendou rothschildi* (**Rothschild's porcupine**): *A. longirostre* (A), *Ha. juxtakochi* (N)

*Sphiggurus insidiosus* (**Bahia hairy dwarf porcupine**): *A. longirostre*

*Sphiggurus mexicanus* (**Mexican hairy dwarf porcupine**): *A. auricularium* (N), *A. longirostre* (A)

*Sphiggurus spinosus* (Paraguay hairy dwarf porcupine): *A. geayi* (A), *A. longirostre* (AN)

*Sphiggurus villosus* (orange-spined hairy dwarf porcupine): *A. cajennense* (A), *A. geayi* (A), *A. longirostre* (A), *A. ovale* (A)

Family Gliridae (dormice)

*Dryomys nitedula* (forest dormouse): *Ha. punctata* (N), *I. ricinus*

*Eliomys melanurus* (Asian garden dormouse): *R. turanicus*

*Eliomys quercinus* (garden dormouse): *Ha. punctata* (L), *Hy. lusitanicum*, *I. hexagonus* (L), *I. ricinus* (NL), *R. bursa*, *R. turanicus*

*Glis glis* (fat dormouse): *I. ricinus* (NL), *R. turanicus*

*Muscardinus avellanarius* (hazel dormouse): *D. reticulatus* (L), *I. ricinus*

Family Heteromyidae (pocket mice, kangaroo rats and kangaroo mice)

*Heteromys anomalus* (Trinidad spiny pocket mouse): *A. ovale* (N)

Family Hystricidae (Old World porcupines)

*Atherurus africanus* (African brush-tailed porcupine): *A. nuttalli* (NL), *A. sparsum*, *Ha. hoodi* (A), *R. senegalensis* (A)

*Hystrix africaeaustralis* (Cape porcupine): *A. latum*, *Hy. truncatum* (A), *R. appendiculatus* (A), *R. evertsi* (N), *R. kochi* (A), R. simus

*Hystrix brachyura* (East Asian porcupine): *A. testudinarium*, *Ha. hystricis* (A)

*Hystrix cristata* (crested porcupine): *A. nuttalli* (L), *A. sparsum*, *I. hexagonus* (L), *R. appendiculatus*, *R. bursa*, *R. evertsi* (A), *R. senegalensis* (A)

*Hystrix indica* (Indian crested porcupine): *A. javanense* (NL), *A. testudinarium* (N), *D. auratus* (NL), *R. turanicus*, *R. (B.) microplus* (A)

*Hystrix pumila* (Palawan porcupine): *A. testudinarium* (N)

Family Muridae (rats, mice and gerbils)

*Acomys cahirinus* (Cairo spiny mouse): *Hy. anatolicum* spp. (N)

*Acomys dimidiatus* (eastern spiny mouse): *Hy. dromedarii* (N), *Hy. marginatum* group (N)

*Acomys russatus* (golden spiny mouse): *Hy. anatolicum* spp. (N)

*Aethomys chrysophilus* (red rock rat): *D. rhinocerinus* (L), *R. evertsi*, *R. simus* (NL), *R. (B.) decoloratus* (L)

*Apodemus agrarius* (striped field mouse): *D. nuttalli*, *D. reticulatus* (NL), *Ha. punctata*, I. ricinus (ANL)

*Apodemus flavicollis* (yellow-necked field mouse): *D. nuttalli*, *D. reticulatus* (NL), *Ha. punctata* (NL), I. ricinus (N), *R. turanicus* (NL)

*Apodemus mystacinus* (broad-toothed field mouse): *R. turanicus*

*Apodemus sylvaticus* (wood mouse): *D. reticulatus* (NL), *Ha. punctata* (NL), *Hy. aegyptium* (ANL), *Hy. anatolicum* spp. (L), *Hy. marginatum* group, I. ricinus (NL), *R. bursa*, *R. turanicus* (AL)

*Apodemus uralensis* (pygmy field mouse): *I. ricinus* (NL)

*Arvicanthis abyssinicus* (Abyssinian grass rat): *R. evertsi* (L)

*Arvicanthis niloticus* (African grass rat): *A. variegatum* (NL), *Ha. muhsamae* (A), *Hy. impressum* (L), *Hy. marginatum* group (L), *Hy. truncatum* (NL), *R. compositus* (N), *R. sulcatus* (N)

*Bandicota bengalensis* (lesser bandicoot-rat): *I. ricinus* (L), *R. turanicus* (NL), *R. (B.) microplus* (N)

*Bandicota indica* (greater bandicoot-rat): *A. varanense*, *Ha. hystricis*

*Berylmys bowersi* (Bowers' white-toothed rat): *A. testudinarium* (L), *D. auratus* (NL)

*Bunomys fratrorum* (fraternal hill rat): *Ha. hystricis* (N)

*Cremnomys cutchicus* (cutch rat): *Hy. dromedarii*

*Desmodillus auricularis* (Cape short-eared gerbil): *Ha. muhsamae*, *Hy. marginatum* group (L)

*Dipodillus campestris* (four-spotted gerbil): *Hy. dromedarii* (N)

*Dipodillus dasyurus* (Wagner's gerbil): *Hy. anatolicum* spp.

*Gerbilliscus afra* (Cape gerbil): *Ha. muhsamae* (A), *Hy. marginatum* group (L), *Hy. truncatum* (N), *R. appendiculatus*

*Gerbilliscus brantsii* (highveld gerbil): *Hy. truncatum* (L)

*Gerbilliscus leucogaster* (bushveld gerbil): *D. rhinocerinus* (NL), *Hy. marginatum* group (N), *Hy. truncatum* (NL), *R. evertsi* (L), *R. simus* (NL)

*Gerbilliscus nigricaudus* (**black-tailed gerbil**): *R. pulchellus* (A)

*Gerbilliscus validus* (**northern savanna gerbil**): *R. simus* (L)

*Gerbillus cheesmani* (**Cheesman's gerbil**): *Hy. dromedarii* (L)

*Gerbillus gerbillus* (**lesser Egyptian gerbil**): *Hy. anatolicum* spp. (N), *Hy. dromedarii* (N), *Hy. marginatum* group (N)

*Gerbillus pyramidum* (**greater Egyptian gerbil**): *Hy. anatolicum* spp. (N), *Hy. dromedarii* (N)

*Lemniscomys griselda* (**Griselda's striped grass mouse**): *R. simus* (N)

*Lemniscomys rosalia* (**eastern single-striped grass mouse**): *R. simus* (NL)

*Lemniscomys striatus* (**common striped grass rat**): *Ha. muhsamae* (A), *R. appendiculatus* (N), *R. senegalensis* (AN)

*Leopoldamys edwardsi* (**Edward's long-tailed giant rat**): *Ha. longicornis*

*Leopoldamys sabanus* (**long-tailed giant rat**): *A. helvolum*, *D. auratus* (NL)

*Lophuromys aquilus* (**blackish brush-furred rat**): *R. kochi* (L)

*Lophuromys flavopunctatus* (**eastern brush-furred rat**): *A. variegatum* (N), *R. appendiculatus* (NL)

*Mastomys coucha* (**southern multimammate mouse**): *R. appendiculatus* (N), *R. pulchellus*, *R. simus* (L)

*Mastomys erythroleucus* (**Guinea multimammate mouse**): *A. variegatum* (L), *Hy. truncatum* (NL)

*Mastomys natalensis* (**Natal multimammate mouse**): *A. hebraeum* (L), *D. rhinocerinus* (N), *Ha. leachi* (AN), *Hy. truncatum* (L), *R. appendiculatus* (L), *R. simus* (NL)

*Maxomys hellwaldii* (**Hellwald's spiny rat**): *Ha. hystricis* (N)

*Maxomys musschenbroekii* (**Musschenbroek's spiny rat**): *Ha. hystricis* (NL)

*Maxomys rajah* (**brown spiny rat**): *D. auratus* (NL)

*Maxomys surifer* (**red spiny rat**): *D. auratus* (AN)

*Meriones crassus* (**little sand jird**): *Hy. anatolicum* spp. (NL), *Hy. dromedarii* (N), *Hy. marginatum* group (N)

*Meriones hurrianae* (**Indian desert jird**): *Hy. anatolicum* spp. (NL), *Hy. dromedarii* (NL), *Hy. marginatum* group (A), *R. turanicus* (A)

*Meriones libycus* (**Libyan jird**): *Hy. anatolicum* spp. (N), *Hy. detritum*, *Hy. dromedarii*, *R. turanicus*

*Meriones meridianus* (**mid-day jird**): *D. nuttalli*, *Hy. anatolicum* spp.

*Meriones persicus* (**Persian jird**): *R. turanicus*

*Meriones rex* (**king jird**): *Hy. anatolicum* spp. (N), *Hy. dromedarii* (N)

*Meriones shawi* (**Shaw's jird**): *Hy. anatolicum* spp. (ANL), *Hy. dromedarii* (N), *Hy. marginatum* group (N), *I. ricinus*

*Meriones tamariscinus* (**tamarisk jird**): *Ha. punctata*, *Hy. anatolicum* spp., *R. turanicus*

*Meriones tristrami* (**Tristram's jird**): *Hy. anatolicum* spp. (NL), *R. turanicus* (N)

*Meriones unguiculatus* (**Mongolian jird**): *D. nuttalli*

*Micaelamys namaquensis* (**Namaqua rock rat**): *Ha. muhsamae*, *Hy. marginatum* group (L), *Hy. truncatum* (L), *R. evertsi* (L), *R. simus* (NL)

*Micromys minutus* (**Eurasian harvest mouse**): *D. nuttalli*, *D. reticulatus* (NL), *I. ricinus*

*Millardia meltada* (**soft-furred field rat**): *Hy. dromedarii*, *R. turanicus* (A)

*Mus cervicolor* (**fawn-colored mouse**): *D. auratus*

*Mus minutoides* (**lesser pygmy mouse**): *Ha. muhsamae*, *R. appendiculatus* (L), *R. simus* (NL)

*Mus musculus* (**house mouse**): *A. variegatum* (L), *D. auratus* (L), *D. reticulatus* (L), *Ha. longicornis*, *Ha. punctata* (NL), *Hy. anatolicum* spp., *Hy. marginatum* group, *I. hexagonus*, *I. ricinus* (ANL), *R. turanicus* (NL)

*Mus platythrix* (**flat-haired mouse**): *R. turanicus*

*Mus triton* (**grey-bellied pygmy mouse**): *R. simus* (L)

*Myotomys unisulcatus* (**bush Karroo rat**): *A. hebraeum*, *A. marmoreum* (L), *Hy. truncatum* (L)

*Nesokia indica* (**short-tailed bandicoot-rat**): *Hy. anatolicum* spp., *Hy. dromedarii*, *I. ricinus* (N), *R. turanicus*

*Niviventer fulvescens* (**chestnut white-bellied rat**): *D. auratus* (N), *Ha. hystricis* (N)

*Niviventer niviventer* (**white-bellied rat**): *R. (B.) microplus*

*Otomys angionensis* (**Angoni groove-toothed rat**): *R. pulchellus*, *R. simus* (NL)

*Otomys irroratus* (**vlei rat**): *Ha. muhsamae*, *I. pilosus* (L), *R. appendiculatus* (L)

***Parotomys brantsii*** (**Brandt's Karroo rat**): *Hy. truncatum*

***Pelomys campanae*** (**bell groove-toothed swamp rat**): *R. simus* (N)

***Pelomys fallax*** (**common creek rat**): *R. appendiculatus* (L), *R. compositus* (N), *R. simus* (NL), *R. (B.) decoloratus* (A)

***Praomys jacksoni*** (**Jackson's forest rat**): *R. compositus* (L), *R. senegalensis*

***Praomys morio*** (**Cameroon soft-furred mouse**): *R. senegalensis*

***Psammomys obesus*** (**fat sand rat**): *Hy. anatolicum* spp. (ANL), *Hy. dromedarii* (N), *Hy. marginatum* group (N)

***Rattus andamanensis*** (**Sladen's rat**): *D. auratus* (N)

***Rattus exulans*** (**Polynesian rat**): *R. (B.) microplus*

***Rattus hoffmanni*** (**Hoffmann's rat**): *R. (B.) microplus*

***Rattus norvegicus*** (**brown rat**): *Ha. longicornis, Ha. punctata, I. hexagonus* (A), *I. ricinus* (ANL), *R. turanicus*

***Rattus pyctoris*** (**Turkestan rat**): *Ha. punctata, Hy. anatolicum* spp., *R. turanicus*

***Rattus rattus*** (**black rat**): *A. triguttatum* (L), *A. varanense* (A), *A. variegatum* (NL), *D. auratus* (ANL), *Ha. elongata* (A), *Ha. hystricis* (N), *Ha. longicornis, Hy. anatolicum* spp. (L), *Hy. dromedarii* (NL), *Hy. lusitanicum, I. hexagonus, I. ricinus* (ANL), *I. schillingsi, R. appendiculatus* (A), *R. pulchellus* (A), *R. simus* (L), *R. turanicus* (NL), *R. (B.) microplus*

***Rattus tiomanicus*** (**Malaysian field rat**): *A. helvolum*

***Rhabdomys pumilio*** (**four-striped grass mouse**): *A. hebraeum* (NL), *A. sylvaticum* (L), *Ha. muhsamae* (A), *Hy. marginatum* group (L), *Hy. truncatum* (NL), *R. appendiculatus* (NL), *R. capensis, R. simus* (NL)

***Rhombomys opimus*** (**great gerbil**): *Hy. anatolicum* spp., *Hy. detritum, Hy. dromedarii, R. turanicus*

***Tatera indica*** (**antelope rat**): *Hy. anatolicum* spp. (N), *Hy. dromedarii, I. ricinus* (N), *R. turanicus*

***Taterillus gracilis*** (**slender gerbil**): *A. variegatum* (NL)

***Thallomys paedulcus*** (**acacia rat**): *R. simus* (NL)

Family Myocastoridae (coypus)

***Myocastor coypus*** (**coypu**): *A. cajennense* (N), *I. hexagonus* (AN), *I. ricinus* (ANL)

Family Nesomyidae (pouched rats and fat mice)

***Cricetomys gambianus*** (**Gambian rat**): *A. variegatum, Ha. muhsamae* (A), *Hy. truncatum* (L), *R. senegalensis* (A)

***Saccostomus campestris*** (**Cape pouched rat**): *Ha. muhsamae* (A), *I. pilosus, R. appendiculatus* (L), *R. evertsi* (L), *R. simus* (L)

***Steatomys pratensis*** (**fat mouse**): *Ha. muhsamae* (A), *R. evertsi, R. sulcatus* (A)

Family Pedetidae (springhares)

***Pedetes capensis*** (**springhare**): *A. hebraeum, Ha. muhsamae* (A), *Hy. marginatum* group (ANL), *R. appendiculatus* (ANL), *R. evertsi* (NL), *R. sulcatus* (A)

Family Sciuridae (squirrels, chipmunks, marmots and prairie dogs)

***Callosciurus erythraeus*** (**Pallas' squirrel**): *A. crassipes, D. auratus* (N)

***Callosciurus quinquestriatus*** (**Anderson's squirrel**): *D. auratus* (AN)

***Dremomys lokriah*** (**orange-bellied Himalayan squirrel**): *D. auratus* (N)

***Dremomys rufigenis*** (**Asian red-cheeked squirrel**): *Ha. hystricis* (N)

***Funambulus palmarum*** (**Indian palm squirrel**): *R. turanicus* (N)

***Funambulus pennantii*** (**northern palm squirrel**): *D. auratus* (L)

***Funambulus tristriatus*** (**jungle palm squirrel**): *R. (B.) microplus* (N)

***Funisciurus congicus*** (**Congo rope squirrel**): *R. simus*

***Funisciurus pyrropus*** (**fire-footed rope squirrel**): *R. senegalensis* (AN)

***Hylopetes alboniger*** (**particolored flying squirrel**): *A. testudinarium*

***Lariscus insignis*** (**three-striped ground squirrel**): *A. testudinarium*

***Marmota bobak*** (**bobak marmot**): *I. hexagonus* (A)

***Marmota caudata*** (**long-tailed marmot**): *R. turanicus*

***Marmota monax*** (**woodchuck**): *I. ricinus* (A)

***Marmota sibirica*** (**Siberian marmot**): *D. nuttalli, I. ricinus* (A)

*Menetes berdmorei* (**Indochinese ground squirrel**): *D. auratus* (N)

*Paraxerus cepapi* (**Smith's bush squirrel**): *Ha. muhsamae* (AN), *R. appendiculatus*, *R. evertsi* (L), *R. muehlensi* (A), *R. simus* (NL)

*Paraxerus ochraceus* (**ochre bush squirrel**): *R. evertsi* (A), *R. pulchellus* (A)

*Paraxerus palliatus* (**red bush squirrel**): *R. appendiculatus*

*Petaurista petaurista* (**red giant flying squirrel**): *A. testudinarium*, *D. auratus* (NL), *R. (B.) microplus* (A)

*Ratufa bicolor* (**black giant squirrel**): *A. testudinarium*, *D. auratus* (N)

*Ratufa indica* (**Indian giant squirrel**): *D. auratus* (N)

*Sciurus aestuans* (**Guianan squirrel**): *Ha. juxtakochi* (N)

*Sciurus anomalus* (**Caucasian squirrel**): *I. ricinus*

*Sciurus carolinensis* (**grey squirrel**): *I. hexagonus*, *I. ricinus* (NL)

*Sciurus granatensis* (**red-tailed squirrel**): *A. longirostre* (NL)

*Sciurus igniventris* (**northern Amazon red squirrel**): *Ha. juxtakochi* (L)

*Sciurus vulgaris* (**Eurasian red squirrel**): *Ha. punctata* (NL), *I. hexagonus*, <u>*I. ricinus*</u> (ANL), *R. bursa*

*Spermophilopsis leptodactylus* (**long-clawed ground squirrel**): *Hy. anatolicum* spp., *Hy. dromedarii*, *R. turanicus*

*Spermophilus citellus* (**European ground squirrel**): *Ha. punctata*, *I. ricinus*

*Spermophilus dauricus* (**daurian ground squirrel**): *D. nuttalli*

*Spermophilus fulvus* (**yellow ground squirrel**): *Hy. anatolicum* spp., *R. turanicus*

*Spermophilus pygmaeus* (**little ground squirrel**): *Ha. punctata*, *I. ricinus*

*Spermophilus relictus* (**Tien Shan ground squirrel**): *Ha. punctata*

*Spermophilus suslicus* (**speckled ground squirrel**): *Ha. punctata*, *I. ricinus*

*Spermophilus undulatus* (**long-tailed ground squirrel**): *D. nuttalli*

*Tamiops swinhoei* (**Swinhoe's striped squirrel**): *A. varanense*

*Xerus erythropus* (**striped ground squirrel**): *A. ex-ornatum*, *A. nuttalli* (NL), *A. variegatum* (ANL), *Ha. hoodi* (A), *Ha. muhsamae* (A), *Hy. marginatum* group (N), *Hy. truncatum* (L)

*Xerus inauris* (**Cape ground squirrel**): *Ha. muhsamae* (A), *R. appendiculatus*

*Xerus rutilus* (**unstriped ground squirrel**): *R. evertsi*

Family Spalacidae (mole-rats)

*Spalax leucodon* (**lesser mole-rat**): *I. ricinus*

Family Thryonomidae (cane rats)

*Thryonomys swinderianus* (**greater cane rat**): *A. hebraeum*, *A. nuttalli* (N), *A. pomposum*, *A. splendidum* (N), *A. variegatum* (ANL), *R. appendiculatus* (ANL), *R. evertsi*, *R. senegalensis* (A), *R. simus*

ORDER SCANDENTIA (TREE SHREWS)

Family Tupaiidae

*Tupaia belangeri* (**northern treeshrew**): *A. testudinarium* (L)

*Tupaia glis* (**Malaysian treeshrew**): *A. crassipes*, *A. testudinarium* (N), *D. auratus* (ANL), *Ha. hystricis* (NL)

ORDER SORICOMORPHA[c]

Family Soricidae (shrews)

*Crocidura flavescens* (**greater musk shrew**): *Ha. muhsamae*

*Crocidura hirta* (**lesser red musk shrew**): *R. simus* (N)

*Crocidura leucodon* (**bicolored shrew**): *I. ricinus* (NL)

*Crocidura russula* (**greater white-toothed shrew**): *I. ricinus*, *R. turanicus* (N)

*Crocidura suaveolens* (**lesser shrew**): *D. reticulatus* (NL), *I. ricinus*

*Crocidura zarudnyi* (**grey shrew**): *R. turanicus*

*Neomys anomalus* (**southern water shrew**): *I. ricinus* (L)

*Neomys fodiens* (**Eurasian water shrew**): *D. reticulatus* (NL), *I. ricinus* (NL)

*Sorex alpinus* (**alpine shrew**): *I. ricinus* (L)

*Sorex araneus* (**common shrew**): *D. reticulatus* (ANL), *Ha. punctata* (NL), <u>*I. ricinus*</u> (NL)

*Sorex caecutiens* (**Laxmann's shrew**): *I. ricinus*

*Sorex coronatus* (**crowned shrew**): *I. ricinus* (L)

*Sorex minutus* (**Eurasian pygmy shrew**): *D. reticulatus* (NL), *I. ricinus* (NL)

*Suncus murinus* (**Asian house shrew**): *Hy. dromedarii, I. ricinus* (L), *R. turanicus* (NL), *R. (B.) microplus* (A)

Family Talpidae (moles)

*Euroscaptor micrura* (**Himalayan mole**): *I. ricinus*

*Talpa europaea* (**European mole**): *D. reticulatus* (NL), *Ha. punctata* (L), *I. hexagonus* (A), I. ricinus (NL)

*Talpa romana* (**Roman mole**): *I. ricinus*

ORDER TUBULIDENTATA (ANT BEARS)

Family Orycteropodidae

*Orycteropus afer* (**aardvark**): *A. lepidum, A. sparsum, A. variegatum* (AN), *Ha. muhsamae* (A), *Hy. impressum, Hy. truncatum* (A), *R. pulchellus* (A), *R. senegalensis* (A), *R. simus* (A), *R. turanicus* (A)

## Class Reptilia (reptiles)

ORDER CROCODYLIA (CROCODILIANS)

Family Alligatoridae (alligators and caimans)

*Caiman crocodilus* (**brown caiman**): *A. dissimile* (A), *A. humerale* (A)

Family Crocodylidae (crocodiles)

*Crocodylus moreletii* (**Belize crocodile**): *A. dissimile* (A)

*Crocodylus niloticus* (**Nile crocodile**): *A. flavomaculatum*

*Osteolaemus tetrapsis* (**African dwarf crocodile**): *A. flavomaculatum*

ORDER SAURIA (LIZARDS)

Family Agamidae (agamid lizards)

*Agama agama* (**common agama**): *A nuttalli* (N)

*Agama caudospinosa* (**Elmenteita rock agama**): *A. sparsum* (A)

*Agama hispida* (**common spiny agama**): *A. sylvaticum* (L)

*Amphibolurus diemensis* (**mountain dragon**): *B. hydrosauri* (L)

*Calotes versicolor* (**variable agama**): *A. varanense* (A)

*Chlamydosaurus kingii* (**frilled dragon**): *A. fimbriatum* (A)

*Gonocephalus sophiae* (**Negros forest dragon**): *A. helvolum*

*Hydrosaurus amboinensis* (**fin-tailed lizard**): *A. varanense*

*Hydrosaurus pustulatus* (**crested lizard**): *A. helvolum* (A)

*Laudakia atricollis* (**blue-headed tree iguana**): *Hy. truncatum*

*Laudakia stellio* (**rough-tailed rock agama**): *Hy. aegyptium* (AN)

*Pogona barbata* (**bearded dragon**): *A. moreliae, B. hydrosauri*

Family Anguidae (glass and alligator lizards)

*Ophisaurus apodus* (**armored glass lizard**): *Ha. punctata*

Family Chamaeleonidae (chameleons)

*Chamaeleo africanus* (**African chameleon**): *A. variegatum*

*Chamaeleo dilepis* (**flap-necked chameleon**): *A. dissimile, Hy. truncatum*

*Chamaeleo gracilis* (**graceful chameleon**): *A. tholloni* (N)

*Chamaeleo quilensis* (**Bocage's chameleon**): *Hy. truncatum* (N)

*Chameoleo senegalensis* (**Senegal chameleon**): *A. nuttalli* (N)

*Furcifer oustaleti* (**Malagasy giant chameleon**): *A. variegatum* (L)

*Furcifer pardalis* (**panther chameleon**): *A. variegatum* (N)

Family Gerrhosauridae (plated lizards)

*Gerrhosaurus major* (**great plated lizard**): *A. exornatum* (N), *A. latum*

*Gerrhosaurus validus* (**giant plated lizard**): *A. exornatum* (A), *A. hebraeum* (A), *A. marmoreum* (N)

Family Iguanidae (iguanas)

*Basiliscus plumifrons* (**green basilisk**): *A. dissimile*

*Corytophanes cristatus* (smooth helmeted iguana): *A. dissimile, A. scutatum* (A)

*Ctenosaura acanthura* (northeastern spiny-tailed iguana): *A. dissimile, A. scutatum, A. varium* (A)

*Ctenosaura pectinata* (Mexican spiny-tailed iguana): *A. dissimile* (A), *A. scutatum*

*Ctenosaura quinquecarinata* (Oaxacan spiny-tailed iguana): *A. dissimile*

*Ctenosaura similis* (black iguana): *A. cajennense, A. dissimile, A. scutatum* (AN)

*Cyclura carinata* (Bahamas rock iguana): *A. albopictum*

*Cyclura cornuta* (rhinoceros iguana): *A. albopictum* (A), *A. dissimile*

*Cyclura cychlura* (Bahamas iguana): *A. dissimile* (ANL)

*Cyclura nubila* (Cayman Islands ground iguana): <u>*A. albopictum*</u> (A), *A. rotundatum*

*Iguana iguana* (green iguana): *A. cajennense,* <u>*A. dissimile*</u> (A), *A. sabanerae* (A), <u>*A. scutatum*</u> (A)

*Leiocephalus carinatus* (northern curly-tailed lizard): *A. albopictum, A. dissimile* (N)

*Plica plica* (iguana): *A. humerale* (N)

*Plica umbra* (iguana): *A. humerale* (N)

*Sceloporus malachiticus* (green spiny lizard): *A. dissimile* (N)

*Sceloporus undulatus* (fence lizard): *A. dissimile* (A)

Family Lacertidae (lacertid lizards)

*Acanthodactylus boskianus* (Bosk's fringe-fingered lizard): *Hy. anatolicum* spp. (N), *Hy. dromedarii* (AN), *Hy. marginatum* group (N)

*Darevskia saxicola* (rock lizard): *Ha. punctata, I. ricinus*

*Eremias arguta* (Steppe racerunner): *Ha. punctata*

*Eremias velox* (rapid racerunner): *Ha. punctata*

*Lacerta agilis* (sand lizard): *Ha. punctata* (L), <u>*I. ricinus*</u> (ANL), *R. bursa*

*Lacerta strigata* (Caspian green lizard): *Ha. punctata, I. ricinus, R. turanicus*

*Lacerta viridis* (green lizard): *Ha. punctata* (L), *Hy. detritum* (NL), <u>*I. ricinus*</u> (NL)

*Lacerta vivipara* (viviparous lizard): <u>*I. ricinus*</u> (NL)

*Meroles knoxii* (Knox's desert lizard): *A. sylvaticum* (L)

*Ophisops elegans* (European snake-eyed lizard): *Ha. punctata*

*Podarcis dugesii* (Madeiran wall lizard): *Ha. punctata, I. ricinus*

*Podarcis hispanica* (Iberian wall lizard): *I. ricinus*

*Podarcis muralis* (wall lizard): *I. ricinus* (NL)

*Podarcis sucula* (ruin lizard): *I. ricinus*

*Podarcis taurica* (meadow lizard): *I. ricinus*

*Psammodromus algirus* (Algerian sandracer): *I. ricinus* (NL)

*Timon lepida* (eyed lizard): *I. ricinus*

*Timon pater* (Lataste's lizard): *I. ricinus*

Family Scincidae (skinks)

*Acontias meleagris* (Cape legless skink): *A. latum*

*Acontias plumbeus* (giant lance skink): *A. exornatum* (A), *A. latum*

*Corucia zebrata* (prehensile-tailed skink): *A. trimaculatum* (AN)

*Dasia grisea* (grey dasia): *A. helvolum* (N)

*Egernia striolata* (tree skink): *A. moreliae*

*Egernia whitii* (White's skink): *B. hydrosauri*

*Mabuya multifasciata* (Indonesian brown mabuya): *A. helvolum* (ANL)

*Mochlus guineensis* (Guinea writhing skink): *A. latum* (AN)

*Niveoscincus metallicus* (metallic skink): *B. hydrosauri*

*Scincus mitranus* (eastern skink): *A. latum* (A)

*Sphenomorphus cumingi* (Cuming's eared skink): *A. helvolum* (N)

*Tiliqua adelaidensis* (Adelaide pygmy blue-tongue skink): *B. hydrosauri*

*Tiliqua nigrolutea* (blotched blue-tongued skink): *A. moreliae,* <u>*B. hydrosauri*</u> (A)

*Tiliqua occipitalis* (western blue-tongued skink): *B. hydrosauri*

*Tiliqua scincoides* (eastern blue-tongued skink): *A. moreliae* (AN), <u>*B. hydrosauri*</u> (A)

*Trachydosaurus rugosus* (shingleback skink): *A. moreliae, A. triguttatum* (N), <u>*B. hydrosauri*</u> (ANL)

*Trachylepis striata* (striped skink): *A. latum* (L)

*Trachylepis varia* (variable skink): *A. latum* (L)

*Tropidophorus grayi* (Gray's keeled skink): *A. helvolum*

Family Teiidae (macroteiids)

*Ameiva amevia* (giant ameiva): *A. dissimile, A. scutatum*

*Kentropyx calcarata* (Spix's kentropyx): *A. humerale* (N)

*Tupinambis teguixin* (banded tegu): *A. dissimile*

Family Varanidae (monitors)

*Varanus acanthurus* (goanna monitor): *A. fimbriatum*

*Varanus albigularis* (white-throated monitor): A. exornatum (ANL), *A. flavomaculatum* (AN), *A. hebraeum* (AN), *A. latum* (A), *A. marmoreum* (AN), *A. nuttalli* (A)

*Varanus bengalensis* (Bengal monitor): *A. crassipes* (A), *A. flavomaculatum* (A), *A. helvolum* (A), *A. trimaculatum* (A), *A. varanense* (AN)

*Varanus doreanus* (blue-tailed monitor): *A. trimaculatum* (AN), *A. varanense* (N)

*Varanus dumerilii* (Dumeril's monitor): *A. fimbriatum* (A), *A. fuscolineatum, A. geoemydae* (N), *A. helvolum* (AN), *A. varanense* (AN)

*Varanus exanthematicus* (African savanna monitor): A. exornatum (ANL), *A. falsomarmoreum* (A), A. flavomaculatum (ANL), *A. latum* (A), *A. marmoreum* (AN), A. nuttalli (AN), *A. sparsum* (A), *A. variegatum* (N)

*Varanus giganteus* (perentie): *A. fimbriatum*

*Varanus gouldii* (Gould's monitor): A. fimbriatum (AN), *A. trimaculatum* (A), B. hydrosauri (A)

*Varanus griseus* (grey monitor): A. crassipes (A), *A. helvolum, A. varanense*

*Varanus indicus* (mangrove monitor): *A. fimbriatum,* A. trimaculatum (ANL), *A. varanense*

*Varanus jobiensis* (peachthroat monitor): *A. trimaculatum*

*Varanus komodoensis* (Komodo dragon): *A. helvolum,* A. komodoense (ANL)

*Varanus macraei* (blue tree monitor): *A. trimaculatum* (NL)

*Varanus nebulosus* (clouded monitor): *A. helvolum, A. varanense*

*Varanus niloticus* (Nile monitor): A. exornatum (ANL), *A. falsomarmoreum,* A. flavomaculatum (ANL), *A. hebraeum, A. latum* (A), *A. marmoreum* (AN), *A. nuttalli* (AN), *A. sparsum* (AN), *A. variegatum* (N)

*Varanus olivaceus* (Gray's monitor): *A. exornatum* (AN), *A. fimbriatum, A. helvolum* (AN), *A. trimaculatum* (A)

*Varanus panoptes* (argus monitor): *A. fimbriatum* (A), *A. moreliae* (A)

*Varanus prasinus* (emerald monitor): *A. trimaculatum*

*Varanus rosenbergi* (Rosenberg's monitor): *A. fimbriatum*

*Varanus rudicollis* (harlequin monitor): *A. fimbriatum* (A), *A. helvolum* (AN), *A. varanense* (AN)

*Varanus salvadorii* (crocodile monitor): *A. fuscolineatum*

*Varanus salvator* (water monitor): *A. crassipes* (ANL), *A. exornatum* (A), A. fimbriatum (AN), *A. flavomaculatum* (A), *A. fuscolineatum, A. geoemydae* (AN), A. helvolum, *A. javanense* (A), *A. komodoense* (A), *A. kraneveldi* (A), *A. latum, A. testudinarium* (A), *A. trimaculatum* (A), A. varanense (AN)

*Varanus timorensis* (Timor tree monitor): *A. trimaculatum* (A)

*Varanus tristis* (freckled monitor): *A. fimbriatum, A. moreliae, A. trimaculatum*

*Varanus varius* (lace monitor): *A. fimbriatum* (A), A. moreliae (A), *A. trimaculatum* (A), *B. concolor, B. hydrosauri*

*Varanus yemenensis* (Yemen monitor): *A. flavomaculatum* (A), *A. sparsum* (AN)

ORDER SERPENTES (SNAKES)

Family Atractaspididae (mole vipers)

*Aparallactus modestus* (western forest centipede-eater): *A. latum* (A)

*Atractaspis aterrima* (slender burrowing asp): *A. latum* (A)

Family Boidae (boas)

*Boa constrictor* (boa constrictor): *A. argentinae* (A), A. dissimile (AN), *A. latum, A. nuttalli* (A), *A. quadricavum,* A. rotundatum (ANL), *A. scutatum, A. sparsum*

*Calabaria reinhardtii* (African burrowing python): *A. latum* (AN)

*Candoia carinata* (bevel-nosed boa): *A. fimbriatum* (A)

*Corallus caninus* (emerald tree boa): *A. rotundatum* (AN), *A. sabanerae* (A)

*Corallus hortulanus* (Amazon tree boa): *A. dissimile*

*Epicrates angulifer* (**Cuban boa**): *A. albopictum, A. quadricavum, D. nitens*

*Epicrates cenchria* (**rainbow boa**): <u>*A. dissimile*</u> (AN), *A. rotundatum*

*Epicrates striatus* (**Haitian boa**): *A. dissimile* (ANL), <u>*A. quadricavum*</u> (A)

*Epicrates subflavus* (**Jamaican boa**): *A. quadricavum*

*Eunectes murinus* (**green anaconda**): *A. dissimile, A. rotundatum*

*Eunectes notaeus* (**yellow anaconda**): *A. argentinae* (A), *A. dissimile* (N)

*Gongylophis colubrinus* (**East African sand boa**): *A. latum* (L)

*Gongylophis conicus* (**baby python**): *A. varanense* (AN)

Family Colubridae (harmless and rear-fanged snakes)

*Alsophis cantherigerus* (**Cuban racer**): *A. quadricavum*

*Alsophis portoricensis* (**Puerto Rican racer**): *A. quadricavum*

*Boiga blandingii* (**Blanding's tree snake**): *A. latum* (AN)

*Boiga dendrophila* (**mangrove cat snake**): *A. helvolum* (A), *A. varanense* (A)

*Boiga pulverulenta* (**powdered tree snake**): *A. latum* (AN)

*Chironius carinatus* (**sipo**): *A. dissimile*

*Chironius laevicollis* (**Brazilian sipo**): *A. rotundatum*

*Coluber constrictor* (**racer**): *A. dissimile, A. rotundatum* (A)

*Coluber gemonensis* (**Balkan racer**): *A. moreliae*

*Coluber ravergieri* (**mountain racer**): *I. ricinus*

*Contia tenuis* (**sharp-tailed snake**): *A. latum*

*Crotaphopeltis hotamboiea* (**herald snake**): <u>*A. latum*</u> (AN)

*Cyclagras gigas* (**Brazilian smooth snake**): *A. dissimile*

*Dasypeltis fasciata* (**western forest egg-eating snake**): *A. latum* (A)

*Dasypeltis medici* (**East African egg-eating snake**): *A. latum* (A)

*Dasypeltis scabra* (**common egg-eating snake**): *A. latum* (A), *A. marmoreum* (N)

*Dendrelaphis calligastra* (**northern bronzeback**): *A. moreliae*

*Dendrelaphis pictus* (**painted bronzeback**): *A. helvolum* (A)

*Dendrelaphis punctulata* (**common bronzeback**): *A. fimbriatum* (A)

*Dispholidus typus* (**boomslang**): *A. latum* (A), *A. marmoreum* (N)

*Dromophis praeornatus* (**African swamp snake**): *A. latum, A. variegatum* (N)

*Drymarchon corais* (**southern indigo snake**): *A. argentinae, A. cajennense, A. dissimile, A. scutatum*

*Elaphe flavolineata* (**black copper rat snake**): *A. helvolum* (AN)

*Elaphe guttata* (**corn snake**): *A. dissimile* (A)

*Elaphe obsoleta* (**North American rat snake**): *A. dissimile* (AN)

*Elaphe radiata* (**radiated rat snake**): *A. helvolum* (A)

*Gonyosoma oxycephalum* (**red-tailed rat snake**): *A. helvolum* (A)

*Grayia smythii* (**Smyth's water snake**): *A. latum* (AN)

*Hapsidophrys lineatus* (**black-lined green snake**): *A. latum* (A)

*Lampropeltis getula* (**common kingsnake**): *A. dissimile* (A)

*Lamprophis aurora* (**aurora house snake**): *A. latum* (A)

*Lamprophis fuliginosus* (**brown house snake**): <u>*A. latum*</u> (ANL)

*Lamprophis lineatus* (**African house snake**): *A. latum* (AL)

*Lamprophis olivaceus* (**olive house snake**): *A. latum* (A)

*Lamprophis virgatus* (**Hallowell's house snake**): *A. latum* (A)

*Leptodeira annulata* (**banded cat-eyed snake**): *A. dissimile* (N)

*Lycophidion capense* (**Cape wolf snake**): *A. latum*

*Macropisthodon flaviceps* (**orange-necked keelback**): *A. helvolum* (N)

*Mehelya capensis* (**Cape file snake**): *A. latum* (ANL)

*Mehelya crossi* (**African file snake**): *A. latum* (A)

*Mehelya nyassae* (**black file snake**): *A. latum* (N)

*Mehelya poensis* (**forest file snake**): *A. latum* (ANL), *A. nuttalli* (NL)

*Natriciteres olivacea* (**olive marsh snake**): *A. latum* (A)

*Natrix natrix* (**grass snake**): *A. argentinae*

*Oxybelis aeneus* (Mexican vine snake): *A. dissimile* (ANL)

*Oxybelis fulgidus* (green vine snake): *A. dissimile*

*Philodryas baroni* (Baron's green racer): *A. dissimile* (N)

*Philothamnus irregularis* (western green snake): *A. latum* (A)

*Pituophis melanoleucus* (pine gopher snake): *A. dissimile* (A)

*Psammophis elegans* (elegant sand racer): *A. latum* (ANL)

*Psammophis mossambicus* (hissing sand snake): *A. latum* (ANL)

*Psammophis phillipsi* (olive grass racer): *A. latum*

*Psammophis schokari* (desert sand snake): *A. latum* (A)

*Psammophis sibilans* (African beauty racer): <u>*A. latum*</u> (ANL)

*Psammophis subtaeniatus* (stripe-bellied sand racer): *A. latum* (AL)

*Psammophylax tritaeniatus* (southern striped skaapsteker): *A. latum* (AN)

*Pseudaspis cana* (mole snake): <u>*A. latum*</u> (A), *A. sylvaticum* (AN)

*Pseudoboa neuwiedii* (Neuwied's false boa): *A. dissimile*

*Pseustes poecilonotus* (puffing snake): *A. dissimile*

*Ptyas dipsas* (Sulawesi black racer): *A. helvolum* (A)

*Ptyas korros* (Chinese rat snake): *A. helvolum* (A)

*Ptyas mucosus* (common rat snake): *A. helvolum*, *A. varanense* (A)

*Rhamnophis aethiopissa* (large-eyed green tree snake): *A. latum* (A)

*Rhamphiophis oxyrhynchus* (western beaked snake): *A. latum* (ANL)

*Sinonatrix trianguligera* (triangle keelback): *A. helvolum* (A)

*Spalerosophis microlepis* (zebra snake): *A. latum* (A)

*Spilotes pullatus* (tropical rat snake): *A. dissimile*, *A. rotundatum*

*Thelotornis kirtlandii* (forest twig snake): *A. latum* (A)

*Waglerophis merremi* (Wagler's snake): *A. rotundatum*

*Xenodon neuwiedii* (Neuwied's false fer-de-lance): *A. rotundatum*

*Xenodon severus* (Amazonian false fer-de-lance): *A. dissimile*

*Zaocys fuscus* (white-bellied rat snake): *A. helvolum* (ANL)

Family Elapidae (cobras, kraits, coral snakes and sea snakes)

*Aspidelaps scutatus* (shieldnose cobra): *A. latum* (A)

*Austrelaps superbus* (Australian copperhead): *B. hydrosauri*

*Bungarus fasciatus* (banded krait): *A. crassipes* (A), *A. fuscolineatum*, *A. varanense* (AN)

*Dendroaspis angusticeps* (common mamba): <u>*A. latum*</u> (A)

*Dendroaspis jamesoni* (Jameson's mamba): *A. flavomaculatum*, *A. latum* (AN), *A. sparsum* (A)

*Dendroaspis polylepis* (black mamba): <u>*A. latum*</u> (ANL)

*Dendroaspis viridis* (green mamba): *A. flavomaculatum*, <u>*A. latum*</u> (ANL), *A. nuttalli* (N)

*Drysdalia coronoides* (white-lipped snake): *B. hydrosauri*

*Elapsoidea sundevalli* (Sundevall's garter snake): *A. latum* (AN)

*Hemachatus haemachatus* (ringneck spitting cobra): *A. exornatum*, *A. latum* (A), *A. marmoreum* (A)

*Micrurus surinamensis* (aquatic coral snake): *A. rotundatum*

*Naja annulifera* (snouted cobra): *A. latum* (ANL)

*Naja atra* (Chinese cobra): *A. varanense* (ANL)

*Naja haje* (Egyptian cobra): *A. exornatum*, *A. flavomaculatum* (A), <u>*A. latum*</u> (ANL)

*Naja melanoleuca* (forest cobra): *A. flavomaculatum* (A), <u>*A. latum*</u> (ANL)

*Naja mossambica* (Mozambique spitting cobra): *A. latum* (ANL)

*Naja naja* (Indian cobra): *A. crassipes*, *A. helvolum* (A), *A. testudinarium*, *A. varanense* (ANL), *R. (B.) microplus* (A)

*Naja nigricollis* (black-necked spitting cobra): *A. exornatum*, *A. flavomaculatum*, <u>*A. latum*</u> (ANL), *A. marmoreum*

*Naja nivea* (Cape cobra): *A. latum* (A)

*Notechis ater* (black tiger snake): *B. hydrosauri*

*Notechis scutatus* (mainland tiger snake): *A. fimbriatum* (A), *A. moreliae* (A), <u>*B. hydrosauri*</u> (AN)

*Ophiophagus hannah* (king cobra): *A. dissimile, A. helvolum* (A), *A. varanense* (A)

*Oxyuranus scutellatus* (taipan): *A. fimbriatum* (AN), *A. moreliae, B. hydrosauri*

*Pseudechis australis* (king brown snake): *A. moreliae*

*Pseudechis porphyriacus* (red-bellied black snake): *A. fimbriatum* (A), *A. moreliae* (L)

*Pseudohaje goldii* (black forest cobra): *A. exornatum, A. flavomaculatum, A. latum* (A)

*Pseudonaja affinis* (dugite): *A. fimbriatum* (A)

*Pseudonaja nuchalis* (western brown snake): *A. fimbriatum*

*Pseudonaja textilis* (eastern brown snake): *A. fimbriatum, A. moreliae, B. hydrosauri*

Family Pythonidae (pythons)

*Antaresia childreni* (Children's python): *A. fimbriatum*

*Aspidites melanocephalus* (black-headed python): *A. fimbriatum, A. moreliae*

*Aspidites ramsayi* (Ramsay's python): *A. fimbriatum*

*Leiopython albertisii* (white-lipped python): *A. varanense*

*Liasis fuscus* (water python): *A. fimbriatum* (A), *A. moreliae* (A)

*Liasis mackloti* (Macklot's python): *A. kraneveldi, A. varanense*

*Liasis olivaceus* (olive python): *A. fimbriatum*

*Morelia amethistina* (amethystine python): *A. fimbriatum* (A), *A. trimaculatum*

*Morelia spilota* (carpet python): *A. fimbriatum* (A), *A. moreliae, A. trimaculatum*

*Morelia viridis* (green tree python): *A. varanense* (A)

*Python curtus* (Sumatran short-tailed python): *A. varanense* (AN)

*Python molurus* (Burmese python): *A. crassipes* (A), *A. dissimile* (AN), *A. fuscolineatum, A. javanense* (A), *A. latum, A. transversale, A. varanense* (ANL)

*Python natalensis* (southern African python): *A. latum* (A), *A. marmoreum* (N), *A. transversale* (N)

*Python regius* (ball python): *A. dissimile, A. exornatum* (A), *A. flavomaculatum* (ANL), *A. latum* (ANL), *A. nuttalli* (A), *A. sparsum* (A), *A. transversale* (ANL), *A. varanense* (A)

*Python reticulatus* (reticulated python): *A. crassipes* (A), *A. dissimile* (A), *A. fimbriatum* (AN), *A. geoemydae* (ANL), *A. helvolum* (A), *A. kraneveldi* (A), *A. latum* (A), *A. testudinarium* (A), *A. transversale, A. varanense* (AN), *D. auratus* (A)

*Python sebae* (African rock python): *A. exornatum* (A), *A. flavomaculatum* (A), *A. latum* (ANL), *A. marmoreum, A. nuttalli* (AN), *A. sparsum* (AN), *A. transversale* (AN), *R. (B.) decoloratus* (A)

*Python timoriensis* (Timor python): *A. kraneveldi* (A)

Family Typhlopidae (blind snakes)

*Ramphotyphlops proximus* (chocolate blind snake): *A. fimbriatum*

Family Viperidae (adders and vipers)

*Agkistrodon piscivorus* (water moccasin): *A. dissimile*

*Bitis arietans* (puff adder): *A. exornatum, A. gemma, A. hebraeum* (A), *A. latum* (AN), *A. marmoreum* (AN), *A. nuttalli* (A), *A. sparsum* (A), *A. variegatum* (A), *R. turanicus* (A)

*Bitis gabonica* (Gabon viper): *A. latum* (A), *A. marmoreum* (L), *A. nuttalli* (AN), *A. variegatum*

*Bothriechis lateralis* (side-striped palm pit viper): *A. rotundatum*

*Bothrops asper* (fer-de-lance): *A. dissimile*

*Bothrops atrox* (common lancehead): *A. dissimile, A. rotundatum* (AN)

*Bothrops jararaca* (jararaca): *A. dissimile* (A), *A. rotundatum*

*Bothrops lanceolatus* (Martinique lancehead): *A. dissimile* (A), *A. rotundatum, A. scutatum* (A)

*Bothrops leucurus* (white-tailed lancehead): *A. rotundatum* (A)

*Bothrops moojeni* (Brazilian lancehead): *A. rotundatum* (A)

*Causus defilippii* (snouted night adder): *A. exornatum, A. latum* (ANL)

*Causus resimus* (green night adder): *A. latum* (A), *R. turanicus*

*Causus rhombeatus* (common night adder): *A. latum* (ANL)

*Crotalus adamanteus* (eastern diamondback rattlesnake): *A. dissimile* (A)

*Crotalus durissus* (cascabel rattlesnake): *A. argenti-*

*nae* (AN), *A. cajennense* (A), *A. dissimile*, *A. rotundatum* (A)

***Echis carinatus* (saw-scaled viper):** *Ha. punctata*

***Lachesis muta* (bushmaster):** *A. dissimile* (A), *A. rotundatum* (ANL)

***Sistrurus miliarius* (pygmy rattlesnake):** *A. dissimile*

***Vipera aspis* (asp viper):** *A. latum* (L), *Ha. punctata*

***Vipera berus* (adder):** *Ha. punctata*

***Vipera russelii* (chain viper):** *A. varanense* (A)

***Vipera ursinii* (field adder):** *I. ricinus* (NL)

ORDER TESTUDINES (CHELONIANS)

Family Chelidae (snake-neck turtles)

***Chelodina longicollis* (common snake-necked turtle):** *B. hydrosauri*

Family Emydidae (New World pond turtles and terrapins)

***Emys orbicularis* (European pond terrapin):** *Hy. aegyptium*

***Terrapene carolina* (common box turtle):** *A. dissimile*, *A. sabanerae* (AN)

***Terrapene ornata* (ornate box turtle):** *A. humerale* (A)

***Trachemys scripta* (common slider):** *A. dissimile*, *A. sabanerae* (A)

Family Geoemydidae (Old World pond turtles)

***Cuora amboinensis* (Malaysian box turtle):** *A. geoemydae*, *A. varanense*

***Cuora flavomarginata* (Chinese box turtle):** *A. geoemydae* (ANL)

***Cuora galbinifrons* (Indochinese box turtle):** *A. geoemydae* (A)

***Cuora mouhotii* (jagged-shelled turtle):** *A. geoemydae* (A)

***Cyclemys dentata* (Asian leaf turtle):** *A. geoemydae*

***Geoemyda splengleri* (black-breasted leaf turtle):** *A. geoemydae* (ANL)

***Heosemys annandalii* (yellow-headed temple turtle):** *A. geoemydae* (AN), *A. rotundatum* (A)

***Heosemys grandis* (giant Asian pond turtle):** *A. geoemydae*, *A. helvolum* (A)

***Heosemys spinosa* (spiny terrapin):** *A. geoemydae* (ANL)

***Mauremys japonica* (Japanese pond turtle):** *A. geoemydae* (AN)

***Mauremys leprosa* (Mediterranean turtle):** *A. dissimile* (A)

***Mauremys reevesii* (Reeves's turtle):** *A. geoemydae*

***Melanochelys tricarinata* (three-keeled land tortoise):** *A. javanense*

***Melanochelys trijuga* (Indian black turtle):** *A. geoemydae* (A)

***Rhinoclemmys annulata* (brown land turtle):** *A. humerale*, *A. sabanerae* (ANL)

***Rhinoclemmys areolata* (furrowed wood turtle):** *A. cajennense* (A), *A. dissimile* (A), *A. rotundatum*, *A. sabanerae* (AN)

***Rhinoclemmys funerea* (black river turtle):** *A. sabanerae* (ANL)

***Rhinoclemmys pulcherrima* (Central American wood turtle):** *A. dissimile* (A), *A. sabanerae* (ANL)

***Rhinoclemmys punctularia* (spot-legged terrapin):** *A. sabanerae* (N)

***Rhinoclemmys rubida* (Mexican spotted terrapin):** *A. dissimile* (A)

Family Kinosternidae (mud and musk turtles)

***Kinosternon angustipons* (narrow-bridged mud turtle):** *A. dissimile*, *A. sabanerae*

***Kinosternon leucostomum* (white-lipped mud turtle):** *A. dissimile*

***Kinosternon scorpioides* (scorpion mud turtle):** *A. dissimile* (A)

Family Pelomedusidae (helmeted side-neck turtles)

***Pelomedusa subrufa* (African helmeted turtle):** *A. nuttalli*

***Pelusios sinuatus* (East African serrated mud turtle):** *A. nuttalli*

Family Podocnemididae

***Podocnemis erythrocephala* (red-headed Amazon river turtle):** *A. cajennense*

Family Testudinidae (tortoises)

***Aldabrachelys gigantea* (Aldabra giant tortoise):** *A. gemma* (A), *A. hebraeum* (A), *A. marmoreum* (ANL), *A. nuttalli* (A), *A. sparsum* (A), *R. (B.) decoloratus* (A)

*Astrochelys radiata* (**radiated tortoise**): *A. chabaudi* (A)

*Chelonoidis carbonaria* (**red-footed tortoise**): *A. humerale* (A), *A. rotundatum* (A), *A. sabanerae* (A)

*Chelonoidis chilensis* (**Argentine tortoise**): <u>*A. argentinae*</u> (A), *A. rotundatum*

*Chelonoidis denticulata* (**yellow-footed tortoise**): *A. cajennense, A. dissimile, A. geayi,* <u>*A. humerale*</u> (A), *A. latum* (A), *A. marmoreum* (AN), *A. rotundatum, A. sabanerae* (A)

*Chelonoidis nigra* (**Galapagos giant tortoise**): *A. marmoreum* (N)

*Chersina angulata* (**angulated tortoise**): *A. hebraeum* (N), *A. marmoreum* (AN), <u>*A. sylvaticum*</u> (ANL)

*Geochelone elegans* (**Indian star tortoise**): <u>*A. clypeolatum*</u> (ANL)

*Geochelone platynota* (**Burmese starred tortoise**): *A. clypeolatum* (AN)

*Geochelone sulcata* (**African spurred tortoise**): *A. marmoreum* (ANL), *A. sparsum* (A)

*Gopherus polyphemus* (**gopher tortoise**): *A. dissimile, A. exornatum, A. rotundatum*

*Homopus areolatus* (**areolated tortoise**): *A. marmoreum* (ANL), *A. sylvaticum* (ANL)

*Homopus boulengeri* (**Boulenger's Cape tortoise**): *A. marmoreum* (NL)

*Homopus femoralis* (**Karoo Cape tortoise**): *A. marmoreum* (ANL)

*Homopus signatus* (**speckled Cape tortoise**): *A. marmoreum* (AL), *A. sylvaticum* (A)

*Indotestudo elongata* (**elongated tortoise**): *A. clypeolatum* (A), *A. geoemydae* (AN)

*Indotestudo forstenii* (**Celebes tortoise**): *A. geoemydae*

*Kinixys belliana* (**Bell's hinged tortoise**): *A. exornatum, A. falsomarmoreum, A. hebraeum* (N), *A. marmoreum* (ANL), <u>*A. nuttalli*</u> (AN), *A. sparsum* (ANL), *A. tholloni, Hy. truncatum* (A)

*Kinixys erosa* (**forest hinged tortoise**): *A. clypeolatum* (A)

*Kinixys homeana* (**Home's hingeback tortoise**): *A. nuttalli*

*Kinixys spekii* (**Speke's hinged tortoise**): *A. nuttalli* (A)

*Malacochersus tornieri* (**pancake tortoise**): *A. falsomarmoreum*

*Manouria emys* (**Asian giant tortoise**): *A. geoemydae, A. testudinarium, A. varanense* (A)

*Manouria impressa* (**impressed tortoise**): *A. geoemydae* (ANL)

*Psammobates geometricus* (**geometric tortoise**): *A. marmoreum* (ANL), *A. sylvaticum* (ANL), *Hy. truncatum* (A)

*Psammobates oculifer* (**African serrated tortoise**): *A. marmoreum* (AN)

*Psammobates tentorius* (**African tent tortoise**): *A. marmoreum* (ANL), *A. sylvaticum* (ANL)

*Pyxis arachnoides* (**spider tortoise**): <u>*A. chabaudi*</u> (ANL)

*Pyxis planicauda* (**flat-backed spider tortoise**): *A. chabaudi*

*Stigmochelys pardalis* (**leopard tortoise**): *A. dissimile,* <u>*A. falsomarmoreum*</u> (A), *A. gemma, A. hebraeum* (AN), *A. latum* (A), <u>*A. marmoreum*</u> (ANL), *A. nuttalli* (A), <u>*A. sparsum*</u> (ANL), *A. tholloni* (A), *Hy. truncatum* (A), *I. ricinus, R. evertsi* (A)

*Testudo graeca* (**Greek tortoise**): <u>*Hy. aegyptium*</u> (ANL), *Hy. anatolicum* spp., *Hy. marginatum* group, *I. ricinus*

*Testudo hermanni* (**Hermann's tortoise**): *A. marmoreum* (AN), *Hy. aegyptium* (ANL), *Hy. anatolicum* spp., *Hy. marginatum* group

*Testudo horsfieldii* (**Afghan tortoise**): *Ha. punctata, Hy. aegyptium* (A), *Hy. anatolicum* spp., *Hy. detritum*

*Testudo kleinmanni* (**Egyptian tortoise**): *Hy. aegyptium* (A)

*Testudo marginata* (**marginated tortoise**): *Hy. aegyptium* (ANL)

Family Trionychidae (softshell turtles)

*Cyclanorbis senegalensis* (**Senegal flapshell turtle**): *A. nuttalli*

*Notes*: a. *Hyalomma isaaci, Hy. marginatum* or *Hy. rufipes* (see section 7.5)

b. *Hyalomma anatolicum* or *Hy. excavatum* (see section 7.3)

c. Insectivores including the Orders Afrosoricida, Erinaceomorpha and Soricomorpha

d. Ungulates including the Orders Artiodactyla and Perissodactyla

e. Xenarthrans including the Orders Cingulata and Pilosa

f. Marsupials including the Orders Dasyuromorphia, Didelphimorphia, Diprotodontia and Peramelemorphia

# Glossary

**abattoir.** A slaughterhouse.
**acaricide.** A pesticide that kills ticks.
**acral.** Relating to or affecting the peripheral parts such as the limbs and ears.
**alopecia.** Loss of hair.
**anorexia.** Loss of appetite for food.
**anthropophilic.** Showing preference for humans.
**anuran.** Any of the order of tailless amphibians with broad bodies and well-developed hind legs, such as frogs and toads.
**arthralgia.** Severe pain in a joint.
**ascites.** Accumulation of serous fluid in the abdominal cavity.
**babesiacide.** A drug that kills *Babesia* sp. parasites.
**bilirubinuria.** The presence of the bile pigment bilirubin in the urine.
**caprine.** Relating to goats.
**carapace.** The upper shell of tortoises and turtles.
**chelonian.** Relating to tortoises and turtles.
**cloaca.** The common chamber into which the hindgut, bladder and genital ducts open, as in reptiles, birds and amphibians.
**conspecific.** Belonging to the same species.
**coronet.** The line of junction between the skin and the hoof.
**croup.** The rump of a horse.
**diapause.** A period of delayed development or growth accompanied by reduced metabolism and inactivity.
**dysphasia.** Impairment of the ability to speak.
**dyspnea.** Difficulty in breathing.
**echidna.** Any of several small, egg-laying, toothless Australasian spiny anteaters belonging to the order Monotremata.
**elephant shrew.** Any of several small African terrestrial mammals belonging to the order Macroscelidea; these are adapted to a diet of insects and take their name from their long mobile snouts.
**encephalitis.** Inflammation of the brain.
**encephalomyelitis.** Inflammation of the brain and spinal cord.
**endemic.** Denoting a temporal pattern of disease occurrence in a population in which the disease occurs with predictable regularity with only relatively minor fluctuations in its frequency over time.
**endocarditis.** Inflammation of the endocardium, the covering lining the cavities of the heart.
**enzootic.** As for endemic but relating only to animal diseases.
**erythema.** Redness of the skin due to capillary dilatation.
**erythrocyte.** A red blood cell.
**erythrophagocytosis.** The process of ingestion and digestion by cells of red blood cells.
**eschar.** A necrotic lesion of the skin covered by a black crust.
**escutcheon.** The region of the skin in cattle and other quadrupeds between the hind legs and above the udder and below the anus.
**fetlock.** The joint of horses and other equines just above the hoof.
**genotype.** A group of organisms each having the same hereditary characteristics.
**hematogenous.** Spread by the bloodstream.
**hemoglobinuria.** The presence of the red blood pigment hemoglobin in the urine.
**hemolysis.** The destruction of red blood cells with liberation of hemoglobin into the surrounding fluid.
**hepatitis.** Inflammation of the liver.
**hepatomegaly.** Enlargement of the liver.
**herpetological.** Relating to reptiles and amphibians.
**homoiothermic.** Warm blooded.
**hyperemia.** The presence of an increased amount of blood in a part of the body.

**hyperesthesia.** An abnormal sensitivity of the skin.

**hyrax.** Any of several small terrestrial or arboreal mammals from Africa and southwestern Asia belonging to the order Hyracoidea, comparable in size and external appearance to rodents.

**indurated.** Hardened.

**infarction.** The development of an infarct, an area of dead tissue resulting from a sudden insufficiency of blood supply.

**insectivore.** Any member of the old order Insectivora, comprising small mammals with long narrow snouts that are primarily insectivorous; examples are shrews and moles (order Soricomorpha), hedgehogs (order Erinaceomorpha) and tenrecs (order Afrosoricida).

**invasive species.** An organism that is introduced into a non-native ecosystem and that causes, or is likely to cause, harm to the economy, environment, and/or animal or human health.

**lagomorph.** Any member of the order Lagomorpha, an order of herbivorous mammals resembling rodents but having two pairs of upper incisor teeth, one behind the other; examples are rabbits, hares and pikas.

**larva.** The second stage in the life cycle of a tick, the stage that hatches from the egg and, following engorgement, will molt into a nymph.

**leukopenia.** A decrease below normal in the number of leukocytes (white blood cells) in the blood.

**lymphadenopathy.** Any disease process affecting a lymph node.

**lymphangitis.** Inflammation of lymphatic vessels.

**lymphoblast.** A primitive cell that matures into a lymphocyte.

**lymphocyte.** A white blood cell formed in lymphatic tissue.

**macrophage.** Any mononuclear phagocytic cell that will ingest and destroy solid substances such as bacteria, bits of necrotic tissue and foreign particles.

**maculopapular.** Having a flat base surrounding a papule (a small solid elevation of the skin) in the center, as in a skin lesion.

**marsupial.** Any member of the old order Marsupialia, comprising mammals that lack a placenta and have an external pouch containing the teats; examples are banded anteaters (order Dasyuromorphia), opossums (order Didelphimorphia), wallabies and kangaroos (order Diprotodontia) and bandicoots (order Peramelemorphia).

**meningitis.** Inflammation of the meninges, the membranes enveloping the brain and spinal cord.

**meningoencephalitis.** Inflammation of the brain and its membranes.

**merozoite.** The motile infective stage of sporozoan protozoa resulting from asexual reproduction.

**monitor.** Any of several very large, flesh-eating lizards of the family Varanidae.

**monocyte.** A large, nongranular white blood cell with a relatively small, kidney-shaped nucleus.

**monotreme.** Any of the lowest order (Monotremata) of mammals that lay eggs and include the platypus and echidnas.

**mucopurulent.** Containing both mucus and pus.

**myalgia.** Muscular pain.

**myiasis.** Any infection due to invasion of tissues or cavities of the body by larvae of dipterous insects.

**myocarditis.** Inflammation of the muscular walls of the heart.

**necrosis.** The death of tissue resulting from irreversible damage.

**neuropathy.** Any disease of the nervous system.

**neutropenia.** A decrease below normal in the number of neutrophils in the blood.

**neutrophil.** A granulocytic white blood cell.

**nosocomial.** Hospital-acquired.

**nymph.** The third stage in the life cycle of a tick, between the larva and the adult.

**nystagmus.** An involuntary rapid movement of the eyeball, usually from side to side.

**ophidian.** Relating to snakes.

**opisthotonus.** A tetanic spasm in which the spine and extremities are bent with convexity forward, the body resting on the head and heels.

**pangolin.** Any of several toothless scaly mammals (order Pholidota) which feed on ants and termites.

**paraplegia.** Paralysis of the entire lower half of the body.

**paresis.** Paralysis or incomplete paralysis.

**pericarditis.** Inflammation of the pericardium, the membranous sac surrounding the heart and roots of the great blood vessels.

**perineum.** The area of the body between the anus and the vulva in the female and between the anus and the scrotum in the male.

**perissodactyl.** Any animal of the order Perissodactyla of hoofed mammals with an uneven number of toes on each foot and with a simple stomach, including horses and tapirs.

**petechiae.** Small hemorrhagic spots in the skin.

**pheromone.** A chemical substance secreted externally by certain animals, such as ticks, to convey information to and produce specific responses (such as attraction, aggregation and attachment) in other individuals of the same species.

**photophobia.** Abnormal fear of light.

**piroplasm.** A stage in the life cycle of many protozoa found in red blood cells.

**plastron.** The under or lower shell of tortoises and turtles.

**poikilothermic.** Cold blooded.

**polyarthritis.** Simultaneous inflammation of several joints.

**protozoa.** Plural form of protozoan.

**protozoan.** Any of the animals of the subkingdom Protozoa of mostly microscopic organisms, including all unicellular forms of animal life.

**rhinitis.** Inflammation of the nasal mucous membrane.

**rickettsia.** Any bacterium of the genus *Rickettsia*.

**saurian.** Relating to lizards.

**schizont.** A stage in the life cycle of many sporozoan protozoa that undergo asexual reproduction by multiple fission.

**scutes.** The external horny plates covering the shell of tortoises and turtles.

**splenomegaly.** Enlargement of the spleen.

**stomatitis.** Inflammation of the mucous membrane of the mouth.

**sutures.** The lines between scutes (external horny plates) on the outside of the shell of tortoises and turtles.

**taxa.** Taxonomic categories or units, such as species or genera.

**thrombocytopenia.** A decrease below normal in the number of platelets in the blood.

**transovarial.** When related to transmission, indicates passage of infectious agents from female ticks to their eggs, producing infected larvae in the next generation.

**transstadial.** When related to transmission, indicates passage of infectious agents from one developmental stage to the next within the same generation.

**ungulate.** Any of the animal species having hoofs, including bovids, camelids, deer, giraffes, hippopotamuses and pigs (order Artiodactyla), and equids, rhinoceroses and tapirs (order Perissodactyla).

**vasculitis.** Inflammation of a blood vessel.

**viremia.** Presence of a virus in the blood.

**withers.** The highest part of the back of a horse located between the shoulder blades.

**xenarthran.** Any member of the old order Xenarthra, comprising mammals unique to the Americas, including armadillos (order Cingulata) and sloths and anteaters (order Pilosa).

# References

Abou-Elela RG, Taher MO, Diab FM (1981) Studies on ticks infesting camels, sheep and goats in Riyadh area (Saudi Arabia). J. Coll. Sci. Univ. Riyadh 12: 385–399.

Aeschlimann A (1963) Observations sur la morphologie, la biologie et la developpement d'*Amblyomma compressum* (Macalister, 1872), la tique des pangolins d'Afrique occidentale. Acta Trop. 20: 154–177.

Aeschlimann A (1967) Biologie et ecologie des tiques (Ixodoidea) de Cote d'Ivoire. Acta Trop. 24: 281–405.

Aeschlimann A (1972) *Ixodes ricinus*, Linne, 1758 (Ixodoidea; Ixodidae): essai preliminaire de synthese sur la biologie de cette espece en Suisse. Acta Trop. 29: 321–340.

Aeschlimann A, Buttiker W (1975) Importations de tiques en Suisse (Acarina: Ixodoidea). Mitt. Schweiz. Entomol. Ges. 48: 69–75.

Aitken THG, Omardeen TA, Gilkes CD (1958) The 1958 Cayenne tick outbreak. J. Agric. Soc. Trinidad Tobago 58: 153–157.

Alani AJ, Herbert IV (1988) Morphology and transmission of *Theileria recondita* (Theileriidae: Sporozoa) isolated from *Haemaphysalis punctata* from north Wales. Vet. Parasitol. 28: 283–291.

Albanese M, Smiraglia CB, Lavagnino A (1971) [On the ticks of Sicily and occurrence of *Hyalomma detritum* and *Amblyomma variegatum*.] Riv. Parassitol. 32: 273–276.

Alekseev AN, Dubinina HV, Semenov AV, Bolshakov CV (2001) Evidence of ehrlichiosis agents found in ticks (Acari: Ixodidae) collected from migratory birds. J. Med. Entomol. 38: 471–474.

Al-Khalifa MS, Hussein HS, Al-Asgah NA, Diab FM (1987) Ticks (Acari: Ixodidae) infesting local domestic animals in western and southern Saudi Arabia. Arab Gulf J. Sci. Res. Agric. Biol. Sci. B5: 301–319.

Al-Khalifa MS, Diab FM, Al-Asgah NA, Hussein HS, Khalil GM (2006) Ticks (Acari: Argasidae, Ixodidae) recorded on wild animals in Saudi Arabia. Fauna Arabia 22: 225–231.

Allan SA, Norval RA, Sonenshine DE, Burridge MJ (1996) Efficacy of tail-tag decoys impregnated with pheromone and acaricide for control of bont ticks on cattle. Ann. N. Y. Acad. Sci. 791: 85–93.

Allan SA, Barre N, Sonenshine DE, Burridge MJ (1998a) Efficacy of tags impregnated with pheromone and acaricide for control of *Amblyomma variegatum*. Med. Vet. Entomol. 12: 141–150.

Allan SA, Simmons LA, Burridge MJ (1998b) Establishment of the tortoise tick *Amblyomma marmoreum* (Acari: Ixodidae) on a reptile-breeding facility in Florida. J. Med. Entomol. 35: 621–624.

Allan SA, Simmons LA, Burridge MJ (2001) Ixodid ticks on white-tailed deer and feral swine in Florida. J. Vector Ecol. 26: 93–102.

Allsopp BA, Bezuidenhout JD, Prozesky L (2004) Heartwater. In: Coetzer JAW, Tustin RC, eds. Infectious Diseases of Livestock. 2nd ed. Oxford: Oxford University Press. pp. 507–535.

Alvarez C., V, Bonilla MR, Chacon GI (2000) Distribucion de la garrapata *Amblyomma cajennense* (Acari: Ixodidae) sobre *Bos taurus* y *Bos indicus* en Costa Rica. Rev. Biol. Trop. 48: 129–135.

Amorim M, Serra-Freire NM (1994a) *Amblyomma nodosum* Neumann, 1899 descricao morfologica do estadio de larva. Rev. Bras. Parasitol. Vet. 3: 131–142.

Amorim M, Serra-Freire NM (1994b) Descricao morfologica do estadio de larva de carrapato (Acari: Ixodidae). 4. *Amblyomma dissimile* Koch, 1844. Bol. Mus. Paraense Emilio Goeldi 10: 273–288.

Amorim M, Serra-Freire NM (1995) Morphological description of tick larval stage (Acari: Ixodidae). 1. *Amblyomma rotundatum* Koch, 1844. Parasitol. Dia 19: 9–19.

Amorim M, Serra-Freire NM (2000) Morphological description of tick larval stage (Acari: Ixodidae). 7. *Amblyomma auricularium* (Conil, 1878). Entomol. Vect. 7: 297–309.

Amorim M, Gazeta GS, Guerim L, Serra-Freire NM (1997) Morphological description of tick larval stage (Acari: Ixodidae). 5. *Anocentor nitens* (Neumann, 1897). Rev. Bras. Parasitol. Vet. 6: 143–156.

Andersen MC, Adams H, Hope B, Powell M (2004) Risk assessment for invasive species. Risk Anal. 24: 787–793.

Anderson JF, Magnarelli LA, Burgdorfer W, Casper EA, Philip RN (1981a) Importation into the United States from Africa of *Rhipicephalus simus* on a boutonneuse fever patient. Am. J. Trop. Med. Hyg. 30: 897–899.

Anderson JF, Magnarelli LA, Keirans JE (1981b) *Aponomma quadricavum* (Acari: Ixodidae) collected from an imported boa, *Epicrates striatus*, in Connecticut. J. Med. Entomol. 18: 123–125.

Anderson JF, Magnarelli LA, Keirans JE (1984) Ixodid and argasid ticks in Connecticut, U.S.A.: *Aponomma latum*, *Amblyomma dissimile*, *Haemaphysalis leachi* group, and *Ornithodoros kelleyi* (Acari: Ixodidae, Argasidae). Int. J. Acarol. 10: 149–151.

Andrew HR, Norval RAI (1989) The carrier status of sheep, cattle, and African buffalo recovered from heartwater. Vet. Parasitol. 34: 261–266.

Anon (1996) Tick-borne infectious diseases. Ohio Vector News 15: 1–6.

Apanaskevich DA, Horak IG (2005) The genus *Hyalomma* Koch, 1844. II. Taxonomic status of *H.* (*Euhyalomma*) *anatolicum* Koch, 1844 and *H.* (*E.*) *excavatum* Koch, 1844 (Acari, Ixodidae) with redescriptions of all stages. Acarina 13: 181–197.

Apanaskevich DA, Horak IG (2007) The genus *Hyalomma* Koch, 1844. III. Redescription of the adults and larva of *H.* (*Euhyalomma*) *impressum* Koch, 1844 (Acari: Ixodidae) with a first description of its nymph and notes on its biology. Folia Parasitol. 54: 51–58.

Apanaskevich DA, Horak IG (2008a) The genus *Hyalomma* Koch, 1844: V. Re-evaluation of the taxonomic rank of taxa comprising the *H.* (*Euhyalomma*) *marginatum* Koch complex of species (Acari: Ixodidae) with redescription of all parasitic stages and notes on biology. Int. J. Acarol. 34: 13–42.

Apanaskevich DA, Horak IG (2008b) The genus *Hyalomma*. VI. Systematics of *H.* (*Euhyalomma*) *truncatum* and the closely related species, *H.* (*E.*) *albiparmatum* and *H.* (*E.*) *nitidum* (Acari: Ixodidae). Exp. Appl. Acarol. 44: 115–136.

Apanaskevich DA, Horak IG, Camicas JL (2007) Redescription of *Haemaphysalis* (*Rhipistoma*) *elliptica* (Koch, 1844), an old taxon of the *Haemaphysalis* (*Rhipistoma*) *leachi* group from East and southern Africa, and of *Haemaphysalis* (*Rhipistoma*) *leachi* (Audouin, 1826) (Ixodida, Ixodidae). Onderstepoort J. Vet. Res. 74: 181–208.

Apanaskevich DA, Santos-Silva MM, Horak IG (2008a) The genus *Hyalomma* Koch, 1844. IV. Redescription of all parasitic stages of *H.* (*Euhyalomma*) *lusitanicum* Koch, 1844 and the adults of *H.* (*E.*) *franchinii* Tonelli Rondelli, 1932 (Acari: Ixodidae) with a first description of its immature stages. Folia Parasitol. 55: 61–74.

Apanaskevich DA, Schuster AL, Horak IG (2008b) The genus *Hyalomma*: VII. Redescription of all parasitic stages of *H.* (*Euhyalomma*) *dromedarii* and *H.* (*E.*) *schulzei* (Acari: Ixodidae) J. Med. Entomol. 45: 817–831.

Arthur DR (1947) Observations on *Ixodes hexagonus* Leach, 1815 (Acarina: Ixodidae). Entomol. Monthly Mag. 83: 69–76.

Arthur DR (1953) The host relationships of *Ixodes hexagonus* Leach in Britain. Parasitology 43: 227–238.

Arthur DR (1957) The *Ixodes schillingsi* group: ticks of Africa and Madagascar, parasitic on primates, with descriptions of two new species (Ixodoidea, Ixodidae). Parasitology 47: 544–559.

Arthur DR (1960) Ticks: A Monograph of the Ixodoidea. Part V. On the Genera *Dermacentor*, *Anocentor*, *Cosmiomma*, *Boophilus* and *Margaropus*. London: Cambridge University Press. 251 pp.

Arthur DR (1962) Ticks and Disease. Evanston, IL: Row, Peterson & Co. 445 pp.

Arthur DR (1963) British Ticks. London: Butterworth. 213 pp.

Arthur DR (1965) Ticks of the Genus *Ixodes* in Africa. London: Athlone Press. 348 pp.

Auffenberg T (1988) *Amblyomma helvolum* (Acarina: Ixodidae) as a parasite of varanid and scincid reptiles in the Philippines. Int. J. Parasitol. 18: 937–946.

Awad FI, Amin MM, Salama SA, Khide S (1981) The role played by *Hyalomma dromedarii* in the transmission of African horse sickness virus in Egypt. Bull. Anim. Health Prod. Afr. 29: 337–340.

Azad AF, Radulovic S (2003) Pathogenic rickettsiae as bioterrorism agents. Ann. N. Y. Acad. Sci. 990: 734–738.

Babudieri B (1959) Q fever: a zoonosis. Adv. Vet. Sci. 5: 81–182.

Baeza CR (1979) Tick paralysis—Canal Zone, Panama. Morbid. Mortal. Weekly Rep. 28: 428/433.

Bailly-Choumara H, Morel PC, Rageau J (1976) Sommaire des donnees actuelles sur les tiques du Maroc (Acari, Ixodoidea). Bull. Inst. Sci. 1: 101–117.

Baker RH (1963) Geographical distribution of terrestrial mammals in Middle America. Am. Midland Nat. 70: 208–249.

Ball GH, Chao J, Telford SR (1969) *Hepatozoon fusifex* sp. n., a hemogregarine from *Boa constrictor* producing marked morphological changes in infected erythrocytes. J. Parasitol. 55: 800–813.

Barnard SM, Durden LA (2000) A Veterinary Guide to the Parasites of Reptiles, Arthropods (Excluding Mites). Malabar, FL: Krieger. 288 pp.

Barnes IR (1955) Cattle egrets colonize a new world. Atlantic Nat. 10: 238–247.

Barre N, Garris GI (1990) Biology and ecology of *Amblyomma variegatum* (Acari: Ixodidae) in the Caribbean: implications for a regional eradication program. J. Agric. Entomol. 7: 1–9.

Barre N, Uilenberg G, Morel PC, Camus E (1987) Danger of introducing heartwater onto the American mainland: potential role of indigenous and exotic *Amblyomma* ticks. Onderstepoort J. Vet. Res. 54: 405–417.

Barre N, Garris GI, Borel G, Camus E (1988) Hosts and population dynamics of *Amblyomma variegatum* (Acari: Ixodidae) on Guadeloupe, French West Indies. J. Med. Entomol. 25: 111–115.

Barre N, Camus E, Borel G, Aprelon R (1991) [Preferential attachment sites of the tick *Amblyomma variegatum*, on its hosts in Guadeloupe (French West Indies).] Rev. Elev. Med. Vet. Pays Trop. 44: 453–458.

Barre N, Garris G, Camus E (1995) Propagation of the tick *Amblyomma variegatum* in the Caribbean. Rev. Sci. Tech. Off. Int. Epizoot. 14: 841–855.

Barre N, Camus E, Fifi J, Fourgeaud P, Numa G, Rose-Rosette F, Borel H (1996) Tropical bont tick eradication campaign in the French Antilles: current status. Ann. N. Y. Acad. Sci. 791: 64–76.

Bauwens D, Strijbosch H, Stumpel AHP (1983) The lizards *Lacerta agilis* and *L. vivipara* as hosts to larvae and nymphs of the tick *Ixodes ricinus*. Holarctic Ecol. 6: 32–40.

Bayless MK, Simmons LA (2000) Tick parasites on the rock monitor lizard (*Varanus albigularis* Daudin, 1802) of Tanzania, Africa. Afr. J. Ecol. 38: 363–364.

Beati L, Keirans JE (2001) Analysis of the systematic relationships among ticks of the genera *Rhipicephalus* and *Boophilus* (Acari: Ixodidae) based on mitochondrial 12S ribosomal DNA gene sequences and morphological characters. J. Parasitol. 87: 32–48.

Becker CAM, Bouju-Albert A, Jouglin M, Chauvin A, Malandrin L (2009) Natural transmission of zoonotic *Babesia* spp. by *Ixodes ricinus* ticks. Emerg. Infect. Dis. 15: 320–322.

Becklund WW (1968) Ticks of veterinary significance found on imports in the United States. J. Parasitol. 54: 622–628.

Bequaert J (1932) *Amblyomma dissimile* Koch, a tick indigenous to the United States (Acarina: Ixodidae). Psyche 39: 45–47.

Bequaert J (1945) Further records of the snake tick, *Amblyomma dissimile* Koch, in Florida. Bull. Brooklyn Entomol. Soc. 40: 129.

Berge TO, Lennette EH (1953) World distribution of Q fever: human, animal and arthropod infection. Am. J. Hyg. 57: 125–143.

Berube LR (2006) Tick collection survey report 2005. Kissimmee, FL: Kissimmee Diagnostic Laboratory, Florida Department of Agriculture & Consumer Services. 22 pp.

Bezuidenhout JD (1987) Natural transmission of heartwater. Onderstepoort J. Vet. Res. 54: 349–351.

Bhattacharyulu Y, Chaudhri RP, Gill BS (1975) Transstadial transmission of *Theileria annulata* through common ixodid ticks infesting Indian cattle. Parasitology 71: 1–7.

Billings AN, Yu XJ, Teel PD, Walker DH (1998) Detection of a spotted fever group rickettsia in *Amblyomma cajennense* (Acari: Ixodidae) in south Texas. J. Med. Entomol. 35: 474–478.

Bilsing SW, Eads RB (1947) An addition to the tick fauna of the United States. J. Parasitol. 33: 85–86.

Birnie EF, Burridge MJ, Camus E, Barre N (1985) Heartwater in the Caribbean: isolation of *Cowdria ruminantium* from Antigua. Vet. Rec. 116: 121–123.

Bishop GC (2004) *Borrelia theileri* infection. In: Coetzer JAW, Tustin RC, eds. Infectious Diseases of Livestock. 2nd ed. Oxford: Oxford University Press. pp. 1435–1436.

Bishopp FC, Trembley HL (1945) Distribution and hosts of certain North American ticks. J. Parasitol. 31: 1–54.

Bjoersdorff A, Bergstrom S, Massung RF, Haemig PD, Olsen B (2001) *Ehrlichia*-infected ticks on migrating birds. Emerg. Infect. Dis. 7: 877–879.

Blake CH (1961) Notes on the history of the cattle egret in the New World. Chat 25: 24–27.

Bloemer SR, Russell CL, Keirans JE (1987) *Amblyomma calcaratum* (Acari, Ixodidae), a Central and South American tick found in Kentucky, USA. J. Med. Entomol. 24: 117.

Bock R, Jackson L, de Vos A, Jorgensen W (2004) Babesiosis of cattle. Parasitology 129: S247–S269.

Bodkin GE (1918–19) The biology of *Amblyomma dissimile* Koch, with an account of its power of reproducing parthogenetically. Parasitology 11: 10–17.

Bokma BH, Shaw JL (1993) Eradication of a new focus of *Amblyomma variegatum* in Puerto Rico. Rev. Elev. Med. Vet. Pays Trop. 46: 355–358.

Borges LMF, Oliveira PR, Ribeiro MFB (2000) Seasonal dynamics of *Anocentor nitens* on horses in Brazil. Vet. Parasitol. 89: 165–171.

Borio L, Inglesby T, Peters CJ, Schmaljohn AL, Hughes JM, Jahrling PB, Ksiazek T, Johnson KM, Meyerhoff A, O'Toole T, Ascher MS, Bartlett J, Breman JG, Eitzen EM, Hamburg M, Hauer J, Henderson DA, Johnson RT, Kwik G, Layton M, Lillibridge S, Nabel GJ, Osterholm MT, Perl TM, Russell P, Tonat K (2008) Hemorrhagic fever viruses as biological weapons: medical and public health management. J. Am. Med. Assoc. 287: 2391–2405.

Boulard Y, Paperna I, Petit G, Landau I (2001) Ultrastructure of developmental stages of *Hemolivia stellata* (Apicomplexa: Haemogregarinidae) in the cane toad *Bufo marinus* and in its vector tick *Amblyomma rotundatum*. Parasitol. Res. 87: 598–604.

Bram RA, George JE (2000) Introduction of nonindigenous arthropod pests of animals. J. Med. Entomol. 37: 1–8.

Bram RA, George JE, Reichard RE, Tabachnick WJ (2002) Threat of foreign arthropod-borne pathogens to livestock in the United States. J. Med. Entomol. 39: 405–416.

Brinck P, Svedmyr A, von Zeipel G (1965) Migrating birds at Ottenby Sweden as carriers of ticks and possible transmitters of tick-borne encephalitis virus. Oikos 16: 88–99.

Brocklesby DW (1965) Evidence that *Rhipicephalus pulchellus* (Gerstacker 1873) may be a vector of some piroplasms. Bull. Epizoot. Dis. Afr. 13: 37–44.

Brouqui P, Parola P, Fournier PE, Raoult D (2007) Spotted fever rickettsioses on southern and eastern Europe. FEMS Immunol. Med. Microbiol. 49: 2–12.

Brown RN, Lane RS, Dennis DT (2005) Geographic distributions of tick-borne diseases and their vectors. In: Goodman JL, Dennis DT, Sonenshine DE, eds. Tick-borne Diseases of Humans. Washington, DC: ASM Press. pp. 363–391.

Bruce WG (1962) Eradication of the red tick from a wild animal compound in Florida. J. Wash. Acad. Sci. 52: 81–85.

Brumpt E (1934) L'ixodine *Amblyomma dissimile* du Venezuela ne presente pas de parthenogenese facultative. Ann. Parasitol. 12: 116–120.

Bruning A (1996) Equine piroplasmosis: an update on diagnosis, treatment and prevention. Brit. Vet. J. 152: 139–151.

Bull CM, Sharrad RD, Smyth M (1977) The pre-molt period of larvae and nymphs of the Australian reptile tick *Aponomma hydrosauri*. Acarologia 19: 593–600.

Burr EW (1983) Tick toxicosis in a crossbred terrier caused by *Hyalomma truncatum*. Vet. Rec. 113: 260–261.

Burridge MJ (1985) Heartwater invades the Caribbean. Parasitol. Today 1: 175–177.

Burridge MJ (1997) Heartwater: an increasingly serious threat to the livestock and deer populations of the United States. In: Proceedings of 101st Annual Meeting of the United States Animal Health Association. Richmond, VA: Spectrum Press. pp. 582–597.

Burridge MJ (2000) Risks of introduction of heartwater into the U.S. associated with importations of reptiles and of wild game animals from Africa. In: Proceedings of 104th Annual Meeting of the United States Animal Health Association. Richmond, VA: Carter Printing Company. pp. 435–442.

Burridge MJ (2001) Ticks (Acari: Ixodidae) spread by the international trade in reptiles and their potential roles in dissemination of diseases. Bull. Entomol. Res. 91: 3–23.

Burridge MJ (2005) Controlling and eradicating tick infestations on reptiles. Compend. Cont. Educ. Pract. Vet. 27: 371–376.

Burridge MJ, Barre N, Birnie EF, Camus E, Uilenberg G (1984) Epidemiological studies on heartwater in the Caribbean, with observations on tick-associated bovine dermatophilosis. In: Coubrough RI, Bertschinger HJ, Button C, van Amstel S, eds. Proceedings of the XIIIth World Congress on Diseases of Cattle, Vol. I. Durban, South Africa: South African Veterinary Association. pp. 542–546.

Burridge MJ, Simmons LA, Allan SA (2000a) Introduction of potential heartwater vectors and other exotic ticks into Florida on imported reptiles. J. Parasitol. 86: 700–704.

Burridge MJ, Simmons LA, Simbi BH, Peter TF, Mahan SM (2000b) Evidence of *Cowdria ruminantium* infection (heartwater) in *Amblyomma sparsum* ticks found on tortoises imported into Florida. J. Parasitol. 86: 1135–1136.

Burridge MJ, Simmons LA, Simbi BH, Mahan SM, Fournier PE, Raoult D (2002a) Introduction of the exotic tick *Amblyomma hebraeum* into Florida on a human host. J. Parasitol. 88: 800–801.

Burridge MJ, Simmons LA, Peter TF, Mahan SM (2002b) Control and eradication of chelonian tick infestations, with particular reference to vectors of heartwater. Ann. N. Y. Acad. Sci. 969: 294–296.

Burridge MJ, Simmons LA, Peter TF, Mahan SM (2002c) Increasing risks of introduction of heartwater onto the American mainland associated with animal movements. Ann. N. Y. Acad. Sci. 969: 269–274.

Burridge MJ, Simmons LA (2003) Exotic ticks introduced into the United States on imported reptiles from 1962 to 2001 and their potential roles in international dissemination of diseases. Vet. Parasitol. 113: 289–320.

Burridge MJ, Simmons LA, Ahrens EH, Naude SA, Malan FS (2004a) Development of a novel self-medicating applicator for control of internal and external parasites of wild and domestic animals. Onderstepoort J. Vet. Res. 71: 41–51.

Burridge MJ, Simmons LA, Condie T (2004b) Control of an exotic tick (*Aponomma komodoense*) infestation in a Komodo dragon (*Varanus komodoensis*) exhibit at a zoo in Florida. J. Zoo Wildl. Med. 35: 248–249.

Burridge MJ, Berube LR, Holt TJ (2006) Invasive ticks: introduction of *Amblyomma kraneveldi* (Anastos) and other exotic ticks (Acari: Ixodidae) into Florida on imported reptiles. Int. J. Acarol. 32: 315–321.

Bushmich SL (1994) Lyme borreliosis in domestic animals. J. Spirochetal Tick-borne Dis. 1: 24–28.

Byrne WR (1997) Q fever. In: Sidell FR, Takafugi ET, Franz DR, eds. Medical Aspects of Chemical and Biological Warfare. Washington, DC: Office of The Surgeon General at TMM Publications. pp. 523–537.

Callow LL, Hoyte HMD (1961) Transmission experiments using *Babesia bigemina*, *Theileria mutans*, *Borrelia* sp. and the cattle tick, *Boophilus microplus*. Aust. Vet. J. 37: 381–390.

Camus E, Barre N, Martinez D, Uilenberg G (1996) Heartwater (Cowdriosis): A Review. 2nd ed. Paris: Office International des Epizooties. 177 pp.

Carelli G, Decaro N, Lorusso E, Paradies P, Elia G, Martella V, Buonavoglia C, Ceci L (2008) First report of bovine anaplasmosis caused by *Anaplasma centrale* in Europe. Ann. N. Y. Acad. Sci. 1149: 107–110.

Charrel RN, Attoui H, Butenko AM, Clegg JC, Deubel V, Frolova TV, Gould EA, Gritsun TS, Heinz FX, Labuda M, Lashkevich VA, Loktev V, Lundkvist A, Lvov DV, Mandl CW, Niedrig M, Papa A, Petrov VS, Plyusnin A, Randolph S, Suss J, Zlobin VI, de Lamballerie X (2004) Tick-borne virus diseases of human interest in Europe. Clin. Microbiol. Infect. 10: 1040–1055.

Chaudhuri RP, Srivastava SC, Naithani RC (1969) On the biology of the ixodid tick, *Hyalomma* (*Hyalomma*) *anatolicum anatolicum* Koch, 1844 (Acarina: Ixodidae). Indian J. Anim. Sci. 39: 257–268.

Chilton NB (1994) Differences in the life cycles of two species of reptile tick: implications for species distributions. Int. J. Parasitol. 24: 791–795.

Chilton NB, Bull CM (1991) A comparison of the reproductive parameters of females of two reptile tick species. Int. J. Parasitol. 21: 907–911.

Chilton NB, Bull CM (1993) A comparison of the off-host survival times of larvae and nymphs of two species of reptile ticks. Int. J. Parasitol. 23: 693–696.

Chilton NB, Bull CM (1994) Influence of environmental factors on oviposition and egg development in *Amblyomma limbatum* and *Aponomma hydrosauri* (Acari: Ixodidae). Int. J. Parasitol. 24: 89–90.

Chilton NB, Andrews RH, Bull CM (2000) Influence of temperature and relative humidity on the moulting success of *Amblyomma limbatum* and *Aponomma hydrosauri* (Acari: Ixodidae) larvae and nymphs. Int. J. Parasitol. 30: 973–979.

Chiszar D, Smith HM, Petkus A, Dougherty J (1993) A fatal attack on a teenage boy by a captive Burmese python (*Python molurus bivittatus*) in Colorado. Bull. Chicago Herpetol. Soc. 28: 261–262.

Christy B (2008) The Lizard King: The True Crimes and Passions of the World's Greatest Reptile Smugglers. New York: Twelve. 241 pp.

Clark LG, Doten EH. (1995) Ticks on imported reptiles into Miami International Airport: November 1994 through January 1995. In: Proceedings for the Veterinary Epidemiology and Economics Symposium. Fort Collins, CO: U.S. Department of Agriculture. pp. 1A17–A25.

Clifford CM, Walker JB, Keirans JE (1983) Clarification of the status of *Rhipicephalus kochi* Donitz, 1905 (Ixodoidea, Ixodidae). Onderstepoort J. Vet. Res. 50: 77–89.

Clinton WJ (1999) Invasive species. Fed. Reg. 64: 6183–6186.

Comstedt P, Bergstrom S, Olsen B, Garpmo U, Marjavaara L, Mejlon H, Barbour AG, Bunikis J (2006) Migratory passerine birds as reservoirs of Lyme borreliosis in Europe. Emerg. Infect. Dis. 12: 1087–1095.

Connell M, Hall WTK (1972) Transmission of *Anaplasma marginale* by the cattle tick *Boophilus microplus*. Aust. Vet. J. 48: 477.

Cooley RA. (1946) The Genera *Boophilus*, *Rhipicephalus*, and *Haemaphysalis* (Ixodidae) of the New World. National Institute of Health bulletin no. 187. Washington, DC: U.S. Government Printing Office. 54 pp.

Cooley RA, Kohls GM (1944) The genus *Amblyomma* (Ixodidae) in the United States. J. Parasitol. 30: 77–111.

Coombs DW, Springer MD (1974) Parasites of feral pig x European wild boar hybrids in southern Texas. J. Wildl. Dis. 10: 436–441.

Corn JL, Barre N, Theibot B, Creekmore TE, Garris GI, Nettles VF (1993) Potential role of cattle egrets, *Bubulcus ibis* (Ciconiformes: Ardeidae), in the dissemination of *Amblyomma variegatum* (Acari: Ixodidae) in the eastern Caribbean. J. Med. Entomol. 30: 1029–1037.

Cornet JP (1995) Contribution a l'etude des tiques (Acarina: Ixodina) de la Republique Centrafricaine. 4. Inventaire et repartition. Acarologia 36: 203–212.

Cotte V, Bonnet S, le Rhun D, le Naour E, Chauvin A, Boulouis HJ, Lecuelle B, Lilin T, Vayssier-Taussat M (2008) Transmission of *Bartonella henselae* by *Ixodes ricinus*. Emerg. Infect. Dis. 14: 1074–1080.

Dantas-Torres F (2007) Rocky Mountain spotted fever. Lancet Infect. Dis. 7: 724–732.

Das HL, Subramanian G (1972a) The biology of *Hyalomma (Hyalomma) marginatum isaaci* (Sharif, 1928) Kaiser & Hoogstraal, 1963 (Acarina: Ixodidae). Acarologia 13: 496–501.

Das HL, Subramanian G (1972b) Biology of *Hyalomma (H.) dromedarii* Koch, 1844 (Acarina: Ixodidae). Indian J. Anim. Sci. 42: 285–289.

Daubney R (1930) Natural transmission of heart-water of sheep by *Amblyomma variegatum* (Fabricius 1794). Parasitology 22: 260–267.

Davies FG, Terpstra C (2004) Nairobi sheep disease. In: Coetzer JAW, Tustin RC, eds. Infectious Diseases of Livestock. 2nd ed. Oxford: Oxford University Press. pp. 1071–1076.

Davis RG (2004a) The ABCs of bioterrorism for veterinarians, focusing on Category A agents. J. Am. Vet. Med. Assoc. 224: 1084–1095.

Davis RG (2004b) The ABCs of bioterrorism for veterinarians, focusing on category B and C agents. J. Am. Vet. Med. Assoc. 224: 1096–1104.

de Moraes-Barros N, Silva JAB, Miyaki CY, Morgante JS (2006) Comparative phylogeography of the Atlantic forest endemic sloth (*Bradypus torquatus*) and the widespread three-toed sloth (*Bradypus variegatus*) (Bradypodidae, Xenarthra). Genetica 126: 189–198.

de Vos AJ (1981) *Rhipicephalus appendiculatus*: cause and vector of diseases in Africa. J. S. Afr. Vet. Assoc. 52: 315–322.

de Vos AJ, de Waal DT, Jackson LA (2004) Bovine babesiosis. In: Coetzer JAW, Tustin RC, eds. Infectious Diseases of Livestock. 2nd ed. Oxford: Oxford University Press. pp. 406–424.

de Waal DT (1990) The transovarial transmission of *Babesia caballi* by *Hyalomma truncatum*. Onderstepoort J. Vet. Res. 57: 99–100.

de Waal DT (2004) Porcine babesiosis. In: Coetzer JAW, Tustin RC, eds. Infectious Diseases of Livestock. 2nd ed. Oxford: Oxford University Press. pp. 435–437.

de Waal DT, van Heerden J (2004) Equine piroplasmosis. In: Coetzer JAW, Tustin RC, eds. Infectious Diseases of Livestock. 2nd ed. Oxford: Oxford University Press. pp. 425–434.

de Waal DT, Lopez Rebollar LM, Potgieter FT (1992) The transovarial transmission of *Babesia trautmanni* by *Rhipicephalus simus* to domestic pigs. Onderstepoort J. Vet. Res. 59: 219–221.

Delpy L, Gouchey SH (1937) Biologie de *Hyalomma dromedarii* (Koch 1844). Ann. Parasitol. 15: 487–499.

Delsing CE, Kullberg BJ (2008) Q fever in the Netherlands: a concise overview and implications of the largest ongoing outbreak. Netherlands J. Med. 66: 365–367.

Dennis DT, Inglesby TV, Henderson DA, Bartlett JG, Ascher MS, Eitzen E, Fine AD, Friedlander AM, Hauer J, Layton M, Lillibridge SR, McDade JE, Osterholm MT, O'Toole T, Parker G, Perl TM, Russell PK, Tonat K (2006) Tularemia as a biological weapon: medical and public health management. J. Am. Med. Assoc. 285: 2763–2773.

Diamant G (1965) The bont tick (*Amblyomma hebraeum*), exotic vector of heartwater found in the United States. Vet. Med. Small Anim. Clin. 60: 847–850.

Dias E, Martins AV (1939) Spotted fever in Brazil: a summary. Am. J. Trop. Med. 19: 103–108.

Donnelly J, Peirce MA (1975) Experiments on the transmission of *Babesia divergens* to cattle by the tick *Ixodes ricinus*. Int. J. Parasitol. 5: 363–367.

Doss MA, Farr MM, Roach KF, Anastos G (1974a) Index-catalogue of Medical and Veterinary Zoology, Special Publication no. 3, Ticks and Tickborne Diseases. I. Genera and Species of Ticks. Part 1. Genera A-G. Washington, DC: U.S. Government Printing Office. 429 pp.

Doss MA, Farr MM, Roach KF, Anastos G (1974b) Index-catalogue of Medical and Veterinary Zoology, Special Publication no. 3, Ticks and Tickborne Diseases. I. Genera and Species of Ticks. Part 3. Genera O-X. Washington, DC: U.S. Government Printing Office. 329 pp.

Dower KM, Petney TN, Horak IG (1988) The developmental success of *Amblyomma hebraeum* and *Amblyomma marmoreum* on the leopard tortoise, *Geochelone pardalis*. Onderstepoort J. Vet. Res. 55: 11–13.

Dumler JS, Bakken JS (1998) Human ehrlichioses: newly recognized infections transmitted by ticks. Annu. Rev. Med. 49: 201–213.

Dumler JS, Choi KS, Garcia-Garcia JC, Barat NS, Scorpio DG, Garyu JW, Grab DJ, Bakken JS (2005) Human granulocytic anaplasmosis and *Anaplasma phagocytophilum*. Emerg. Infect. Dis. 11: 1828–1834.

Dumpis U, Crook D, Oksi J (1999) Tick-borne encephalitis. Clin. Infect. Dis. 28: 882–890.

Duncan IM (1989) The use of flumethrin pour-on for de-ticking black rhinoceros (*Diceros bicornis*) prior to translocation in Zimbabwe. J. S. Afr. Vet. Assoc. 60: 195–197.

Duncan IM (1992) Tick control on cattle with flumethrin pour-on through a Duncan applicator. J. S. Afr. Vet. Assoc. 63: 125–127.

Dunn LH (1915) Observations on the preoviposition, oviposition and incubation periods of *Dermacentor nitens* in Panama (Arach., Acar.). Entomol. News 26: 214–219.

Dunn LH (1918) Studies on the iguana tick, *Amblyomma dissimile*, in Panama. J. Parasitol. 5: 1–10.

Dunn LH (1923) The ticks of Panama, their hosts, and diseases they transmit. Am. J. Trop. Med. 3: 91–104.

Durden LA, Kollars TM (1992) An annotated list of the ticks (Acari: Ixodoidea) of Tennessee, with records of four exotic species for the United States. Bull. Soc. Vector Ecol. 17: 125–131.

Durden LA, Klompen JSH, Keirans JE (1993) Parasitic arthropods of sympatric opossums, cotton rats, and cotton mice from Merritt Island, Florida. J. Parasitol. 79: 283–286.

Eads RB, Menzies GC, Hightower BG (1956) The ticks of Texas, with notes on their medical significance. Texas J. Sci. 8: 7–24.

Eads RB, Campos EG, Trevino HA (1966) Quarantine problems associated with the importation of bananas from Mexico. J. Econ. Entomol. 59: 896–899.

Eddy GW, Joyce CR (1942) Ticks collected on the Tama (Iowa) Indian Reservation with notes on other species. Iowa State Coll. J. Sci. 16: 539–543.

Edwards EE, Arthur DR (1947) The seasonal activity of the tick, *Ixodes ricinus* L., in Wales. Parasitology 38: 72–85.

Elbl A, Anastos G (1966a) Ixodid Ticks (Acarina, Ixodidae) of Central Africa. Volume I. General Introduction. Genus *Amblyomma* Koch, 1844. Tervuren, Belgium: Musee Royal de l'Afrique Centrale. 275 pp.

Elbl A, Anastos G (1966b) Ixodid Ticks (Acarina, Ixodidae) of Central Africa. Volume III. Genus *Rhipicephalus* Koch, 1844. Tervuren, Belgium: Musee Royal de l'Afrique Centrale. 555 pp.

Elbl A, Anastos G (1966c) Ixodid Ticks (Acarina, Ixodidae) of Central Africa. Volume IV. Genera *Aponomma* Neumann, 1899, *Boophilus* Curtice, 1891, *Dermacentor* Koch, 1844, *Haemaphysalis* Koch, 1844, *Hyalomma* Koch, 1844, and *Rhipicentor* Nuttall and Warburton, 1908. Lists and Bibliography. Tervuren, Belgium: Musee Royal de l'Afrique Centrale. 412 pp.

Ellis J, Oyston PCF, Green M, Titball RW (2002) Tularemia. Clin. Microbiol. Rev. 15: 631–646.

Enge KM, Krysko KL, Hankins KR, Campbell TS, King FW (2004) Status of the Nile monitor (*Varanus niloticus*) in southwestern Florida. Southeast. Nat. 3: 571–582.

Engeman R, Woolard JW, Perry ND, Witmer G, Hardin S, Brashears L, Smith H, Muiznieks B, Constantin B (2006) Rapid assessment for a new invasive species threat: the case of the Gambian giant pouched rat in Florida. Wildl. Res. 33: 439–448.

Erasmus LD (1952) Regional tick paralysis: sensory and motor changes caused by a male tick, genus *Hyalomma*. S. Afr. Med. J. 26: 985–987.

Ergonul O (2006) Crimean-Congo haemorrhagic fever. Lancet Infect. Dis. 6: 203–214.

Estrada-Pena A (2001) Forecasting habitat suitability for ticks and prevention of tick-borne diseases. Vet. Parasitol. 98: 111–132.

Estrada-Pena A, Jongejan F (1999) Ticks feeding on humans: a review of records on human-biting Ixodoidea with special reference to pathogen transmission. Exp. Appl. Acarol. 23: 685–715.

Estrada-Pena A, Guglielmone AA, Mangold AJ, Castella J (1993) A description of *Amblyomma tigrinum* Koch, *A. neumanni* Ribaga, and *A. testudinis* (Conil) immatures (Acarina: Ixodidae). Folia Parasitol. 40: 147–153.

Estrada-Pena A, Bouattour A, Camicas JL, Walker AR (2004a) Ticks of Domestic Animals in the Mediterranean Region: A Guide to Identification of Species. Zaragoza, Spain: University of Zaragoza. 131 pp.

Estrada-Pena A, Guglielmone AA, Mangold AJ (2004b) The distribution and ecological 'preferences' of the tick *Amblyomma cajennense* (Acari: Ixodidae), an ectoparasite of humans and other mammals in the Americas. Ann. Trop. Med. Parasitol. 98: 283–292.

Estrada-Pena A, Acedo CS, Quilez J, del Cacho E (2005) A

retrospective study of climatic suitability for the tick *Rhipicephalus* (*Boophilus*) *microplus* in the Americas. Global Ecol. Biogeogr. 14: 565–573.

Estrada-Pena A, Bouattour A, Camicas JL, Guglielmone AA, Horak I, Jongejan F, Latif A, Pegram R, Walker AR (2006) The known distribution and ecological preferences of the tick subgenus *Boophilus* (Acari: Ixodidae) in Africa and Latin America. Exp. Appl. Acarol. 38: 219–235.

Estrada-Pena A, Pegram RG, Barre N, Venzal JM (2007) Using invaded range data to model the climate suitability for *Amblyomma variegatum* (Acari: Ixodidae) in the New World. Exp. Appl. Acarol. 41: 203–214.

Evans GO (1951a) The seasonal incidence of *Ixodes ricinus* (L.) on cattle in mid-Wales. Bull. Entomol. Res. 41: 459–468.

Evans GO (1951b) The distribution and economic importance of *Ixodes ricinus* (L.) in Wales and the Welsh border counties with special reference to N. W. Cardiganshire. Bull. Entomol. Res. 41: 469–485.

Evans GO (1951c) The distribution of *Ixodes ricinus* (L.) on the body of cattle and sheep. Bull. Entomol. Res. 41: 709–723.

Falotico T, Labruna MB, Verderane MP, de Resende BD, Izar P, Ottoni EB (2007) Repellent efficacy of formic acid and the abdominal secretion of carpenter ants (Hymenoptera: Formicidae) against *Amblyomma* ticks (Acari: Ixodidae). J. Med. Entomol. 44: 718–721.

Faye O, Cornet JP, Camicas JL, Fontenille D, Gonzalez JP (1999) [Experimental transmission of Crimean-Congo haemorrhagic fever virus: role of three vectorial species in maintenance and transmission cycles in Senegal.] Parasite 6: 27–32.

Feldman-Muhsam B (1953) *Rhipicephalus bursa* in Israel. Bull. Res. Counc. Israel 3: 201–206.

Feldman-Muhsam B (1955) On two rare genera of ticks of domestic stock in Israel. Bull. Res. Counc. Israel 5B: 193–194.

Feldman-Muhsam B, Saturen IM (1961) Notes on the ecology of ixodid ticks of domestic stock in Israel. Bull. Res. Counc. Israel 10B: 53–61.

Fielden LJ, Rechav Y (1994) Attachment sites of the tick *Amblyomma marmoreum* on its tortoise host, *Geochelone pardalis*. Exp. Appl. Acarol. 18: 339–349.

Fielden LJ, Magano S, Rechav Y (1992) Laboratory studies on the life cycle of *Amblyomma marmoreum* (Acari: Ixodidae) on two different hosts. J. Med. Entomol. 29: 750–756.

Floch H, Fauran P (1958) Ixodides de la Guyane et des Antilles Francaises. Arch. Inst. Pasteur Guyane Francaise Inini 446: 1–94.

Floch H, Fauran P (1959) Sur les ixodides, autres que ceux du genre *Amblyomma* en Guyane et aux Antilles Francaises. Acarologia 1: 393–407.

Forlano M, Scofield A, Elisei C, Fernandes KR, Ewing SA, Massard CL (2005) Diagnosis of *Hepatozoon* spp. in *Amblyomma ovale* and its experimental transmission in domestic dogs in Brazil. Vet. Parasitol. 134: 1–7.

Foster GW, Moler PE, Kinsella JM, Terrell SP, Forrester DJ (2000) Parasites of eastern indigo snakes (*Drymarchon corais couperi*) from Florida, USA. Comp. Parasitol. 67: 124–128.

Fourie LJ, Horak IG (2000) Status of Dorper sheep as hosts of ectoparasites. Small Rumin. Res. 36: 159–164.

Fournier PE, Raoult D (2005) Mediterranean spotted fever and other tick-borne rickettsioses. In: Goodman JL, Dennis DT, Sonenshine DE, eds. Tick-borne Diseases of Humans. Washington, DC: ASM Press. pp. 302–327.

Fournier PE, Zhu Y, Yu X, Raoult D (2006) Proposal to create subspecies of *Rickettsia sibirica* and an amended description of *Rickettsia sibirica*. Ann. N. Y. Acad. Sci. 1078: 597–606.

Franke J, Telecky TM (2001) Reptiles as Pets: An Examination of the Trade in Live Reptiles in the United States. Washington, DC: Humane Society of the United States. 146 pp.

Friedhoff KT (1988) Transmission of *Babesia*. In: Ristic M, ed. Babesiosis of Domestic Animals and Man. Boca Raton, FL: CRC Press. pp. 23–52.

Friedhoff KT (1997) Tick-borne diseases of sheep and goats caused by *Babesia*, *Theileria* or *Anaplasma* spp. Parassitologia 39: 99–109.

Garben AFM, Vos H, van Bronswijk JEMH (1981) *Haemaphysalis punctata* Canestrini and Fanzago 1877, a tick of pastured seadunes on the island of Texel (The Netherlands). Acarologia 23: 19–25.

Garris GI (1984) Colonization and life cycle of *Amblyomma variegatum* (Acari: Ixodidae) in the laboratory in Puerto Rico. J. Med. Entomol. 21: 86–90.

Garris GI (1987) *Amblyomma variegatum* (Acari: Ixodidae): population dynamics and hosts used during an eradication program in Puerto Rico. J. Med. Entomol. 24: 82–86.

Garris GI, Bokma BH, Strickland RK, Combs GP (1989) Evaluation of the eradication program for *Amblyomma variegatum* (Acari: Ixodidae) on Puerto Rico. Exp. Appl. Acarol. 6: 67–76.

Gear JHS (1992) Tick-bite fever (tick typhus) in southern Africa. In: Fivaz B, Petney T, Horak I, eds. Tick Vector Biology, Medical and Veterinary Aspects. Berlin: Springer-Verlag. pp. 135–142.

Gear J, de Meillon B (1941) The hereditary transmission of the rickettsiae of tick-bite fever through the common dog-tick, *Haemaphysalis leachi*. S. Afr. Med. J. 15: 389–392.

George JE (1990) Wildlife as a constraint to the eradication of *Boophilus* spp. (Acari: Ixodidae). J. Agric. Entomol. 7: 119–125.

George JE, Davey RB, Pound JM (2002) Introduced ticks and tick-borne diseases: the threat and approaches to eradication. Vet. Clin. Food Anim. Pract. 18: 401–416.

Gilot B, Robin Y, Pautou G, Moncada E, Vigny F, Guichard JP (1974) Ecologie et role pathogene de *Dermacentor reticulatus* (Fabricius, 1794) (Ixodoidea) dans le sud-est de la France. Acarologia 16: 220–249.

Gilot B, Pautou G, Moncada E, Ain G (1975) Premiere contribution a l'etude ecologique d'*Ixodes ricinus* (Linne, 1758) (Acarina, Ixodoidea) dans le sud-est de la France. Acta Trop. 32: 232–258.

Goddard J, Norment BR (1983) Notes on the geographical distribution of the Gulf Coast tick, *Amblyomma maculatum* (Koch) [Acari: Ixodidae]. Entomol. News 94: 103–104.

Gomes, AF (1993) The tick vectors of cowdriosis in Angola. Rev. Elev. Med. Vet. Pays Trop. 46: 237–243.

Gonzalez-Acuna D, Guglielmone AA (2005) Ticks (Acari: Ixodoidea: Argasidae, Ixodidae) of Chile. Exp. Appl. Acarol. 35: 147–163.

Goodman JL (2005) Human granulocytic anaplasmosis (ehrlichiosis). In: Goodman JL, Dennis DT, Sonenshine DE, eds. Tick-borne Diseases of Humans. Washington, DC: ASM Press. pp. 218–238.

Gorenflot A, Moubri K, Precogout E, Carcy B, Schetters TPM (1998) Human babesiosis. Ann. Trop. Med. Parasitol. 92: 489–501.

Gothe R, Neitz AWH (1991) Tick paralyses: pathogenesis and etiology. Adv. Dis. Vector Res. 8: 177–204.

Gothe R, Gold Y, Bezuidenhout JD (1986) Investigations into the paralysis-inducing ability of *Rhipicephalus evertsi mimeticus* and that of hybrids between this subspecies and *Rhipicephalus evertsi evertsi*. Onderstepoort J. Vet. Res. 53: 25–29.

Graham, OH, Hourrigan JL (1977) Eradication programs for the arthropod parasites of livestock. J. Med. Entomol. 13: 629–658.

Granstrom M (1997) Tick-borne zoonoses in Europe. Clin. Microbiol. Infect. 3: 156–169.

Graves S, Stenos J (2003) *Rickettsia honei*: a spotted fever group rickettsia on three continents. Ann. N. Y. Acad. Sci. 990: 62–66.

Gray JS (1984) Studies on the dynamics of active populations of the sheep tick, *Ixodes ricinus* L. in Co. Wicklow, Ireland. Acarologia 25: 167–178.

Gray JS (1991) The development and seasonal activity of the tick *Ixodes ricinus*: a vector of Lyme borreliosis. Rev. Med. Vet. Entomol. 79: 323–333.

Gray JS (1998) The ecology of ticks transmitting Lyme borreliosis. Exp. Appl. Acarol. 22: 249–258.

Gray JS, de Vos AJ (1981) Studies on a bovine *Babesia* transmitted by *Hyalomma marginatum rufipes* Koch, 1844. Onderstepoort J. Vet. Res. 48: 215–223.

Gray JS, Potgieter FT (1982) Studies on the infectivity of *Boophilus decoloratus* males and larvae infected with *Babesia bigemina*. Onderstepoort J. Vet. Res. 49: 1–2.

Gray JS, Turley T, Strickland KL (1978) Studies on the ecology of sheep tick, *Ixodes ricinus* in Co. Wicklow, Ireland. Irish Vet. J. 32: 25–34.

Gray J, von Stegingk LV, Gurtelschmid M, Granstrom M (2002) Transmission studies of *Babesia microti* in *Ixodes ricinus* ticks and gerbils. J. Clin. Microbiol. 40: 1259–1263.

Greca H, Langoni H, Souza LC (2008) Brazilian spotted fever: a reemergent zoonosis. J. Venom. Anim. Toxins Trop. Dis. 14: 3–18.

Gresikova M, Kaluzova M (1997) Biology of tick-borne encephalitis virus. Acta Virol. 41: 115–124.

Gritsun, TS, Lashkevich VA, Gould EA (2003) Tick-borne encephalitis. Antiviral Res. 57: 129–146.

Guglielmone AA (1990) Sites of attachment in *Amblyomma triguttatum triguttatum* Koch (Acari: Ixodidae) on natural hosts. Ann. Parasitol. Hum. Comp. 65: 145–148.

Guglielmone AA (1994) The seasonal occurrence of *Amblyomma triguttatum triguttatum* Koch (Acari: Ixodidae). Acarologia 35: 107–113.

Guglielmone AA, Moorhouse DE (1986) Reproduction in *Amblyomma triguttatum triguttatum*. Acarologia 27: 235–239.

Guglielmone AA, Mangold AJ, Keirans JE (1990) Redescription of the male and female of *Amblyomma parvum* Aragao, 1908, and description of the nymph and larva, and description of all stages of *Amblyomma pseudoparvum* sp. n. (Acari: Ixodida: Ixodidae). Acarologia 31: 143–159.

Guglielmone AA, Mangold AJ, Garcia MD (1991) The life cycle of *Amblyomma parvum* Aragao, 1908 (Acari: Ixodidae) under laboratory conditions. Exp. Appl. Acarol. 13: 129–136.

Guglielmone AA, Luciani CA, Mangold AJ (2001) Aspects of the ecology of *Amblyomma argentinae* Neumann, 1904 [=*Amblyomma testudinis* (Conil, 1877)] (Acari: Ixodidae). Syst. Appl. Acarol. Spec. Publ. 8: 1–12.

Guglielmone AA, Estrada-Pena A, Keirans JE, Robbins RG (2003a) Ticks (Acari: Ixodida) of the Neotropical Zoogeographic Region. Houten, The Netherlands: Atalanta. 173 pp.

Guglielmone AA, Estrada-Pena A, Luciani CA, Mangold AJ, Keirans JE (2003b) Hosts and distribution of *Amblyomma auricularium* (Conil 1878) and *Amblyomma pseudoconcolor* Aragao, 1908 (Acari: Ixodidae). Exp. Appl. Acarol. 29: 131–139.

Guglielmone AA, Estrada-Pena A, Mangold AJ, Barros-Battesti DM, Labruna MB, Martins JR, Venzal JM, Arzua M, Keirans JE (2003c) *Amblyomma aureolatum* (Pallas, 1772) and *Amblyomma ovale* Koch, 1844 (Acari: Ixodidae): hosts, distribution and 16S rDNA sequences. Vet. Parasitol. 113: 273–288.

Guglielmone AA, Robbins RG, Apanaskevich DA, Petney TN, Estrada-Pena A, Horak IG (2009) Comments on controversial tick (Acari: Ixodida) species names and species described or resurrected from 2003 to 2008. Exp. Appl. Acarol. 48: 311–327.

Hadrill DJ, Boid R, Jones TW, Bell-Sakyi L (1990) Bovine babesiosis on Nevis—implications for tick control. Vet. Rec. 126: 403–404.

Hamel HD, Gothe R (1978) Influence of infestation rate on tick-paralysis in sheep induced by *Rhipicephalus evertsi evertsi* Neumann, 1897. Vet. Parasitol. 4: 183–191.

Hammond DL, Dorsett WA (1988) Tick infestation in a ball python (*Python regius*). Companion Anim. Pract. 2: 39–40.

Hanson BA, Frank PA, Mertins JW, Corn JL (2007) Tick paralysis of a snake caused by *Amblyomma rotundatum* (Acari: Ixodidae). J. Med. Entomol. 44: 155–157.

Hengge UR, Tannapfel A, Tyring SK, Erbel R, Arendt G, Ruzicka T (2003) Lyme borreliosis. Lancet Infect. Dis. 3: 489–500.

Hightower BG, Lehman VW, Eads RB (1953) Ectoparasites

from mammals and birds on a quail preserve. J. Mammal. 34: 268–271.

Hillman-Smith AKK, Groves CP (1994) *Diceros bicornis*. Mammal. Spec. 455: 1–8.

Hixson H (1940) Field biology and environmental relationships of the Gulf Coast tick in southern Georgia. J. Econ. Entomol. 33: 179–189.

Hoogstraal H (1953) Ticks (Ixodoidea) of the Malagasy faunal region (excepting the Seychelles): their origins and host relationships, with descriptions of five new *Haemaphysalis* species. Bull. Mus. Comp. Zool. 111: 36–113.

Hoogstraal H (1956) African Ixodoidea. I. Ticks of the Sudan (with Special Reference to Equatorial Province and with Preliminary Reviews of the Genera *Boophilus*, *Margaropus*, and *Hyalomma*). Research report NM 005 050.29.07. Washington, DC: Bureau of Medicine & Surgery, U.S. Department of the Navy. 1101 pp.

Hoogstraal H (1967) Ticks in relation to human diseases caused by *Rickettsia* species. Annu. Rev. Entomol. 12: 377–420.

Hoogstraal H (1970) Identity, distribution, and hosts of *Haemaphysalis* (*Rhipistoma*) *indica* Warburton (resurrected) (Ixodoidea: Ixodidae), a carnivore parasite of the Indian subregion. J. Parasitol. 56: 1013–1022.

Hoogstraal H, Kaiser MN (1958) Observations on Egyptian *Hyalomma* ticks (Ixodoidea, Ixodidae). 2. Parasitism of migrating birds by immature *H. rufipes* Koch. Ann. Entomol. Soc. Am. 51: 12–16.

Hoogstraal H, Kaiser MN (1959a) Observations on Egyptian *Hyalomma* ticks (Ixodoidea, Ixodidae). 5. Biological notes and differences in identity of *H. anatolicum* and its subspecies *anatolicum* Koch and *excavatum* Koch among Russian and other workers. Identity of *H. lusitanicum* Koch. Ann. Entomol. Soc. Am. 52: 243–261.

Hoogstraal H, Kaiser MN (1959b) Ticks (Ixodoidea) of Arabia with special references to the Yemen. Fieldiana Zool. 39: 297–322.

Hoogstraal H, Kaiser MN (1960) Observations on ticks (Ixodoidea) of Libya. Ann. Entomol. Soc. Am. 53: 445–457.

Hoogstraal H, Kaiser MN (1961a) Ticks from European-Asiatic birds migrating through Egypt into Africa. Science 133: 277–278.

Hoogstraal H, Kaiser MN (1961b) Records of *Hunterellus theilerae* Fiedler (Encyrtidae, Chalcidoidea) parasitizing *Hyalomma* ticks on birds migrating through Egypt. Ann. Entomol. Soc. Am. 54: 616–617.

Hoogstraal H, Wassef HY (1985) *Dermacentor* (*Indocentor*) *auratus* (Acari: Ixodoidea: Ixodidae): hosts, distribution, and medical importance in tropical Asia. J. Med. Entomol. 22: 170–177.

Hoogstraal H, Kaiser MN, Traylor MA, Gaber S, Guindy E (1961) Ticks (Ixodoidea) on birds migrating from Africa to Europe and Asia. Bull. World Health Org. 24: 197–212.

Hoogstraal H, Kaiser MN, Traylor MA, Guindy E, Gaber S (1963) Ticks (Ixodidae) on birds migrating from Europe and Asia to Africa, 1959–1961. Bull. World Health Org. 28: 235–262.

Hoogstraal H, Traylor MA, Gaber S, Malakatis G, Guindy E, Helmy I (1964) Ticks (Ixodoidea) on migrating birds in Egypt, spring and fall 1962. Bull. World Health Org. 30: 355–367.

Hoogstraal H, Trapido H, Kohls GM (1965) Studies on southeast Asian *Haemaphysalis* ticks (Ixodoidea, Ixodidae). The identity, distribution, and hosts of *H.* (*Kaiseriana*) *hystricis* Supino. J. Parasitol. 51: 467–480.

Hoogstraal H, Roberts FHS, Kohls GM, Tipton VJ (1968) Review of *Haemaphysalis* (*Kaiseriana*) *longicornis* Neumann (resurrected) of Australia, New Zealand, New Caledonia, Fiji, Japan, Korea, and northeastern China and USSR, and its parthenogenetic and bisexual populations (Ixodoidea, Ixodidae). J. Parasitol. 54: 1197–1213.

Hoogstraal H, Wassef HY, Uilenberg G (1974) *Haemaphysalis* (*Elongiphysalis*) *elongata* Neumann subgen. n. (Ixodoidea: Ixodidae): structural variation, hosts, and distribution in Madagascar. J. Parasitol. 60: 480–498.

Hoogstraal H, Wassef HY, Buttiker W (1981) Ticks (Acarina) of Saudi Arabia, fam. Argasidae, Ixodidae. Fauna Saudi Arabia 3: 25–110.

Hooshmand-Rad P, Hawa NJ (1973) Transmission of *Theileria hirci* in sheep by *Hyalomma anatolicum anatolicum*. Trop. Anim. Health Prod. 5: 103–109.

Hoover C (1998) The U.S. Role in the International Live Reptile Trade: Amazon Tree Boas to Zululand Dwarf Chameleons. Washington, DC: TRAFFIC North America. 59 pp.

Horak IG, MacIvor KMF, Petney TN, de Vos V (1987) Some avian and mammalian hosts of *Amblyomma hebraeum* and *Amblyomma marmoreum* (Acari: Ixodidae). Onderstepoort J. Vet. Res. 54: 397–403.

Horak IG, Camicas JL, Keirans JE (2002) The Argasidae, Ixodidae and Nuttalliellidae (Acari: Ixodida): a world list of valid tick names. Exp. Appl. Acarol. 28: 27–54.

Horak IG, Golezardy H, Uys AC (2006a) The host status of African buffaloes, *Syncerus caffer*, for *Rhipicephalus* (*Boophilus*) *decoloratus*. Onderstepoort J. Vet. Res. 73: 193–198.

Horak IG, McKay IJ, Heyne H, Spickett AM (2006b) Hosts, seasonality and geographic distribution of the South African tortoise tick, *Amblyomma marmoreum*. Onderstepoort J. Vet. Res. 74: 13–25.

Hourrigan JL, Strickland RK, Kelsey OL, Knisely BE, Crago CC, Whittaker S, Gilhooly DJ (1969) Eradication efforts against tropical bont tick, *Amblyomma variegatum*, in the Virgin Islands. J. Am. Vet. Med. Assoc. 154: 540–545.

Hubalek Z (2004) An annotated checklist of pathogenic microorganisms associated with migratory birds. J. Wildl. Dis. 40: 639–659.

Hubalek Z, Sixl W, Halouzka J, Mikulaskova M (1997) Prevalence of *Francisella tularensis* in *Dermacentor reticulatus* ticks collected in adjacent areas of the Czech and Austrian Republics. Central Eur. J. Public Health 4: 199–201.

Hyslop NSG (1980) Dermatophilosis (streptothricosis) in animals and man. Comp. Immunol. Microbiol. Infect. Dis. 2: 389–404.

Ioffe-Uspensky I, Mumcuoglu KY, Uspensky I, Galun R (1997) *Rhipicephalus sanguineus* and *R. turanicus* (Acari: Ixodidae): closely related species with different biological characteristics. J. Med. Entomol. 34: 74–81.

Jakowska S (1972) Lesions produced by ticks, *Amblyomma dissimile*, in *Bufo marinus* toads from the Dominican Republic. Am. Zool. 12: 731.

Jenkins PT, Genovese K, Ruffler H (2007) Broken Screens: the Regulation of Live Animal Imports in the United States. Washington, DC: Defenders of Wildlife. 56 pp.

Jensenius M, Fournier PE, Kelly P, Myrvang B, Raoult D (2003) African tick bite fever. Lancet Infect. Dis. 3: 557–564.

Jiang J, Sangkasuwan V, Lerdthusnee K, Sukwit S, Chuenchitra T, Rozmajzl PJ, Eamsila C, Jones JW, Richards AL (2005) Human infection with *Rickettsia honei*, Thailand. Emerg. Infect. Dis. 11: 1473–1475.

Johnson-Delaney CA (1996) Reptile zoonosis and threats to public health. In: Mader DR, ed. Reptile Medicine and Surgery. Philadelphia: W. B. Saunders. pp. 20–33.

Jones EK, Clifford CM, Keirans JE, Kohls GM (1972) The Ticks of Venezuela (Acarina: Ixodoidea) with a Key to the Species of *Amblyomma* in the Western Hemisphere. Provo, UT: Brigham Young University. 40 pp.

Jongejan F (1992) Experimental transmission of *Cowdria ruminantium* (Rickettsiales) by the American reptile tick *Amblyomma dissimile* Koch, 1844. Exp. Appl. Acarol. 15: 117–121.

Jongejan F, Uilenberg G (2004) The global importance of ticks. Parasitology 129: S3–S14.

Jongejan F, Zivkovic D, Pegram RG, Tatchell RJ, Fison T, Latif AA, Paine G (1987) Ticks (Acari: Ixodidae) of the Blue and White Nile ecosystems in the Sudan with particular reference to the *Rhipicephalus sanguineus* group. Exp. Appl. Acarol. 3: 331–346.

Jonsson NN (2006) The productivity effects of cattle tick (*Boophilus microplus*) infestation on cattle, with particular reference to *Bos indicus* cattle and their crosses. Vet. Parasitol. 137: 1–10.

Kaiser MN, Hoogstraal H (1958) *Hunterellus theileri* Fiedler (Encyrtidae, Chalcidoidea) parasitizing an African *Hyalomma* tick on a migrant bird in Egypt. J. Parasitol. 44: 392.

Kaiser MN, Hoogstraal H (1964) The *Hyalomma* ticks (Ixodoidea, Ixodidae) of Pakistan, India and Ceylon, with keys to subgenera and species. Acarologia 6: 257–286.

Kaiser MN, Hoogstraal H, Watson GE (1974) Ticks (Ixodoidea) on migrating birds in Cyprus, fall 1967 and spring 1968, and epidemiological considerations. Bull. Entomol. Res. 64: 97–110.

Karrar G (1965) Further studies on the epizootiology of heartwater in The Sudan. Sudan J. Vet. Sci. Anim. Husb. 6: 83–85.

Kaufman TS (1972) A Revision of the Genus *Aponomma* Neumann, 1899 (Acarina: Ixodidae). College Park, MD: University of Maryland. 238 pp.

Keirans JE (1984) George Henry Falkiner Nuttall and the Nuttall Tick Catalogue. Washington, DC: Agricultural Research Service, U.S. Department of Agriculture. 1785 pp.

Keirans JE (1993) *Dermacentor rhinocerinus* (Denny 1843) (Acari: Ixodida: Ixodidae): redescription of the male, female and nymph and first description of the larva. Onderstepoort J. Vet. Res. 60: 59–68.

Keirans JE, Durden LA (2001) Invasion: exotic ticks (Acari: Argasidae, Ixodidae) imported into the United States. A review and new records. J. Med. Entomol. 38: 850–861.

Keirans JE, Klompen JSH (1996) *Amblyomma quadricavum* (Schulze) (new combination), and *Amblyomma arianae* Keirans and Garris, a new junior synonym of *Amblyomma quadricavum* (Acari: Ixodidae). Proc. Entomol. Soc. Wash. 98: 164–165.

Keirans JE, Oliver JH (1993) First description of the male and redescription of the immature stages of *Amblyomma rotundatum* (Acari: Ixodidae), a recently discovered tick in the U.S.A. J. Parasitol. 79: 860–865.

Keirans JE, Restifo RA (1993) *Haemaphysalis juxtakochi* Cooley (Acari: Ixodidae), a neotropical tick species, found in Ohio. J. Med. Entomol. 30: 1074–1075.

Kelly PJ (2004) Q fever. In: Coetzer JAW, Tustin RC, eds. Infectious Diseases of Livestock. 2nd ed. Oxford: Oxford University Press. pp. 565–572.

Kelly PJ (2006) *Rickettsia africae* in the West Indies. Emerg. Infect. Dis. 12: 224–226.

Kelly PJ, Mason PR (1991) Transmission of a spotted fever group rickettsia by *Amblyomma hebraeum* (Acari: Ixodidae). J. Med. Entomol. 28: 598–600.

Kelly PJ, Beati L, Mason PR, Matthewman LA, Roux V, Raoult D (1996) *Rickettsia africae* sp. nov., the etiological agent of African tick bite fever. Int. J. Syst. Bacteriol. 46: 611–614.

Kenny MJ, Shaw SE, Hillyard PD, Forbes AB (2004) Ectoparasite and haemoparasite risks associated with imported exotic reptiles. Vet. Rec. 154: 434–435.

Ketchum HR, Teel PD, Strey OF, Longnecker MT (2005) Feeding predilection of the Gulf Coast tick, *Amblyomma maculatum* Koch, nymphs on cattle. Vet. Parasitol. 133: 349–356.

Khan MH (1993) Studies on host range and incidence of *Hyalomma* (H.) *marginatum isaaci*. Indian J. Anim. Health 32: 155–158.

Khan MH, Srivastava SC (1994) Studies on the host-range and incidence of *Hyalomma* (H.) *dromedarii* Koch, 1844. J. Vet. Parasitol. 8: 21–25.

Kim S, Guirgis S, Harris D, Keelan T, Mayer M, Zaki M, Steinert L, Benach J, White D, Lyman DO, Ormsbee R (1978) Q fever—New York. Morbid. Mortal. Weekly Rep. 27: 321–322/327.

Kjemtrup AM, Conrad PA (2000) Human babesiosis: an emerging tick-borne disease. Int. J. Parasitol. 30: 1323–1337.

Klushkina EA (1972) [The occurrence of *Amblyomma gemma* Don. (Ixodidae) in the Crimea.] Parazitologiia 6: 306.

Knight MM, Norval RAI, Rechav Y (1978) The life cycle of the tick *Hyalomma marginatum rufipes* Koch (Acarina: Ixodidae) under laboratory conditions. J. Parasitol. 64: 143–146.

Kocan KM, de la Fuente J, Guglielmone AA, Melendez RD (2003) Antigens and alternatives for control of *Anaplasma marginale* infection in cattle. Clin. Microbiol. Rev. 16: 698–712.

Kohler G, Hoffmann G, Janitschke K, Wiesenhutter E (1967) Untersuchungen zur Kenutris der Zeckenfauna Syriens. Z. Tropenmed. Parasitol. 18: 375–381.

Kohls GM (1960) Records and new synonymy of New World *Haemaphysalis* ticks, with descriptions of the nymph and larva of *H. juxtakochi* Cooley. J. Parasitol. 46: 355–361.

Kok DJ, Fourie LJ (1995) The role of *Hyalomma* ticks in foot infestations and temporary lameness of sheep in a semi-arid region of South Africa. Onderstepoort J. Vet. Res. 62: 201–206.

Kolonin GV (1995) Review of the ixodid fauna (Acari: Ixodidae) of Vietnam. J. Med. Entomol. 32: 276–282.

Kolonin GV (2009) Fauna of Ixodid Ticks of the World (Acari, Ixodidae). Moscow. http://www.kolonin.org/1.html.

Kozuch O, Nosek J (1971) Transmission of tick-borne encephalitis (TBE) virus by *Dermacentor marginatus* and *D. reticulatus* ticks. Acta Virol. 15: 334.

Krysko KL, Nifong JC, Snow RW, Enge KM, Mazzotti FJ (2008) Reproduction of the Burmese python (*Python molurus bivittatus*) in southern Florida. Appl. Herpetol. 5: 93–95.

Labruna MB, Camargo LM, Terrassini FA, Schumaker TT, Camargo EP (2002) Notes on parasitism by *Amblyomma humerale* (Acari: Ixodidae) in the state of Rondonia, western Amazon, Brazil. J. Med. Entomol. 39: 814–817.

Labruna MB, Keirans JE, Camargo LMA, Ribeiro AF, Soares RM, Camargo EP (2005) *Amblyomma latepunctatum*, a valid tick species (Acari: Ixodidae) long misidentified with both *Amblyomma incisum* and *Amblyomma scalpturatum*. J. Parasitol. 91: 527–541.

Labruna MB, Naranjo V, Mangold AJ, Thompson C, Estrada-Pena A, Guglielmone AA, Jongejan F, de la Fuente J (2009) Allopatric speciation in ticks: genetic and reproductive divergence between geographic strains of *Rhipicephalus (Boophilus) microplus*. BMC Evolut. Biol. 9: 46.

Labuda M, Nuttall PA (2004) Tick-borne viruses. Parasitology 129: S221–S245.

Landau I, Paperna I (1997) The assignment of *Hepatozoon mauritanicum*, a tick-transmitted parasite of tortoise, to the genus *Hemolivia*. Parasite 4: 365–367.

Lawrence JA (2004a) Theileriosis of sheep and goats. In: Coetzer JAW, Tustin RC, eds. Infectious Diseases of Livestock. 2nd ed. Oxford: Oxford University Press. pp. 498–499.

Lawrence JA (2004b) *Theileria buffeli/orientalis* infection. In: Coetzer JAW, Tustin RC, eds. Infectious Diseases of Livestock. 2nd ed. Oxford: Oxford University Press. pp. 500–501.

Lawrence JA, Norval RAI (1979) A history of ticks and tick-borne diseases of cattle in Rhodesia. Rhodesia Vet. J. 10: 28–40.

Lawrence JA, Williamson SM (2004a) Turning sickness. In: Coetzer JAW, Tustin RC, eds. Infectious Diseases of Livestock. 2nd ed. Oxford: Oxford University Press. pp. 475–477.

Lawrence JA, Williamson SM (2004b) *Theileria taurotragi* infection. In: Coetzer JAW, Tustin RC, eds. Infectious Diseases of Livestock. 2nd ed. Oxford: Oxford University Press. pp. 478–479.

Lawrence JA, Williamson SM (2004c) *Theileria mutans* infection. In: Coetzer JAW, Tustin RC, eds. Infectious Diseases of Livestock. 2nd ed. Oxford: Oxford University Press. pp. 480–482.

Lawrence JA, MacKenzie PKI, Norval RAI (1981) Isolation of *Theileria mutans* in Zimbabwe. Zimbabwe Vet. J. 12: 27–30.

Lawrence JA, Perry BD, Williamson SM (2004a) East Coast fever. In: Coetzer JAW, Tustin RC, eds. Infectious Diseases of Livestock. 2nd ed. Oxford: Oxford University Press. pp. 448–467.

Lawrence JA, Perry BD, Williamson SM (2004b) Corridor disease. In: Coetzer JAW, Tustin RC, eds. Infectious Diseases of Livestock. 2nd ed. Oxford: Oxford University Press. pp. 468–471.

Lawrence JA, Perry BD, Williamson SM (2004c) Zimbabwe theileriosis. In: Coetzer JAW, Tustin RC, eds. Infectious Diseases of Livestock. 2nd ed. Oxford: Oxford University Press. pp. 472–474.

Lawrence JA, de Vos AJ, Irvin AD (2004d) *Theileria separata* infection. In: Coetzer JAW, Tustin RC, eds. Infectious Diseases of Livestock. 2nd ed. Oxford: Oxford University Press. p. 485.

Lehmann HD, Roth B, Schneider CC (1969) Die Zecke *Amblyomma testudinis* (Conil 1877), ihre Entwicklung und ihre Wirkung auf den Wirt. Z. Tropenmed. Parasitol. 20: 247–259.

Lewis D, Young ER (1980) The transmission of a human strain of *Babesia divergens* by *Ixodes ricinus* ticks. J. Parasitol. 66: 359–360.

L'Hostis M, Bureaud A, Gorenflot A (1996) Female *Ixodes ricinus* (Acari, Ixodidae) in cattle of western France: infestation level and seasonality. Vet. Res. 27: 589–597.

Li Y, Luo J, Liu Z, Guan G, Gao J, Ma M, Dang Z, Liu A, Ren Q, Lu B, Liu J, Zhao J, Li J, Liu G, Bai Q, Yin H (2007) Experimental transmission of *Theileria* sp. (China 1) infective for small ruminants by *Haemaphysalis longicornis* and *Haemaphysalis qinghaiensis*. Parasitol. Res. 101: 533–538.

Li Y, Luo J, Guan G, Ma M, Liu J, Ren Q, Niu Q, Lu B, Gao J, Liu Z, Dang Z, Tian Z, Zhang B, He Z, Bai Q, Yin H (2009) Experimental transmission of *Theileria uilenbergi* infective for small ruminants by *Haemaphysalis longicornis* and *Haemaphysalis qinghaiensis*. Parasitol. Res. 104: 1227–1231.

Lightfoot CJ, Norval RAI (1981) Tick problems in wildlife in Zimbabwe. I. The effects of tick parasitism on wild ungulates. S. Afr. J. Wildl. Res. 11: 41–45.

Ligon BL (2004) Monkeypox: a review of the history and emergence in the Western Hemisphere. Sem. Pediatr. Infect. Dis. 15: 280–287.

Lindgren E, Talleklint L, Polfeldt T (2000) Impact of climatic change on the northern latitude limit and population density of the disease-transmitting European tick *Ixodes ricinus*. Environ. Health Perspec. 108: 119–123.

Lindquist L, Vapalahti O (2008) Tick-borne encephalitis. Lancet 371: 1861–1871.

Lord CC, Day JF (2000) First record of *Amblyomma auricularium* (Acari: Ixodidae) in the United States. J. Med. Entomol. 37: 977–978.

Lounsbury CP (1900) Tick-heartwater experiment. Agric. J. Cape Town 16: 682–687.

Lounsbury CP (1906) Ticks and African Coast fever. Agric. J. Cape Good Hope 28: 634–654.

Lucas JMS (1954) Fatal anaemia in poultry caused by a heavy tick infestation. Vet. Rec. 66: 573–574.

Mackenzie PKI, Norval RAI (1980) The transmission of *Cowdria ruminantium* by *Amblyomma tholloni*. Vet. Parasitol. 7: 265–268.

MacLeod J (1932) The bionomics of *Ixodes ricinus* L., the "sheep tick" of Scotland. Parasitology 24: 382–400.

MacLeod J (1940) The seasonal and annual incidence of the sheep tick, *Ixodes ricinus*, in Britain. Bull. Entomol. Res. 30: 103–118.

MacLeod J, Gordon WS (1932) Studies in louping-ill (an encephalomyelitis of sheep). II.—Transmission by the sheep tick, *Ixodes ricinus* L. J. Comp. Pathol. Therapeut. 45: 240–256.

Madder M, Thys E, Geysen D, Baudoux C, Horak I (2007) *Boophilus microplus* ticks found in West Africa. Exp. Appl. Acarol. 43: 233–234.

Mahadev PVM (1977) Life cycle, feeding behaviour and ovipositional ability of *Rhipicephalus sanguineus* and *R. turanicus* (Acarina: Ixodidae). Indian J. Acarol. 2: 12–20.

Mahan SM, Peter TF, Simbi BH, Kocan K, Camus E, Barbet AF, Burridge MJ (2000) Comparison of efficacy of American and African *Amblyomma* ticks as vectors of heartwater (*Cowdria ruminantium*) infection by molecular analyses and transmission trials. J. Parasitol. 86: 44–49.

Mahara F (1997) Japanese spotted fever: report of 31 cases and review of the literature. Emerg. Infect. Dis. 3: 105–111.

Mahara F (2006) Rickettsioses in Japan and the Far East. Ann. N. Y. Acad. Sci. 1078: 60–73.

Mans BJ, Gothe R, Neitz AWH (2004) Biochemical perspectives on paralysis and other forms of toxicoses caused by ticks. Parasitology 129: S95–S111.

Marques S, Barros-Battesti DM, Faccini JLH, Onofrio VC (2002) Brazilian distribution of *Amblyomma varium* Koch, 1844 (Acari: Ixodidae), a common parasite of sloths (Mammalia: Xenarthra). Mem. Inst. Oswaldo Cruz 97: 1141–1146.

Matsumoto K, Ogawa M, Brouqui P, Raoult D, Parola P (2005) Transmission of *Rickettsia massiliae* in the tick, *Rhipicephalus turanicus*. Med. Vet. Entomol. 19: 263–270.

Matthysse JG, Colbo MH (1987) The Ixodid Ticks of Uganda Together with Species Pertinent to Uganda Because of Their Present Known Distribution. College Park, MD: Entomological Society of America. 426 pp.

Matuschka FR, Fischer P, Musgrave K, Richter D, Spielman A (1991) Hosts on which nymphal *Ixodes ricinus* most abundantly feed. Am. J. Trop. Med. Hyg. 44: 100–107.

Maurin M, Raoult D (1999) Q fever. Clin. Microbiol. Rev. 12: 518–553.

Mazzoni R, Cunningham AA, Daszak P, Apolo A, Perdomo E, Speranza G (2003) Emerging pathogen of wild amphibians in frogs (*Rana catesbeiana*) farmed for international trade. Emerg. Infect. Dis. 9: 995–998.

McBee K, Baker RJ (1982) *Dasypus novemcinctus*. Mammal. Spec. 162: 1–9.

McCarthy PH (1960) Observations on the infestation of the larger domestic animals and the kangaroo, by the ornate kangaroo tick (*Amblyomma triguttatum*). Aust. Vet. J. 36: 436–437.

McCulloch B, Kalaye WJ, Tungaraza R, Suda BJ, Mbasha EM (1968) A study of the life history of the tick *Rhipicephalus appendiculatus*—the main vector of East Coast fever—with reference to its behaviour under field conditions and with regard to its control in Sukumaland, Tanzania. Bull. Epizoot. Dis. Afr. 16: 477–500.

McDiarmid L, Petney T, Dixon B, Andrews R (2000) Range expansion of the tick *Amblyomma triguttatum triguttatum*, an Australian vector of Q fever. Int. J. Parasitol. 30: 791–793.

Mehl R, Michaelsen J, Lid G (1984) Ticks (Acari, Ixodides) on migratory birds in Norway. Fauna Norv. Ser. B 31: 46–58.

Mehlhorn H, Schein E (1998) Redescription of *Babesia equi* Laveran, 1901 as *Theileria equi* Mehlhorn, Schein 1998. Parasitol. Res. 84: 467–475.

Mertins JW, Schlater JL (1991) Exotic ectoparasites of ostriches recently imported into the United States. J. Wildl. Dis. 27: 180–182.

Meshaka WE, Butterfield BP, Hauge JB (2004) The Exotic Amphibians and Reptiles of Florida. Malabar, FL: Krieger. 155 pp.

Michel JC (1973) *Hepatozoon mauritanicum* (Et. et Ed. Sergent, 1904) n. comb., parasite de *Testudo graeca*: redescription de la sporogonie chez *Hyalomma aegyptium* et de la schizogonie tissulaire d'apres le materiel d'E. Brumpt. Ann. Parasitol. 48: 11–21.

Milne A (1945) The ecology of the sheep tick, *Ixodes ricinus* L.: the seasonal activity in Britain with particular reference to northern England. Parasitology 36: 142–152.

Milne A (1947) The ecology of the sheep tick, *Ixodes ricinus* L.: the infestations of hill sheep. Parasitology 38: 34–50.

Milne A (1949a) The ecology of the sheep tick, *Ixodes ricinus* L.: host relationships of the tick. Part 1. Review of previous work in Britain. Parasitology 39: 167–172.

Milne A (1949b) The ecology of the sheep tick, *Ixodes ricinus* L.: host relationships of the tick. Part 2. Observations on hill and moorland grazings in northern England. Parasitology 39: 173–197.

Minami T, Ishihara T (1980) *Babesia ovata* sp. n. isolated from cattle in Japan. Nat. Inst. Anim. Health Quart. 20: 101–113.

Mitchell CJ, Spillett JJ (1968) Ecological notes on *Rhipicephalus turanicus* Pomerantzev in West Bengal, India (Acarina: Ixodidae). J. Med. Entomol. 5: 5–8.

Montgomery E (1917) On a tick-borne gastro-enteritis of sheep and goats occurring in British East Africa. J. Comp. Pathol. Therapeut. 30: 28–57.

Morel PC (1961) V. Tiques (Acarina, Ixodoidea) (deuxieme note). Mem. Inst. Franc. Afr. Noire 62: 83–90.

Morel PC (1978) [Wild animal ticks in Upper Volta.] Rev. Elev. Med. Vet. Pays Trop. 31: 69–78.

Morel PC, Finelle P (1961) Les tiques des animaux domestiques du Centrafrique. Rev. Elev. Med. Vet. Pays Trop. 14: 191–197.

Morshed MG, Scott JD, Fernando K, Beati L, Mazerolle DF, Geddes G, Durden LA (2005) Migratory songbirds disperse ticks across Canada, and first isolation of the Lyme disease spirochete, *Borrelia burgdorferi*, from the avian tick, *Ixodes auritulus*. J. Parasitol. 91: 780–790.

Morzaria SP, Brocklesby DW, Harradine DL (1977) Experimental transmission of *Babesia major* by *Haemaphysalis punctata*. Res. Vet. Sci. 23: 261–262.

Mungall EC, Sheffield WJ (1994) Exotics on the Range: the Texas Example. College Station: Texas A&M University Press. 265 pp.

Nadchatram M (1960) Malaysian parasites. XLVIII. Notes on rearing Malayan ixodid ticks (Acarina, Ixodidae) with special reference to *Ixodes granulatus* Supino and *Amblyomma geoemydae* (Cantor). Stud. Inst. Med. Res. Malaya 29: 217–224.

Nakamura-Uchiyama F, Komuro Y, Yoshii A, Nawa Y (2000) *Amblyomma testudinarium* tick bite: one case of engorged adult and a case of extraordinary number of larval tick infestation. J. Dermatol. 27: 774–777.

Nava S, Mangold AJ, Guglielmone AA (2006) The natural hosts for larvae and nymphs of *Amblyomma neumanni* and *Amblyomma parvum* (Acari: Ixodidae). Exp. Appl. Acarol. 40: 123–131.

Nava S, Lareschi M, Rebollo C, Benitez Usher C, Beati L, Robbins RG, Durden LA, Mangold AJ, Guglielmone AA (2007) The ticks (Acari: Ixodida: Argasidae, Ixodidae) of Paraguay. Ann. Trop. Med. Parasitol. 101: 255–270.

Nava S, Szabo MPJ, Mangold AJ, Guglielmone AA (2008a) Distribution, hosts, 16S rDNA sequences and phylogenetic position of the neotropical tick *Amblyomma parvum* (Acari: Ixodidae). Ann. Trop. Med. Parasitol. 102: 409–425.

Nava S, Mangold AJ, Guglielmone AA (2008b) Aspects of the life cycle of *Amblyomma parvum* (Acari: Ixodidae) under natural conditions. Vet. Parasitol. 156: 270–276.

Neilson FJA, Mossman DH (1982) Anaemia and deaths in red deer (*Cervus elaphus*) fawns associated with heavy infestations of cattle tick (*Haemaphysalis longicornis*). New Zealand Vet. J. 30: 125–126.

Neitz WO (1933) The blesbuck (*Damaliscus albifrons*) as a carrier of heartwater and blue tongue. J. S. Afr. Vet. Med. Assoc. 4: 24–26.

Neitz WO (1935) The blesbuck (*Damaliscus albifrons*) and the black-wildebeest (*Connochaetes gnu*) as carriers of heartwater. Onderstepoort J. Vet. Sci. Anim. Ind. 5: 35–40.

Neitz WO (1956) Classification, transmission, and biology of piroplasms of domestic animals. Ann. N. Y. Acad. Sci. 64: 56–111.

Neitz WO (1964) Tick-borne diseases as a hazard in the rearing of calves in Africa. Bull. Off. Int. Epizoot. 62: 607–625.

Neitz WO (1972) The experimental transmission of *Theileria ovis* by *Rhipicephalus evertsi mimeticus* and *R. bursa*. Onderstepoort J. Vet. Res. 39: 83–85.

Norval RAI (1974) The life cycle of *Amblyomma hebraeum* Koch, 1844 (Acarina: Ixodidae). J. Entomol. Soc. S. Afr. 37: 357–367.

Norval RAI (1975) Studies on the ecology of *Amblyomma marmoreum* Koch 1844 (Acarina: Ixodidae). J. Parasitol. 61: 737–742.

Norval RAI (1977) Ecology of the tick *Amblyomma hebraeum* Koch in the Eastern Cape Province of South Africa. I. Distribution and seasonal activity. J. Parasitol. 63: 734–739.

Norval RAI (1982) The ticks of Zimbabwe. IV. The genus *Hyalomma*. Zimbabwe Vet. J. 13: 2–10.

Norval RAI (1985) The ticks of Zimbabwe. XI. The genus *Aponomma*. Zimbabwe Vet. J. 16: 5–8.

Norval RAI, Horak IG (2004) Vectors: ticks. In: Coetzer JAW, Tustin RC, eds. Infectious Diseases of Livestock. 2nd ed. Oxford: Oxford University Press. pp. 3–42.

Norval RAI, Mackenzie PKI (1981) The transmission of *Cowdria ruminantium* by *Amblyomma sparsum*. Vet. Parasitol. 8: 189–191.

Norval RAI, Mason CA (1981) The ticks of Zimbabwe. II. The life cycle, distribution and hosts of *Rhipicephalus simus* Koch, 1844. Zimbabwe Vet. J. 12: 2–9.

Norval RAI, Tebele N (1984) The ticks of Zimbabwe. VIII. *Rhipicephalus compositus*. Zimbabwe Vet. J. 15: 3–8.

Norval RAI, Colborne J, Tannock J, Mackenzie PKI (1980) The life cycle of *Amblyomma tholloni* Neumann, 1899 (Acarina: Ixodidae) under laboratory conditions. Vet. Parasitol. 7: 255–263.

Norval RAI, Andrew HR, Meltzer MI (1991) Seasonal occurrence of the bont tick *Amblyomma hebraeum* in the southern lowveld of Zimbabwe. Exp. Appl. Acarol. 13: 81–96.

Norval RAI, Sonenshine DE, Allan SA, Burridge MJ (1996) Efficacy of pheromone-acaricide-impregnated tail-tag decoys of controlling the bont tick, *Amblyomma hebraeum* (Acari: Ixodidae), on cattle in Zimbabwe. Exp. Appl. Acarol. 20: 31–46.

Nosek J (1971) The ecology, bionomics, and behaviour of *Haemaphysalis* (*Aboimisalis*) *punctata* tick in central Europe. Z. Parasitenk. 37: 198–210.

Nosek J (1972) The ecology and public health importance of *Dermacentor marginatus* and *D. reticulatus* ticks in central Europe. Folia Parasitol. 19: 93–102.

Nosek J, Balat F (1982) Infestation of migratory birds with *Ixodes ricinus* and *I. arboricola* ticks. Biologia 37: 215–219.

Ntiamoa-Baidu Y, Carr-Saunders C, Matthews BE, Preston PM, Walker AR (2004) An updated list of the ticks of Ghana and an assessment of the distribution of the ticks of Ghanaian wild mammals in different vegetation zones. Bull. Entomol. Res. 94: 245–260.

Nuorteva P, Hoogstraal H (1963) Incidence of ticks (Ixodoidea, Ixodidae) on migratory birds arriving in Finland during spring of 1962. Ann. Med. Exp. Biol. Fenn. 41: 457–468.

Oba MSP, Schumaker TTS (1983) Estudo de biologia de *Amblyomma rotundatum* (Koch, 1844), em infestacoes experimentais de *Bufo marinus* (L., 1758) sob condicoes variadas de umidade relativa e de temperatura do ar. Mem. Inst. Butantan 47: 195–204.

Ogden NH, Hailes RS, Nuttall PA (1998) Interstadial variation in the attachment sites of *Ixodes ricinus* ticks on sheep. Exp. Appl. Acarol. 22: 227–232.

Ohman C (1961) The geographical and topographical distribution of *Ixodes ricinus* in Finland. Acta Soc. Pro Fauna Flora Fenn. 76: 1–38.

Okaeme AN (1986) Ectoparasites and gastrointestinal parasites of nomadic cattle infiltrating into Kainji Lake National Park, Nigeria. Int. J. Zoon. 13: 40–44.

Okanga S, Rebelo C (2006) Tick prevalence and species diversity on Aldabran giant tortoises (*Dipsochelys dussumieri*) in relation to host range and host size in a restored ecosystem, Kenya. Afr. J. Ecol. 44: 395–400.

Olenev NO (1927) On the geographical distribution in the Palaearctic Region of the ticks *Dermacentor reticulatus* Fabr. and *D. niveus* Neum. (Ixodoidea). Parasitology 19: 451–454.

Oliveira PR, Borges LMF, Lopes CML, Leite RC (2000) Population dynamics of the free-living stages of *Amblyomma cajennense* (Fabricius, 1787) (Acari: Ixodidae) on pastures of Pedro Leopoldo, Minas Gerais State, Brazil. Vet. Parasitol. 92: 295–301.

Oliveira PR, Borges LMF, Leite RC, Freitas CMV (2003) Seasonal dynamics of the Cayenne tick, *Amblyomma cajennense* on horses in Brazil. Med. Vet. Entomol. 17: 412–416.

Oliver JH (1974) Symposium on reproduction of arthropods of medical and veterinary importance. IV. Reproduction in ticks (Ixodoidea). J. Med. Entomol. 11: 26–34.

Oliver JH, Owsley MR, Claiborne CM (1988) Chromosomes, reproductive biology, and developmental stages of *Aponomma varanensis* (Acari: Ixodidae). J. Med. Entomol. 25: 73–77.

Oliver JH, Hayes MP, Keirans JE, Lavender DR (1993) Establishment of the foreign parthenogenetic tick *Amblyomma rotundatum* (Acari: Ixodidae) in Florida. J. Parasitol. 79: 786–790.

Olsen B, Jaenson TGT, Bergstrom S (1995) Prevalence of *Borrelia burgdorferi* sensu lato-infected ticks on migrating birds. Appl. Environ. Microbiol. 61: 3082–3087.

Onofrio VC, Barros-Battesti DM, Marques S, Faccini JLH, Labruna MB, Beati L, Guglielmone AA (2008) Redescription of *Amblyomma varium* Koch, 1844 (Acari: Ixodidae) based on light and scanning electron microscopy. Syst. Parasitol. 69: 137–144.

Otsuka H (1974) [Studies on transmission of *Babesia gibsoni* Patton (1910) by *Haemaphysalis longicornis* Neumann (1901)]. Bull. Fac. Agric. Miyazaki Univ. 21: 359–367.

Ouhelli H (1994) Comparative development of *Hyalomma marginatum* (Koch, 1844), *H. detritum* (Schulze, 1919), *H. anatolicum excavatum* (Koch, 1844), *H. lusitanicum* (Koch, 1884) and *H. dromedarii* (Koch, 1844) under laboratory conditions. Acta Parasitol. 39: 153–157.

Ouhelli H, Pandey VS (1984) Development of *Hyalomma lusitanicum* under laboratory conditions. Vet. Parasitol. 15: 57–66.

Ouhelli H, Pandey VS, Benzaouia T, Belkasmi A (1982) Seasonal prevalence of *Rhipicephalus turanicus* on sheep in Morocco. Trop. Anim. Health Prod. 14: 247–248.

Padilla M, Dowler RC (1994) *Tapirus terrestris*. Mammal. Spec. 481: 1–8.

Paling RW, Grootenhuis JG, Young AS (1981) Isolation of *Theileria mutans* from Kenyan buffalo, and transmission by *Amblyomma gemma*. Vet. Parasitol. 8: 31–37.

Papadopoulos B, Nuncio MS, Filipe AR (1992) The occurrence of *Rhipicephalus turanicus* Pomerantzev, Matikashvily & Lototsky, 1940, a species of *R. sanguineus* group, in Portugal. Acarologia 33: 331–333.

Papadopoulos B, Humair PF, Aeschlimann A, Vaucher C, Buttiker W (2001) Ticks on birds in Switzerland. Acarologia 42: 3–19.

Paperna I, Kremer-Mecabell T, Finkelman S (2002) *Hepatozoon kisrae* n. sp. infecting the lizard *Agama stellio* is transmitted by the tick *Hyalomma* cf. *aegyptium*. Parasite 9: 17–27.

Parola P (2006) Rickettsioses in sub-Saharan Africa. Ann. N. Y. Acad. Sci. 1078: 42–47.

Parola P, Raoult D (2001) Ticks and tickborne bacterial diseases in humans: an emerging infectious threat. Clin. Infect. Dis. 32: 897–928.

Parola P, Inokuma H, Camicas JL, Brouqui P, Raoult D (2001) Detection and identification of spotted fever group rickettsiae and ehrlichiae in African ticks. Emerg. Infect. Dis. 7: 1014–1017.

Parola P, Davoust B, Raoult D (2005) Tick- and flea-borne rickettsial emerging zoonoses. Vet. Res. 36: 469–492.

Pavlov P (1963) Recherches sur la "tick paralysis" observee chez des poulets en Bulgarie et provoquee par des nymphes d'*Haemaphysalis punctata* Can. et Fanz. Ann. Parasitol. Hum. Comp. 38: 459–461.

Pegram R (2006) End of the Caribbean *Amblyomma* Programme. International Consortium on Ticks and Tick Borne Diseases (ICTTD-3) Newsletter 30: 4–6.

Pegram RG, Oosterwijk GPM (1990) The effect of *Amblyomma variegatum* on liveweight gain of cattle in Zambia. Med. Vet. Entomol. 4: 327–330.

Pegram RG, Walker JB, Clifford CM, Keirans JE (1987a) Comparison of populations of the *Rhipicephalus simus* group: *R. simus*, *R. praetextatus*, and *R. muhsamae* (Acari: Ixodidae). J. Med. Entomol. 24: 666–682.

Pegram RG, Clifford CM, Walker JB, Keirans JE (1987b) Clarification of the *Rhipicephalus sanguineus* group (Acari, Ixodoidea, Ixodidae). I. *R. sulcatus* Neumann,1908 and *R. turanicus* Pomerantsev, 1936. Syst. Parasitol. 10: 3–26.

Pegram RG, Rota A, Onkelinx R, Wilson DD, Bartlette P, Nisbett BS, Swanston G, Vanterpool P, de Castro JJ (1996) Eradicating the tropical bont tick from the Caribbean. World Anim. Rev. 87: 56–65.

Pegram R, Indar L, Eddi C, George J (2004) The Caribbean *Amblyomma* Program: some ecologic factors affecting its success. Ann. N. Y. Acad. Sci. 1026: 302–311.

Perez-Eid C, Cabrita J (2003) La larve et la nymphe de *Hyalomma (Hyalomma) lusitanicum* Koch, 1844 (Acari: Ixodida): description morphologique, habitats, hotes. Acarologia 43: 327–335.

Perreau P, Morel PC, Barre N, Durand P (1980) [Cowdriosis (heartwater) by *Cowdria ruminantium* in ruminants of French Indies (Guadeloupe) and Mascarene Islands (La Reunion and Mauritius).] Rev. Elev. Med. Vet. Pays Trop. 33: 21–22.

Perry ND, Hanson B, Hobgood W, Lopez RL, Okraska CR, Karem K, Damon IK, Carroll DS (2006) New invasive species in southern Florida: Gambian rat (*Cricetomys gambianus*). J. Mammal. 87: 262–264.

Peter TF, Anderson EC, Burridge MJ, Mahan SM (1998) Demonstration of a carrier state for *Cowdria ruminantium* in wild ruminants from Africa. J. Wildl. Dis. 34: 567–575.

Peter TF, Anderson EC, Burridge MJ, Perry BD, Mahan SM (1999) Susceptibility and carrier status of impala, sable and tsessebe for *Cowdria ruminantium* infection (heartwater). J. Parasitol. 85: 468–472.

Peter TF, Burridge MJ, Mahan SM (2000) Competence of the African tortoise tick, *Amblyomma marmoreum* (Acari: Ixodidae), as a vector of the agent of heartwater (*Cowdria ruminantium*). J. Parasitol. 86: 438–441.

Peter TF, Burridge MJ, Mahan SM (2002) *Ehrlichia ruminantium* infection (heartwater) in wild animals. Trends Parasitol. 18: 214–218.

Petit G, Landau I, Baccam D, Lainson R (1990) [Description and life cycle of *Hemolivia stellata* n. g., n. sp., a haemogregarine of Brazilian toads.] Ann. Parasitol. Hum. Comp. 65: 3–15.

Petney TN, Al-Yaman F (1985) Attachment sites of the tortoise tick *Hyalomma aegyptium* in relation to tick density and physical condition of the host. J. Parasitol. 71: 287–289.

Petney TN, Horak IG (1988) Comparative host usage by *Amblyomma hebraeum* and *Amblyomma marmoreum*, (Acari, Ixodidae) the South African vectors of the disease heartwater. J. Appl. Entomol. 105: 490–495.

Petney TN, Horak IG, Rechav Y (1987) The ecology of the African vectors of heartwater, with particular reference to *Amblyomma hebraeum* and *Amblyomma variegatum*. Onderstepoort J. Vet. Res. 54: 381–395.

Pfafle M, Petney T, Elgas M, Skuballa J, Taraschewski H (2009) Tick-induced blood loss leads to regenerative anaemia in the European hedgehog (*Erinaceus europaeus*). Parasitology 136: 443–452.

Piercy PL (1956) Transmission of anaplasmosis. Ann. N. Y. Acad. Sci. 64: 40–48.

Pipano E, Shkap V (2004) *Theileria annulata* theileriosis. In: Coetzer JAW, Tustin RC, eds. Infectious Diseases of Livestock. 2nd ed. Oxford: Oxford University Press. pp. 486–497.

Potgieter FT (1979) Epizootiology and control of anaplasmosis in South Africa. J. S. Afr. Vet. Assoc. 50: 367–372.

Potgieter FT, Stoltsz WH (2004) Bovine anaplasmosis. In: Coetzer JAW, Tustin RC, eds. Infectious Diseases of Livestock. 2nd ed. Oxford: Oxford University Press. pp. 594–616.

Potgieter FT, van Rensburg L (1987) Tick transmission of *Anaplasma centrale*. Onderstepoort J. Vet. Res. 54: 5–7.

Pound JM, Miller JA, George JE, Lemeilleur CA (2000) The '4-poster' passive topical treatment device to apply acaricide for controlling ticks (Acari: Ixodidae) feeding on white-tailed deer. J. Med. Entomol. 37: 588–594.

Poupon MA, Lommano E, Humair PF, Douet V, Rais O, Schaad M, Jenni L, Gern L (2006) Prevalence of *Borrelia burgdorferi* sensu lato in ticks collected from migratory birds in Switzerland. Appl. Environ. Microbiol. 72: 976–979.

Prieto R (1974) Distribucion geografica, influencia ecologica y selectividad hospedera de *Amblyomma cajennense* (Fabricius, 1787) en Cuba. Rev. Cubana Cienc. Vet. 5: 47–51.

Psaroulaki A, Germanakis A, Gikas A, Scoulica E, Tselentis Y (2005) Simultaneous detection of "*Rickettsia mongolotimonae*" in a patient and in a tick in Greece. J. Clin. Microbiol. 43: 3558–3559.

Rahman MH, Mondal MMH (1985) Tick fauna of Bangladesh. Indian J. Parasitol. 9: 145–149.

Reaser JK, Clark EE, Meyers NM (2008) All creatures great and minute: a public policy primer for companion animal zoonoses. Zoonoses Public Health 55: 385–401.

Rechav Y (1992) The role of ticks in the epidemiology of Crimean-Congo haemorrhagic fever in southern Africa. In: Fivaz B, Petney T, Horak I, eds. Tick Vector Biology, Medical and Veterinary Aspects. Berlin: Springer-Verlag. pp. 149–157.

Rechav Y, Knight MM, Norval RAI (1977) Life cycle of the tick *Rhipicephalus evertsi evertsi* Neumann (Acarina: Ixodidae) under laboratory conditions. J. Parasitol. 63: 575–579.

Reed RN, Rodda GH (2009) Giant constrictors: biological and management profiles and an establishment risk assessment for nine large species of pythons, anacondas, and the boa constrictor. Open-file report 2009-1202. Reston, VA: U.S. Geological Survey. 302 pp.

Reeves W, Durden L, Wills W (2002) New records of ticks (Acari: Argasidae, Ixodidae) from South Carolina. J. Agric. Urban Entomol. 19: 197–204.

Reeves WK, Durden LA, Dasch GA (2006) A spotted fever group *Rickettsia* from an exotic tick species, *Amblyomma exornatum* (Acari: Ixodidae), in a reptile breeding facility in the United States. J. Med. Entomol. 43: 1099–1101.

Riek RF (1966) The life cycle of *Babesia argentina* (Lignieres, 1903) (Sporozoa: Piroplasmidea) in the tick vector *Boophilus microplus* (Canestrini). Aust. J. Agric. Res. 17: 247–254.

Robbins RG (1990) 961st regular meeting—November 1, 1990. Proc. Entomol. Soc. Wash. 93: 518–520.

Roberts FHS (1953) The Australian species of *Aponomma* and *Amblyomma* (Ixodoidea). Aust. J. Zool. 1: 111–161.

Roberts FHS (1962) On the status of morphologically divergent tick populations of *Amblyomma triguttatum* Koch (Acarina: Ixodidae). Aust. J. Zool. 10: 367–381.

Roberts FHS (1964a) Further observations on the Australian species of *Aponomma* and *Amblyomma* with descriptions of the nymphs of *Amblyomma moreliae* (L. Koch) and *Amb. loculosum* Neumann (Acarina: Ixodidae). Aust. J. Zool. 12: 288–313.

Roberts FHS (1964b) The tick fauna of Tasmania. Rec. Queen Victoria Mus. Launceston New Ser. 17: 1–8.

Roberts FHS (1969) The larvae of Australian Ixodidae (Acarina: Ixodoidea). J. Aust. Entomol. Soc. 8: 37–78.

Roberts FHS (1970) Australian Ticks. Melbourne, Australia: Commonwealth Scientific & Industrial Research Organization. 267 pp.

Robinson LE (1926) The Genus *Amblyomma*. Nuttall GHF, Warburton C, Robinson LE, eds. Cambridge: University Press. 302 pp.

Robson J, Robb JM, Hawa NJ, Al-Wahayyib T (1969) Ticks (Ixodoidea) of domestic animals in Iraq. 6. Distribution. J. Med. Entomol. 6: 125–127.

Roby TO, Anthony DW (1963) Transmission of equine piroplasmosis by *Dermacentor nitens* Neumann. J. Am. Vet. Med. Assoc. 142: 768–769.

Roby TO, Anthony DW, Thornton CW, Holbrook AA (1964) The hereditary transmission of *Babesia caballi* in the tropical horse tick, *Dermacentor nitens* Neumann. Am. J. Vet. Res. 25: 494–499.

Rodda GH, Jarnevich CS, Reed RN (2009) What parts of the US mainland are climatically suitable for invasive alien pythons spreading from the Everglades National Park? Biol. Invasions 11: 241–252.

Rubina M, Hadani A, Ziv M (1982) The life cycle of the tick *Hyalomma anatolicum excavatum* Koch, 1844, maintained under field conditions in Israel. Rev. Elev. Med. Vet. Pays Trop. 35: 255–264.

Ruiz-Carus R, Matheson RE, Roberts DE, Whitfield PE (2006) The western Pacific red lionfish, *Pterois volitans* (Scorpaenidae), in Florida: evidence for reproduction and parasitism in the first exotic marine fish established in state waters. Biol. Conserv. 128: 384–390.

Saikku P, Ulmanen I, Brummer-Korvenkontio M (1971) Ticks (Ixodidae) on migratory birds in Finland. Acta Entomol. Fenn. 28: 46–51.

Salisbury LE (1959) Ticks in the South African zoological survey collection—part X—*Rhipicephalus muehlensi*. Onderstepoort J. Vet. Res. 28: 125–132.

Samuel WM, Trainer DO (1970) *Amblyomma* (Acarina: Ixodidae) on white-tailed deer, *Odocoileus virginianus* (Zimmermann), from south Texas with implications for theileriasis. J. Med. Entomol. 7: 567–574.

Sang R, Onyango C, Gachoya J, Mabinda E, Konongoi S, Ofula V, Dunster L, Okoth F, Coldren R, Tesh R, Travassos da Rosa A, Finkbeiner S, Wang D, Crabtree M, Miller B (2006) Tick-borne arbovirus surveillance in market livestock, Nairobi, Kenya. Emerg. Infect. Dis. 12: 1074–1080.

Santos Dias JAT (1993) Contribuicao para o Estudo da Sistematica e Taxonomia das Especies do Genero *Aponomma* Neumann, 1899 (Acarina-Ixodoidea). Lisbon, Portugal: Ministerio do Planeamento e da Administracao do Territorio. 204 pp.

Saratsiotis A (1972) Contribution a l'etude morphologique et biologique du genre *Aponomma* Neumann, 1899 (Acariens, Ixodidea). I. *Ap. flavomaculatum* (Lucas, 1846). II. *Ap. latum* (Koch,1844). Acarologia 13: 476–495.

Saratsiotis A, Battelli C (1975) *Rhipicephalus turanicus* Pomerantzev, Metkashilli et Lotozki 1940, en Italie. Comparation morphologique avec *Rhipicephalus sanguineus* S. Str. (Latreille, 1908) (Acariens, Ixodidae). Riv. Parassitol. 36: 207–214.

Schetters T (2005) Vaccination against canine babesiosis. Trends Parasitol. 21: 179–184.

Schloegel LM, Picco AM, Kilpatrick AM, Davies AJ, Hyatt AD, Daszak P (2009) Magnitude of the US trade in amphibians and presence of *Batrachochytrium dendrobatidis* and ranavirus infection in imported North American bullfrogs (*Rana catesbeiana*). Biol. Conserv. 142: 1420–1426.

Schofield PJ (2009) Geographic extent and chronology of the invasion of non-native lionfish (*Pterois volitans* [Linnaeus 1758] and *P. miles* [Bennett 1828]) in the western North Atlantic and Caribbean Sea. Aquat. Invasions 4: 473–479.

Schumaker TT, Barros DM (1994) Notes on the biology of *Amblyomma dissimile* Koch, 1844 (Acari: Ixodida) on *Bufo marinus* (Linnaeus, 1758) from Brazil. Mem. Inst. Oswaldo Cruz 89: 29–31.

Scott JD, Fernando K, Banerjee SN, Durden LA, Byrne SK, Banerjee M, Mann RB, Morshed MG (2001) Birds disperse ixodid (Acari: Ixodidae) and *Borrelia burgdorferi*-infected ticks in Canada. J. Med. Entomol. 38: 493–500.

Sementelli A, Smith HT, Meshaka WE, Engeman RM (2008) Just green iguanas? The associated costs and policy implications of exotic invasive wildlife in south Florida. Public Works Manage. Policy 12: 599–606.

Sergent E, Donatien A, Parrot L, Lestoquard F (1928) Tiques et piroplasmoses bovines d'Algerie. Bull. Soc. Pathol. Exot. 21: 846–849.

Serra Freire NM (1983) Tick paralysis in Brazil. Trop. Anim. Health Prod. 15: 124–126.

Sharif M (1928) A revision of the Indian Ixodidae with special reference to the collection in the Indian Museum. Rec. Indian Mus. 30: 217–344.

Shaw SE, Day MJ, Birtles RJ, Breitschwerdt EB (2001) Tick-borne infectious diseases of dogs. Trends Parasitol. 17: 74–80.

Shepherd AJ, Swanepoel R, Shepherd SP, Leman PA, Mathee O (1991) Viraemic transmission of Crimean-Congo haemorrhagic fever virus to ticks. Epidemiol. Infect. 106: 373–382.

Simmons LA, Burridge MJ (2000) Introduction of the exotic ticks *Amblyomma humerale* Koch and *Amblyomma geoemydae* (Cantor) (Acari: Ixodidae) into the United States on imported reptiles. Int. J. Acarol. 26: 239–242.

Simmons LA, Burridge MJ (2002) Introduction of the exotic tick *Amblyomma chabaudi* Rageau (Acari: Ixodidae) into Florida on imported tortoises. Fla. Entomol. 85: 288–289.

Simmons LA, Stadler CK, Burridge MJ (2002) Introduction of the exotic tick *Amblyomma helvolum* Koch (Acari: Ixodidae) into the United States on imported cobras (Squamata: Elapidae). Int. J. Acarol. 28: 45–48.

Singh KRP, Bhatt PN (1968) Transmission of Kyasanur Forest disease virus by *Hyalomma marginatum isaaci*. Indian J. Med. Res. 56: 610–613.

Singh KRP, Dhanda V (1965) Description and keys of immature stages of some species of Indian *Hyalomma* Koch, 1844 (Ixodoidea, Ixodidae). Acarologia 7: 636–651.

Sippel WL, Cooperrider DE, Gainer JH, Allen RW, Mouw JEB, Teigland MB (1962) Equine piroplasmosis in the United States. J. Am. Vet. Med. Assoc. 141: 694–698.

Siroky P, Petrzelkova KJ, Kamler M, Mihalca AD, Modry D (2006) *Hyalomma aegyptium* as dominant tick in tortoises of the genus *Testudo* in Balkan countries, with notes on its host preferences. Exp. Appl. Acarol. 40: 279–290.

Siroky P, Mikulicek P, Jandzik D, Kami H, Mihalca AD, Rouag R, Kamler M, Schneider C, Zaruba M, Mogry D (2009) Co-distribution pattern of a haemogregarine *Hemolivia mauritanica* (Apicomplexa: Haemogregarinidae) and its vector *Hyalomma aegyptium* (Metastigmata: Ixodidae). J. Parasitol. 95: 728–733.

Siuda K, Dutkiewicz J (1979) [*Hyalomma marginatum* Koch, 1844 (Acarina, Ixodidae) in Poland—an example for transport of exogenous tick by migratory birds.] Wiad. Parazytol. 25: 333–338.

Sjostedt A (2007) Tularemia: history, epidemiology, pathogen physiology, and clinical manifestations. Ann. N. Y. Acad. Sci. 1105: 1–29.

Smallridge CJ, Bull CM (1999) Transmission of the blood parasite *Hemolivia mariae* between its lizard and tick hosts. Parasitol. Res. 85: 858–863.

Smith MW (1974) A survey of the distribution of the ixodid ticks *Boophilus microplus* (Canestrini, 1888) and *Amblyomma cajennense* (Fabricius, 1787) in Trinidad and Tobago and the possible influence of the survey results on planned livestock development. Trop. Agric. 51: 559–567.

Smith MW (1975) Some aspects of ecology and lifecycle of *Amblyomma cajennense* (Fabricius 1787) in Trinidad and their influence on tick control measures. Ann. Trop. Med. Parasitol. 69: 121–129.

Smyth M (1973) The distribution of three species of reptile ticks, *Aponomma hydrosauri* (Denny), *Amblyomma albolimbatum* Neumann, and *Amb. limbatum* Neumann. Aust. J. Zool. 21: 91–101.

Snetsinger R (1968) Distribution of ticks and tick-borne diseases in Pennsylvania. Penn. State Univ. Progr. Rep. 288: 1–8.

Sonenshine DE (1991) Biology of Ticks. New York: Oxford University Press. 447 pp.

Sonenshine DE, Allan SA, Peter TF, McDaniel R, Burridge MJ (2000) Does geographic range affect the attractant-aggregation-attachment pheromone of the tropical bont tick, *Amblyomma variegatum*? Exp. Appl. Acarol. 24: 283–299.

Sreenivasan MA, Bhat HR, Naik SV (1979) Experimental transmission of Kyasanur forest disease virus by *Dermacentor auratus* Supino. Indian J. Med. Res. 69: 701–707.

Stanek G, Strle F (2003) Lyme borreliosis. Lancet 362: 1639–1647.

Steere A (2001) Lyme disease. New England J. Med. 345: 115–125.

Stenos J, Graves S, Popov VL, Walker DH (2003) *Aponomma hydrosauri*, the reptile-associated tick reservoir of *Rickettsia honei* on Flinders Island, Australia. Am. J. Trop. Med. Hyg. 69: 314–317.

Stewart CG (1992) Bovine ehrlichiosis. In: Fivaz B, Petney T, Horak I, eds. Tick Vector Biology: Medical and Veterinary Aspects. Berlin: Springer-Verlag. pp. 101–107.

Stewart R (1991) Flinders Island spotted fever: a newly recognised endemic focus of tick typhus in Bass Strait. Part 1. Clinical and epidemiological features. Med. J. Aust. 154: 94–99.

Stoltsz WH (2004) Ovine and caprine anaplasmosis. In: Coetzer JAW, Tustin RC, eds. Infectious Diseases of Livestock. 2nd ed. Oxford: Oxford University Press. pp. 617–624.

Strickland RK, Gerrish RR (1964) Distribution of the tropical horse tick in the United States, with notes on associated cases of equine piroplasmosis. J. Am. Vet. Med. Assoc. 144: 875–878.

Stuen S (2007) *Anaplasma phagocytophilum*—the most widespread tick-borne infection in animals in Europe. Vet. Res. Commun. 31 (suppl. 1): 79–84.

Sumption KJ, Scott GR (2004) Lesser-known rickettsias infecting livestock. In: Coetzer JAW, Tustin RC, eds. Infectious Diseases of Livestock. 2nd ed. Oxford: Oxford University Press. pp. 536–549.

Sutherst RW, Maywald GF (1985) A computerized system for matching climates in ecology. Agric. Ecosyst. Environ. 13: 281–299.

Swanepoel A (1959) Tick paralysis: regional neurological involvement caused by *Hyalomma truncatum*. S. Afr. Med. J. 33: 909–911.

Swanepoel R, Burt FJ (2004) Crimean-Congo haemorrhagic fever. In: Coetzer JAW, Tustin RC, eds. Infectious Diseases of Livestock. 2nd ed. Oxford: Oxford University Press. pp. 1077–1085.

Swanepoel R, Laurenson MK (2004) Louping ill. In: Coetzer JAW, Tustin RC, eds. Infectious Diseases of Livestock. 2nd ed. Oxford: Oxford University Press. pp. 995–1003.

Sweatman GK (1968) Temperature and humidity effects on the oviposition of *Hyalomma aegyptium* ticks of different engorgement weights. J. Med. Entomol. 5: 429–439.

Szabo MPJ, Mangold AJ, Joao CF, Bechara GH, Guglielmone AA (2005) Biological and DNA evidence of two dissimilar populations of the *Rhipicephalus sanguineus* tick group (Acari: Ixodidae) in South America. Vet. Parasitol. 130: 131–140.

Szabo MPJ, Pereira LF, Castro MB, Garcia MV, Sanches GS, Labruna MB (2009) Biology and life cycle of *Amblyomma incisum* (Acari: Ixodidae). Exp. Appl. Acarol. 48: 263–271.

Szymanski S (1987) Seasonal activity of *Dermacentor reticulatus* (Fabricius, 1794) (Ixodidae) in Poland. III. Larvae and nymphs. Acta Parasitol. Polonica 32: 265–280.

Talleklint L, Jaenson TGT (1998) Increasing geographical distribution and density of *Ixodes ricinus* (Acari: Ixodidae) in central and northern Sweden. J. Med. Entomol. 35: 521–526.

Taulman JF, Robbins LW (1996) Recent range expansion and distributional limits of the nine-banded armadillo (*Dasypus novemcinctus*) in the United States. J. Biogeogr. 23: 635–648.

Teel PD (1991) Applications of modeling to the ecology of *Boophilus microplus* (Say) (Acari: Ixodidae). J. Agric. Entomol. 8: 291–296.

Teel PD, Hopkins SW, Donahue WA, Strey OF (1998) Population dynamics of immature *Amblyomma maculatum* (Acari: Ixodidae) and other ectoparasites on meadowlarks and northern bobwhite quail resident to the coastal prairie of Texas. J. Med. Entomol. 35: 483–488.

Telecky TM (2001) United States import and export of live turtles and tortoises. Turtle Tortoise Newsletter 4: 8–13.

Telford SR (1984) Haemoparasites of reptiles. In: Hoff GL, Frye FL, Jacobson ER, eds. Diseases of Amphibians and Reptiles. New York: Plenum. pp. 385–517.

Theiler A (1906) Transmission of equine piroplasmosis by ticks in South Africa. J. Comp. Pathol. Therapeut. 19: 283–292.

Theiler A (1912) The transmission of gall-sickness by ticks. Agric. J. Union S. Afr. 3: 173–181.

Theiler G (1949a) Zoological survey of the Union of South Africa: tick survey. Part II.—Distribution of *Boophilus* (*Palpoboophilus*) *decoloratus*, the blue tick. Onderstepoort J. Vet. Sci. Anim. Indust. 22: 255–268.

Theiler G (1949b) Zoological survey of the Union of South Africa: tick survey. Part III.—Distribution of *Rhipicephalus appendiculatus*, the brown tick. Onderstepoort J. Vet. Sci. Anim. Indust. 22: 269–284.

Theiler G (1950) Zoological survey of the Union of South Africa. Tick survey—part V. Distribution of *Rhipicephalus evertsi*, the red tick. Onderstepoort J. Vet. Sci. Anim. Indust. 24: 33–36.

Theiler G (1962) The Ixodoidea Parasites of Vertebrates in Africa South of the Sahara. Project S.9958. Onderstepoort, South Africa: Veterinary Services. 260 pp.

Theiler G, Salisbury LE (1959) Ticks in the South African zoological survey collection, part IX—"the *Amblyomma marmoreum* group." Onderstepoort J. Vet. Res. 28: 47–124.

Theiler G, Walker JB, Wiley AJ (1956) Ticks in the South African zoological survey collection: part VIII: two East African ticks. Onderstepoort J. Vet. Res. 27: 83–99.

Thorner AR, Walker DH, Petri WA (1998) Rocky Mountain spotted fever. Clin. Infect. Dis. 27: 1353–1360.

Timoney PJ (1998) Louping-ill. In: Buisch WW, Hyde JL, eds. Foreign Animal Diseases. 6th ed. Richmond, VA: Carter Printing Company. pp. 267–272.

Townsend JH, Krysko KL, Enge KM (2003) Introduced iguanas in southern Florida: a history of more than 35 years. Iguana 10: 111–118.

Trees AJ (1978) The transmission of *Borrelia theileri* by *Boophilus annulatus* (Say, 1821). Trop. Anim. Health Prod. 10: 93–94.

Ueckermann EA, van Harten A, Smith Meyer MKP (2006) The mites and ticks (Acari) of Yemen: an annotated check-list. Fauna Arabia 22: 243–286.

Uilenberg G (1965) *Amblyomma chabaudi* Rageau, 1964 (Ixodidae). Description de la femelle et de la nymphe. Variations morphologiques du male. Ann. Parasitol. Hum. Comp. 40: 681–692.

Uilenberg G (1967) *Amblyomma chabaudi* Rageau, 1964 (Ixodidae). Elevage au laboratoire. Description de la larve. Observations complaementaires sur la nymphe. Ann. Parasitol. Hum. Comp. 42: 343–351.

Uilenberg G (1982) Experimental transmission of *Cowdria ruminantium* by the Gulf Coast tick *Amblyomma maculatum*: dangers of introducing heartwater and benign African theileriasis onto the American mainland. Am. J. Vet. Res. 43: 1279–1282.

Uilenberg G (1983a) [New acquisitions concerning the vector role of the tick *Amblyomma* (Ixodidae).] Rev. Elev. Med. Vet. Pays Trop. 36: 61–66.

Uilenberg G (1983b) Heartwater (*Cowdria ruminantium* infection): current status. Adv. Vet. Sci. Comp. Med. 27: 427–480.

Uilenberg, G (1997) General review of tick-borne diseases of sheep and goats world-wide. Parassitologia 39: 161–165.

Uilenberg G, Hashemi-Fesharki R (1984) *Theileria orientalis* in Iran. Vet. Quart. 6: 1–4.

Uilenberg G, Schreuder BEC (1976) Studies on Theileriidae (Sporozoa) in Tanzania. 1. Tick transmission of *Haematoxenus veliferus*. Tropenmed. Parasitol. 27: 106–111.

Uilenberg G, Robson J, Pedersen V (1974) Some experiments on the transmission of *Theileria mutans* (Theiler, 1906) and *Theileria parva* (Theiler, 1904) by the ticks *Amblyomma variegatum* (Fabricius, 1794) and *Rhipicephalus appendiculatus* Neumann, 1901, in Uganda. Tropenmed. Parasitol. 25: 207–216.

Uilenberg G, Hoogstraal H, Klein JM (1979) Les tiques (Ixodoidea) de Madagascar et leur role vecteur. Arch. Inst. Pasteur Madagascar numero special: 7–153.

Uilenberg G, Rombach MC, Perie NM, Zwart D (1980) Blood parasites of sheep in the Netherlands. II. *Babesia motasi* (Sporozoa, Babesiidae). Vet. Quart. 2: 3–14.

Uilenberg G, Barre N, Camus E, Burridge MJ, Garris GI (1984) Heartwater in the Caribbean. Prev. Vet. Med. 2: 255–267.

Uilenberg G, Perie NM, Spanjer AAM, Franssen FFJ (1985) *Theileria orientalis*, a cosmopolitan blood parasite of cattle: demonstration of the schizont stage. Res. Vet. Sci. 38: 352–357.

USAHA (2005) Report of the Committee on Parasitic Diseases. In: Proceedings of the One Hundred and Ninth Annual Meeting of the United States Animal Health Association. Montgomery, AL: Skinner Printing Company. pp. 498–518.

USAHA (2006) Report of the Committee on Parasitic Diseases. In: Proceedings of the One Hundred and Tenth Annual Meeting of the United States Animal Health Association. Richmond, VA: Pat Campbell & Associates. pp. 519–532.

USAHA (2007) Report of the Committee on Parasitic Diseases. In: Proceedings of the One Hundred and Eleventh Annual Meeting of the United States Animal Health Association. Kansas City, MO: Meghan Richey & Richardson Printing. pp. 561–576.

USDA (1962a) Reports of ticks collected calendar year 1962. Washington, DC: Animal Disease Eradication Division, Agricultural Research Service, U.S. Department of Agriculture. 6 pp.

USDA (1962b) Report of cooperative cattle fever tick eradication activities, fiscal year 1962. Washington, DC: Animal Disease Eradication Division, Agricultural Research Service, U.S. Department of Agriculture. 14 pp.

USDA (1963) Report of cooperative tick eradication activities, fiscal year 1963. Hyattsville, MD: Agricultural Research Service, Animal Disease Eradication Division, U.S. Department of Agriculture. 7 pp.

USDA (1964a) National tick survey, calendar year 1963. Hyattsville, MD: Agricultural Research Service, Animal Disease Eradication Division, U.S. Department of Agriculture. 8 pp.

USDA (1964b) Report of cooperative tick eradication activities, fiscal year 1964. Hyattsville, MD: Agricultural Research Service, Animal Disease Eradication Division, U.S. Department of Agriculture. 6 pp.

USDA (1965a) National tick survey—CY 1964. Hyattsville, MD: Agricultural Research Service, Animal Disease Eradication Division, U.S. Department of Agriculture. 12 pp.

USDA (1965b) Report of cooperative tick eradication activities, fiscal year 1965. Hyattsville, MD: Agricultural Research Service, Animal Health Division, U.S. Department of Agriculture. 6 pp.

USDA (1966a) National tick surveillance program, calendar year 1965. Hyattsville, MD: Agricultural Research Service, Animal Health Division, U.S. Department of Agriculture. 15 pp.

USDA (1966b) Report of cooperative tick eradication activities, fiscal year 1966. Hyattsville, MD: Agricultural Research Service, Animal Health Division, U.S. Department of Agriculture. 4 pp.

USDA (1967a) National tick surveillance program, calendar year 1966. Hyattsville, MD: Agricultural Research Service, Animal Health Division, U.S. Department of Agriculture. 15 pp.

USDA (1967b) Report of cooperative tick eradication activities, fiscal year 1967. Hyattsville, MD: Agricultural Research Service, Animal Health Division, U.S. Department of Agriculture. 8 pp.

USDA (1968a) National tick surveillance program, calendar year 1967. Hyattsville, MD: Agricultural Research Service, Animal Health Division, U.S. Department of Agriculture. 15 pp.

USDA (1968b) Report of cooperative tick eradication activities, fiscal year 1968. Hyattsville, MD: Agricultural Research Service, Animal Health Division, U.S. Department of Agriculture. 7 pp.

USDA (1969) National tick surveillance program, calendar year 1968. Hyattsville, MD: Agricultural Research Service, Animal Health Division, U.S. Department of Agriculture. 15 pp.

USDA (1970a) National tick surveillance program, calendar year 1969. Hyattsville, MD: Agricultural Research Service, Animal Health Division, U.S. Department of Agriculture. 15 pp.

USDA (1970b) Cooperative state-federal tick eradication activities, fiscal year 1969. Report no. ARS 91-83. Hyattsville, MD: Agricultural Research Service, Animal Health Division, U.S. Department of Agriculture. 11 pp.

USDA (1970c) Cooperative state-federal tick eradication activities, fiscal year 1970. Report no. ARS 91-96. Hyattsville, MD: Agricultural Research Service, Animal Health Division, U.S. Department of Agriculture. 11 pp.

USDA (1971) National tick surveillance program, calendar year 1970. Report no. ARS 91-102. Hyattsville, MD: Agricultural Research Service, Animal Health Division, U.S. Department of Agriculture. 15 pp.

USDA (1972) National tick surveillance program, calendar year 1971. Report no. APHIS 91-8. Hyattsville, MD: Animal & Plant Health Inspection Service, Veterinary Services, U.S. Department of Agriculture. 14 pp.

USDA (1973a) National tick surveillance program, calendar year 1972. Report no. APHIS 91-12. Hyattsville, MD: Animal & Plant Health Inspection Service, Veterinary Services, U.S. Department of Agriculture. 15 pp.

USDA (1973b) Cooperative state-federal tick eradication activities, fiscal year 1972. Report no. APHIS 91-14. Hyattsville, MD: Animal & Plant Health Inspection Service, Veterinary Services, U.S. Department of Agriculture. 9 pp.

USDA (1974a) National tick surveillance program, calendar year 1973. Report no. APHIS 91-22. Hyattsville, MD: Animal & Plant Health Inspection Service, Veterinary Services, U.S. Department of Agriculture. 15 pp.

USDA (1974b) Cooperative state-federal tick eradication activities, fiscal year 1973. Report no. APHIS 91-24. Hyattsville, MD: Animal & Plant Health Inspection Service, Veterinary Services, U.S. Department of Agriculture. 8 pp.

USDA (1975) National tick surveillance program, calendar year 1974. Report no. APHIS 91-28. Hyattsville, MD: Animal & Plant Health Inspection Service, Veterinary Services, U.S. Department of Agriculture. 17 pp.

USDA (1977a) National tick surveillance program, calendar year 1975. Report no. APHIS 91-32. Hyattsville, MD: Animal & Plant Health Inspection Service, Veterinary Services, U.S. Department of Agriculture. 15 pp.

USDA (1977b) National tick surveillance program, calendar year 1976. Report no. APHIS 91-33. Hyattsville, MD: Animal & Plant Health Inspection Service, Veterinary Services, U.S. Department of Agriculture. 17 pp.

USDA (1978) National tick surveillance program, calendar year 1977. Report no. APHIS 91-35 October 1978. Hyattsville, MD: Animal & Plant Health Inspection Service, Veterinary Services, U.S. Department of Agriculture. 17 pp.

USDA (1980) National tick surveillance program, calendar year 1978. Report no. APHIS 91-35 May 1980. Hyattsville, MD: Animal & Plant Health Inspection Service, Veterinary Services, U.S. Department of Agriculture. 21 pp.

USDA (1981) National tick surveillance program, calendar year 1979. Report no. APHIS 91-39 July 1981. Hyattsville, MD: Animal & Plant Health Inspection Service, Veterinary Services, U.S. Department of Agriculture. 23 pp.

USDA (1982) National tick surveillance program, calendar years 1980 and 1981. Report no. APHIS 91-39 Sept. 1982. Washington, DC: U.S. Government Printing Office. 35 pp.

USDA (1983) National tick surveillance program, calendar year 1982. Report no. APHIS 91-39 August 1983. Hyattsville, MD: Animal & Plant Health Inspection Service, Veterinary Services, U.S. Department of Agriculture. 17 pp.

USDA (1985a) National tick surveillance program, calendar year 1983. Report no. APHIS 91-39 February 1985. Hyattsville, MD: Animal & Plant Health Inspection Service, Veterinary Services, U.S. Department of Agriculture. 19 pp.

USDA (1985b) National tick surveillance program, calendar year 1984. Report no. APHIS 91-39 October 1985. Washington, DC: U.S. Government Printing Office. 17 pp.

USDA (1987a) National tick surveillance program, calendar year 1985. Report no. APHIS 91-39 June 1987. Hyattsville, MD: Animal & Plant Health Inspection Service, Veterinary Services, U.S. Department of Agriculture. 19 pp.

USDA (1987b) National tick surveillance program, calendar year 1986. Report no. APHIS 91-39 October 1987. Hyattsville, MD: Animal & Plant Health Inspection Service, Veterinary Services, U.S. Department of Agriculture. 19 pp.

USDA (1988) National tick surveillance program, calendar year 1987. Report no. APHIS 91-39. Hyattsville, MD: Animal & Plant Health Inspection Service, Veterinary Services, U.S. Department of Agriculture. 19 pp.

USDA (1989) Importation of ostriches and other ratites. Fed. Reg. 54: 34485–34487.

USDA (1990) Foreign animal disease report no. 18-4. Hyattsville, MD: Emergency Programs, Veterinary Services, Animal & Plant Health Inspection Service, U.S. Department of Agriculture. 27 pp.

USDA (1994) National tick surveillance program, calendar years 1988–89. Report no. APHIS 91-45-005. Hyattsville, MD: Animal & Plant Health Inspection Service, U.S. Department of Agriculture. 43 pp.

USDA (2000) Importation and interstate movement of certain land tortoises. Fed. Reg. 65: 15216–15218.

van der Borght-Elbl A (1977) Ixodid Ticks (Acarina, Ixodidae) of Central Africa. Volume V. The Larval and Nymphal Stages of the More Important Species of the Genus *Amblyomma* Koch, 1844. Tervuren, Belgium: Musee Royal de l'Afrique Centrale. 158 pp.

Verderane MP, Falotico T, Resende BD, Labruna MB, Izar P, Ottoni EB (2007) Anting in a semifree-ranging group of *Cebus apella*. Int. J. Primatol. 28: 47–53.

Viseras J, Hueli LE, Adroher FJ, Garcia-Fernandez P (1999) Studies on the transmission of *Theileria annulata* to cattle by the tick *Hyalomma lusitanicum*. J. Vet. Med. B 46: 505–509.

Voltzit OV, Keirans JE (2002) A review of Asian *Amblyomma* species (Acari, Ixodida, Ixodidae). Acarina 10: 95–136.

Voltzit OV, Keirans JE (2003) A review of African *Amblyomma* species (Acari, Ixodida, Ixodidae). Acarina 11: 135–214.

Waag DM, Williams JC, Peacock MG, Raoult D (1991) Methods of isolation, amplification, and purification of *Coxiella burnetii*. In: Williams JC, Thompson HA, eds. Q Fever: the Biology of *Coxiella burnetii*. Boca Raton, FL: CRC Press. pp. 73–115.

Wagner GG, Holman P, Waghela S (2002) Babesiosis and heartwater: threats without boundaries. Vet. Clin. Food Anim. Pract. 18: 417–430.

Waldenstrom J, Lundkvist A, Falk KI, Garpmo U, Bergstrom S, Lindegren G, Sjostedt A, Mejlon H, Fransson T, Haemig PD, Olsen B (2007) Migrating birds and tickborne encephalitis virus. Emerg. Infect. Dis. 13: 1215–1218.

Walker AR, Lloyd CM (1993) Experiments on the relationship between feeding of the tick *Amblyomma variegatum* (Acari: Ixodidae) and dermatophilosis skin disease in sheep. J. Med. Entomol. 30: 136–143.

Walker AR, Bouattour A, Camicas JL, Estrada-Pena A, Horak IG, Latif AA, Pegram RG, Preston PM (2003) Ticks of Domestic Animals in Africa: A Guide to Identification of Species. Edinburgh: Bioscience Reports. 221 pp.

Walker ED, Stobierski MG, Poplar ML, Smith TW, Murphy AJ, Smith PC, Schmitt SM, Cooley TM, Kramer CM (1998) Geographic distribution of ticks (Acari: Ixodidae) in Michigan, with emphasis on *Ixodes scapularis* and *Borrelia burgdorferi*. J. Med. Entomol. 35: 872–882.

Walker JB (1955) *Rhipicephalus pulchellus* Gerstacker 1873: a description of the larva and nymph with notes on the adults and on its biology. Parasitology 45: 95–98.

Walker JB (1974) The Ixodid Ticks of Kenya: A Review of Present Knowledge of Their Hosts and Distribution. London: Commonwealth Institute of Entomology. 220 pp.

Walker JB, Olwage A (1987) The tick vectors of *Cowdria ruminantium* (Ixodoidea, Ixodidae, genus *Amblyomma*) and their distribution. Onderstepoort J. Vet. Res. 54: 353–379.

Walker JB, Parsons BT (1964) The laboratory rearing of *Amblyomma sparsum* Neumann, 1899. Parasitology 54: 173–175.

Walker JB, Keirans JE, Horak IG (2000) The Genus *Rhipicephalus* (Acari, Ixodidae): A Guide to the Brown Ticks of the World. Cambridge: Cambridge University Press. 643 pp.

Wanzala W, Okanga S (2006) Ticks (Acari: Ixodidae) associated with wildlife and vegetation of Haller Park along the Kenyan coastline. J. Med. Entomol. 43: 789–794.

Warner RD, Marsh WW (2002) Rocky Mountain spotted fever. J. Am. Vet. Med. Assoc. 221: 1413–1417.

Wassef HY, Hoogstraal H (1984) *Dermacentor* (*Indocentor*) *auratus* (Acari: Ixodoidea: Ixodidae): identity of male and female. J. Med. Entomol. 21: 169–173.

Waudby HP, Petit S (2007) Seasonal density fluctuations of the exotic ornate kangaroo tick, *Amblyomma triguttatum triguttatum* Koch, and its distribution on Yorke Peninsula, South Australia. Parasitol. Res. 101: 1203–1208.

Waudby HP, Petit S, Dixon B, Andrews RH (2007) Hosts of the exotic ornate kangaroo tick, *Amblyomma triguttatum triguttatum* Koch, on southern Yorke Peninsula, South Australia. Parasitol. Res. 101: 1323–1330.

Whitehouse CA (2004) Crimean-Congo hemorrhagic fever. Antiviral Res. 64: 145–160.

Whitworth T, Popov V, Han V, Bouyer D, Stenos J, Graves S, Ndip L, Walker D (2003) Ultrastructural and genetic evidence of a reptilian tick, *Aponomma hydrosauri*, as a host of *Rickettsia honei* in Australia. Ann. N. Y. Acad. Sci. 990: 67–74.

Wilcocks C, Manson-Bahr PEC (1972) Manson's Tropical Diseases. 17th ed. London: Bailliere Tindall. 1164 pp.

Williams R, Bayless M (1998) Tick removal behavior by a white-throated monitor lizard (*Varanus albigularis*) (Sauria: Varanidae). Bull. Chicago Herpetol. Soc. 33: 101–102.

Williams RJ, Al-Busaidy S, Mehta FR, Maupin GO, Wagoner KD, Al-Awaidy S, Suleiman AJM, Khan AS, Peters CJ, Ksiazek TG (2000) Crimean-Congo haemorrhagic fever: a seroepidemiological and tick survey in the Sultanate of Oman. Trop. Med. Int. Health 5: 99–106.

Wilson DD, Bram RA (1998) Foreign pests and vectors of arthropod-borne diseases. In: Buisch WW, Hyde JL, eds. Foreign Animal Diseases. 6th ed. Richmond, VA: Carter Printing Company. pp. 205–222.

Wilson DD, Richard RD (1984) Interception of a vector of heartwater, *Amblyomma hebraeum* Koch (Acari: Ixodidae) on black rhinoceroses imported into the United States. In: Proceedings of 88th Annual Meeting of the United States Animal Health Association. Richmond, VA: Carter Printing Co. pp. 303–311.

Wilson N, Barnard SM (1985) Three species of *Aponomma* (Acari: Ixodidae) collected from imported reptiles in the United States. Fla. Entomol. 68: 478–480.

Wilson N, Kale HW (1972) Ticks collected from Indian River County, Florida (Acari: Metastigmata). Fla. Entomol. 55: 53–57.

Yamane I, Conrad PA, Gardner I (1993) *Babesia gibsoni* infections in dogs. J. Protozool. Res. 3: 111–125.

Yeoman GH, Walker JB, Ross JPJ, Docker TM (1967) The Ixodid Ticks of Tanzania: A Study of the Zoogeography of the Ixodidae of an East African Country. London: Commonwealth Institute of Entomology. 215 pp.

Yeruham I, Hadani A (2004) Ovine babesiosis. In: Coetzer JAW, Tustin RC, eds. Infectious Diseases of Livestock. 2nd ed. Oxford: Oxford University Press. pp. 438–445.

Yeruham I, Hadani A, Galker (Kronthal) F, Mauer E, Rubina M, Rosen S (1985) The geographical distribution and animal hosts of *Rhipicephalus bursa* (Canestrini and Fanzago, 1877) in Israel. Rev. Elev. Med. Vet. Pays Trop. 38: 173–179.

Yeruham I, Hadani A, Galker F, Rosen S (1989) Notes on the biology of the tick *Rhipicephalus bursa* (Canestrini and Fanzago, 1877) in Israel. Rev. Elev. Med. Vet. Pays Trop. 42: 233–235.

Yeruham I, Hadani A, Galker F, Rosen S (1996) The seasonal occurrence of ticks (Acari: Ixodidae) on sheep and in the field in the Judean area of Israel. Exp. Appl. Acarol. 20: 47–56.

Yeruham I, Hadani A, Galker F (1998) Some epizootiological and clinical aspects of ovine babesiosis caused by *Babesia ovis*—a review. Vet. Parasitol. 74: 153–163.

Yin H, Luo J (2007) Ticks of small ruminants in China. Parasitol. Res. 101 (suppl. 2): S187–S189.

Yonow T (1995) The life-cycle of *Amblyomma variegatum* (Acari: Ixodidae): a literature synthesis with a view to modelling. Int. J. Parasitol. 25: 1023–1060.

Young AS, Grootenhuis JG, Kimber CD, Kanhai GK, Stagg DA (1977) Isolation of a *Theileria* species from eland (*Taurotragus oryx*) infective for cattle. Tropenmed. Parasitol. 27: 185–194.

Young E (1965) *Aponomma exornatum* (Koch) as a cause of mortality among monitors. J. S. Afr. Vet. Med. Assoc. 36: 579.

Zaria LT (1993) *Dermatophilus congolensis* infection (dermatophilosis) in animals and man! An update. Comp. Immunol. Microbiol. Infect. Dis. 16: 179–222.

Zaria LT, Amin JD (2004) Dermatophilosis. In: Coetzer JAW, Tustin RC, eds. Infectious Diseases of Livestock. 2nd ed. Oxford: Oxford University Press. pp. 2026–2041.

Zintl A, Mulcahy G, Skerrett HE, Taylor SM, Gray JS (2003) *Babesia divergens*, a bovine blood parasite of veterinary and zoonotic importance. Clin. Microbiol. Rev. 16: 622–636.

Zumpt F, Glajchen D (1950) Tick paralysis in man: a suspected case due to *Rhipicephalus simus* Koch. S. Afr. Med. J. 24: 1092–1094.

# Index

Amblyomma
  albopictum, 12, 105–6, 166
  americanum, 158
  arcanum, 16
  argentinae, 12, 106–8, 166
  astrion, 146, 156
  auricularium, 108–9, 162, 163, 164, 165, 166
  cajennense, 9–10, 11, 12, 109–12, 146, 152, 157, 161, 162, 164, 165, 166, 168, 169, 170, 173
  calcaratum, 112–13, 168
  chabaudi, 12, 14–15
  clypeolatum, 12, 86
  coelebs, 12, 113–14, 166, 168
  cohaerens, 146, 156
  compressum, 11, 14–16
  crassipes, 12, 87, 166
  cyprium, 127
  dissimile, 12, 106, 114–16, 146, 147, 160, 161, 162, 163, 164, 165, 166, 168, 173, 177, 184
  dubitatum, 110
  echidnae, 100
  exornatum, 12, 16–17, 160, 161, 162, 163, 164, 165, 166, 169, 177
  falsomarmoreum, 12, 18, 166
  fimbriatum, 12, 96–97
  flavomaculatum, 12, 18–19, 160, 161, 166
  fuscolineatum, 12, 87–88, 166
  geayi, 11, 116–17
  gemma, 11, 12, 20–21, 146, 156, 168, 171, 173
  geoemydae, 12, 88, 168
  hebraeum, 11, 12, 21–22, 146, 151, 156, 166, 168, 169, 170, 171, 173, 175, 176, 191
  helvolum, 12, 97–98, 166
  humerale, 12, 13, 117, 162, 164, 166
  incisum, 11, 117–18, 168
  javanense, 11, 89, 166
  komodoense, 12, 89–90, 163, 165, 186
  kraneveldi, 12, 91
  latepunctatum, 117, 118
  latum, 11, 12, 22–24, 160, 161, 166, 168
  lepidum, 12, 13, 24–25, 146, 168, 170, 171, 173, 176
  longirostre, 11, 12–13, 118–19, 168
  maculatum, 147, 171, 172, 173, 174
  marmoreum, 12, 25–27, 146, 160, 161, 162, 164, 165, 166, 168, 172, 173, 177, 184, 188
  moreliae, 12, 100–101, 168
  multipunctum, 11, 119
  nodosum, 11, 12, 120
  nuttalli, 12, 13, 27–29, 150, 161, 162, 166, 168
  oblongoguttatum, 11, 120–21, 166, 168
  ovale, 11, 121, 157, 162, 164, 166, 168, 169
  parvum, 11, 122, 166, 168
  pictum, 11, 123, 166
  pomposum, 11, 29, 146, 166, 168, 170, 171, 173, 176
  pseudoconcolor, 11, 123–24, 166, 168
  quadricavum, 11, 124, 166
  rhinocerotis, 11, 29–30
  rotundatum, 12, 125–26, 147, 157, 161, 162, 165, 166, 168, 169, 177, 184
  sabanerae, 12, 13, 126–27, 166
  scalpturatum, 117, 118
  scutatum, 12, 127, 166
  sparsum, 11, 12, 25, 30–31, 146, 162, 165, 166, 168, 172, 173, 177, 184–85, 188
  splendidum, 12, 32, 150
  subleave, see Amblyomma javanense
  sylvaticum, 12, 32–33
  testudinarium, 11, 91–92, 166, 168
  testudinis, see Amblyomma argentinae
  tholloni, 11, 12, 33–34, 146, 168, 173
  transversale, 12, 34–35, 162, 166
  triguttatum, 11, 101, 150, 168, 169
  trimaculatum, 12, 98–99
  varanense, 12, 87, 92–93, 162, 163, 165, 166
  'variegatum, 11, 12, 13, 57–61, 138, 144, 145, 146, 148, 150, 151, 152, 156, 157, 162, 166, 167, 168, 169, 170, 171, 173, 175, 176, 189–93
  varium, 11, 127–28, 166
Anaplasma
  bovis, 45, 60, 70, 138, 171
  centrale, 54, 137–38, 171
  marginale, 45, 49, 53, 56, 65, 71, 74, 78, 133, 136, 137, 170, 171
  ovis, 78, 80, 94, 138, 176
  phagocytophilum, 77, 138–39, 169, 171, 176
Anaplasmosis,
  bovine, 137–38, 170, 171
  caprine, 138, 176
  granulocytic, 139, 168, 169
  ovine, 138, 176
Anocentor nitens, see Dermacentor nitens
Aponomma
  concolor, see Bothriocroton concolor
  crassipes, see Amblyomma crassipes
  exornatum, see Amblyomma exornatum
  fimbriatum, see Amblyomma fimbriatum
  flavomaculatum, see Amblyomma flavomaculatum
  fuscolineatum, see Amblyomma fuscolineatum
  hydrosauri, see Bothriocroton hydrosauri
  komodoense, see Amblyomma komodoense
  kraneveldi, see Amblyomma kraneveldi
  latum, see Amblyomma latum

*lucasi*, see *Amblyomma varanense*
*ochraeum*, see *Amblyomma latum*
*quadricavum*, see *Amblyomma quadricavum*
*transversale*, see *Amblyomma transversale*
*trimaculatum*, see *Amblyomma trimaculatum*
*varanense*, see *Amblyomma varanense*
*varanensis*, see *Amblyomma varanense*
Applicator, self-medicating, 192

*Babesia*
*argentina*, see *Babesia bovis*
*bigemina*, 48, 56, 67, 133, 136, 139, 170, 171
*bovis*, 133, 136, 139–40, 170, 171
*caballi*, 41, 48, 63, 70, 74, 78, 84, 94, 130, 149–50, 175, 176
*canis*, 84, 141, 175, 176
*divergens*, 77, 140, 169, 170, 171
*equi*, see *Theileria equi*
*gibsoni*, 99, 141, 175, 176
*major*, 67, 140, 171
*merionis*, 63, 70, 176
*microti*, 77, 142, 168, 169
*motasi*, 67, 78, 140–41, 176
*occultans*, 74, 171
*ovata*, 99
*ovis*, 70, 78, 80, 141
*rossi*, 38, 141, 175, 176
sp. EU1, 77, 169
*trautmanni*, 54, 80, 176
*volgeli*, 141
Babesiosis,
bovine, 133, 139–40, 170, 171, 173–75, 193
canine, 141–42, 175, 176
caprine, 140–41, 176
cerebral, see Sickness, turning
human, 140, 142, 168, 168
ovine, 140–41, 176
porcine, 141, 176
rodent, 142, 176
*Bartonella henselae*, 77
Bioterrorism, 145, 149, 151, 153, 159, 169
*Boophilus*
*annulatus*, see *Rhipicephalus (Boophilus) annulatus*
*decoloratus*, see *Rhipicephalus (Boophilus) decoloratus*
*microplus*, see *Rhipicephalus (Boophilus) microplus*
*Borrelia*
*afzelii*, 142
*burgdorferi*, 77, 82, 142, 169
*garinii*, 142
*theileri*, 48, 56, 133, 136, 143, 171
Borreliosis,
bovine, 143–44, 171
Lyme, 142–43, 168, 169
*Bothriocroton*
*concolor*, 12, 101–2
*hydrosauri*, 12, 102–4, 147, 152, 168, 169, 176–77

Caribbean Amblyomma Program, 190–91
*Cowdria ruminantium*, see *Ehrlichia ruminantium*
Cowdriosis, see Heartwater
*Coxiella burnetii*, 17, 28, 60, 65, 71, 80, 84, 101, 150, 169
*Cytauxzoon taurotragi*, see *Theileria taurotragi*

Decoy, tick, 191
*Dermacentor*
*albipictus*, 137
*andersoni*, 137, 152, 158
*auratus*, 11, 12, 93–94, 153, 157, 166, 168, 169
*marginatus*, 141, 153
*nitens*, 11, 128–30, 149, 165, 168, 176
*nuttalli*, 12, 94, 138, 149, 153, 157, 168, 169, 176
*reticulatus*, 11, 12, 83–84, 141, 148, 149, 150, 153, 158, 166, 168, 169, 175, 176
*rhinocerinus*, 11, 35–36, 168
*variabilis*, 137, 152, 158
Dermatophilosis, 60, 145–46, 171, 190
*Dermatophilus congolensis*, 60, 145, 171
Disease,
Corridor, 154–55, 171
January, see Theileriosis, Zimbabwe
lumpy wool, see Dermatophilosis
Lyme, see Borreliosis, Lyme
Nairobi sheep, 20, 148, 175, 176
Senkobo, see Dermatophilosis

*Ehrlichia*
*bovis*, see *Anaplasma bovis*
*ovina*, 48, 74, 78, 176
*phagocytophila*, see *Anaplasma phagocytophilum*
*ruminantium*, 20, 22, 25, 27, 29, 31, 34, 60, 112, 116, 146, 171, 172, 173, 176
Ehrlichiosis, ovine, 176
Encephalitis
central European, 157
Russian spring/summer, 157
tickborne, 157–58, 168, 169
Eradication,
*A. komodoense*, 186
*A. sparsum*, 184–85
*A. variegatum*, 189–93
*R. evertsi*, 183–84
*R. (B.) annulatus*, 133, 193
*R. (B.) microplus*, 133, 193

Fever,
African tick-bite, 151–52, 169, 190
biliary, see Piroplasmosis, equine
boutonneuse, see Typhus, tick
Brazilian spotted, 152–53, 168, 169
Crimean-Congo hemorrhagic, 144–45, 168, 169
East Coast, 154–55, 171
Eyach, 169
Flinders Island spotted, 152, 169, 177
Japanese spotted, 152, 169
Mediterranean spotted, see Typhus, tick
Omsk hemorrhagic, 148–49, 168, 169
oriental spotted, see Fever, Japanese spotted
pasture, 138, 171
Q, 17, 28, 150–51, 168, 169
rabbit, see Tularemia
Rocky Mountain spotted, 152–53, 168
Texas cattle, see Babesiosis, bovine
tickborne, 138, 176
Footrot, strawberry, see Dermatophilosis
*Francisella tularensis*, 77, 84, 158, 169

Gallsickness, see Anaplasmosis, bovine

*Haemaphysalis*
*bancrofti*, 156

*bispinosa*, 141
*concinna*, 156
*elliptica*, 37, 38
*elongata*, 12, 36, 168
*flava*, 152
*hoodi*, 12, 36–37
*humerosa*, 156
*hystricis*, 11, 94–95, 152, 166, 168, 169
*indica*, 38
*intermedia*, 141
*japonica*, 156
*juxtakochi*, 12, 130–31, 168
*leachi*, 11, 12, 37–38, 141, 150, 152, 166, 168, 169, 175, 176
*leachi muhsamae*, see *Haemaphysalis muhsamae*
*longicornis*, 11, 99, 141, 156, 166, 168, 171, 175, 176
*muhsamae*, 11, 12, 38, 166
*muhsami*, see *Haemaphysalis muhsamae*
*punctata*, 11, 12, 13, 66–67, 140, 141, 150, 153, 156, 157, 166, 168, 169, 171, 176
Heartwater, 146–47, 170, 173, 175, 176, 177, 180, 184, 188, 189
Hemogregarines, 147
*Hemolivia*
  *mariae*, 104, 147
  *mauritanicum*, 68, 147
  *stellata*, 126, 147
*Hepatozoon*
  *fusifex*, 116, 147
  *kisrae*, 68, 147
Hepatozoonosis, canine, 121
*Hyalomma*
  *aegyptium*, 12, 67–68, 147, 168
  *albiparmatum*, 11, 12, 39, 152, 168, 169
  *anatolicum*, 68–70, 138, 144, 149, 150, 155, 169, 171, 175, 176
  *anatolicum* spp., 11, 68–70, 147, 150, 168
  *anatolicum anatolicum*, 68
  *anatolicum excavatum*, 68
  *asiaticum*, 138, 153
  *detritum*, 11, 70–71, 137, 150, 155, 166, 168, 169, 171, 176
  *dromedarii*, 11, 62–63, 149, 150, 155, 166, 168, 171, 176
  *excavatum*, 68–70, 138, 150, 153, 155, 171, 176

*impressum*, 12, 39–40
*isaaci*, 71, 72, 73
*lusitanicum*, 12, 64–65, 137, 150, 155, 166, 168, 169, 171
*marginatum*, 71, 72, 74, 144, 169, 171, 176
*marginatum* group, 11, 12, 13, 71–74, 149, 150, 155, 166, 168
*marginatum isaaci*, 71
*marginatum marginatum*, 71
*marginatum rufipes*, 71
*rufipes*, 71, 72, 74, 137, 144, 152, 169, 170, 171
*scupense*, 70–71
*transiens*, see *Hyalomma truncatum*
*truncatum*, 11, 12, 40–41, 144, 149, 152, 153, 157, 168, 169, 170, 175, 176
Hyalomma,
  camel, see *Hyalomma dromedarii*
  shiny, see *Hyalomma truncatum*
  small, see *Hyalomma anatolicum* spp.
  tortoise, see *Hyalomma aegyptium*

*Ixodes*
  *angustus*, 142
  *dentatus*, 143
  *granulatus*, 152
  *hexagonus*, 12, 81–82, 143, 152, 157, 166, 168, 169
  *luciae*, 12, 130–31, 168
  *muris*, 142
  *neotomae*, 143
  *ovatus*, 142
  *pacificus*, 138, 143
  *persulcatus*, 143, 147, 158
  *pilosus*, 12, 41–42, 166, 168
  *ricinus*, 11, 12, 13, 74–77, 138, 140, 142, 143, 147, 150, 157, 158, 162, 166, 168, 169, 170, 171, 175, 176
  *scapularis*, 138, 142, 143
  *schillingsi*, 11, 42–43, 168
  *spinipalpis*, 142, 143
  *trianguliceps*, 142
  *uriae*, 143

Louping ill, 147–48, 175, 176
Lymphadenopathy,
  *Dermacentor*-borne necrosis-erythema, 153
  tickborne, 153, 169

Paralysis,
  spring lamb, 157
  tick, 36, 41, 48, 49, 53, 54, 61, 66, 77, 78, 82, 94, 112, 121, 126, 156–57, 168, 170, 177
Pheromone, 22, 191
*Pholeoixodes hexagonus*, see *Ixodes hexagonus*
Piroplasmosis, equine 149–50, 175–76, 180

Redwater, see Babesiosis, bovine
*Rhipicephalus*
  *appendiculatus*, 11, 12, 43–45, 138, 147, 148, 152, 154, 156, 157, 166, 167, 168, 169, 170, 171, 175, 176
  *bursa* 11, 77–78, 137, 138, 141, 149, 150, 157, 166, 168, 169, 171, 175, 176
  *capensis*, 12, 45–46, 154, 171
  *compositus*, 12, 46–47, 168
  *evertsi*, 11, 12, 47–49, 152, 157, 162, 165, 166, 168, 169, 183–84
  *evertsi evertsi*, 47–49, 139, 144, 149, 150, 154, 156, 157, 169, 170, 171, 176
  *evertsi mimeticus*, 47, 49, 144, 150, 157, 169, 176
  *follis*, 45
  *gertrudae*, 45
  *kochi*, 12, 49–50
  *longus*, 46
  *muehlensi*, 11, 12, 50–51, 168
  *praetextatus*, 52
  *pseudolongus*, 46
  *pulchellus*, 11, 12, 51, 144, 148, 152, 154, 156, 168, 169, 171, 176
  *pumilio*, 138
  *sanguineus*, 78, 141, 152
  *senegalensis*, 12, 52, 150, 168
  *simus*, 11, 12, 52–54, 137, 141, 148, 150, 152, 154, 157, 168, 169, 170, 171, 176
  *sulcatus*, 12, 54–55, 168
  *turanicus*, 12, 54, 78–80, 150, 152, 166, 168, 169, 176
  *zambeziensis*, 43
*Rhipicephalus (Boophilus)*
  *annulatus*, 11, 12, 132–34, 137, 139, 140, 144, 162, 165, 166, 168, 170, 171, 173–75, 193
  *decoloratus*, 11, 12, 55–56, 139, 144, 150, 162, 168, 171

*geigyi,* 139, 140
*microplus,* 11, 12, 56, 133, 134–36, 137, 139, 140, 144, 150, 165, 166, 167, 168, 170, 171, 173–75, 176, 183, 193

*Rickettsia*
*aeschlimannii,* 74, 169
*africae,* 22, 25, 60, 151, 169
*conorii,* 38, 39, 41, 45, 49, 51, 54, 60, 74, 80, 82, 152, 169
*helvetica,* 77, 169
*honei,* 104, 152, 169, 177
*japonica,* 95, 152, 169
*massiliae,* 80
*rickettsii,* 111, 152–53, 169
*sibirica mongolitimonae,* 41, 70, 153
*sibirica sibirica,* 67, 84, 94, 153, 169
*slovaca,* 84, 153, 169

Rickettsiosis, lymphangitis-associated 153

Sickness,
　sweating, 41, 157
　turning, 154–55
Spirochetosis, see Borreliosis, bovine
Streptothricosis, see Dermatophilosis

*Theileria*
*annulata,* 63, 65, 70, 71, 74, 153, 155, 170, 171
*buffeli,* 67, 99, 153, 156, 171
*equi,* 48, 49, 63, 70, 71, 74, 78, 80, 84, 136, 149–50, 175, 176
*hirci,* see *Theileria lestoquardi*
*lawrencei,* see *Theileria parva*
*lestoquardi,* 70, 153, 155–56, 176
*mutans,* 20, 22, 60, 153, 156, 171
*orientalis,* 67, 99, 153, 156, 171
*ovis,* 49, 67, 78, 154, 176
*parva,* 45, 46, 48, 51, 53, 153, 154–55, 170, 171
*parva bovis,* see *Theileria parva*
*parva lawrencei,* see *Theileria parva*
*parva parva,* see *Theileria parva*

*separata,* 48, 49, 78, 154, 176
*taurotragi,* 45, 49, 51, 154, 156, 171
*uilenbergi,* 99
*velifera,* 22, 25, 60, 154, 171

Theileriosis,
　benign bovine, 154, 156, 171
　benign caprine, 154, 176
　benign ovine, 154, 176
　bovine, 156, 170, 171
　cerebral, see Sickness, turning
　equine, see Piroplasmosis, equine
　malignant, 155–56, 175, 176
　tropical, 155, 171
　Zimbabwe, 154–55, 171

Tick,
　Anatolian brown, see *Rhipicephalus bursa*
　blue, see *Rhipicephalus (Boophilus) decoloratus*
　brown ear, see *Rhipicephalus appendiculatus*
　castor bean, see *Ixodes ricinus*
　cattle, see *Rhipicephalus (Boophilus) annulatus*
　cattle fever, 173
　Cayenne, see *Amblyomma cajennense*
　glossy, see *Rhipicephalus simus*
　Gulf Coast, see *Amblyomma maculatum*
　hedgehog, see *Ixodes hexagonus*
　iguana, see *Amblyomma dissimile*
　ornate dog, see *Dermacentor reticulatus*
　ornate kangaroo, see *Amblyomma triguttatum*
　red-legged, see *Rhipicephalus evertsi*
　red sheep, see *Haemaphysalis punctata*
　rotund toad, see *Amblyomma rotundatum*
　sheep, see *Ixodes ricinus*
　sourveld, see *Ixodes pilosus*
　South African bont, see *Amblyomma hebraeum*
　South African tortoise, see *Amblyomma marmoreum*
　southern cattle, see *Rhipicephalus (Boophilus) microplus*
　tropical bont, see *Amblyomma variegatum*
　tropical horse, see *Dermacentor nitens*
　yellow dog, see *Haemaphysalis leachi*
　zebra, see *Rhipicephalus pulchellus*

Toxicosis,
　brown tick, 157
　tick, 41, 45, 157, 170
*Trypanosoma theileri,* 70, 171
Trypanosomiasis, benign, 171
Tularemia, 158–59, 168, 169
Typhus,
　Israel tick, 80
　North Asian tick, 153, 169
　Siberian tick, see Typhus, North Asian tick
　tick, 152, 168, 169

Virus,
　African horse sickness, 63
　Crimean-Congo hemorrhagic fever, 41, 48, 49, 51, 60, 70, 74, 144, 168, 169
　Dugbe, 20, 25, 60
　Eyach fever, 77, 169
　Jos, 60
　Kyasanur Forest disease, 74, 94
　louping ill, 45, 77, 147, 176
　Nairobi sheep disease, 45, 51, 54, 60, 148, 176
　Omsk hemorrhagic fever, 84, 148, 168, 169
　Thogoto, 20, 25, 45, 60
　tickborne encephalitis, 77, 84, 157, 169

Warfare, biological, 145, 151
Weapons, biological, 149, 159

Michael Burridge is a professor emeritus of infectious diseases at the University of Florida, Gainesville. His research on ticks has been published in many scientific journals, including the *Journal of Parasitology*, the *International Journal of Acarology*, the *Bulletin of Entomological Research*, *Veterinary Parasitology*, the *Journal of Medical Entomology*, the *Journal of Zoo and Wildlife Medicine* and the *Florida Entomologist*.